D1264032

FLUID MECHANICS

TIDAL BORE ON THE CHIEN TANG RIVER, BY LI SUNG (1166–1243).

FLUID MECHANICS

A CONCISE INTRODUCTION TO THE THEORY

CORRECTED EDITION

CHIA-SHUN YIH

STEPHEN P. TIMOSHENKO UNIVERSITY PROFESSOR OF FLUID MECHANICS
DEPARTMENT OF APPLIED MECHANICS AND ENGINEERING SCIENCE
THE UNIVERSITY OF MICHIGAN

WEST RIVER PRESS
3530 WEST HURON RIVER DRIVE
ANN ARBOR, MICHIGAN 48103
U.S.A.

TO

YUAN-CHENG FUNG

AS A TOKEN OF FRIENDSHIP

FLUID MECHANICS

Library of Congress Catalog Card Number 78-65697

0-9602190-0-5

PREFACE

More than anything else, this is a textbook to introduce an interested person, or one whose calling obliges him to learn something about the subject, to the theory of fluid mechanics. As an introduction it is not intended to be exhaustive in its choice of material. Exhaustiveness would be impossible even in a much more ambitious endeavor; in an introduction to the subject, it is also undesirable. What I have set out to do is to give a concise and connected account of the theoretical aspects of classical fluid mechanics and its modern developments in a systematic way and at such a level that any serious student with modest mathematical preparation can use it to become acquainted with the fundamentals, the theory, and the main results of the subject. I have endeavored to make the book self-contained. No previous acquaintance is assumed with any fluid-mechanics subject under discussion in this book, and it is only very seldom that a subject, once chosen for discussion, is not pursued to the final solution. The few exceptions arise from necessity: the context seems to call for mentioning a study or a result, but the limitation of space definitely forbids its thorough treatment. In all such cases, as indeed throughout the book, references are given. Thus I hope the book will be useful to anyone who wishes to instruct himself on the subject.

This book is primarily intended for first- or second-year graduate students in American universities, although portions of it can be taught to qualified undergraduates by an experienced instructor. My experience in teaching engineering students has naturally made me aware of what they are expected to learn about fluid mechanics, but my own interest in fluid mechanics is somewhat wider than that motivated by engineering applications alone and this book is not intended exclusively for engineering students. A student of meteorology may find here what he needs as an adequate foundation for his subsequent studies, and a physicist, if an accident should arouse his interest in fluid mechanics, may also find here things worth his while to learn.

In practice the material contained in this book may be more than can be presented in a course consisting of three lectures (or discussions, since it is the American university that I have in mind) per week for one school year. If so, portions may be omitted according to the priority of interest. It seems that Chaps. 1 to 4, 7, and 10 ought to be included in any choice for such a course. In Chaps. 4 to 10, supplemental material can be inserted for instruction according to the wishes and judgment of the instructor.

The level of mathematics assumed of the reader is not very high. Aside from advanced calculus, some knowledge of Fourier series, theory of functions of a complex variable, and differential equations would be very helpful, but in the majority of cases whatever mathematical results I need I endeavor to develop ab initio. This practice is illustrated in Chap. 1, where Cartesian tensors are used and some results in linear algebra are presented; in Chap. 4, where conformal mapping is discussed; in Chaps. 5 and 6, where the method of characteristics for partial differential equations is used; in Chap. 7, where Fourier series is used; and in Appendix 2, where general tensors are discussed.

I have written Appendix 2 for a good reason. Students are seldom taught how to transform a set of partial differential equations in Cartesian coordinates to their forms in curvilinear coordinates. In most books all they find are some recipes. Surely for those who wish to learn the transformations in a systematic way there should be a source to learn from. A lucid source is L. Brillouin, "Les tenseurs en mécanique et en elasticité," Dover Publications Inc., New York, 1946, on which Appendix 2 relies heavily. For the sake of brevity, however, I have taken one important advantage. Since I am interested only in transforming the rate-of-deformation and vorticity tensors, the equation of continuity, the Navier-Stokes equations, and so on, from Cartesian coordinates to other coordinates, *the space involved is always Euclidean*. I am therefore spared the task of defining parallelism in curved spaces, and this makes it possible to give the bare essentials in a rather short appendix.

Problems are given in each chapter. Their order follows the text approximately, and the degree of difficulty varies. Some are mere exercises to help the student grasp the material in the text, but some require a good deal of thought and a measure of skill for their solution. I have tried to make the problems as interesting as possible without putting them out of reach of the student or turning them into topics for long research.

I am happy to avail myself of this opportunity to thank many people for their help. The first three chapters and Appendix 1 were read by Prof. Yuan-Cheng Fung, whose criticisms and suggestions have been helpful. Professor T. Y. T. Wu carefully read the entire manuscript, offered many valuable criticisms, and caught some errors. To him I owe my sincere gratitude. In a book published in 1965, I acknowledged the benefit I had received from the many who had been responsible for my initial interest in fluid mechanics or for intensifying it. In addition to these people, I must now mention Prof. Sydney Goldstein, from whom I learned the theory of the flow of compressible fluids in 1947 and whose warm friendship I have the good fortune to enjoy. This book was started in 1956, when my friend and former teacher, Prof. John S. McNown, succeeded in bringing me to The University of Michigan. The 12 years since 1956, during which chapters of this book have been written and taught to classes of graduate students, have been happy and useful years of my life. For this happiness I am indebted to my colleagues at the University of Michigan. The stimulation of discussions with them, the comfort of their friendship, and the encouragement of their generosity have been responsible for the interest and the peace of mind necessary for writing this book.

I should like to express my indebtedness to Prof. Walter R. Debler, of the University of Michigan, Prof. S. Taneda, of Kyushu University, Prof. Benjamin Gebhart, of Cornell University, and Prof. Stanley Corrsin, of Johns Hopkins University, all of whom supplied photographs acknowledged in the legends to illustrations. My sincere thanks are due to my friend and student, Dr. Chin-Hsiu Li, for making initial drawings for the figures, to Mrs. Ruth Haeussler for typing the manuscript, and to the McGraw-Hill staff for their patient and splendid cooperation.

Foreword for the Corrected Edition

I have taken advantage of the occasion of the new printing of this book to make corrections and to improve the text at several places. Many readers and friends have kindly pointed out the misprints and mistakes in the old edition, and Mrs. Beverly Pyle has painstakingly typed the corrigenda. To them I want to express my sincere appreciation.

I am especially grateful to Professor Milton van Dyke for his encouragement and generous help, without which this edition would never have seen the light of day. Finally, I take this opportunity to thank the Alexander von Humboldt Foundation of the Federal Republic of Germany for a Humboldt Award, which allowed me to work at leisure in Karlsruhe, where much of the final revision of this book was done.

CHIA-SHUN YIH, 1978

CONTENTS

CHAPTER ONE

FUNDAMENTALS

CHAPTER TWO

THE BASIC EQUATIONS

CHAPTER THREE

GENERAL THEOREMS FOR THE FLOW OF AN INVISCID FLUID

CHAPTER FOUR

IRROTATIONAL FLOWS OF AN INVISCID FLUID OF CONSTANT DENSITY

CHAPTER FIVE

WAVES IN AN INCOMPRESSIBLE FLUID

CHAPTER SIX

THE DYNAMICS OF INVISCID COMPRESSIBLE FLUIDS

CHAPTER SEVEN

EFFECTS OF VISCOSITY

CHAPTER EIGHT

HEAT TRANSFER AND BOUNDARY LAYERS OF A GAS

CHAPTER NINE

HYDRODYNAMIC STABILITY

CHAPTER TEN

TURBULENCE

APPENDIX ONE

BASIC THERMODYNAMICS

APPENDIX TWO

CURVILINEAR COORDINATES

GUIDE TO READER

Equation numbers are continuous within each chapter. If an equation is referred to in another chapter, its number is prefixed by the number of the chapter in which it is to be found. For instance, if Eq. (100) in Chap. 4 is referred to in the same chapter, it is identified simply as (100), but the same equation referred to in Chap. 5 will be identified as (4.100). The same applies to figure numbers and problem numbers. When equations in the appendixes are referred to in the text, the prefix A1 or A2 is used. For instance, Eq. (20) in Appendix 2 is referred to in the text as (A2.20). Section numbers, however, are always referred to explicitly, e.g., "Sec. 3.5.1 of Chap. 5."

It is quite impossible to preserve the customary usage of symbols and at the same time maintain a one-to-one correspondence. Usage takes precedence in this book, so that from time to time a symbol represents different things. I often define a quantity more than once, so that one rarely has to search far to find the definition of any symbol. In my opinion, a list of symbols is more a nuisance than a necessity, let alone a help.

References are given at the end of each chapter, which also serve as an author index. Like the material treated, these references are not exhaustive. Works of authors explicitly referred to in the text are listed, and in addition, standard works (mostly books) related to the chapter are listed.   From all these references one can find other references, and the process can be repeated.   (This method of informing oneself was called the *star method* by André Maurois.   Each ray of a star is a reference, at the end of which is another star with its own rays.)   No other author index will be given.   A subject index is provided at the end of the book.

CHAPTER ONE
FUNDAMENTALS

1. DEFINITION OF A FLUID

The defining property of a fluid is that it cannot withstand shearing forces, however small, without sustained motion. Since both gases and liquids have this property, they both are fluids and subject to a unified treatment as far as their macroscopic motion is concerned. Fluids may, of course, be in a state of equilibrium under the action of surface forces applied normal to their boundary. Indeed, the entire subject of hydrostatics deals with fluids in such a state.

In this book, only fluids which have no privileged directions are considered. Such fluids are called *isotropic*.

2. VELOCITIES

If matter were infinitely divisible, it would be meaningful to define the velocity of a material point as its time rate of displacement; but matter in general and fluids in particular are not infinitely divisible. Strictly speaking, we can comprehend the velocity only of a molecule, an atom, a nucleus, or an electron; the "velocity" of a geometrical point in the empty space between the electrons and the nucleus in an atom, between atom and atom in a molecule, or between the molecules themselves is physically meaningless.

It would be a hopeless situation indeed if in order to study fluid motion we had to deal with the molecules directly. Fortunately, although there is much empty space between molecules, the number of molecules per unit volume of a liquid or of a gas under ordinary conditions is extremely great. A mole of gas has approximately 6.024×10^{23} molecules (Loschmidt number) and under normal conditions occupies a volume of 22.4 liters, so that 1 cm^3 contains 2.687×10^{19} molecules. The number of molecules in 1 μ^3 (1 $\mu = 1/1{,}000$ mm) is about 2.687×10^7. For such a small volume this is an exceedingly great number. Thus gases under ordinary conditions—and, a fortiori, liquids—can be considered for all practical purposes to be continuous. The velocity of a fluid particle of a very small volume (1 μ^3, say) can be defined to be the average of the momenta of the vast number of molecules contained therein divided by the total mass of the particle. Since the volume of this particle is very small, the velocity so defined can be considered to be the velocity of the material point situated at the center of mass of the fluid particle, as if the fluid were indefinitely divisible. This approach is valid unless a highly rarefied gas is being studied.

According to this definition, the velocity $\mathbf{v}$ of a fluid particle consisting of N_i molecules $(i = 1, 2, \ldots, n)$ of the ith substance present in the particle is

$$\mathbf{v} = \frac{\sum_{i=1}^{n} \left[m_i \sum_{j=1}^{N_i} \mathbf{v}(i,j) \right]}{\sum_{i=1}^{n} N_i m_i}, \tag{1}$$

in which m_i is the molecular weight of the ith substance and $\mathbf{v}(i,j)$ is the velocity of the jth of the N_i molecules of molecular weight m_i.

The velocity of a fluid particle can also be defined as

$$\mathbf{v} = \frac{\sum_{j=1}^{N} \mathbf{v}(j)}{N}, \tag{2}$$

in which
$$N = \sum_{i=1}^{n} N_i$$

and $\mathbf{v}(j)$ is the velocity of the jth of the N molecules of the n substances taken together without regard to the *kind* of molecules. The particle velocity defined by (1) is a mean velocity of the molecules weighted by the molecular weights, whereas that defined by (2) is an unweighted mean velocity. In Sec. 5, when the equation of continuity is discussed, it will be evident that the two definitions do not correspond to the same form of the equation of continuity if the fluid is a mixture. In Chap. 8, the matter will be further discussed in connection with the diffusion equation. In order to facilitate later discussions, Eq. (1) will be put into a slightly more convenient form. If for each constituent of a fluid mixture the mean velocity is defined to be

$$\mathbf{v}(i) = \sum_{j=1}^{N_i} \frac{\mathbf{v}(i,j)}{N_i}$$

Eq. (1) can be written as

$$\mathbf{v} = \frac{\sum\limits_{i=1}^{n} N_i m_i \mathbf{v}(i)}{\sum\limits_{i=1}^{n} N_i m_i} \tag{1a}$$

Whatever the volume of the fluid particle, the density ρ_i of the ith constituent must be proportional to $N_i m_i$. Thus $(1a)$ can be written as

$$\mathbf{v} = \frac{\sum\limits_{i=1}^{n} \rho_i \mathbf{v}(i)}{\rho}, \tag{1b}$$

in which ρ is the total density, or

$$\rho = \sum_{i=1}^{n} \rho_i. \tag{3}$$

The full meaning of the definition of the velocity of a particle must be understood. The average of the momenta of the molecules contained in a fluid particle may be zero, even though these molecules are moving with great speeds individually. In fact, it is zero if the molecules are in completely random motion. The average of the molecular momenta is therefore a measure of the ordered part of molecular motion. The intensity of random molecular motion is manifested in a property called *temperature*.

If the fluid is considered as a continuum, the velocity at any point in the fluid is postulated as a function of time t and the coordinates x_i ($i = 1, 2, 3$) of the point. It is evident that without some modification the continuum concept is incompatible with the concept of diffusion; this basic incompatibility, however, presents no insurmountable difficulties.

Other properties, kinematic or dynamic, can also be defined either from the molecular or the continuum point of view. In this book a fluid will be considered as a continuum, although wherever mass diffusion is involved, the appropriateness of the continuum approach must be judged in the light of molecular considerations.

3. ACCELERATION

To find the acceleration of a particle, we must see how its velocity changes with time and therefore must keep track of its identity—at least for a short interval of time. There are two different descriptions of fluid motion. In the one, the coordinates of fluid particles are considered to be a function of time and of their permanent identifications, such as their initial coordinates. In the other, the velocities and other properties of fluid particles are considered to be functions of time and fixed spatial coordinates independent of time. The former is called the *material* or *Lagrangian* description, after Joseph Louis Lagrange (1736–1813), and the latter the *spatial* or *Eulerian* description, after Leonard Euler (1707–1783), although historians argue that both descriptions should be attributed to Euler.

With the Lagrangian description, the Cartesian coordinates (c_1,c_2,c_3) of the position of a fluid particle at the initial time serve to identify the particle. The subsequent (Cartesian) coordinates of the position of the same particle will be denoted by (X_1,X_2,X_3). These coordinates are functions of c_i and time t. For a definite particle the identifying coordinates are fixed, and the coordinates X_i are functions of time alone. If the coordinates are Cartesian, the velocity components are

$$u_i = \frac{\partial X_i}{\partial t}, \qquad i = 1, 2, 3,$$

and the acceleration components are simply

$$a_i = \frac{\partial^2 X_i}{\partial t^2}, \qquad i = 1, 2, 3.$$

With the Eulerian description, the velocity and the acceleration of a fluid in motion are considered to be functions of time and position. The Cartesian coordinates (x_1,x_2,x_3) describing the position are now independent of time. However, to find the velocity or the acceleration we still have to follow the particle for a short interval of time dt. During this interval, the coordinates of the particle followed have changed by the amounts dX_i. The corresponding change of the velocity consists of two parts: the first part is a local change with time, and the second part is a change due to the change of position of the particle. If the velocity component in the direction of x_i is again denoted by u_i, then since u_i is a function of t and x_i,

$$du_i = \frac{\partial u_i}{\partial t} dt + \frac{\partial u_i}{\partial x_j} dx_j$$

for all increments dt and dx_j. The physical process of following the particles corresponds to the identification of dx_j with the particle displacements dX_j. Thus, when a particle is followed,

$$du_i = \frac{\partial u_i}{\partial t}\, dt + \frac{\partial u_i}{\partial x_j}\, dX_j, \qquad i = 1, 2, 3, \tag{3a}$$

in which, as in the preceding equation, the last term stands for the sum of three terms, with j ranging over 1, 2, and 3. This summation convention will always be used unless otherwise stated. If the coordinates are Cartesian, the displacement components are

$$dX_i = u_i\, dt, \qquad i = 1, 2, 3. \tag{3b}$$

Dividing (3a) by dt and taking the limit, we have

$$a_i = \frac{\partial u_i}{\partial t} + u_j \frac{\partial u_i}{\partial x_j}, \qquad i = 1, 2, 3. \tag{4}$$

The operator

$$\frac{D}{Dt} \equiv \frac{\partial}{\partial t} + u_j \frac{\partial}{\partial x_j} \tag{5}$$

stands for what is commonly called the *substantial differentiation*, which means differentiation with respect to time by following the substance. From (5) it follows that

$$a_i = \frac{Du_i}{Dt}, \qquad i = 1, 2, 3. \tag{4a}$$

A flow is *steady* if all dependent variables, such as velocity, acceleration, density, temperature, and pressure, are independent of time at any fixed point.

3.1. The Proper Definition of Acceleration When Diffusion Is Present

In employing the continuum approach special care must be taken when diffusion is present, as can be illustrated by considering the force acting on a fluid particle of volume V. Suppose that there are two gases, one of molecular weight m_1 and the other of molecular weight m_2, and that there are N_1 molecules of the first gas and N_2 of the second gas in ΔV. The density in the cgs system is

$$\rho = \frac{N_1 M_1 + N_2 M_2}{\Delta V} \qquad \text{in g/cm}^3, \tag{6}$$

in which M_1 and M_2 are the molecular masses of the gases, which are related to m_1 and m_2 by

$$M_1 = \frac{m_1}{L}, \qquad M_2 = \frac{m_2}{L},$$

L being the Loschmidt number, i.e., the number of molecules in m g of a gas of molecular weight m. Now, because of diffusion, the two gases in the particle do not have the same mean velocity or acceleration.

If the acceleration of the particle is defined by a formula similar to (2), i.e., as the unweighted mean of the accelerations of all the molecules in the particle of volume ΔV, again $a_i \rho \, \Delta V$ will not give the components of the net force acting on the particle. Only if $\mathbf{a}$ (or a_i) is defined by a formula like (1b), i.e., as a mean acceleration weighted by the density, will it be obvious that $\mathbf{a}\rho \, \Delta V$ is the net force acting on the particle. Fortunately, the differences in accelerations (or velocities) of the various gas components in a mixture are usually small compared with the acceleration of any one gas or with the force acting on the entire fluid particle. Hence the error caused by using an inadequately defined acceleration to compute the net force on a particle of a gas mixture is small. The same is true of liquids. But even the conceptual difficulty disappears if there is only one substance or if the mixture is homogeneous and the flow is steady and the velocity is the same for all components at one section (and therefore everywhere).

4. PATH LINES AND STREAMLINES

The coordinates X_i used in the Lagrangian description are functions of the identifying coordinates c_i and t. For fixed c_i, we have

$$X_i = F_i(t), \qquad i = 1, 2, 3,$$

as the parametric equations describing the locus of the particle under consideration. This locus is called a *path line*.

Although the Eulerian description is less convenient for describing path lines, it is superior to the Lagrangian method for describing streamlines, or lines to which the velocity vectors of the fluid are tangent at a particular *instant*. The differential equations for such lines are, in Cartesian coordinates,

$$\frac{dx_1}{u_1} = \frac{dx_2}{u_2} = \frac{dx_3}{u_3}. \tag{7}$$

As mentioned before, a flow is called steady if at any fixed point in the fluid the velocity does not change with time. For steady flows path lines are coincident with streamlines, for when a particle reaches the position of its predecessor on a streamline, because of the steadiness of the motion it has the same velocity as its predecessor and therefore goes the way its predecessor went, and so on. But if the flow is unsteady, path lines and streamlines may (though not necessarily) differ. A simple case of unsteady flow in which path lines and streamlines are coincident is rectilinear parallel flow with a velocity varying with time.

The streamlines passing through a closed curve which does not lie on a surface generated by streamlines form a tubular surface. The fluid contained in such a surface is called a *stream tube*.

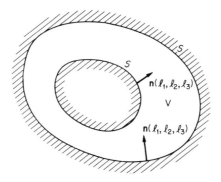

FIGURE 1. DEFINITION SKETCH FOR GREEN'S THEOREM.

5. CONTINUITY

Since the principle of mass conservation must not be violated, the velocity distribution in a fluid in motion must satisfy a certain condition. Obviously, since mass is involved, the density (mass per unit volume) of the fluid ρ must be considered. If S is an arbitrary fixed surface enclosing a fluid volume V and u_n is the velocity along the normal to S drawn *into* V, the net amount of mass flowing into V per unit time is

$$\int_S \rho u_n \, dS.$$

This must be equal to the rate at which the mass in V is increasing. Thus

$$\frac{\partial}{\partial t} \int_V \rho \, dV = \int_S \rho u_n \, dS. \tag{8}$$

Equation (8) is an integral equation of continuity and can be reduced to a differential form. Green's theorem states that for any three single-valued and differentiable functions (U_1, U_2, U_3) in Cartesian coordinates,

$$\int_S l_i U_i \, dS = -\int_V \frac{\partial U_i}{\partial x_i} \, dV, \tag{9}$$

in which the l's are the direction cosines of the normal to S drawn into V (Fig. 1). Taking $U_i = \rho u_i$, we have

$$\int_S \rho u_n \, dS = \int_S l_i \rho u_i \, dS = -\int_V \frac{\partial (\rho u_i)}{\partial x_i} \, dV. \tag{10}$$

From (8) and (10) it follows that

$$\int_V \left[\frac{\partial \rho}{\partial t} + \frac{\partial (\rho u_i)}{\partial x_i} \right] dV = 0. \tag{11}$$

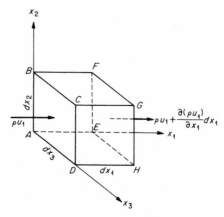

FIGURE 2. DERIVING THE EQUATION OF CONTINUITY.

Since V is arbitrary,

$$\frac{\partial \rho}{\partial t} + \frac{\partial(\rho u_i)}{\partial x_i} = 0, \tag{12}$$

which is the general differential equation of continuity.

The differential form of the equation of continuity, Eq. (12), has been derived from its integral form by use of Green's theorem in order to provide a desirable link between these two forms. Since many readers may not be familiar with Green's theorem, we shall derive (12) directly. Consider a rectangular box (Fig. 2) with sides of lengths dx_1, dx_2, and dx_3. The rate of mass inflow at the surface $ABCD$ is

$$\rho u_1 \, dx_2 \, dx_3.$$

The rate of mass outflow at the surface $EFGH$ is

$$\left[\rho u_1 + \frac{\partial(\rho u_1)}{\partial x_1} \, dx_1 \right] dx_2 \, dx_3.$$

Thus the net outflow at the two surfaces normal to the x_1 axis is

$$\frac{\partial(\rho u_1)}{\partial x_1} \, dx_1 \, dx_2 \, dx_3.$$

Similarly, the net outflows at the two pairs of surfaces normal to the two other axes are, respectively,

$$\frac{\partial(\rho u_2)}{\partial x_2} \, dx_1 \, dx_2 \, dx_3 \quad \text{and} \quad \frac{\partial(\rho u_3)}{\partial x_3} \, dx_1 \, dx_2 \, dx_3.$$

Thus the rate of total outflow is

$$\frac{\partial(\rho u_i)}{\partial x_i} \, dx_1 \, dx_2 \, dx_3,$$

which must be equal to the rate of decrease of mass in the box, i.e., to

$$- \frac{\partial}{\partial t} (\rho \, dx_1 \, dx_2 \, dx_3).$$

Since $dx_1 \, dx_2 \, dx_3$ is independent of t (and of x_i), the equality can be written in the form of (12).

A fluid is *incompressible* if its particles maintain their densities along their paths, i.e., if the substantial derivative of the density is zero:

$$\frac{D\rho}{Dt} = 0. \tag{13}$$

From (12) and (13) it follows that for incompressible fluids the equation of continuity is simply

$$\frac{\partial u_i}{\partial x_i} = 0, \tag{14}$$

whether the flow is steady or not, and whether the fluid is homogeneous or not. The integral form of (14) is

$$\int_S u_n \, dS = 0. \tag{15}$$

Equation (14) can also be directly derived from the requirement that the volume of a particle of an incompressible fluid remain unchanged with time. Consider such a particle in the form of a rectangular box, with sides of lengths dx_1, dx_2, and dx_3. The difference between the u_1 at one end of the side dx_1 and that at the other end is

$$\frac{\partial u_1}{\partial x_1} dx_1.$$

Thus after a time dt has elapsed, the x_1 projection of the side dx_1 will become

$$\left(1 + \frac{\partial u_1}{\partial x_1} dt\right) dx_1.$$

Similarly, the corresponding projections of the other two sides will be

$$\left(1 + \frac{\partial u_2}{\partial x_2} dt\right) dx_2 \quad \text{and} \quad \left(1 + \frac{\partial u_3}{\partial x_3} dt\right) dx_3.$$

Thus the volume of the particle is, after a time dt,

$$\left(1 + \frac{\partial u_1}{\partial x_1} dt\right)\left(1 + \frac{\partial u_2}{\partial x_2} dt\right)\left(1 + \frac{\partial u_3}{\partial x_3} dt\right) dx_1 \, dx_2 \, dx_3$$

if terms of higher orders in dt and dx_i are neglected in computing the volume. With higher-order terms again neglected, the increase in volume is

$$\frac{\partial u_i}{\partial x_i} dt \, dx_1 \, dx_2 \, dx_3.$$

For an incompressible fluid, this is zero. Since $dt\, dx_1\, dx_2\, dx_3$ is not zero, (14) results. The advantage of this derivation lies in its independence of considerations of mass conservation and density conservation. One may also obtain (14) by using Fig. 2 but considering volume flux through the surface of the box instead of mass flux.

For steady flow of a compressible fluid or for flows (steady or unsteady) of a fluid with homogeneous density, (8) becomes

$$\int_S \rho u_n\, dS = 0, \tag{16}$$

which has the important consequence that streamlines for such flows or fluids cannot end in the interior of the fluid if it has no mass sources or sinks. For we can consider the space occupied by the fluid to consist of stream tubes of a constant rate of mass flow. As expressed in (16), just as many tubes must enter S as leave it. Thus stream tubes, and, in the limit, streamlines, cannot terminate in the interior of the fluid but must either form closed curves or terminate on the boundary of a fluid domain free from sources and sinks. The same is true also for an incompressible fluid of variable density, whether or not its flow is steady, as can be shown by a consideration of the volume flux.

5.1. The Form of the Continuity Equation When Diffusion Is Present

For the sake of brevity a mixture of only two substances in the fluid state will be considered. Evidently the continuum approach can be applied to each substance. Denoting the density of the first and second substances at any point by ρ_1 and by ρ_2, respectively, and the mean velocity components of the first and second substances by u_i and v_i, respectively, the two equations of continuity are

$$\frac{\partial \rho_1}{\partial t} + \frac{\partial(\rho_1 u_i)}{\partial x_i} = 0 \quad \text{and} \quad \frac{\partial \rho_2}{\partial t} + \frac{\partial(\rho_2 v_i)}{\partial x_i} = 0.$$

Addition of these two equations produces

$$\frac{\partial \rho}{\partial t} + \frac{\partial(\rho_1 u_i + \rho_2 v_i)}{\partial x_i} = 0, \tag{17}$$

or

$$\frac{\partial \rho}{\partial t} + \frac{\partial(\rho w_i)}{\partial x_i} = 0 \tag{12a}$$

if $\rho = \rho_1 + \rho_2$ and w_i is the ith component of the mean velocity defined by

$$w_i = \frac{\rho_1 u_i + \rho_2 v_i}{\rho}.$$

Thus if (12) is to be used when diffusion is present, the velocity components in it should be defined by (1) or (1b) rather than by (2). When

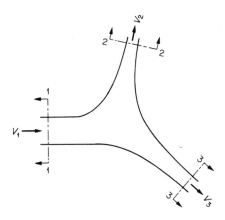

FIGURE 3. APPLICATION OF THE EQUATION
OF CONTINUITY IN INTEGRAL FORM.

(2) is used to define the mean velocity and if (12) is nevertheless used, it will be a different equation from (17). Yet this is often what is done in the literature. This is because u_i and v_i are only slightly different as a result of diffusion, and as far as mass conservation is concerned the error is usually small. [For a fluid mixture that can be considered incompressible, (14) rather than (12) is used with the diffusion equation, in which the diffusive terms take care of the inequality of u_i and v_i. This point will be further discussed in Chap. 8.] It is well to keep in mind. however, the conceptual inconsistency of the practice.

5.2. Application of the Integral Form of the Equation of Continuity

Equation (8) and the more restrictive equation (16) are equations of continuity in integral forms. The surface S is often called a *control surface*. Application of these equations to one-dimensional flows will now be illustrated.

With reference to Fig. 3, the control surface consists of the wall of the branching pipe and of the sections 1-1, 2-2, and 3-3. If the fluid is incompressible and the average velocities across the sections are denoted by V_1, V_2, and V_3, (15) states that

$$-V_1A_1 + V_2A_2 + V_3A_3 = 0. \qquad (18)$$

The negative sign before the first term follows from the convention that normal velocities are counted positive if directed outward from the control volume. The symbols A_1, A_2, and A_3 denote the cross-sectional areas of the three sections. Given the areas and any two velocities, (18) enables one to find the third velocity.

For a compressible fluid in steady flow, if the average densities across the three sections are denoted by ρ_1, ρ_2, and ρ_3, (18) is replaced by

$$-\rho_1 V_1 A_1 + \rho_2 V_2 A_2 + \rho_3 V_3 A_3 = 0.$$

If the flow is unsteady, the rate of increase of mass in the control volume is

$$\rho_1 V_1 A_1 - \rho_2 V_2 A_2 - \rho_3 V_3 A_3.$$

6. STREAM FUNCTIONS

The integrals of the differential equations of the streamlines, Eqs. (7), can be represented by a system of equations of the form

$$f(x,y,z) = a, \tag{19}$$

$$g(x,y,z) = b \tag{20}$$

if (x_1, x_2, x_3) are replaced by (x,y,z) for convenience. For any value of the constant a or b, (19) or (20) represents a stream surface in which streamlines are embedded, and the intersections of the surfaces represented by (19) with those represented by (20) are precisely the streamlines. Along any streamline the values of the functions f and g are both constant. These functions will then be called *stream functions*.

Of course the choice of these functions is not unique, and a function $h(f,g)$ can always be chosen such that on a solid boundary h remains constant. That the surface

$$h(f,g) = \text{const} \tag{21}$$

is a stream surface follows from the fact that (21) can be substituted for either (19) or (20) to furnish the solution of (7), a standard result in the theory of partial differential equations. Since a fixed boundary must be a stream surface, (21) will represent a fixed boundary if, for instance, it passes through the trace of that boundary for $x = c$. If this trace is the curve

$$C(y,z) = 0, \tag{22}$$

which may be open or closed, we can solve for y and z in terms of a and b from the equations

$$a = f(c,y,z) \qquad \text{and} \qquad b = g(c,y,z)$$

and obtain

$$y = y(c,a,b) \qquad \text{and} \qquad z = z(c,a,b).$$

Substituting these in $C(y,z)$, we obtain a function $h(a,b)$ which, upon substitution of (19) and (20), leads to a function $h(f,g)$ that vanishes on the curve C at $x = c$, and hence on the solid boundary.

The velocity components can be expressed in terms of the stream functions. Since the f and g surfaces contain the streamlines, their normals must be perpendicular to the streamlines. If subscripts denote partial differentiation, the direction numbers of the normals to the f surfaces are (f_x, f_y, f_z) and those of the normals to the g surfaces are (g_x, g_y, g_z). The direction numbers of the streamlines are (u_1, u_2, u_3) or, for convenience, (u, v, w). Because of the stated perpendicularity,

$$uf_x + vf_y + wf_z = 0, \tag{23}$$

and

$$ug_x + vg_y + wg_z = 0, \tag{24}$$

from which it follows that

$$u = \lambda(f_y g_z - f_z g_y), \qquad v = \lambda(f_z g_x - f_x g_z), \qquad w = \lambda(f_x g_y - f_y g_x), \tag{25}$$

in which λ is an arbitrary function of the coordinates. In vector form, (25) can be written as

$$\mathbf{u} = \lambda(\mathbf{grad}\, f) \times (\mathbf{grad}\, g). \tag{25a}$$

So far we have not investigated whether the equation of continuity is satisfied by (25). If the fluid in question is incompressible, it follows from (25) that

$$\frac{\partial u_i}{\partial x_i} = u_x + v_y + w_z$$

$$= \lambda_x(f_y g_z - f_z g_y) + \lambda_y(f_z g_x - f_x g_z) + \lambda_z(f_x g_y - f_y g_x), \tag{26}$$

since

$$\mathrm{div}\,[(\mathbf{grad}\, f) \times (\mathbf{grad}\, g)] = 0.$$

Again, with (25), (26) can be written as

$$u_x + v_y + w_z = \frac{1}{\lambda}(u\lambda_x + v\lambda_y + w\lambda_z) = u(\ln \lambda)_x + v(\ln \lambda)_y + w(\ln \lambda)_z.$$

For the equation of continuity to be satisfied, λ must be such that

$$u(\ln \lambda)_x + v(\ln \lambda)_y + w(\ln \lambda)_z = 0;$$

that is, $\ln \lambda$ must satisfy the same equation as f and g. Thus $\ln \lambda$, and therefore λ, must be a function of f and g only.

Without loss of generality λ can be taken to be unity, as will now be shown. If F and G are functions of f and g,

$$(\mathbf{grad}\, F) \times (\mathbf{grad}\, G) = (F_f G_g - F_g G_f)(\mathbf{grad}\, f) \times (\mathbf{grad}\, g).$$

Thus, if

$$\lambda = F_f G_g - F_g G_f, \tag{27}$$

then

$$(u, v, w) = (\mathbf{grad}\, F) \times (\mathbf{grad}\, G), \tag{28}$$

in which the new λ is simply 1. For a given function $\lambda(f, g)$, $F(f, g)$, and $G(f, g)$ can always be found such that (27) is satisfied. In fact, F can be chosen so that it vanishes on a solid boundary, (27) can then be solved for G, and there still is a good deal of latitude in the choice of G, since

(27) does not determine G uniquely. Because λ can be taken to be 1 without loss of generality, (25) and (25a) can assume the simple forms

$$u = f_y g_z - f_z g_y, \qquad v = f_z g_x - f_x g_z, \qquad w = f_x g_y - f_y g_x, \qquad (29)$$

and
$$\mathbf{u} = (\mathbf{grad}\,f) \times (\mathbf{grad}\,g). \qquad (29a)$$

The rate of flow through a stream tube made of two f surfaces and two g surfaces bears a simple relationship to the values of the stream functions f and g on these surfaces. A surface intersecting the streamlines may be called a γ *surface*. Two families of curves orthogonal to each other can be introduced on such a surface. If these serve as coordinate curves and the coordinates are denoted by α and β, the square of an infinitesimal distance ds on the surface can be expressed by

$$ds^2 = h^2(d\alpha)^2 + k^2(d\beta)^2,$$

in which h and k are functions of α and β and determine the metric for the γ surface. The rate of flow (or discharge) through a portion S of the γ surface bounded by the traces of

$$f = f_1, \qquad f = f_2, \qquad g = g_1, \qquad g = g_2$$

is, from (29a),

$$Q = \int_S \left(\frac{f_\alpha}{h} \frac{g_\beta}{k} - \frac{f_\beta}{k} \frac{g_\alpha}{h} \right) hk \, d\alpha \, d\beta$$

$$= \int_S \frac{(f,g)}{(\alpha,\beta)} \, d\alpha \, d\beta,$$

in which $(f_\alpha g_\beta - f_\beta g_\alpha)/hk$ is the velocity component normal to the γ surface and

$$\frac{\partial(f,g)}{\partial(\alpha,\beta)} \equiv \begin{vmatrix} f_\alpha & f_\beta \\ g_\alpha & g_\beta \end{vmatrix}$$

is the Jacobian of transformation between the coordinates (α,β) and (f,g) on the γ surface. Hence, if the directions of increasing (α,β,γ) and those of (f,g,γ) are all right-handed or all left-handed (otherwise a negative sign would have to be added),

$$Q = \int_{g_1}^{g_2} \int_{f_1}^{f_2} df \, dg = (f_2 - f_1)(g_2 - g_1). \qquad (30)$$

Thus, the discharge through a stream tube (Fig. 4) bounded by two f surfaces and two g surfaces is equal to the product of the difference of the f values and that of the g values.

For two-dimensional flows parallel to the z plane, g can be chosen to be z and f can be denoted by ψ, which is Lagrange's stream function (apart from a change of sign). Equations (29) then become

$$u = \psi_y, \qquad v = -\psi_x, \qquad w = 0, \qquad (31)$$

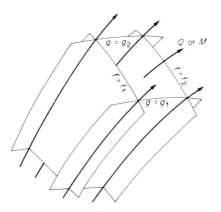

FIGURE 4. VOLUME DISCHARGE Q OR MASS
DISCHARGE M THROUGH A TUBE BOUNDED BY
SURFACES ON WHICH ONE OR THE OTHER
STREAM FUNCTION IS CONSTANT.

and (30) becomes

$$q = \psi_2 - \psi_1, \tag{32}$$

in which q is the discharge per unit distance in the z direction across an
arc with end values ψ_1 and ψ_2 for the stream function (see Fig. 5a).

For flows which are symmetric with respect to an axis, g can be
chosen to be φ for either cylindrical coordinates (r,φ,z) or spherical
coordinates (R,θ,φ). The f is precisely Stokes' stream function ψ, apart
from a difference in sign. If (u,v,w) denote the velocity components in
the directions of (r,φ,z) or (R,θ,φ), we have, for cylindrical coordinates,

$$u = -\frac{1}{r}\psi_z, \qquad v = v(r,z), \qquad w = \frac{1}{r}\psi_r \tag{33}$$

and, for spherical coordinates,

$$u = \frac{1}{R^2 \sin \theta}\psi_\theta, \qquad v = -\frac{1}{R \sin \theta}\psi_R, \qquad w = w(R,\theta). \tag{34}$$

The discharge through an annular space formed by revolving an arc with
end values of ψ equal to ψ_2 and ψ_1 about the axis of symmetry is then
(Fig. 5b)

$$Q = 2\pi(\psi_2 - \psi_1), \tag{35}$$

whether cylindrical or spherical coordinates are used.

If the fluid is compressible but the motion is steady, it can be shown
that in order to satisfy the equation of continuity, λ (which is no longer
a function of f and g only) should be equal to ρ_0/ρ, in which ρ_0 is a

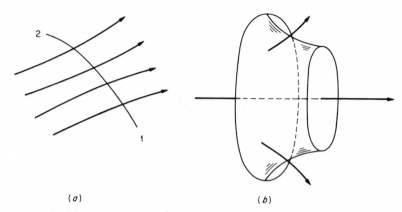

FIGURE 5   (a).   THE DISCHARGE PER UNIT LENGTH PERPENDICULAR TO THE PLANE
OF A TWO-DIMENSIONAL FLOW ACROSS A CURVE 1-2.   (b).   THE DISCHARGE OF AN
AXISYMMETRIC FLOW THROUGH A SURFACE GENERATED BY REVOLVING AN ARC ABOUT
THE AXIS OF SYMMETRY.

reference density.   The appropriate expression for the velocity is

$$\mathbf{u} = \frac{\rho_0}{\rho} \, (\mathbf{grad}\, f) \times (\mathbf{grad}\, g), \tag{36}$$

which, aside from the factor ρ_0/ρ, reduces to (31), (33), or (34) for two-dimensional or axisymmetric flows.   The mass discharge through a stream tube described in Fig. 4 is

$$M = \int_S \rho_0 \frac{\partial(f,g)}{\partial(\alpha,\beta)} \, d\alpha \, d\beta = \int_{g_1}^{g_2} \int_{f_1}^{f_2} \rho_0 \, df \, dg = \rho_0(f_2 - f_1)(g_2 - g_1), \tag{37}$$

which, for two-dimensional and for axisymmetric flows, reduces to equations identical in form to (32) and (35), respectively, except for the factor ρ_0.

7.   RATE OF RELATIVE DISPLACEMENT

The rate of displacement of a fluid particle relative to that of a neighboring one is determined by the differences in velocity components between these two particles du_j.   Since du_j can be computed from

$$du_j = \frac{\partial u_j}{\partial x_i} \, dx_i,$$

in which the dx's are the differences between the coordinates (assumed Cartesian throughout this chapter) of the two particles, the nine quantities $\partial u_j/\partial x_i$ completely describe the rate of relative displacement.   Before exploring the physical significance of these nine quantities further, an important mathematical property they possess in common should be studied first.

The question may be asked: How do these nine quantities transform for a change of the coordinates? For the Cartesian coordinates considered here the answer is straightforward. The original Cartesian coordinates will be denoted by (x_1, x_2, x_3) and the new ones by (x_1', x_2', x_3'), and both sets of coordinates will be assumed to be right-handed. If the cosine of the angle between the x_i' axis and the x_j axis is denoted by a_{ij}, so that the a's are direction cosines, the coordinate transformation for a rotation of the axes can be summarized as follows:

	x_1	x_2	x_3
x_1'	a_{11}	a_{12}	a_{13}
x_2'	a_{21}	a_{22}	a_{23}
x_3'	a_{31}	a_{32}	a_{33}

The six equations

$$x_i = a_{\alpha i} x_\alpha' \qquad \text{and} \qquad x_i' = a_{i\alpha} x_\alpha, \qquad \text{with } i = 1, 2, \text{ and } 3, \qquad (38)$$

then follow from the fact that the projection of a line on another is equal to the sum of the projections of a broken line with the same starting and end points. According to analytic geometry, the cosine of the angle between two lines is equal to the sum of the three products of the direction cosines. Since the coordinates are Cartesian, it follows that

$$a_{i\alpha} a_{j\alpha} = a_{\alpha i} a_{\alpha j} = \delta_{ij}, \qquad (39)$$

in which δ_{ij} is the Kronecker delta, defined by

$$\delta_{ij} = \begin{cases} 0 & \text{for } i \neq j, \\ 1 & \text{for } i = j. \end{cases}$$

The displacement dX_i transforms as x_i, since as far as the transformation is concerned, it is immaterial whether the particle is followed. From (3b) it then follows that the velocity components transform as the coordinates:

$$u_i = a_{\alpha i} u_\alpha', \qquad u_i' = a_{i\alpha} u_\alpha. \qquad (40)$$

Sets of quantities like (x_1, x_2, x_3) and (u_1, u_2, u_3) which obey the law of transformation given in (38) and (40) are called *vectors*.

We are now in a position to investigate the law of transformation of the nine quantities $\partial u_j / \partial x_i$. It follows from

$$\frac{\partial}{\partial x_i} = \frac{\partial x_\beta'}{\partial x_i} \frac{\partial}{\partial x_\beta'} = a_{\beta i} \frac{\partial}{\partial x_\beta'}$$

and the first of (40) that

$$\frac{\partial u_j}{\partial x_i} = a_{\beta i} \frac{\partial (a_{\alpha j} u_\alpha')}{\partial x_\beta'} = a_{\alpha j} a_{\beta i} \frac{\partial u_\alpha'}{\partial x_\beta'} \qquad (41)$$

and conversely that

$$\frac{\partial u_j'}{\partial x_i'} = a_{j\alpha} a_{i\beta} \frac{\partial u_\alpha}{\partial x_\beta}, \tag{42}$$

the right-hand sides of which have nine terms. Sets of quantities obeying the transformation laws (41) and (42) are called *tensors of the second order*. There are tensors of higher orders defined by a similar but generalized law. Furthermore, the coordinates for which this generalized law is valid are not restricted to Cartesian ones, although a_{ij} has to be generalized and the u's may no longer be the physical components of the velocity. A detailed discussion of general coordinates is given in Appendix 2.

8. VORTICITY AND RATE OF DEFORMATION

The physical significance of the tensor $\partial u_j/\partial x_i$ is best explained by resolving it into two parts:

$$\frac{\partial u_j}{\partial x_i} = \frac{1}{2}(e_{ij} + v_{ij}), \tag{43}$$

in which

$$e_{ij} = \frac{\partial u_j}{\partial x_i} + \frac{\partial u_i}{\partial x_j}, \tag{44}$$

$$v_{ij} = \frac{\partial u_j}{\partial x_i} - \frac{\partial u_i}{\partial x_j}. \tag{45}$$

If i and j are all equal to 1,

$$e_{11} = 2\frac{\partial u_1}{\partial x_1},$$

which is twice the rate at which a fluid element is lengthening in the x_1 direction, per unit time and per unit length along that direction. The quantities e_{22} and e_{33} have similar meanings. If i and j are not equal, the physical meanings of e_{ij} and v_{ij} can be seen with the aid of Fig. 6. Suppose the motion is independent of x_3 and a fluid particle $ABCD$ has moved and deformed, after a time interval dt, to $A'B'C'D'$. Since the velocity at B in the x_2 direction is in excess of that at A by the amount $(\partial u_2/\partial x_1) dx_1$, the coordinate x_2 of B' exceeds that of A' by the amount $(\partial u_2/\partial x_1) dx_1 dt$. Similarly, the coordinate x_1 of D' exceeds that of A' by $(\partial u_1/\partial x_2) dx_2 dt$. The lengths AB and AD have changed some during the deformation, but the effect of these changes on the angles $d\theta_1$ and $d\theta_2$ is secondary and vanishes in the limit as the lengths themselves are made smaller and smaller. Since dt is very small, $d\theta_1$ and $d\theta_2$ are also very small and can be used in place of their sines. Thus, if $d\theta_1$ and $d\theta_2$ are measured counterclockwise,

$$d\theta_1 = \frac{\partial u_2}{\partial x_1} dt, \qquad d\theta_2 = -\frac{\partial u_1}{\partial x_2} dt.$$

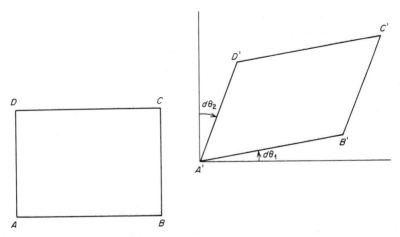

FIGURE 6.   THE ROTATIONS OF ADJACENT SIDES.   HERE $d\theta_1$ IS POSITIVE AND $d\theta_2$
NEGATIVE.

Thus, $$e_{12} = \frac{d}{dt}(\theta_1 - \theta_2), \qquad v_{12} = \frac{d}{dt}(\theta_1 + \theta_2).$$

It is clear that e_{12} is the rate at which the angle BAD is changing with time
and that v_{12} is twice the rate at which the fluid particle is rotating about
the x_3 axis. Although to arrive at this conclusion the motion has been
assumed, for convenience, to be independent of x_3, any dependence on
it has only a secondary effect on the quantities involved, which vanishes
in the limit as the fluid particle is made smaller and smaller. Therefore
the conclusion is true in general, and similar physical interpretations can
be given to e_{23}, v_{32}, e_{31}, v_{13}, etc.

From the definitions of e_{ij} and v_{ij} and with (41) or (42) it can readily
be shown that they are second-order Cartesian tensors, with their transformations governed by formulas identical in structure to (41) and (42).
Since the former measures the rate of deformation, it can be called the
rate-of-deformation tensor. The latter measures twice the rate of rotation,
and is called the *vorticity tensor*.

Since
$$e_{ij} = e_{ji} \qquad \text{and} \qquad v_{ij} = -v_{ji},$$

e_{ij} is symmetric and v_{ij} is antisymmetric. It is the antisymmetry of the
vorticity tensor that accounts for the fact that vorticity (or rotation) can
be considered as a vector. The vorticity tensor has only three essentially
distinct elements, namely, v_{23}, v_{31}, and v_{12}, which will be denoted by ξ_1,
ξ_2, and ξ_3, respectively. According to (42) and for a change of coordinates from x_i to x_i',
$$v_{ij}' = a_{i\alpha}a_{j\beta}v_{\alpha\beta}.$$

Because of the antisymmetry of v_{ij}, only the cases $i \neq j$ and $\alpha \neq \beta$ need be considered, and the last equation can be replaced by

$$\xi'_k = A_{k\gamma}\xi_\gamma, \tag{46}$$

in which k and γ are such that (i,j,k) and (α,β,γ) are cyclic permutations of (1,2,3) and

$$A_{k\gamma} = a_{i\alpha}a_{j\beta} - a_{i\beta}a_{j\alpha} \tag{47}$$

is the cofactor corresponding to the element $a_{k\gamma}$ in the transformation matrix. If the equality of $a_{k\gamma}$ and $A_{k\gamma}$ can be established, the vector nature of ξ_i will be proved.

Consider the determinant

$$D = |a_{ij}| = \begin{vmatrix} a_{11} & a_{12} & a_{13} \\ a_{21} & a_{22} & a_{23} \\ a_{31} & a_{32} & a_{33} \end{vmatrix}. \tag{48}$$

Because of (39),

$$D^2 = \begin{vmatrix} a_{11} & a_{12} & a_{13} \\ a_{21} & a_{22} & a_{23} \\ a_{31} & a_{32} & a_{33} \end{vmatrix} \begin{vmatrix} a_{11} & a_{21} & a_{31} \\ a_{12} & a_{22} & a_{32} \\ a_{13} & a_{23} & a_{33} \end{vmatrix} = \begin{vmatrix} 1 & 0 & 0 \\ 0 & 1 & 0 \\ 0 & 0 & 1 \end{vmatrix} = 1.$$

Thus $$D = \pm 1.$$

The positive sign should be taken for transformation from a right-handed system to a right-handed one or from a left-handed one to a left-handed one, and the negative sign should be taken in the other cases. Since these Cartesian systems have been considered to be right-handed, D is equal to 1. It follows then from the theory of determinants that

$$a_{i\alpha}A_{j\alpha} = \delta_{ij}. \tag{49}$$

Multiplication of this equation by $a_{i\beta}$ and summation over i produce, with (39),

$$\delta_{\alpha\beta}A_{j\alpha} = \delta_{ij}a_{i\beta},$$

or $$A_{j\beta} = a_{j\beta}, \tag{50}$$

which is what needs to be proved. Equation (46) can now be written as

$$\xi'_k = a_{k\gamma}\xi_\gamma, \tag{51}$$

which shows that ξ_i indeed behaves like a vector. Since the three essentially distinct components of v_{ij} are the three components of the vorticity vector $\boldsymbol{\xi}$, the vorticity tensor v_{ij} may be considered as a vector. From the definition of ξ_i it follows, in vector notation, that

$$\boldsymbol{\xi} = \text{curl } \mathbf{u} \tag{52}$$

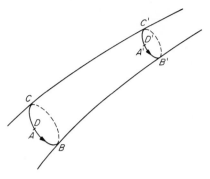

FIGURE 7. TWO CIRCUITS ON A VORTEX
TUBE. THE POINTS A AND D ARE ONE
POINT, AND SO ARE A' AND D'.

and that $\boldsymbol{\xi}$ is just twice the rate-of-rotation vector $\boldsymbol{\omega}$ of the fluid. From (52) it follows immediately that

$$\frac{\partial \xi_i}{\partial x_i} = 0, \tag{53}$$

which is entirely kinematic in nature. The validity of (53) does not depend on either the steadiness of the motion or the properties of the fluid.

9. VORTEX LINES AND CIRCULATION

A line to which vorticity vectors are tangent at all its points is called a *vortex line*. The differential equations of a vortex line are

$$\frac{dx_1}{\xi_1} = \frac{dx_2}{\xi_2} = \frac{dx_3}{\xi_3}. \tag{54}$$

A tube with a wall consisting of vortex lines is called a *vortex tube*.

The circulation Γ along any line is defined by the line integral

$$\Gamma = \int u_i \, dx_i. \tag{55}$$

There is a simple relationship between the circulation around a closed curve and the vorticity of the fluid over any surface bounded by that curve. Consider a vortex tube with circuits (Fig. 7) drawn on its lateral surface. The Stokes theorem states that

$$\int u_i \, dx_i = \int_S l_i \xi_i \, dS = \int_S \xi_n \, dS, \tag{56}$$

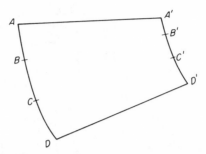

FIGURE 8. THE LATERAL SURFACE OF THE
VORTEX TUBE IN FIG. 7 AFTER CUTTING
ALONG AA', OR DD'.

in which the line integral is performed along a curve enclosing the surface
S and the l's are the direction cosines of the right-handed (with respect
to the circuit) normal to S. If S is the lateral surface of the vortex tube
in Fig. 8, split along AA',

$$\Gamma_{A'B'C'D'} + \Gamma_{D'D} + \Gamma_{DCBA} + \Gamma_{AA'} = \int_S \xi_n \, dS.$$

Because the lateral surface is generated by vortex lines, the vorticity
component ξ_n normal to that surface is zero. Since

$$\Gamma_{D'D} = -\Gamma_{AA'} \quad \text{and} \quad \Gamma_{DCBA} = -\Gamma_{ABCD},$$

it follows that[1]

$$\Gamma_{A'B'C'D'} = \Gamma_{ABCD},$$

or, from (56) again,

$$\xi_n \sigma = \xi'_n \sigma', \tag{57}$$

in which σ and σ' represent the areas of the surfaces $ABCD$ and $A'B'C'D'$,
respectively, and ξ_n and ξ'_n are now the respective *mean* values of the
vorticity components normal to them. If σ and σ' are infinitesimal, ξ_n
and ξ'_n are the local values of the normal component of the vorticities.

If U_i in (9) is taken to be ξ_i and that equation is applied to a closed
surface S, it follows from (53) that

$$\int_S \xi_n \, dS = \int_S l_i \xi_i \, dS = 0, \tag{58}$$

in which ξ_n is the vorticity component normal to S and the l's now have
the same meaning as they have in (9). From (58) we can conclude, as

---

[1] This result was proved by Helmholtz (1858), who noted that the divergence of the
vorticity vector is zero and used Green's theorem, or (9), to reach the result. The
present proof, due to Lord Kelvin (1824–1907) [W. Thomson (1869)], is really just a
paraphrase of Helmholtz'.

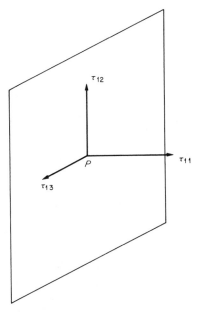

FIGURE 9. THE STRESS COMPONENT τ_{11}
ACTS IN A DIRECTION NORMAL TO THE
PLANE IN WHICH τ_{12} AND τ_{13} ACT.

we did with streamlines, that vortex lines cannot terminate in the interior
of the fluid. In fact, the similarity of (14) and (53) and the similarity
of (7) and (54) enable us to apply the results obtained for streamlines and
velocity to vorticity lines and vorticity, respectively. Thus, if we choose,
we may speak of vorticity functions in terms of which the vorticity
components can be expressed, just as the velocity components are expressed in terms of the stream functions in (29). Then it follows from an
equation similar to (30) that the circulation around a vortex tube made of
vorticity surfaces (which is merely a vorticity discharge, as we have seen)
is precisely the product of the differences of the vorticity functions
pertaining to these vorticity surfaces.

10. THE STRESS TENSOR

So far we have dealt only with continuity and the fundamental kinematics
of fluid motion. Now a set of dynamical quantities, the stresses, will
be considered, to find their law of transformation as the coordinates are
transformed. Imagine a plane perpendicular to the x_1 direction passing
through a point P in the fluid, with an outward (away from the part of
the fluid under consideration) normal in the x_1 direction. The force
acting on a small element of this plane (Fig. 9) containing the point P

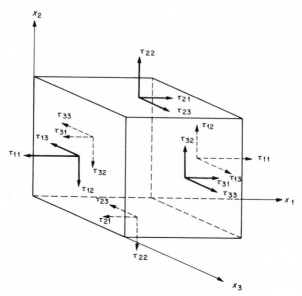

FIGURE 10.   DEFINITION SKETCH FOR STRESS COMPONENTS.

can be resolved into three components, one in each coordinate direction. The limits of the ratios of these components to the area of the plane element as the area approaches zero are called the *normal stress* (τ_{11}) and the *shearing stresses* (τ_{12} and τ_{13}) at the point P.   By passing planes perpendicular to the other two directions of the coordinate axes we can define the remaining stress components.   The nine components of the stress will be denoted by τ_{ij}, with i and j ranging over 1, 2, 3.   The first subscript indicates the axis which is normal to the plane on which the stress component acts (Fig. 10), and the second subscript indicates the direction in which the stress component is acting (if the outward normal to the surface is in the x_i direction) or the opposite of that direction (if the outward normal is in the negative x_i direction) when that stress component is positive.

To find the law of transformation of τ_{ij}, a fluid element (Fig. 11) bounded by the coordinate planes and a plane ABC perpendicular to the x_1' axis of the new coordinate system may be considered.   The areas of the four triangles of the tetrahedron satisfy the following relationships:

$$\triangle BCO = a_{11}\triangle ABC, \qquad \triangle CAO = a_{12}\triangle ABC, \qquad \triangle ABO = a_{13}\triangle ABC.$$

Although the fluid element is in general not in a state of equilibrium, the term for the effective or the body force contains as a factor the volume of the fluid element and hence is an infinitesimal of higher order and can be neglected in comparison with the surface forces, which can thus be

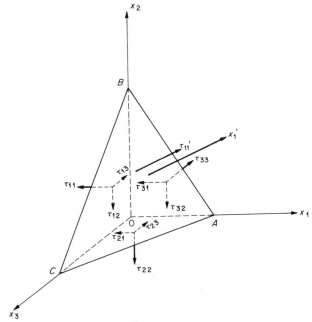

FIGURE 11. THE TETRAHEDRON ILLUSTRATING THE TRANSFORMA-
TION OF STRESS COMPONENTS. ONLY τ'_{11} IS SHOWN ON ABC.

considered to be in equilibrium. Resolving the forces acting on the
fluid element in the directions x'_i and imposing the condition of equilibrium produce, after the common factor $\triangle ABC$ is canceled out,

$$\tau'_{1i} = a_{1\alpha}a_{i\beta}\tau_{\alpha\beta}.$$

If tetrahedrons similar to $OABC$ are considered, it follows that in general

$$\tau'_{ij} = a_{i\alpha}a_{j\beta}\tau_{\alpha\beta}, \tag{59}$$

and conversely [by applying (39)]

$$\tau_{ij} = a_{\alpha i}a_{\beta j}\tau'_{\alpha\beta}. \tag{60}$$

Thus the set of quantities τ_{ij}, like e_{ij}, transforms as a tensor. If the
inertial effects and other effects of a higher order are again neglected, the
rotational equilibrium of the shearing stresses (Fig. 12) demands that

$$\tau_{ij} = \tau_{ji},$$

so that, like e_{ij}, τ_{ij} is a symmetric tensor.

From the condition of equilibrium of the tetrahedron in Fig. 11
it follows also that the x_i component of the force acting on the area
ABC is

$$F_i = a_{1\alpha}\tau_{\alpha i}\triangle$$

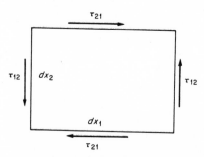

FIGURE 12.   THE SIDE dx_3 IS NORMAL TO
THE PLANE OF PAPER.   THE COUPLE OF
THE STRESS PAIR τ_{12} IS $\tau_{12}\ dx_2\ dx_3\ dx_1$,
AND THE COUPLE OF THE STRESS PAIR τ_{21}
IS $-\tau_{21}\ dx_1\ dx_3\ dx_2$.   HENCE $\tau_{12} = \tau_{21}$.
SIMILARLY $\tau_{ij} = \tau_{ji}$.

if we use $\triangle$ to denote $\triangle ABC$.   The direction of x_1' is the direction of the normal n to the surface ABC.   Therefore $F_i/\triangle$ can be denoted by τ_{ni}, and the above equation can be written

$$\tau_{ni} = a_{n\alpha}\tau_{\alpha i} = l_\alpha \tau_{\alpha i}, \tag{61}$$

in which l_1, l_2, and l_3 are the direction cosines of the normal n.

11.   PRINCIPAL DIRECTIONS

Of the nine elements of e_{ij} or τ_{ij}, six are distinct.   Several questions then present themselves.   Are there some directions for which the pertinent rates of shearing deformation or the pertinent shear stresses are zero? If so, how many and are they somehow related?   Are these directions (called *principal directions*) the same for the rates of deformation as for the stresses?   The answer to the last question is essential to the formulation of the constitutive equations and hence the equations of motion of a fluid.

To answer the first question, the direction of x_j' (j fixed) is assumed to be a principal direction for the rates of deformation (so that e_{ij}' is zero if i is not equal to j).   Whether such a direction really exists can then be tested.   The conditions on the rates of deformation pertinent to this direction are

$$e_{ij}' = e\delta_{ij}. \tag{62}$$

Multiplication of the equation

$$e_{ij}' = a_{i\alpha}a_{j\beta}e_{\alpha\beta} \tag{63}$$

by a_{ik} and summation over i produce, by virtue of (39) and (62),

$$a_{ik}e_{ij}' = a_{ik}\delta_{ij}e = a_{jk}e = a_{ik}a_{i\alpha}a_{j\beta}e_{\alpha\beta} = a_{j\beta}\delta_{k\alpha}e_{\alpha\beta} = a_{j\beta}e_{k\beta} \tag{64}$$

for a definite j and for $k = 1, 2,$ and 3. Written out in full, the three equations

$$a_{jk}e = a_{j\beta}e_{k\beta} \tag{64a}$$

contained in (64) are

$$(e_{11} - e)a_{j1} + e_{12}a_{j2} + e_{13}a_{j3} = 0,$$

$$e_{21}a_{j1} + (e_{22} - e)a_{j2} + e_{23}a_{j3} = 0, \tag{65}$$

$$e_{31}a_{j1} + e_{32}a_{j2} + (e_{33} - e)a_{j3} = 0.$$

Since Eqs. (65) are linear and homogeneous in a_{j1}, a_{j2}, and a_{j3}, which are direction cosines of the x'_j axis referred to the coordinate system (x_1, x_2, x_3) and therefore must not all vanish, it is necessary and sufficient for the existence of the direction we are looking for that the secular equation (in e)

$$D(e) \equiv \begin{vmatrix} e_{11} - e & e_{12} & e_{13} \\ e_{21} & e_{22} - e & e_{23} \\ e_{31} & e_{32} & e_{33} - e \end{vmatrix} = 0 \tag{66}$$

be satisfied and have real roots. The sufficiency follows from the fact that, with real roots of e, two of (65) can be solved with

$$a_{j1}^2 + a_{j2}^2 + a_{j3}^2 = 1 \tag{39a}$$

to furnish real solutions for a_{j1}, a_{j2}, and a_{j3}.

That (66) does have real roots follows from the fact that e_{ij} is symmetric. The proof is simple. If the roots were complex, the a's would be complex. Multiplication of (64a) by a_{jk}^* (the complex conjugate of a_{jk}) produces

$$ea_{jk}a_{jk}^* = a_{j\beta}a_{jk}^*e_{k\beta}. \tag{67}$$

The complex conjugate of the left-hand side is,[1] because $e_{\beta k} = e_{k\beta}$,

$$e^*a_{jk}^*a_{jk} = a_{j\beta}^*a_{jk}e_{k\beta} = a_{j\beta}a_{jk}^*e_{\beta k} = a_{j\beta}a_{jk}^*e_{k\beta}, \tag{68}$$

which is precisely $ea_{jk}a_{jk}^*$. Consequently $e = e^*$, and any root of (66) is real. By virtue of (65) and (39a), the reality of these roots implies the reality of the direction cosines of the principal directions, and hence the existence of these directions. The proof for the reality of the roots of (66) is valid for the secular equation of any second-order symmetric tensor, no matter what the number of dimensions is, and in particular is valid for the secular equation of the stress tensor. Consequently, principal directions for stresses exist also.

Equation (66) has three roots (e_1, e_2, and e_3), all of which are now known to be real. Corresponding to each root e_i there is therefore a set of direction cosines a_{ik} ($k = 1, 2, 3$) indicating one of the principal directions. Are the three principal directions orthogonal? This is a

---

[1] Remember that the components $e_{k\beta}$ are real.

special form of our second question. Now (64) or (65) can be written as

$$e_1 a_{1k} = a_{1\beta} e_{k\beta}, \qquad e_2 a_{2k} = a_{2\beta} e_{k\beta}, \qquad e_3 a_{3k} = a_{3\beta} e_{k\beta}, \qquad k = 1, 2, 3.$$
(69)

Multiplying the first of these by a_{2k} and the second by a_{1k}, summing over k, and taking the difference of the two resulting equations, we have, since the repeated indices can be interchanged,

$$(e_1 - e_2) a_{1k} a_{2k} = a_{1\beta} a_{2k} e_{k\beta} - a_{1k} a_{2\beta} e_{k\beta}$$

$$= a_{1\beta} a_{2k} e_{k\beta} - a_{1\beta} a_{2k} e_{\beta k}.$$
(70)

But $e_{k\beta}$ is equal to $e_{\beta k}$, so that

$$(e_1 - e_2) a_{1k} a_{2k} = 0.$$
(71)

If e_1 is not equal to e_2, it follows that

$$a_{1k} a_{2k} = 0,$$

and the second principal direction is perpendicular to the first. By taking the principal directions by pairs, we can show that if the roots of the secular equation are all different, the principal directions are orthogonal.

If two of the roots are equal, the principal direction corresponding to the remaining root is, aside from the sense of the direction, uniquely determined. The other two principal directions can be any direction in the plane normal to the first principal direction and in particular can be any two orthogonal directions in such a plane. If all three of the roots of (66) are equal, any direction is a principal direction, and in particular any three orthogonal directions can be taken to be the principal directions. The mathematical proofs of these statements involve a straightforward discussion of the rank of the determinant in (66) for multiple roots of e and will be given in Prob. 14.

The third question asked at the beginning of this section was: Are the principal directions the same for the rates of deformation as for stresses? The affirmative answer follows from the assumed isotropy of the fluid. If the rates of shearing corresponding to an orthogonal set of directions are zero, in which directions will the corresponding shear stresses act if they are not zero? They may act in one direction as well as in the opposite one if the fluid is isotropic, so that there are no privileged directions. Since it is reasonable to assume that the state of stress must be unique for given rates of deformation, it follows that these shear stresses are zero and that the principal directions for the stresses are identical with those for the rates of deformation.

12. INVARIANTS

Equation (66) can be written in the form

$$e^3 - I_1 e^2 + I_2 e - I_3 = 0,$$
(72)

in which $\qquad I_1 = e_{ii},$ $\qquad\qquad\qquad\qquad\qquad\qquad\qquad\qquad\qquad$ (73)

$$I_2 = \begin{vmatrix} e_{11} & e_{12} \\ e_{21} & e_{22} \end{vmatrix} + \begin{vmatrix} e_{22} & e_{23} \\ e_{32} & e_{33} \end{vmatrix} + \begin{vmatrix} e_{11} & e_{13} \\ e_{31} & e_{33} \end{vmatrix},$$ (74)

$$I_3 = \begin{vmatrix} e_{11} & e_{12} & e_{13} \\ e_{21} & e_{22} & e_{23} \\ e_{31} & e_{32} & e_{33} \end{vmatrix}.$$ (75)

The I's are invariants; i.e., they do not change with a rotational change of axes to which the rates of deformation are referred. Therefore, the roots of (72) do not change with a rotation of the coordinate axes, and the principal rates of deformation and principal directions are essentially unique (apart from senses of the directions and the degenerate cases), whatever the original rates are from which the principal rates are determined. The proof of the invariance of the I's follows.

Any sum of the form

$$S = e_{\alpha\beta}e_{\beta\gamma}e_{\gamma\delta}\cdots e_{\nu\alpha}$$

is invariant. For instance, by the law of transformation of e_{ij},

$$e'_{ij}e'_{ji} = a_{i\alpha}a_{j\beta}a_{j\gamma}a_{i\delta}e_{\alpha\beta}e_{\gamma\delta} = \delta_{\alpha\delta}\delta_{\beta\gamma}e_{\alpha\beta}e_{\gamma\delta} = e_{\alpha\beta}e_{\beta\alpha},$$

which shows that $e_{\alpha\beta}e_{\beta\alpha}$ is an invariant. The demonstration for higher orders is entirely similar. Indeed, any sum of the form of S or a combination of such sums is invariant. Since I_1 is $e_{\alpha\alpha}$ and

$$I_2 = \tfrac{1}{2}(I_1{}^2 - e_{\alpha\beta}e_{\beta\alpha}),$$ (76)

I_1 and I_2 are invariant.

If the x' directions are the principal directions, however determined, and e'_{11} is denoted by e_1, e'_{22} by e_2, and e'_{33} by e_3, we have

$$e'_{ij}e'_{jk}e'_{kl} = e_i{}^3\delta_{il}$$

$$= [e_1e_2e_3 - (e_1e_2 + e_2e_3 + e_3e_1)e_i + (e_1 + e_2 + e_3)e_i{}^2]\delta_{il},$$
$$\text{not summed over } i,$$

no matter what i is. Thus,

$$e'_{ij}e'_{jk}e'_{kl} = (I_3 - I_2e_i + I_1e_i{}^2)\delta_{il}$$

$$= I_3\delta_{il} - I_2e'_{il} + I_1e'_{ij}e'_{jl}, \qquad \text{not summed over } i.$$ (77)

In the x system it follows from (77) by the law of transformation that

$$e_{\alpha\beta}e_{\beta\gamma}e_{\gamma\delta} = a_{i\alpha}a_{j\beta}a_{m\beta}a_{k\gamma}a_{n\gamma}a_{l\delta}e'_{ij}e'_{mk}e'_{nl} = a_{i\alpha}a_{l\delta}e'_{ij}e'_{jk}e'_{kl}$$

$$= a_{i\alpha}a_{l\delta}(I_3\delta_{il} - I_2e'_{il} + I_1e'_{ij}e'_{jl})$$

$$= a_{i\alpha}a_{i\delta}I_3 - I_2e_{\alpha\delta} + I_1a_{i\alpha}a_{j\beta}a_{p\beta}a_{l\delta}e'_{ij}e'_{vl}$$

$$= I_3\delta_{\alpha\delta} - I_2e_{\alpha\delta} + I_1e_{\alpha\beta}e_{\beta\delta}.$$ (78)

From this it follows that

$$e_{\alpha\beta}e_{\beta\gamma}e_{\gamma\alpha} = I_3 - I_2 e_{\alpha\alpha} + I_1 e_{\alpha\beta}e_{\beta\alpha},$$

so that, in view of (73) and (76), I_3 is a combination of sums of the form of S and is therefore invariant.

Furthermore, from (78) it follows that

$$e_{\alpha\beta}e_{\beta\gamma}e_{\gamma\delta}e_{\delta\alpha} = I_3\delta_{\alpha\delta}e_{\delta\alpha} - I_2 e_{\alpha\delta}e_{\delta\alpha} + I_1 e_{\alpha\beta}e_{\beta\delta}e_{\delta\alpha}$$

$$= I_3 e_{\alpha\alpha} - I_2 e_{\alpha\delta}e_{\delta\alpha} + I_1 e_{\alpha\beta}e_{\beta\delta}e_{\delta\alpha},$$

so that all sums of the form of S and of order 4 can be expressed in terms of those of order less than 4. The process can be repeated indefinitely, and we arrive at the conclusion that all sums of the form of S (and hence all invariants of e_{ij}) can be expressed in terms of those of order less than 4, or, equivalently, in terms of I_1, I_2, and I_3, which are, therefore, the three essential invariants.

13. RELATIONSHIP BETWEEN STRESSES AND RATES OF DEFORMATION

Since

$$e'_{ik}e'_{kj} = (a_{i\alpha}a_{k\gamma}e_{\alpha\gamma})(a_{k\delta}a_{j\beta}e_{\delta\beta}) = a_{i\alpha}a_{j\beta}\delta_{\gamma\delta}e_{\alpha\gamma}e_{\delta\beta} = a_{i\alpha}a_{j\beta}e_{\alpha\gamma}e_{\gamma\beta},$$

the sum $e_{ik}e_{kj}$ transforms like e_{ij} and is called a *contracted tensor*. Other contracted tensors are

$$e_{ik}e_{kl}e_{lj}, \qquad e_{ik}e_{kl}e_{lm}e_{mj}, \qquad \text{etc.}$$

If the stress tensor τ_{ij} is expressed in terms of the rate of deformation e_{ij}, the expression must be a sum of terms each of which transforms like e_{ij}—or τ_{ij}. The most general relationship between stresses and the rates of deformation is then, for the isotropic pure[1] fluids considered in this book,

$$\tau_{ij} = F_0(I)\delta_{ij} + F_1(I)e_{ij} + F_2(I)e_{ik}e_{kj} + F_3(I)e_{ik}e_{kl}e_{lj} + \cdots, \quad (79)$$

in which the functions $F_n(I)$ are functions of the invariants and therefore functions of I_1, I_2, and I_3 only; the symbol I is used to denote these three essential invariants collectively. Because of (78) and its extensions, (79) can be written as [Reiner (1945)]

$$\tau_{ij} = f_0(I)\delta_{ij} + f_1(I)e_{ij} + f_2(I)e_{ik}e_{kj}, \quad (80)$$

in which, of course, the functions f_0, f_1, and f_2 must be such that the equation is dimensionally homogeneous. All the mechanical properties of the fluid pertinent to its motion are contained in the three functions f.

---

[1] A pure fluid is defined in this book as a fluid in which the stresses depend only on the present rates of deformation, not on deformation itself or on the history of motion.

The form of (79) or (80) implies the assumption that the principal directions of the stresses are identical with those of the rates of deformation.
It also implies that the stresses are independent of the vorticity, according
to the so-called *principle of material indifference* (to rotation).

A Newtonian fluid may be defined as a fluid for which the relationship
between the stresses and the rates of deformation is linear. Therefore,
for a Newtonian fluid, $f_2(I)$ must be zero, $f_1(I)$ must be a constant (independent of I_1, I_2, and I_3), and $f_0(I)$ can contain I_1 only and that to the
first power. Hence (80) must have the form [Stokes (1845)]

$$\tau_{ij} = (-p + \lambda\theta)\delta_{ij} + \mu e_{ij}, \tag{81}$$

in which p is the pressure corresponding to the local density and temperature if the effects of the motion are neglected,

$$\theta = \frac{e_{\alpha\alpha}}{2} = \frac{\partial u_\alpha}{\partial x_\alpha}$$

is the *dilatation*, or rate of volume expansion, and λ and μ are physical
constants. Equation (81) can be explained in the following way. If
there were no motion, θ and e_{ij} would be zero and τ_{ij} would be (as it
should) the isotropic stress tensor $-p\delta_{ij}$. The effect of motion, under the
assumption of linearity of the relationship between stresses and rates
of deformation, is twofold. First, the isotropic part of the stress tensor
will be modified by an amount $\lambda\theta$, proportional to the rate of volume
expansion. Second, an additional modification is effected by the tensor
μe_{ij}.

If preferred, (81) can be obtained by considering the relationship
between principal stresses τ_1, τ_2, and τ_3 and principal rates of deformation
e_1, e_2, and e_3. Because of symmetry this relationship (for Newtonian
fluids) must be

$$\tau_i = -p + \lambda\theta + \mu e_i, \qquad i = 1, 2, 3. \tag{82}$$

A transformation of coordinates then results in (81).

From (81) it follows by "contraction" that

$$\tau_{ii} = -3p + 3\lambda\theta + \mu e_{ii} = -3p + (3\lambda + 2\mu)\theta. \tag{83}$$

Since p has been assumed, with good reason, to be the thermodynamic
pressure at equilibrium condition and not the mean value of the normal
stresses when the fluid is in motion, $3\lambda + 2\mu$ is not zero in general.
(The mean value of the normal stresses is $\tau_{ii}/3$ and can be expected to
depend on the motion.) In fact one-third of this quantity is denoted by
μ_v and is called the *bulk* or *volume viscosity*. Had p been defined to be
$\tau_{ii}/3$, then $3\lambda + 2\mu$ would indeed be zero. But p can no longer be identified with the thermodynamic pressure and calculated from the local
temperature and density, for surely the mean normal stress should depend

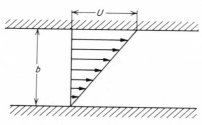

FIGURE 13. VELOCITY DISTRIBUTION IN
THE FLUID BETWEEN TWO PARALLEL
PLATES THE LOWER OF WHICH IS STATION-
ARY AND THE UPPER OF WHICH IS MOVING
WITH VELOCITY U IN ITS OWN PLANE.

on the state of motion. The following quotation from Stokes (1845) shows that he was quite aware of the difficulty involved in the rheological aspect of hydrodynamics:

> ... we may at once put $\mu_v = 0$, if we assume that in the case of a uniform motion of dilatation the pressure at any instant depends only on the actual density and temperature at that instant and not on the rate at which the former changes with the time. In most cases to which it would be interesting to apply the theory of the friction of fluids, the density of the fluid is either constant or may without sensible error be regarded as constant, or else changes slowly with time. In the first two cases, the results would be the same and in the third nearly the same, whether μ_v were equal to zero or not. Consequently, if theory and experiments should in such cases agree, the experiments must not be regarded as confirming that part of the theory which relates to supposing μ_v to be equal to zero.

Although it is known that Stokes' assumption does not always give valid results for ultrasonic absorption, it is commonly accepted that it causes little error in most cases of interest, especially for monatomic gases. In the following chapters μ_v will be assumed to be zero even if p is regarded as the thermodynamic pressure at equilibrium condition. After this assumption, (81) becomes

$$\tau_{ij} = \left(-p - \frac{2\mu}{3}\theta\right)\delta_{ij} + \mu e_{ij}. \tag{84}$$

Of course, for a strictly incompressible fluid, (82) reduces to $\tau_{ii} = -3p$, and, as Stokes indicated, the question of whether the volume viscosity vanishes or not is then immaterial.

The physical meaning of μ will now be investigated. Imagine a layer of incompressible fluid between an infinite fixed plate and a parallel one moving in its own plane with velocity U in the direction of x_1 (Fig. 13). Even without the equations of motion it is possible to see that for constant temperature and constant ρ the motion of the fluid described by

$$u_1 = \frac{Ux_2}{b}, \qquad u_2 = u_3 = 0, \tag{85}$$

$$p = \text{const} - g\rho x_2 \qquad g = \text{grav. accel.} \tag{86}$$

is both kinematically and dynamically possible. It is kinematically possible because the equation of continuity is satisfied and the boundary conditions for u_1 are satisfied if it is assumed that there is no slip between fluid and solid. It is dynamically possible because there is no acceleration according to (85) and there is no unbalanced force on the fluid particle (Fig. 13) according to (84) and (86). In fact, as will be shown later, (85) and (86) constitute the solution to the problem of steady fluid flow under the prescribed conditions if the flow is assumed to be laminar, i.e., to have smooth streamlines. Thus, the shear stress can be computed from (84) and (85) to be

$$\tau_{12} = \mu \frac{du_1}{dx_2} = \mu \frac{U}{b},$$

which states that μ is the "constant" of proportionality between the shear stress and the relevant rate of deformation, which here happens to be the velocity gradient U/b. This local constant of proportionality is called *viscosity*, which can vary in space and with time.

In the analogous case of heat conduction between two horizontal plates where the lower plate has a constant temperature taken as the datum and the upper plate has a temperature T, the heat transferred per unit area of any hoizontal plane through the fluid is kT/b, k being the thermal conductivity. Thus μ corresponds to k and can be considered as a coefficient of momentum transfer. Since $\alpha = k/\rho c$ (c is the specific heat which may be for constant volume or for constant pressure) is the *thermal diffusivity*, the corresponding quantity $\nu = \mu/\rho$ (called the *kinematic viscosity*) can be regarded as the momentum diffusivity of the fluid. The ratio ν/α is an index of the capacity of the fluid to diffuse momentum as compared with its capacity to diffuse heat. It is usually denoted by σ and called the *Prandtl number*. For most gases the Prandtl number is of the magnitude of 1. Another analogous phenomenon is mass diffusion, which can be compared with momentum diffusion in the same way.

PROBLEMS

1 Show from the meaning of D/Dt that the equation of incompressibility (13) can be written as

$$\frac{\partial \rho}{\partial t} + \mathbf{v} \cdot \mathbf{grad}\ \rho = 0$$

in any coordinates whatever. Write out $D\rho/Dt$ for cylindrical and spherical coordinates.

2 For two-dimensional flows of an incompressible fluid the equation of continuity is

$$\frac{\partial u}{\partial x} + \frac{\partial v}{\partial y} = 0.$$

Show that this is necessary and sufficient for the existence of a stream function ψ, in terms of which

$$u = \frac{\partial \psi}{\partial y}, \qquad v = -\frac{\partial \psi}{\partial x}.$$

Hint: The necessity part is simple. To prove sufficiency, use the Stokes theorem.

3 Does the solution to the previous problem make it obvious that the following equations hold for steady two-dimensional flows of a compressible fluid?

$$u = \frac{\rho_0}{\rho} \frac{\partial \psi}{\partial y} \qquad \text{and} \qquad v = -\frac{\rho_0}{\rho} \frac{\partial \psi}{\partial x}.$$

Explain. Why must the two-dimensional flow be steady for a stream function to exist for a compressible fluid but not for an incompressible fluid?

4 Use a figure in the form of a curvilinear block to show that the equation of continuity in cylindrical coordinates is

$$\frac{\partial \rho}{\partial t} + \frac{1}{r} \frac{\partial (\rho r u)}{\partial r} + \frac{1}{r} \frac{\partial (\rho v)}{\partial \varphi} + \frac{\partial (\rho w)}{\partial z} = 0.$$

5 Show that if the fluid is incompressible, the equation of continuity in the preceding problem becomes

$$\frac{\partial (r u)}{\partial r} + \frac{\partial v}{\partial \varphi} + \frac{\partial (r w)}{\partial z} = 0.$$

6 Show that for steady axisymmetric flows of a compressible fluid the equation of continuity,

$$\frac{\partial (\rho r u)}{\partial r} + \frac{\partial (\rho r w)}{\partial z} = 0,$$

is necessary and sufficient for the existence of a stream function (the Stokes stream function) ψ, in terms of which

$$u = -\frac{\rho_0}{\rho r} \frac{\partial \psi}{\partial z}, \qquad w = \frac{\rho_0}{\rho r} \frac{\partial \psi}{\partial r}.$$

7 Use a figure in the form of a curvilinear block to show that the equation of continuity in spherical coordinates is

$$\frac{\partial \rho}{\partial t} + \frac{1}{R^2} \frac{\partial}{\partial R}(\rho R^2 u) + \frac{1}{R \sin \theta} \frac{\partial}{\partial \theta}(\rho v \sin \theta) + \frac{1}{R \sin \theta} \frac{\partial(\rho w)}{\partial \varphi} = 0.$$

8 Show that if the fluid is incompressible, the equation of continuity in the preceding problem becomes

$$\frac{\partial}{\partial R}(R^2 u \sin \theta) + \frac{\partial}{\partial \theta}(Rv \sin \theta) + \frac{\partial}{\partial \varphi}(Rw) = 0.$$

9 Show that for steady axisymmetric flows of a compressible fluid the equation of continuity

$$\frac{\partial(\rho R^2 u \sin \theta)}{\partial R} + \frac{\partial(\rho Rv \sin \theta)}{\partial \theta} = 0$$

is necessary and sufficient for the existence of the Stokes stream function ψ, in terms of which

$$u = -\frac{\rho_0}{\rho R^2 \sin \theta} \frac{\partial \psi}{\partial \theta}, \qquad v = \frac{\rho_0}{\rho R \sin \theta} \frac{\partial \psi}{\partial R}.$$

10 Consider two-dimensional flows of an incompressible fluid. Show directly from the equation for streamlines and from

$$u = \frac{\partial \psi}{\partial y}, \qquad v = -\frac{\partial \psi}{\partial x}$$

that $\psi = $ const on a streamline. Similarly, show directly from the equation for streamlines that the Stokes stream function, whenever it exists, is constant along a streamline.

11 A torus is formed if a circle with center at point C of the line OC lying in the plane of the circle is revolved about a line BO perpendicular to OC and situated in that plane. The toroidal coordinates are the angle θ made by OC and any arbitrarily chosen reference position of OC, the magnitude r of the radius vector measured from C, and its angular distance φ measured from the direction of OC. If there is symmetry of flow about BO, what is the equation of continuity in toroidal coordinates? If, in addition, the fluid is incompressible, does there exist a stream function? If so, how can the velocity components be expressed in terms of the stream function?

12 Prove (42) from (41). *Hint:* Use (39).

13 From a physical point of view, the principal rates of deformation, which are the roots of (72) for e, should be independent of the choice of the Cartesian coordinates to which the components e_{ij} correspond. Does this fact indicate that I_1, I_2, and I_3 are invariants? Explain.

14 Discuss the cases in which (66) has a double or a triple root. The following discussion is given in full because of the rather difficult nature of the problem. The reader may produce his own discussion.

If two of the roots are equal, the principal direction corresponding to the remaining root is (aside from the sense of the direction) uniquely determined. The other two principal directions can be any direction in the plane normal to the first principal direction and in particular can be any two orthogonal directions in such a plane. If all three of the roots of (66) are equal, any direction is a principal direction, and in particular any three orthogonal directions can be taken to be the principal directions. To prove these statements, the rank of the determinant in (66) for multiple roots of e must be considered first.

The rank of a determinant of n rows and n columns is n if the determinant does not vanish and m if *all* its minors with $n - m$ rows and $n - m$ columns vanish. If e is equal to a root of (66), $D(e)$ vanishes and its rank is (as will presently be shown) 2, 1, or 0, according as the root is a simple, double, or triple root. Consider the equations

$$D(e + x) = 0, \qquad D(e - x) = 0 \tag{a}$$

as equations in x. If e is a multiple root of (66), the number zero (for x) is a multiple root of these equations, with exactly the same multiplicity, as is immediately clear upon factorization of the left-hand sides. Thus, if e is equal to a root of (66), zero is a multiple root of the equation (considered as one in x)

$$D(e + x)D(e - x) = 0 \tag{b}$$

with multiplicity 2, 4, or 6 according as e is a simple, double, or triple root of (66). If, for convenience, the elements of the determinant $D(e)$ are denoted by b_{ij}, from the rule of multiplication of determinants it follows that

$$D(e + x)D(e - x) = \begin{vmatrix} c_{11} - x^2 & c_{12} & c_{13} \\ c_{21} & c_{22} - x^2 & c_{23} \\ c_{31} & c_{32} & c_{33} - x^2 \end{vmatrix}, \tag{c}$$

in which

$$c_{ij} = b_{i\alpha}b_{j\alpha} + (b_{ji} - b_{ij})x,$$

or, since b_{ij} is equal to b_{ji}, thanks to symmetry,

$$c_{ij} = b_{i\alpha}b_{j\alpha}. \tag{d}$$

Equation (b) can then be written as

$$x^6 - \sigma_1 x^4 + \sigma_2 x^2 - \sigma_3 = 0, \tag{e}$$

in which σ_n ($n = 1, 2, 3$) is the sum of the principal minors of n rows and n columns of the determinant

$$\begin{vmatrix} c_{11} & c_{12} & c_{13} \\ c_{21} & c_{22} & c_{23} \\ c_{31} & c_{32} & c_{33} \end{vmatrix}.$$

It is easy to see that, from (d),

$$\sigma_1 = c_{11} + c_{22} + c_{33} = c_{ii} = b_{i\alpha}b_{i\alpha} \qquad (f)$$

and that, from (c),

$$\sigma_3 = [D(e)]^2. \qquad (g)$$

Furthermore,

$$\sigma_2 = \begin{vmatrix} c_{11} & c_{12} \\ c_{21} & c_{22} \end{vmatrix} + \begin{vmatrix} c_{22} & c_{23} \\ c_{32} & c_{33} \end{vmatrix} + \begin{vmatrix} c_{11} & c_{13} \\ c_{31} & c_{33} \end{vmatrix}. \qquad (h)$$

The first determinant on the right-hand side of (h) is, written out in full according to (d),

$$\begin{vmatrix} b_{11}b_{11} + b_{12}b_{12} + b_{13}b_{13} & b_{11}b_{21} + b_{12}b_{22} + b_{13}b_{23} \\ b_{21}b_{11} + b_{22}b_{12} + b_{23}b_{13} & b_{21}b_{21} + b_{22}b_{22} + b_{23}b_{23} \end{vmatrix},$$

which can be resolved into nine determinants, of which three are zero because of the proportionality of rows or columns. Thus,

$$\begin{vmatrix} c_{11} & c_{12} \\ c_{21} & c_{22} \end{vmatrix} = \begin{vmatrix} b_{11} & b_{12} \\ b_{21} & b_{22} \end{vmatrix} (b_{11}b_{22} - b_{12}b_{21})$$

$$+ \begin{vmatrix} b_{11} & b_{13} \\ b_{21} & b_{23} \end{vmatrix} (b_{11}b_{23} - b_{13}b_{21})$$

$$+ \begin{vmatrix} b_{12} & b_{13} \\ b_{22} & b_{23} \end{vmatrix} (b_{12}b_{23} - b_{13}b_{22})$$

$$= \begin{vmatrix} b_{11} & b_{12} \\ b_{21} & b_{23} \end{vmatrix}^2 + \begin{vmatrix} b_{11} & b_{13} \\ b_{21} & b_{23} \end{vmatrix}^2 + \begin{vmatrix} b_{12} & b_{13} \\ b_{22} & b_{23} \end{vmatrix}^2,$$

and the first determinant on the right-hand side of (h) is equal to the sum of the squares of the minors taken from the first and second rows of $D(e)$. Similar statements can be made concerning the other two determinants in (h). Thus, σ_2 is the sum of the squares of all the 2×2 minors of $D(e)$. Consequently, all the σ's are sums of squares.

The case of simple roots of the secular equation has already been considered. We need mention here only that for a simple root

$$\sigma_1 \neq 0, \qquad \sigma_2 \neq 0, \qquad \text{and} \qquad \sigma_3 = 0,$$

so that the rank of $D(e)$ is 2. As shown before, the principal directions are thus uniquely determined except for the sense.

If e is a double root of (66), zero is a root of multiplicity 4 of (e), and

$$\sigma_2 = 0, \qquad \sigma_3 = 0, \qquad \sigma_1 \neq 0,$$

so that the rank of $D(e)$ is 1. In that case the rows of $D(e)$ are proportional, and the three equations in (65) are essentially one equation, which, solved in conjunction with (39a), has infinitely many solutions corresponding to infinitely many directions. All these directions are perpendicular to the first principal direction corresponding to the distinct root of (66). Two orthogonal directions can be chosen to form with the first principal direction an orthogonal set of principal directions.

If e is a triple root of (66), zero is a root of multiplicity 6 of (e), and

$$\sigma_1 = 0, \qquad \sigma_2 = 0, \qquad \sigma_3 = 0,$$

so that, from (f), all the elements of $D(e)$ are zero and (65) are automatically satisfied, whatever values the direction cosines may have. Thus any direction is a principal direction, and, in particular, any set of three orthogonal directions may be taken to be the principal directions. The existence of three principal directions and the orthogonality (by choice in degenerate cases) of these directions are now completely proved.

15 Assuming an isothermal atmosphere obeying the ideal-gas law

$$\frac{p}{\rho} = RT_0,$$

find p and ρ as functions of the height z above sea level, at which $p = p_0$ and $\rho = \rho_0$.

16 Assuming an isentropic atmosphere obeying the law

$$\frac{p}{\rho^\gamma} = \text{const},$$

find p and ρ as functions of the height z above sea level, at which $p = p_0$ and $\rho = \rho_0$.

17 If a fluid moves in concentric circles, with a speed $v = f(r)$, r being the radial distance from the axis of symmetry, find the magnitude of the acceleration of each particle in terms of v and r, using (4) and Cartesian coordinates.

18 The density ρ of the fluid in a pail does not change at any point fixed *with respect to the pail*, and at time $t = 0$ the density distribution is given by

$$\rho = \rho_0 - \beta z,$$

in which ρ_0 and β are constants and z increases in the direction of the vertical. The coordinates x, y, and z are *fixed* in space. The pail moves with a vertical velocity

$$w = -gt,$$

g being the gravitational acceleration. Calculate, at any $t > 0$, the quantities

$$\frac{\partial \rho}{\partial t}, \quad \frac{\partial \rho}{\partial z}, \quad \text{and} \quad \frac{D\rho}{Dt}.$$

REFERENCES

HELMHOLTZ, H. L. F. VON, 1858: Über Integrale der hydrodynamischen Gleichungen welche den Wirbelbewegungen entsprechen, *Crelle*, **55**.

REINER, M., 1945: A Mathematical Theory of Dilatancy, *Am. J. Math.*, **67**: 350.

STOKES, G. G., 1845: On the Theory of Internal Friction of Fluids in Motion, etc., *Cambridge Trans.*, **8**: 287.

THOMSON, W. (LORD KELVIN), 1869: On Vortex Motion, *Edinburgh Trans.*, **25**: 217–260; reprinted in "Mathematical and Physical Papers," vol. 4, p. 13. Cambridge University Press, London.

ADDITIONAL READING

FREDRICKSON, A. G., 1964: "Principles and Applications of Rheology," Prentice-Hall Inc., Englewood Cliffs, N.J. (From this book readers interested in the rheological aspects of hydrodynamics can obtain a systematic account and an extensive bibliography on the subject.)

REINER, M., 1951: The Rheological Aspect of Hydrodynamics, *Quart. Appl. Math.*, **8**: 341–349.

TRUESDELL, C., 1954: The Present Status of the Controversy Regarding the Bulk Viscosity of Fluids, *Proc. Roy. Soc. London Ser. A.*, **226**: 59–65.

CHAPTER TWO

THE BASIC
EQUATIONS

1. THE NAVIER-STOKES EQUATIONS

An arbitrary mass M of a fluid in motion is of constant volume throughout the motion if the fluid is incompressible but of variable volume if the fluid can expand or be compressed. In any case let its volume at any instant be denoted by V. If Newton's second law is applied to the mass M for motion in the direction of x_i,

$$\int_V \rho \frac{Du_i}{Dt} \, dV = \int_S \tau_{ni} \, dS + \int_V \rho X_i \, dV, \tag{1}$$

in which S is the surface of V, $\tau_{ni} \, dS$ is the force acting on dS in the x_i direction, n is the outward-drawn normal to S, and X_i is now the body force in the x_i direction per unit mass. From (1.61) and Green's theorem[1]

---

[1] Note that the outward-drawn normal is used here for convenience. This accounts for the absence of the minus sign before the volume integral in (2), which otherwise would be present, as in (1.9).

it follows that

$$\int_S \tau_{ni}\, dS = \int_S l_\alpha \tau_{\alpha i}\, dS = \int_V \frac{\partial \tau_{\alpha i}}{\partial x_\alpha}\, dV, \tag{2}$$

in which the l's are the direction cosines of the outward-drawn normal n. Hence (1) becomes

$$\int_V \rho \frac{Du_i}{Dt}\, dV = \int_V \frac{\partial \tau_{\alpha i}}{\partial x_\alpha}\, dV + \int_V \rho X_i\, dV. \tag{3}$$

Since V is arbitrary, it follows from (3) that

$$\rho \frac{Du_i}{Dt} = \rho X_i + \frac{\partial \tau_{\alpha i}}{\partial x_\alpha}. \tag{4}$$

But the relationship between stresses and rates of deformation for a Newtonian fluid is given by (1.81). Thus (4) can be written as

$$\rho \frac{Du_i}{Dt} = \rho X_i + \frac{\partial}{\partial x_i}(-p + \lambda\theta) + \frac{\partial}{\partial x_\alpha}\mu e_{\alpha i}. \tag{5}$$

By virtue of (1.44), (5) can be written as

$$\rho \frac{Du_i}{Dt} = \rho X_i + \frac{\partial}{\partial x_i}(-p + \lambda\theta) + \frac{\partial}{\partial x_\alpha}\left[\mu\left(\frac{\partial u_\alpha}{\partial x_i} + \frac{\partial u_i}{\partial x_\alpha}\right)\right]. \tag{6}$$

These equations assume the form, if λ and μ are constant,

$$\rho \frac{Du_i}{Dt} = \rho X_i - \frac{\partial p}{\partial x_i} + (\lambda + \mu)\frac{\partial \theta}{\partial x_i} + \mu\, \nabla^2 u_i, \tag{7}$$

in which ∇^2 is the Laplacian operator $\partial^2/(\partial x_\alpha\, \partial x_\alpha)$, consisting of three terms. Equations (7) are called the *Navier-Stokes equations* after Louis Marie Henri Navier (1785–1836) and George Gabriel Stokes (1819–1903), and are the fundamental equations of motion. For incompressible fluids, the Navier-Stokes equations are

$$\rho \frac{Du_i}{Dt} = \rho X_i - \frac{\partial p}{\partial x_i} + \mu\, \nabla^2 u_i. \tag{7a}$$

(See Appendix 2 for the forms of the Navier-Stokes equations in curvilinear coordinates.)

These equations can be put in vector form by the use of the vorticity components defined by the equations

$$\xi_1 = \frac{\partial u_3}{\partial x_2} - \frac{\partial u_2}{\partial x_3}, \qquad \xi_2 = \frac{\partial u_1}{\partial x_3} - \frac{\partial u_3}{\partial x_1}, \qquad \xi_3 = \frac{\partial u_2}{\partial x_1} - \frac{\partial u_1}{\partial x_2}, \tag{8}$$

which can be written in brief as

$$\xi_k = \epsilon_{\alpha\beta k}\frac{\partial u_\beta}{\partial x_\alpha} \qquad \text{or} \qquad \frac{\partial u_\beta}{\partial x_\alpha} - \frac{\partial u_\alpha}{\partial x_\beta} = \epsilon_{\alpha\beta\gamma}\xi_\gamma, \tag{9}$$

in which ϵ_{ijk} is zero if i, j, and k are not all different, 1 if they are of the same cyclic order as 1, 2, and 3, and -1 otherwise. We shall do this for the incompressible and homogeneous fluid for which the Navier-Stokes equations are

$$\frac{\partial u_i}{\partial t} + u_\alpha \frac{\partial u_i}{\partial x_\alpha} = -\frac{\partial \chi}{\partial x_i} + \nu \, \nabla^2 u_i, \tag{10}$$

in which
$$\chi = \frac{p}{\rho} + \Omega \tag{11}$$

if the body force per unit mass can be derived from a potential Ω by

$$X_i = -\frac{\partial \Omega}{\partial x_i}. \tag{12}$$

Equations (10) can be written as

$$\frac{\partial u_i}{\partial t} + u_\alpha \left(\frac{\partial u_i}{\partial x_\alpha} - \frac{\partial u_\alpha}{\partial x_i} \right) = -\frac{\partial \chi'}{\partial x_i} + \nu \, \nabla^2 u_i, \tag{13}$$

in which
$$\chi' = \frac{p}{\rho} + \Omega + \frac{u_\alpha u_\alpha}{2}. \tag{14}$$

By virtue of (8) or (9),

$$u_\alpha \left(\frac{\partial u_i}{\partial x_\alpha} - \frac{\partial u_\alpha}{\partial x_i} \right) = u_\alpha \epsilon_{\alpha i \beta} \xi_\beta = -\epsilon_{\alpha \beta i} u_\alpha \xi_\beta, \tag{15}$$

so that (13) can be written as

$$\frac{\partial u_i}{\partial t} - \epsilon_{\alpha \beta i} u_\alpha \xi_\beta = -\frac{\partial \chi'}{\partial x_i} + \nu \, \nabla^2 u_i, \tag{16}$$

or, in vector form,

$$\frac{\partial \mathbf{u}}{\partial t} - \mathbf{u} \times \boldsymbol{\xi} = -\mathbf{grad} \, \chi' + \nu \nabla \cdot \nabla \mathbf{u}. \tag{17}$$

For a compressible fluid the Navier-Stokes equations (7) can be put in a form similar to (17). The details are omitted here.

2. THE MOMENTUM EQUATIONS IN INTEGRAL FORM

Since the quantity $\rho \, dV$ in (3) is the mass of a fluid particle of density ρ and volume dV and hence does not change with time as the particle is followed in its motion,

$$\frac{D(\rho \, dV)}{Dt} = 0, \tag{18}$$

and (1) can be written as

$$\frac{D}{Dt} \int_V \rho u_i \, dV = \int_S \tau_{ni} \, dS + \int_V \rho X_i \, dV. \tag{19}$$

These equations are the equation of motion in integral form but written in such a way that its physical meaning can be simply stated. It means that the rate of change of momentum of a mass of fluid in motion is equal to the sum of the surface force and the body force—with momentum and forces considered as vectors, of course. Equations (19) are the basis of what may be called the *momentum principle* in fluid dynamics, and are often applied even in elementary books on the subject. But to apply (19) the stresses must be known, or if only the resultant surface forces are wanted, the momentum fluxes must be known. In elementary applications the fluid is usually assumed to be inviscid, the pressure is obtained from the Bernoulli equation (see Chap. 3), and the velocity distribution across a section is assumed uniform.

3. DISSIPATION OF ENERGY

If the body force (per unit mass) possesses a potential Ω, (4) gives

$$\rho u_i \frac{Du_i}{Dt} = -\rho u_i \frac{\partial \Omega}{\partial x_i} + u_i \frac{\partial \tau_{\alpha i}}{\partial x_\alpha}. \tag{20}$$

If, furthermore, Ω is supposed to be independent of time,

$$u_i \frac{\partial \Omega}{\partial x_i} = \frac{D\Omega}{Dt},$$

and (20) can be written as

$$\rho \frac{D}{Dt}\left(\frac{u_i u_i}{2} + \Omega\right) = \frac{\partial(u_i \tau_{\alpha i})}{\partial x_\alpha} - \tau_{\alpha i} \frac{\partial u_i}{\partial x_\alpha}. \tag{21}$$

Since $D(\rho \, dV)/Dt = 0$ for the fluid mass $\rho \, dV$ as it moves about, integration over V and utilization of (18) and Green's theorem produces

$$\frac{D}{Dt} \int_V \rho\left(\frac{u_i u_i}{2} + \Omega\right) dV = \int_S l_\alpha u_i \tau_{\alpha i} \, dS - \int_V \tau_{\alpha i} \frac{\partial u_i}{\partial x_\alpha} \, dV, \tag{22}$$

in which the l's have the same meaning as in (2). By virtue of (1.61), (22) can be written as

$$\frac{D}{Dt}(K + P) = \int_S u_i \tau_{ni} \, dS - \int_V \tau_{\alpha i} \frac{\partial u_i}{\partial x_\alpha} \, dV, \tag{23}$$

in which K is the kinetic energy and P the potential energy of the fluid mass M and the surface integral is the rate of work done by surface forces. It follows that the volume integral on the right-hand side of (23) is the difference between W, the rate of work done by surface forces, and the rate of change of mechanical energy $(K + P)$ and is therefore the rate of increase of internal energy.   After the stress components are expressed in terms of the rates of deformation according to 1.81), this integral is

$$\int_V \left[-p\theta + \lambda\theta^2 + \mu \frac{\partial u_i}{\partial x_\alpha} \left(\frac{\partial u_\alpha}{\partial x_i} + \frac{\partial u_i}{\partial x_\alpha} \right) \right] dV,$$

of which, for example, $$-\int_V p\theta \, dV = \frac{D}{Dt} I$$

is the rate at which work is being done to compress the fluid and therefore the rate at which the internal energy is increasing by compression.   The rest of the integral is positive definite and is the rate at which the total available energy is being lost by viscous dissipation.   Equation (23) can then be written

$$\frac{D}{Dt} (K + P + I) = W - \int_V \Phi \, dV, \tag{24}$$

in which the volume integral is the rate of energy dissipation and Φ is the dissipation function

$$\Phi = \lambda\theta^2 + \mu \frac{\partial u_i}{\partial x_\alpha} \left(\frac{\partial u_\alpha}{\partial x_i} + \frac{\partial u_i}{\partial x_\alpha} \right)$$

$$= \mu \left[\frac{\lambda}{\mu} \theta^2 + \tfrac{1}{2}(e_{11}^2 + e_{22}^2 + e_{33}^2) + (e_{12}^2 + e_{23}^2 + e_{31}^2) \right]. \tag{25}$$

For special case of uniform and isotropic expansion,

$$\frac{e_{11}}{2} = \frac{e_{22}}{2} = \frac{e_{33}}{2} = \frac{\theta}{3}, \qquad e_{ij} = 0 \text{ for } i \neq j,$$

and (25) gives

$$\Phi = (\lambda + \tfrac{2}{3}\mu)\theta^2.$$

But in this case there is no shear deformation, only volume change (with time).   For this reason the quantity $\lambda + \tfrac{2}{3}\mu$ is called the *volume viscosity*. If the fluid is incompressible,

$$\Phi = \mu[\tfrac{1}{2}(e_{11}^2 + e_{22}^2 + e_{33}^2) + (e_{12}^2 + e_{23}^2 + e_{31}^2)], \tag{26}$$

in which, as in (25), e_{ij} is given by (1.44).

4. VORTICITY EQUATION
FOR A FLUID OF CONSTANT DENSITY AND VISCOSITY

The equations governing the vorticity components have a simple form if the density and the viscosity of the fluid are constant.   The simplest way to derive these equations governing the vorticity components is to derive one and write the other two by symmetry. Since

$$\xi_3 = \frac{\partial u_2}{\partial x_1} - \frac{\partial u_1}{\partial x_2},$$

the equation governing ξ_3 for a fluid of constant density and viscosity can be obtained from (10) by assigning the indices 1 and 2 to i in turn and eliminating χ by cross differentiation. The result is

$$\frac{\partial \xi_3}{\partial t} + u_\alpha \frac{\partial \xi_3}{\partial x_\alpha} + \frac{\partial u_\alpha}{\partial x_1}\frac{\partial u_2}{\partial x_\alpha} - \frac{\partial u_\alpha}{\partial x_2}\frac{\partial u_1}{\partial x_\alpha} = \nu \nabla^2 \xi_3. \tag{27}$$

But

$$\frac{\partial u_\alpha}{\partial x_1}\frac{\partial u_2}{\partial x_\alpha} - \frac{\partial u_\alpha}{\partial x_2}\frac{\partial u_1}{\partial x_\alpha} = \left(\frac{\partial u_\alpha}{\partial x_1} - \frac{\partial u_1}{\partial x_\alpha}\right)\frac{\partial u_2}{\partial x_\alpha} - \left(\frac{\partial u_\alpha}{\partial x_2} - \frac{\partial u_2}{\partial x_\alpha}\right)\frac{\partial u_1}{\partial x_\alpha}$$

$$= \xi_3 \frac{\partial u_2}{\partial x_2} - \xi_2 \frac{\partial u_2}{\partial x_3} - \xi_1 \frac{\partial u_1}{\partial x_3} + \xi_3 \frac{\partial u_1}{\partial x_1}$$

$$= -\xi_\alpha \frac{\partial u_\alpha}{\partial x_3} = -\xi_x \frac{\partial u_3}{\partial x_x}, \tag{28}$$

since $\quad \dfrac{\partial u_1}{\partial x_1} + \dfrac{\partial u_2}{\partial x_2} = -\dfrac{\partial u_3}{\partial x_3} \quad$ and[1] $\quad \xi_\alpha \dfrac{\partial u_\alpha}{\partial x_3} = \xi_\alpha \dfrac{\partial u_3}{\partial x_\alpha}.$

Thus (27) becomes

$$\left(\frac{\partial}{\partial t} + u_\alpha \frac{\partial}{\partial x_\alpha}\right)\xi_3 = \xi_\alpha \frac{\partial u_3}{\partial x_\alpha} + \nu \nabla^2 \xi_3. \tag{29}$$

The vorticity equations for an incompressible fluid with constant viscosity are therefore

$$\frac{D\xi_i}{Dt} = \xi_\alpha \frac{\partial u_i}{\partial x_\alpha} + \nu \nabla^2 \xi_i. \tag{30}$$

For two-dimensional flows only one vorticity component, say ξ_3 or ζ, is different from zero, and (30) becomes

$$\frac{D\zeta}{Dt} = \nu \nabla^2 \zeta. \tag{30a}$$

From (30) the most important observation to be made is that irrotational motion (for which the vorticity components are zero) is a possible motion even for a viscous fluid, as far as the equations of motion are concerned and as long as viscosity is constant, since

$$\xi_1 = 0, \qquad \xi_2 = 0, \qquad \xi_3 = 0$$

[1] $\xi_\alpha\left(\dfrac{\partial u_\alpha}{\partial x_3} - \dfrac{\partial u_3}{\partial x_\alpha}\right) = \xi_1 \xi_2 - \xi_2 \xi_1 = 0.$

satisfy (30). Strictly irrotational flows are usually unrealistic for a viscous fluid because for such flows there is always slip between the fluid and a solid boundary, to which a viscous fluid must adhere. Thus it is the boundary conditions imposed on a viscous fluid and not the equations governing its motion that are not satisfied by irrotational flows. In particular cases (see Chap. 7) solid boundaries can be realistically made to move according to the dictates of irrotational motion. In such cases irrotational flows become possible flows even for viscous fluids. The flow caused by a body moving with speed U in a large body of an otherwise tranquil liquid would be irrotational if the boundary of the body could be made to move tangentially with the (variable) velocity required by the solution for the irrotational flow. Such a flow is equivalent to an irrotational flow past the body, with uniform velocity $-U$ at infinity and with the boundary of the body moving tangentially as needed. In the latter flow, the work done at a closed surface at infinity on the fluid contained between that surface and the body is zero because the velocity and $p + \rho\Omega$ at infinity are uniform. If the flow is steady, the left-hand side of (24) is zero, and W represents the rate of work done by the moving boundary of the body. It follows that the rate of dissipation of energy is exactly equal to the rate of work done by the moving boundary. If the boundary could be made to move with the velocity demanded by the solution for irrotational flow past the body, the drag force would be zero, as will be shown in Chap. 4, and the power expenditure would be exactly W, which is much less than the power required to propel the same body with a fixed surface through an otherwise tranquil fluid at a speed U. In this paragraph, it has been tacitly assumed that the dissipation of energy through viscosity does not make the temperature of the fluid so nonuniform as to cause the viscosity to vary significantly. If the viscosity varies, irrotational flows are in general no longer dynamically possible, whether or not the boundary of the body is allowed to move.

The terms involving the kinematic viscosity in (30) represent the effect of ν on the diffusion of vorticity and are responsible for transmitting the effect of the solid boundary into the fluid. If such a boundary is suddenly placed in an originally irrotational flow, the vorticity will be concentrated at the boundary initially. Through viscosity this concentrated vorticity will diffuse into the fluid, to render a part of the flow predominantly rotational. If the viscosity of the fluid is small,[1] and if the fluid does not separate from the boundary, this part is limited to a thin layer over the solid boundary, outside of which the flow is predominantly irrotational, with the steep change of velocity in this thin layer providing in fact the sort of slip demanded by irrotational motion.

A detailed discussion of (30) is rewarding. The left-hand side stands for the rate of change of a vorticity component, and the last term on the

[1] By this we mean the Reynolds number is large compared with unity (see Sec. 5).

right represents its diffusion by viscosity, which diffuses the velocity components in the same way, as can be seen from (7a). It will now be shown that the first three terms (written as one term) on the right-hand side represent the rate of change of the vorticity component ξ_i of a fluid particle due to the stretching and turning of the vortex line passing through that particle.

The distance of a point (x_1, x_2, x_3) along the vortex line from a fluid particle situated on it will be denoted by s, and the length of an infinitesimal segment of such a line starting from the particle will be denoted by ds. The direction numbers of the line segment are therefore ξ_1, ξ_2, and ξ_3, so that

$$\frac{\xi_\alpha}{\xi} = \frac{\partial x_\alpha}{\partial s}, \tag{31}$$

with

$$\xi = (\xi_\alpha \xi_\alpha)^{\frac{1}{2}} \tag{32}$$

indicating the magnitude of the vorticity vector. Thus

$$\xi_\alpha \frac{\partial u_i}{\partial x_\alpha} = \xi \frac{\partial u_i}{\partial x_\alpha} \frac{\partial x_\alpha}{\partial s} = \xi \frac{\partial u_i}{\partial s}. \tag{33}$$

If we identify the first direction ($i = 1$) with the direction of increasing s, then $\xi_1 = \xi$, and for an *inviscid* fluid of constant density, (30) and (33) give

$$\frac{D\xi}{Dt} = \xi \frac{\partial u_s}{\partial s}. \tag{34}$$

Now $\partial u_s / \partial s$ is the rate of stretching in the direction of s per unit time per unit distance, or

$$\frac{\partial u_s}{\partial s} = \frac{1}{ds} \frac{D}{Dt} ds, \tag{35}$$

in which ds is the length of a material segment taken along the vortex line at the point in question. Thus (34) can be written as

$$\frac{D}{Dt} \frac{\xi}{ds} = 0, \tag{36}$$

which states that the strength of vorticity of a fluid particle increases or decreases in direct proportion to the length of a material segment of the vortex line[1] passing through it as the fluid moves. In other words, the change of the strength of vorticity for an inviscid fluid of constant density is due entirely to stretching or shrinking of the vortex line.

But the vorticity equations for an inviscid fluid of constant density

---

[1] We have derived (35) and (36) by considering the change of ds in a short time interval dt, with ds originally taken along a vortex line. In Chap. 3 it will be shown that for an inviscid fluid with constant density or entropy, vortex lines move with the fluid.

are

$$\frac{D\xi_i}{Dt} = \xi_\alpha \frac{\partial u_i}{\partial x_\alpha}. \tag{37}$$

By identifying the direction of i with that of s, we have extracted some very interesting and useful information concerning the strength of vorticity. But the three equations in (37) say rather more than (34), for they give the rates of change of the three vorticity *components*. The remaining information can be extracted from (37) most easily if we use the *local* Cartesian coordinates s, n, and m, with s measured along the vortex line, n measured along its normal and away from the center of curvature (in the plane of the osculating circle of the vortex line, which is in general a spatial curve), and m along its binormal. Equations (31) and (37) then give, aside from (34),

$$\frac{D\xi_n}{Dt} = \xi \frac{\partial u_n}{\partial s} \quad \text{and} \quad \frac{D\xi_m}{Dt} = \xi \frac{\partial u_m}{\partial s}. \tag{38}$$

But $\partial u_n/\partial s$ and $\partial u_m/\partial s$ are the rates of turning (per unit time per unit length) of the element ds about the m axis and the n axis, respectively. Now the element ds is measured along the vortex line, and the strength of vorticity is ξ. Hence the right-hand sides of (38) indicate that the rates of change of ξ_m and ξ_n are due entirely to the turning of the vortex line passing through it if the fluid has constant density and no viscosity. Going back to (30), we see then that the rate of change of any component of vorticity of a fluid particle is due to the stretching and turning of the vortex line on the one hand and to viscous diffusion on the other. The Laplacian of a quantity is always multiplied by the diffusivity in the diffusion equation governing that quantity. Hence, ν, the kinematic viscosity, can be considered as a vorticity diffusivity on inspection of (30) or a momentum diffusivity on inspection of (7).

For a compressible fluid or an incompressible fluid with variable density, the viscosity μ and the kinematic viscosity ν cannot both be constant, since ρ is not. Furthermore, in general p is not a function of ρ alone. Under such circumstances, the exact vorticity equations are so complicated that it is not worthwhile to derive them. However, if the effects of viscosity and volume viscosity are neglected, and if the gas is homentropic, i.e., its density is a function of its pressure alone as indicated by (A1.16), the vorticity equations are relatively simple in form (see Chap. 3).

5. SIMILARITY OF FLOWS

Similarity of flows demands first of all similarity in boundary geometry, which means that the boundary of one flow can be made to coincide with that of another if its linear dimensions are multiplied by a constant. If geometrical similarity exists, it is meaningful to speak of corresponding points of two flows. Dynamic similarity exists if the variables[1] of one flow are proportional to those of another at the corresponding points. What conditions are sufficient for dynamic similarity to exist, aside from geometrical similarity of the boundaries? To answer this question without undue encumbrance, the dynamics of an incompressible fluid with constant viscosity but variable density may be considered.

The governing equations of motion are (7a), in addition to which the equation of incompressibility

$$\frac{D\rho}{Dt} \equiv \frac{\partial \rho}{\partial t} + u_\alpha \frac{\partial \rho}{\partial x_\alpha} = 0 \tag{39}$$

and the equation of continuity

$$\frac{\partial u_\alpha}{\partial x_\alpha} = 0 \tag{40}$$

must be satisfied. The following substitutions are now made:

$$u_i' = \frac{u_i}{U}, \qquad \rho' = \frac{\rho}{\rho_0}, \qquad p' = \frac{p}{\rho_0 U^2}, \qquad x_i' = \frac{x_i}{L}, \qquad t' = \frac{tU}{L}, \tag{41}$$

in which U is a reference velocity, ρ_0 a reference density, and L a reference length, and x_i' no longer indicates coordinates after a rotation of the coordinate axes. The quantities with a prime are dimensionless quantities, in terms of which (7a) and (39) can be expressed. The dimensionless equations read

$$\rho'\left(\frac{\partial}{\partial t'} + u_\alpha' \frac{\partial}{\partial x_\alpha'}\right) u_i' = -\frac{\partial p'}{\partial x_i'} + \frac{\rho' L}{U^2} X_i + \frac{\nu}{UL} \frac{\partial^2}{\partial x_\alpha' \partial x_\alpha'} u_i', \tag{42}$$

$$\frac{\partial \rho'}{\partial t'} + u_\alpha' \frac{\partial \rho'}{\partial x_\alpha'} = 0, \quad \text{and} \quad \frac{\partial u_\alpha'}{\partial x_\alpha'} = 0. \tag{43}$$

Gravity will be assumed to be the only body force. Thus, if x_3 is measured vertically upward,

$$X_1 = 0, \qquad X_2 = 0, \qquad X_3 = -g. \tag{44}$$

With the boundary geometry understood to be similar, dynamic similarity demands that the dimensionless quantities pertinent to one flow be identical with the corresponding ones pertinent to another, for which the reference quantities U, ρ_0, and L may be different. Inspection of (42) and (43) reveals that dynamic similarity is possible only if the dimensionless parameters gL/U^2 and ν/UL are the same for the two flows under consideration and if the initial values of u_i', ρ', and p' everywhere—as well

---

[1] For an incompressible fluid, p may be changed by a constant. If proportionality of all variables exists after this change, the flows are dynamically similar.

as their subsequent values at the (closed) boundary of the domain of flow—be respectively the same for one flow as for another at the corresponding points. With the similarity of initial and boundary conditions taken for granted, the flows are similar if the two numbers

$$F \equiv \frac{U}{\sqrt{gL}} \quad \text{and} \quad R \equiv \frac{UL}{\nu}$$

are the same for the two flows. The first number is called the *Froude number*, after William Froude (1810–1879), an English engineer, and the second is called the *Reynolds number*, after Osborne Reynolds (1842–1912), a professor of the Victoria University (then Owen College), Manchester, England.

Equation (42) can be written in the more familiar form

$$\rho' \frac{D}{Dt'} u_i' = - \frac{\partial p'}{\partial x_i'} - \frac{\rho'}{F^2} \delta_{i3} + \frac{1}{R} \nabla^2 u_i', \qquad (42a)$$

in which D/Dt' and ∇^2 are now the substantial-differentiation and Laplacian operators, respectively, in terms of t', x_i', and u_i'.

If the density is constant, it can be taken as the reference density, so that $\rho' = 1$, and the second term on the right-hand side of (42) can be absorbed into the first if p' is changed to $p' + F^{-2}x_3'$. If, in addition, the boundaries are fixed, the Froude number ceases to be a significant parameter of the flow, and the only effect of gravity is to contribute a hydrostatic part to the pressure. If there is a free surface, the boundary condition on that surface may involve g, and the Froude number will be a significant parameter for dynamic similarity, in spite of the fact it can be made to disappear from the equations of motion. However, there are cases in which the boundary condition on the free surface does not involve g, for example, the case of steady laminar flow in an open channel with *rectilinear* and *strictly parallel* streamlines. In fact, in such a case even the Reynolds number does not matter, because the left-hand side of (42) is zero and the parameter ν/UL can be absorbed into the term $\partial p'/\partial x_i'$.

From a physical point of view U^2/gL is the ratio of an inertia force per unit mass to the gravity force per unit mass, because U^2/L is really a representative convective acceleration when such an acceleration exists. Similarly, UL/ν, being equal to the ratio of U^2/L to $\nu U/L^2$, is the ratio of an inertia force per unit mass to a viscous force per unit mass if convective acceleration exists. In steady parallel flows there is no convective acceleration, and neither F nor R can be significant parameters, even if there is a free surface. This is not surprising, because steady parallel flows are obviously similar, whatever the values of F and R. However, if waves are present or possible, or if the surface is not parallel to the straight bottom, F and R will be important. Even for flows which are steady, rectilinear, and parallel *in the mean*, the Reynolds number is important if the flows are turbulent, so that fluid particles are actually accelerated.

For the flows of a compressible fluid, (39) is not valid, and the general equation of continuity has to be used. Furthermore, the energy equation or equation of thermal diffusion and the equation of state must be considered in addition, and other dimensionless parameters must be kept the same in order to ensure similarity between two flows. Except in meteorological problems, the effect of gravity can usually be neglected, so that the Froude number need not be kept the same to ensure similarity.

6. DIMENSIONAL ANALYSIS

As shown in the last section, the similarity of two flows depends on the equality of certain dimensionless parameters for one flow to the corresponding parameters of the other. In other words, if these parameters are kept the same for any two flows, they are geometrically and dynamically similar even though the scale, the velocity at corresponding points, and the physical properties of the fluid may be different for the two flows. The dimensionless parameters governing similarity have been obtained by making the governing equations and boundary conditions dimensionless. This is the best procedure[1] when these equations and boundary conditions are known.

In the literature of fluid mechanics, an alternative procedure finding these dimensionless parameters is often used. Briefly, it consists of

1 guessing, by experience, perception, or intuition, the n physical variables a, b, c, d, etc., that have a bearing on the problem and

2 choosing among them m repeating variables, m being the number of fundamental dimensions involved, and combining powers of these with each of the remaining physical variables in turn to form the dimensionless parameters sought.

For instance, if D is the drag on a body with a given shape and a representative length d moving with speed U in an infinite fluid of constant viscosity μ and density ρ, the variables are simply D, d, U, μ, and ρ. The fundamental dimensions are those of mass, length, and time, or M, L, and T. Choosing ρ, d, and U as the repeating variables, one forms the combinations

$$\rho^\alpha d^\beta U^\gamma D \qquad \text{and} \qquad \rho^\alpha d^\beta U^\gamma \mu$$

and demands that the exponents of M, L, and T be zero in each combination. The results are

$$\frac{D}{\rho U^2 d^2} \qquad \text{and} \qquad \frac{\mu}{\rho U d}.$$

---

[1] It is best even if the flow is turbulent.

The first parameter is often defined to be twice the coefficient of drag C_D, and the second is the reciprocal of the Reynolds number R. In this example $n = 5$, $m = 3$, and the number of dimensionless parameters is 2. In general, this number is $n - m$, which we shall denote by N.

With the dimensionless parameters denoted by $\Pi_1, \Pi_2, \ldots, \Pi_N$, the dimensional analysis is completed by the statement

$$F(\Pi_1, \Pi_2, \Pi_3, \ldots, \Pi_N) = 0. \tag{45}$$

This statement is based implicitly on the realization that any equation in physics must be dimensionally homogeneous, and hence can be made dimensionless by division by the common dimension of its terms. The result is (45), which can be in differential form, as (42) and (43), or in "integrated" form, whenever dimensional analysis is applied. For the example just given, (45) assumes the familiar form

$$C_D = f(R). \tag{46}$$

Whether the Π's have been obtained from the dimensionless forms of the governing equations and boundary conditions or by the procedure of dimensional analysis just outlined, (45) can serve as a guide for systematic experimentation, provided the Π's have been correctly obtained. Of course the Π's will not be correct if the wrong equations or boundary conditions have been used to obtain them or if they have been incorrectly chosen in step 1.

7. EQUATIONS OF MOTION RELATIVE TO A ROTATING SYSTEM

A rotating system is not an inertial system, since Newton's first and second laws are not valid relative to a rotating system. In other words, the expression for acceleration is not invariant with respect to a transformation corresponding to a rotation. To an observer rotating with the system, the additional terms in the expression for the acceleration appear as body forces. Since the viscous terms are unaffected by the rotation of the system, we shall ignore them for the time being and put them back in place later. For an inviscid fluid the equations of motion are, in cylindrical coordinates,

$$\rho\left(\frac{Du}{Dt} - \frac{v^2}{r}\right) = -\frac{\partial p}{\partial r} - \rho\frac{\partial \Omega}{\partial r}, \tag{47}$$

$$\rho\left(\frac{Dv}{Dt} + \frac{uv}{r}\right) = -\frac{1}{r}\frac{\partial p}{\partial \varphi} - \frac{\rho}{r}\frac{\partial \Omega}{\partial \varphi}, \tag{48}$$

$$\rho\frac{Dw}{Dt} = -\frac{\partial p}{\partial z} - \rho\frac{\partial \Omega}{\partial z}, \tag{49}$$

in which
$$\frac{D}{Dt} = \frac{\partial}{\partial t} + u\frac{\partial}{\partial r} + \frac{v}{r}\frac{\partial}{\partial\varphi} + w\frac{\partial}{\partial z}, \tag{50}$$

Ω being the body-force potential, which is assumed to exist.

Suppose that a frame of reference is rotating with constant angular velocity ω (with respect to an inertial system) about the z axis and that the cylindrical coordinates (r',φ',z') are used in it. The velocity components with respect to this rotating system are denoted by u', v', and w'. Then

$$u' = u, \qquad v' = v - \omega r, \qquad w' = w, \qquad t' = t, \tag{51}$$

and
$$r' = r, \qquad \varphi' = \varphi - \omega t, \qquad z' = z. \tag{52}$$

The unprimed quantities are for the inertial system, and the two sets of coordinates coincide at $t = 0$.

Substituting (51) and (52) into (47) to (49), we have

$$\rho\left(\frac{Du'}{Dt'} - \frac{v'^2}{r'} - 2\omega v' - \omega^2 r'\right) = -\frac{\partial p}{\partial r'} - \rho\frac{\partial\Omega}{\partial r'},$$

$$\rho\left(\frac{Dv'}{Dt'} + \frac{u'v'}{r'} + 2\omega u'\right) = -\frac{1}{r'}\frac{\partial p}{\partial\varphi'} - \frac{\rho}{r}\frac{\partial\Omega}{\partial\varphi'},$$

$$\rho\frac{Dw'}{Dt'} = -\frac{\partial p}{\partial z'} - \rho\frac{\partial\Omega}{\partial z'},$$

in which
$$\frac{D}{Dt'} = \frac{\partial}{\partial t'} + u'\frac{\partial}{\partial r'} + \frac{v'}{r'}\frac{\partial}{\partial\varphi'} + w'\frac{\partial}{\partial z'}.$$

Dropping the primes, we have

$$\rho\left(\frac{Du}{Dt} - \frac{v^2}{r} - 2\omega v - \omega^2 r\right) = -\frac{\partial p}{\partial r} - \rho\frac{\partial\Omega}{\partial r}, \tag{53}$$

$$\rho\left(\frac{Dv}{Dt} + \frac{uv}{r} + 2\omega u\right) = -\frac{1}{r}\frac{\partial p}{\partial\varphi} - \frac{\rho}{r}\frac{\partial\Omega}{\partial\varphi}, \tag{54}$$

$$\rho\frac{Dw}{Dt} = -\frac{\partial p}{\partial z} - \rho\frac{\partial\Omega}{\partial z}, \tag{55}$$

in which u, v, and w are now relative to the rotating frame of reference.

It is evident that the terms $-2\omega v$ and $2\omega u$ are components of the Coriolis acceleration $2\omega \times v$, whose components in Cartesian coordinates are $-2\omega v$ and $2\omega u$, u and v now being Cartesian components, provided the z axis is still the axis of rotation. Hence the corresponding equations in Cartesian coordinates (x, y, z) are

$$\rho\left(\frac{Du}{Dt} - 2\omega v - \omega^2 x\right) = -\frac{\partial p}{\partial x} - \rho\frac{\partial \Omega}{\partial x}, \tag{56}$$

$$\rho\left(\frac{Dv}{Dt} + 2\omega u - \omega^2 y\right) = -\frac{\partial p}{\partial y} - \rho\frac{\partial \Omega}{\partial y}, \tag{57}$$

$$\rho\frac{Dw}{Dt} = -\frac{\partial p}{\partial z} - \rho\frac{\partial \Omega}{\partial z}, \tag{58}$$

in which u, v, and w are Cartesian components of the velocity relative to the rotating frame of reference and D/Dt is in Cartesian coordinates.

If ρ is constant or a function of p alone, (53) to (55) can be written as

$$\frac{Du}{Dt} - \frac{v^2}{r} - 2\omega v = -\frac{\partial P}{\partial r}, \tag{59}$$

$$\frac{Dv}{Dt} + \frac{uv}{r} + 2\omega u = -\frac{1}{r}\frac{\partial P}{\partial \varphi}, \tag{60}$$

$$\frac{Dw}{Dt} = -\frac{\partial P}{\partial z}, \tag{61}$$

and (56) to (58) can be written as

$$\frac{Du}{Dt} - 2\omega v = -\frac{\partial P}{\partial x}, \tag{62}$$

$$\frac{Dv}{Dt} + 2\omega u = -\frac{\partial P}{\partial y}, \tag{63}$$

$$\frac{Dw}{Dt} = -\frac{\partial P}{\partial z}, \tag{64}$$

in which $\qquad P = \int \frac{dp}{\rho} + \Omega - \frac{\omega^2 r^2}{2}, \qquad r^2 = x^2 + y^2, \tag{65}$

and the operator D/Dt is in terms of the particular coordinates used.

If viscous effects are to be considered, the viscous terms can be added to Eqs. (53) to (58). For instance, if the fluid is incompressible and the viscosity is constant but ρ is not, then (53) to (55) should be replaced by

$$\rho\left(\frac{Du}{Dt} - \frac{v^2}{r} - 2\omega v - \omega^2 r\right) = -\frac{\partial p}{\partial r} - \rho\frac{\partial \Omega}{\partial r} + \mu\left(\nabla^2 u - \frac{u}{r^2} - \frac{2}{r^2}\frac{\partial v}{\partial \varphi}\right), \tag{66}$$

$$\rho\left(\frac{Dv}{Dt} + \frac{uv}{r} + 2\omega u\right) = -\frac{1}{r}\frac{\partial p}{\partial \varphi} - \frac{\rho}{r}\frac{\partial \Omega}{\partial \varphi} + \mu\left(\nabla^2 v - \frac{v}{r^2} + \frac{2}{r^2}\frac{\partial u}{\partial \varphi}\right), \tag{67}$$

$$\rho\frac{Dw}{Dt} = -\frac{\partial p}{\partial z} - \rho\frac{\partial \Omega}{\partial z} + \mu\,\nabla^2 w, \tag{68}$$

the Laplacian operator ∇^2 in cylindrical coordinates being given in (A2.68)
If both ρ and μ are constant, (59) to (61) should be replaced by

$$\frac{Du}{Dt} - \frac{v^2}{r} - 2\omega v = -\frac{\partial P}{\partial r} + \nu\left(\nabla^2 u - \frac{u}{r^2} - \frac{2}{r^2}\frac{\partial v}{\partial \varphi}\right), \tag{69}$$

$$\frac{Dv}{Dt} + \frac{uv}{r} + 2\omega u = -\frac{1}{r}\frac{\partial P}{\partial \varphi} + \nu\left(\nabla^2 v - \frac{v}{r^2} + \frac{2}{r^2}\frac{\partial u}{\partial \varphi}\right), \tag{70}$$

$$\frac{Dw}{Dt} = -\frac{\partial P}{\partial z} + \nu\,\nabla^2 w. \tag{71}$$

For Cartesian coordinates the viscous terms which must be added to (56) to (58) are, respectively, $\mu\,\nabla^2 u$, $\mu\,\nabla^2 v$, and $\mu\,\nabla^2 w$, and those which must be added to (62) to (64) are $\nu\,\nabla^2 u$, $\nu\,\nabla^2 v$, $\nu\,\nabla^2 w$, respectively. The equations for more complicated cases involving compressibility or variable viscosity (or both) need not be presented in full. For the form of the viscous terms, see (6) for Cartesian coordinates or (A2.80) for general coordinates. Note that if viscosity is to be considered for a compressible fluid, (59) to (64) must not be used as the inviscid parts of the equations of motion, since the entropy is no longer constant, so that ρ is no longer a function of p alone.

7.1. The Linear Acceleration of the Frame of Reference

We have so far considered only the rotation of the frame of reference. A linear translation of the frame with acceleration $b_i(t)$ ($i = 1, 2, 3$) has the sole effect of changing the body force X_i (per unit mass) by the amount $-b_i(t)$. This is seen in the following way. Let the acceleration of any fluid particle with respect to the linearly accelerating frame be a_i' and that with respect to an inertial frame a_i. Then

$$a_i = a_i' + b_i(t). \tag{72}$$

If $b_i(t)$ is moved to the right-hand side of the equation of motion, it has the sole effect of reducing X_i by $-b_i$.

If the frame of reference has a linear acceleration as well as a constant rotation, we can take care of the linear acceleration first, in the simple way mentioned above, and then take care of the rotation as described in the preceding section.

8. THE ENERGY EQUATION

For the investigation of flows of a compressible fluid or of heat transfer, we need an equation representing the first law of thermodynamics as applied to a moving and deforming fluid. This is called the energy equation because it is an equation of the energy "budget."

Consider a control volume V bounded by a surface S *fixed* in space. (Note that the surface S is no longer a material surface as in Secs. 1 to 3.) Since V is now a fixed volume, (18) does not hold, and the operator D/Dt can no longer be moved forward as in (22). Otherwise the development in Sec. 3 remains valid, and, instead of (24), we have

$$\int_V \left[\rho \frac{D}{Dt}\left(\frac{u_i u_i}{2} + \Omega\right) - p\theta \right] dV = \int_S u_i \tau_{ni}\, dS - \int_V \Phi\, dV. \tag{73}$$

This, like (24), reveals the meaning of the dissipation function and enables us to evaluate the rate of work done at S by surface forces, represented by the first term on the right-hand side and denoted by W. Now the sum of the kinetic energy and internal energy per unit mass is

$$\epsilon = \frac{u_i u_i}{2} + c_v T, \tag{74}$$

where c_v is the specific heat at constant volume in energy units and T the absolute temperature. If c_v is not constant, the last term, representing the internal energy per unit mass, should be replaced by $\int c_v\, dT$ [see (A1.10)]. If

W_g = rate of work done by body forces on fluid in V,

E_1 = energy flux into V through S by conduction,

E_2 = energy flux into V through S by convection,

then the energy equation is

$$\int_V \frac{\partial}{\partial t} \rho\epsilon\, dV = W + W_g + E_1 + E_2. \tag{75}$$

Now since n is the outward normal to S, the energy flux by heat conduction into V per unit area of S is $k\, \partial T/\partial n$, k being the thermal conductivity in energy units, and the energy convected into V per unit area of S is $-\rho u_n \epsilon$. Thus, with l_α indicating the direction cosines of the direction of increasing n,

$$E_1 + E_2 = \int_S \left(k \frac{\partial T}{\partial n} - \rho u_n \epsilon \right) dS = \int_S \left(k l_\alpha \frac{\partial T}{\partial x_\alpha} - \rho l_\alpha u_\alpha \epsilon \right) dS$$

$$= \int_V \left[\frac{\partial}{\partial x_\alpha}\left(k \frac{\partial T}{\partial x_\alpha}\right) - \frac{\partial}{\partial x_\alpha} \rho u_\alpha \epsilon \right] dV. \tag{76}$$

As we have said before,

$$W = \int_S u_i \tau_{ni}\, dS. \tag{77}$$

The rate of work done by body forces is

$$W_g = \int_V \rho u_i X_i\, dV = -\int_V \rho u_i \frac{\partial \Omega}{\partial x_i}\, dV = -\int_V \rho \frac{D\Omega}{Dt}\, dV, \tag{78}$$

provided the total body force per unit mass has a potential Ω and provided Ω is independent of time.

Calculating W from (73) and substituting both the W so obtained and (76) and (78) into (75), we finally have

$$\int_V \left[\rho c_v \frac{DT}{Dt} + p\theta - \frac{\partial}{\partial x_\alpha}\left(k \frac{\partial T}{\partial x_\alpha} \right) - \Phi \right] dV = 0.$$

In obtaining this, the equation of continuity

$$\frac{\partial \rho}{\partial t} + \frac{\partial(\rho u_\alpha)}{\partial x_\alpha} = 0$$

has been used. Since V is arbitrary,

$$\rho c_v \frac{DT}{Dt} + p\theta = \frac{\partial}{\partial x_\alpha}\left(k \frac{\partial T}{\partial x_\alpha} \right) + \Phi, \tag{79}$$

which is the energy equation sought. Note that c_v and k are in energy units. If they are in thermal units,

$$\rho J c_v \frac{DT}{Dt} + p\theta = J \frac{\partial}{\partial x_\alpha}\left(k \frac{\partial T}{\partial x_\alpha} \right) + \Phi, \tag{80}$$

where J is Joule's value. If k is constant or its variation negligible, the energy equation takes the form

$$\rho J c_v \frac{DT}{Dt} + p\theta = kJ \nabla^2 T + \Phi, \tag{81}$$

thermal units being used for c_v and k.

Since, in thermal units,

$$J(c_p - c_v) = R \quad \text{and} \quad \frac{p}{\rho} = RT,$$

we can write

$$\epsilon = \tfrac{1}{2} u_i u_i + \left(J c_p T - \frac{p}{\rho} \right)$$

and

$$\rho \frac{D\epsilon}{Dt} = \frac{1}{2} \rho \frac{D}{Dt} u_i u_i + \rho J c_p \frac{DT}{Dt} - \frac{Dp}{Dt} - p\theta, \tag{82}$$

since

$$\frac{D}{Dt} \rho^{-1} = \frac{\theta}{\rho} \tag{83}$$

by the equation of continuity. Thus the energy equation can be written as

$$\rho J c_p \frac{DT}{Dt} - \frac{Dp}{Dt} = J \frac{\partial}{\partial x_\alpha}\left(k \frac{\partial T}{\partial x_\alpha} \right) + \Phi. \tag{84}$$

The J in this equation is to be dropped if energy units for c_p and k are used.

For liquids the θ in (79) to (81) can be equated to zero. For heat transfer in gases with negligible pressure variation in the field, such as in jets of preheated air issuing from an opening into colder air, (84) is more convenient to use, for the θ in (81) is not zero but the Dp/Dt in (84) can be equated to zero.

In arriving at (79) and (84), c_v and c_p have been assumed constant. If they are not constant, the first term in (79) should be replaced by $\rho \, D(c_v T)/Dt$ and the first term in (84) by $\rho J \, D(c_p T)/Dt$.

PROBLEMS

1 A sphere of specific weight $g\rho_s$ and radius a moves slowly with constant velocity U in a viscous fluid of viscosity μ and specific weight $g\rho$. Inertia effects are negligible.   Perform the dimensional analysis, aiming at the determination of U in terms of the other variables, and produce the equation involving the dimensionless variables (or variable).

2 Consider several problems involving fluid flow. For each problem perform the dimensional analysis, obtain the equation (in general form only, of course) governing the dimensionless parameters, and check the form of the parameters by starting from the fundamental equations governing the phenomenon.

3 If heat conduction and dissipation of energy are neglected, show that for an ideal gas (80) or (84) can be reduced to

$$\frac{D}{Dt}\frac{p}{\rho^{\gamma}} = 0, \qquad \text{or} \qquad \frac{D}{Dt}S = 0,$$

γ being c_p/c_v and S being the entropy (see Appendix 1, Sec. 8.1).

4 Show that (79) can be written as

$$\rho T \frac{DS}{Dt} = \frac{\partial}{\partial x_\alpha}\left(k\,\frac{\partial T}{\partial x_\alpha}\right) + \Phi$$

and, if radiation effects are included, as

$$\rho T \frac{DS}{Dt} = -\frac{\partial q_\alpha}{\partial x_\alpha} + \Phi,$$

in which (q_1, q_2, q_3) is the total heat-flux vector.

ADDITIONAL READING

NAVIER, L. M. H., 1822: Mémoire sur les lois du mouvement des fluides, *Mém. Acad. Sci.*, **6**: 389.

POISSON, S. D., 1829: Mémoire sur les équations générales de l'équilibre et du mouvement des corps solides élastique et des fluides, *J. École Polytech.*, **13**: 1. (This paper and the one above are of historical interest. The derivation of the Navier-Stokes equation, as it is known today, was given by Saint-Venant and Stokes.)

SAINT-VENANT, J.-C. B. DE, 1843: Note à joindre au mémoire sur la dynamique des fluides, *Comp. Rend.*, **17**: 1240. (In this note Saint-Venant gave the same constitutive equations for Newtonian fluids as given by Stokes and specifically stated the coincidence of the principal directions for stresses with those for the rates of deformation, which is the underlying assumption for these constitutive equations. The dynamical equations of motion in terms of velocity and stress components were known before 1843. With the constitutive equations proposed by Saint-Venant, these dynamical equations become the Navier-Stokes equations.)

STOKES, G. G., 1845: On the Theory of Internal Friction of Fluids in Motion, etc., *Cambridge Trans.*, **8**: 287.

CHAPTER THREE

GENERAL THEOREMS
FOR THE FLOW
OF AN
INVISCID FLUID

1. INTRODUCTION

An *inviscid* fluid is a fluid with no viscosity. For a gas, inviscidness implies that both the viscosity and the volume viscosity are zero. Since all fluids (except helium II at very low temperatures) are viscous, the study of the motion of inviscid fluids needs justification. It will be given for an incompressible fluid only, that for a compressible fluid being similar. For an incompressible fluid, inspection of (2.42a) reveals that except in a region in which $\nabla^2 u_i'$ is very large compared with 1, the last term can be neglected if R is very large compared with 1. The exceptional regions are near the boundaries, where the nonslip condition (or sometimes the stress condition), being incompatible with the solutions of the equations of motion for an inviscid fluid, gives rise to a layer of large vorticity and rates of shearing deformation. These conditions call for the use of the equations of motion without truncation of the last term. The last term in (2.42a) can be neglected outside of these regions and regions where turbulence supplies the mechanism for macroscopic mixing, e.g., wakes, or where (as in a long pipe) viscosity, however small, has a long time to exercise its influence.   The study of the flow of inviscid fluids is therefore not without physical significance.

2. CONSERVATION OF CIRCULATION

For an inviscid fluid with constant density or an inviscid barotropic[1] gas, for which a relationship $F(\rho,p,t)$ exists, circulation around a material circuit is conserved as that circuit moves about.   If a line element along this circuit is denoted by ds, with components dx_i, these components will change with time as a result of the stretching and turning of the line element ds.   They are therefore functions of time and not merely differentials of the coordinates independent of time.   Therefore the symbol dx_i is a little misleading, but it would not be adequate to use dX_i either, because dX_i might mean the infinitesimal displacement of the *same* particle, in the notation of Chap. 1.   We choose to use dx_i and to keep its meaning in mind.   According to the definition of dx_i,

$$\frac{D}{Dt}\,dx_i = du_i, \tag{1}$$

in which du_i is the difference of the velocity components u_i at the two ends of ds.

If the fluid is inviscid, Eqs. (2.5) become, for $i = 1, 2, 3$,

$$\frac{Du_i}{Dt} = -\frac{1}{\rho}\frac{\partial p}{\partial x_i} + X_i, \tag{2}$$

which are called the *Euler equations*.   If, in addition, the density is either constant (as for a homogeneous liquid) or a function of pressure and time only (as for a barotropic[2] gas), the equations of motion are

$$\frac{Du_i}{Dt} = -\frac{\partial}{\partial x_i}\left(\int \frac{dp}{\rho} + \Omega\right), \tag{2a}$$

in which Ω is a body-force potential which is assumed to exist.

From (1) and (2a) it follows that

$$\frac{D}{Dt}(u_i\,dx_i) = \frac{Du_i}{Dt}\,dx_i + u_i\,du_i = d\left(-\int \frac{dp}{\rho} - \Omega + \tfrac{1}{2}u_iu_i\right). \tag{3}$$

Integration of (3) around the closed circuit C produces

$$\frac{D}{Dt}\Gamma = \frac{D}{Dt}\int_C u_i\,dx_i = \int_C \frac{D}{Dt}(u_i\,dx_i) = 0, \tag{4}$$

---

[1] In this book, results valid for barotropic gases are applied to homogeneous gases with uniform entropy only.   The latter are more special but more often encountered.
[2] A homentropic ideal gas is a barotropic gas.

in which Γ is the circulation around the circuit C and (3) has been utilized. The last integral in (4) is equal to zero because the quantity in the last parenthesis in (3) is single-valued. The reversal of the order of substantial differentiation and integration is permissible because the integration is performed along the material circuit C.

Equation (4) states the persistence of circulation and is a theorem[1] due to Lord Kelvin [William Thomson, 1824–1907] (1869). Since the circulation along a closed circuit is equal to the integral of the normal component of the vorticity over a surface bounded by that circuit, an initially irrotational state is a state in which the circulation along *any* closed circuit is zero. Equation (4) then guarantees that, under the stated conditions, the circulation along *any* closed circuit will remain zero subsequently, since any circuit at a subsequent moment corresponds to an initial circuit consisting of the same fluid particles. Thus the vorticity everywhere must be zero subsequently, and the persistence of irrotationality is a consequence of (4), which states more than the persistence of irrotationality, because the circulation, which is conserved, may be different from zero.

3. THE VORTICITY EQUATIONS FOR AN INVISCID FLUID WITH VARIABLE DENSITY

If the body force per unit mass possesses a potential Ω, so that X_i is given by (2.12), Ω can be eliminated from any two of the three equations in (2) by cross differentiation. Working with the first two equations in (2), recalling Eqs. (2.27) and (2.28), and remembering that now ρ is assumed to be neither constant nor a function of p and t only, we have

$$\frac{D\xi_3}{Dt} + M = \frac{1}{\rho^2}\left(\frac{\partial\rho}{\partial x_1}\frac{\partial p}{\partial x_2} - \frac{\partial\rho}{\partial x_2}\frac{\partial p}{\partial x_1}\right), \tag{5}$$

in which M is equal to the quantity between the second and the third equality signs in (2.28). Since

$$\frac{D\rho}{Dt} + \rho\frac{\partial u_\alpha}{\partial x_\alpha} = 0,$$

M is now equal to

$$-\xi_\alpha\frac{\partial u_\alpha}{\partial x_3} - \frac{\xi_3}{\rho}\frac{D\rho}{Dt} \quad \text{or} \quad -\xi_\alpha\frac{\partial u_3}{\partial x_\alpha} - \frac{\xi_3}{\rho}\frac{D\rho}{Dt}, \tag{6}$$

and (5) becomes, after division by ρ,

$$\frac{D}{Dt}\frac{\xi_3}{\rho} = \frac{\xi_\alpha}{\rho}\frac{\partial u_3}{\partial x_\alpha} + \frac{1}{\rho^3}\left(\frac{\partial\rho}{\partial x_1}\frac{\partial p}{\partial x_2} - \frac{\partial\rho}{\partial x_2}\frac{\partial p}{\partial x_1}\right).$$

[1] Truesdell (1954, p. 92) notes in his scholarly work that Hankel (1861) had already published the theorem before Lord Kelvin did in 1869.

In general,

$$\frac{D}{Dt}\frac{\xi_k}{\rho} = \frac{\xi_\alpha}{\rho}\frac{\partial u_k}{\partial x_\alpha} + \frac{1}{\rho^3}\left(\frac{\partial \rho}{\partial x_i}\frac{\partial p}{\partial x_j} - \frac{\partial \rho}{\partial x_j}\frac{\partial p}{\partial x_i}\right),\tag{7}$$

in which i, j, and k are in the cyclic order of 1, 2, and 3. There are three equations in (7), and they are the vorticity equations for an inviscid fluid with variable density, which can be written in vector form as

$$\left(\frac{\partial}{\partial t} + \mathbf{u}\cdot\mathbf{grad}\right)\frac{\boldsymbol{\xi}}{\rho} = \left(\frac{\boldsymbol{\xi}}{\rho}\cdot\mathbf{grad}\right)\mathbf{u} + \frac{1}{\rho^3}\,(\mathbf{grad}\,\rho \times \mathbf{grad}\,p).\tag{7a}$$

If ρ is constant, (7) reduces to (2.37). For a gas, if ρ is a function of p alone (or, more generally, of p and t only), the quantity in the last parenthesis of (7) is zero, and (7) becomes

$$\frac{D}{Dt}\frac{\xi_k}{\rho} = \frac{\xi_\alpha}{\rho}\frac{\partial u_k}{\partial x_\alpha}.\tag{8}$$

If ρ is neither constant nor a function of p and t only, the last parenthesis in (7) will be different from zero, except in two very special cases in wave motion (see Chap. 5) and in some trivial cases.

If the fluid is incompressible, then since $D\rho/Dt = 0$, (7) can be written as

$$\frac{D\xi_k}{Dt} = \xi_\alpha\frac{\partial u_\alpha}{\partial x_\alpha} + \frac{1}{\rho^2}\left(\frac{\partial \rho}{\partial x_i}\frac{\partial p}{\partial x_j} - \frac{\partial \rho}{\partial x_j}\frac{\partial p}{\partial x_i}\right).\tag{7b}$$

The last term is then seen as representing the rate of creation of vorticity by density variation. The corresponding term in (7) has a similar meaning, except that it contributes to the rate of creation of ξ_k/ρ. For an inviscid and nonhomentropic gas, the rate of change of the ratio of vorticity to the density is in part due to the turning and stretching of the vortex lines and in part to the nonhomentropy, which can create vorticity.

The physical meaning of the last parenthesis in (7) can be made clear by considering two superposed horizontal layers of liquids of density ρ_1 and ρ_2, both being accelerated by a horizontal pressure gradient. The accelerations must be inversely proportional to the densities, for otherwise the pressure would not be continuous across the interface. In this case the gradient of ρ is vertical, that of p horizontal, and a vortex sheet is being created at the interface. If the density variation is continuous, distributed vorticity will be created throughout the fluid, unless the pressure gradient is also in the direction of the vertical.

Note that when dp/ρ is not an exact differential, (4) should be replaced by

$$\frac{D\Gamma}{Dt} = -\int_C \frac{dp}{\rho}.\tag{9}$$

Since

$$dp = \frac{\partial p}{\partial x_\alpha}dx_\alpha,$$

application of Stokes' theorem to (9) yields

$$\frac{D\Gamma}{Dt} = \iint_S \frac{1}{\rho^2} \, (\mathbf{grad} \, \rho \times \mathbf{grad} \, p)_n \, dS, \tag{10}$$

in which the surface S is bounded by the curve C in (9). Comparing (10) with (1.56), we see again that the integrand (without n) is the rate of creation of vorticity. Equation (10) is due to Bjerknes (1918).

4. EFFECT OF STRETCHING OF VORTEX LINES ON THE VORTICITY OF A HOMENTROPIC GAS

For a homentropic gas the vorticity is governed by (8). Comparing (8) with (2.37), we see that ξ_i/ρ for a homentropic gas corresponds to ξ_i for a liquid of constant density. The same development for a fluid of constant density leading to (2.36) can be applied to a homentropic gas, leading to

$$\frac{D}{Dt} \frac{\xi}{\rho \, ds} = 0, \tag{11}$$

in which ξ and ds has the same meaning as in (2.36). This equation states that the magnitude of vorticity in a homentropic gas is proportional to the length ds of the vortex-line segment and to the density of the fluid.

For two-dimensional flows, the only vorticity component that does not vanish is $\xi_3 = \zeta$, say, and (2.37) and (8) become, respectively,

$$\frac{D\zeta}{Dt} = 0 \quad \text{and} \quad \frac{D}{Dt} \frac{\zeta}{\rho} = 0, \tag{12}$$

the first of which is valid for constant ρ and the second for homentropy. These results are exactly the same as (2.38) and (11) for two-dimensional flows, because for such flows the vortex lines are straight lines in the direction of x_3 or z which do not stretch or shrink, and therefore ds is constant.

There is a connection between (4) and (12). If the flow is two-dimensional and ρ is constant, the area dA occupied by a (two-dimensional) fluid particle is constant. Imagine the area to be infinitesimally small; then $dA \cdot \zeta$ is Γ, which by virtue of (4) is conserved. Therefore ζ is conserved, as stated by the first of Eqs. (12). If ρ is not constant, dA is inversely proportional to ρ. Therefore ζ/ρ is conserved, as stated by the second of Eqs. (12).

For axisymmetric flows the vortex lines are concentric circles with centers on the axis of symmetry. Thus ds, which is along these circles, is proportional to r if r, φ, and z are cylindrical coordinates and if the z axis is the axis of symmetry. The only nonzero component of the vorticity is

(see Appendix 2)

$$\eta = \frac{\partial u}{\partial z} - \frac{\partial w}{\partial r},$$

in which u and w are the velocity components in the directions of increasing r and z, respectively. Thus (2.36) and (11) respectively become,

$$\frac{D}{Dt}\frac{\eta}{r} = 0 \quad \text{and} \quad \frac{D}{Dt}\frac{\eta}{\rho r} = 0, \tag{13}$$

the first of which is valid for constant density and the second for constant entropy.

5. STEADY TWO-DIMENSIONAL OR AXISYMMETRIC FLOWS OF AN INVISCID FLUID OF CONSTANT DENSITY

If ρ is constant, two-dimensional flow of an inviscid fluid is governed by the first equation in (12). If, in addition, the flow is *steady*, the equation states that

$$\zeta = \text{a function of } \psi \text{ only}, \tag{14}$$

in which ψ is the stream function. Since

$$\zeta = \frac{\partial v}{\partial x} - \frac{\partial u}{\partial y} = -\nabla^2\psi, \tag{15}$$

in which x and y are Cartesian coordinates and u and v the corresponding velocity components, (14) can be written as

$$\frac{\partial^2\psi}{\partial x^2} + \frac{\partial^2\psi}{\partial y^2} = f(\psi). \tag{14a}$$

Similarly, for steady axisymmetric flow of an inviscid fluid of constant density, the first equation in (13) can be written as

$$\frac{\partial^2\psi}{\partial r^2} - \frac{1}{r}\frac{\partial\psi}{\partial r} + \frac{\partial^2\psi}{\partial z^2} = r^2 f(\psi), \tag{16}$$

in which ψ is now the Stokes stream function, the vorticity being given by

$$\eta = -\frac{1}{r}\left(\frac{\partial^2\psi}{\partial r^2} - \frac{1}{r}\frac{\partial\psi}{\partial r} + \frac{\partial^2\psi}{\partial z^2}\right). \tag{17}$$

6. MOVEMENT OF THE VORTEX LINES

In fluid-mechanics literature, the statement can often be found that "the vortex lines in an inviscid fluid move with the fluid." The precise meaning of this statement is that the fluid particles constituting a vortex line will continue to constitute a vortex line as the fluid flows provided it is inviscid and either a homogeneous liquid or a barotropic gas. This is the theorem of Hermann Ludwig Ferdinand von Helmholtz (1821–1894).

The original proof of this theorem [Helmholtz (1858)] is plausible but not rigorous, as pointed by Lamb (1932, p. 206). A rigorous proof follows readily from equations provided by Augustin Louis de Cauchy (1789–1857) in the introduction of his memoir on waves (1827). A derivation of Cauchy's equations by the use of Weber's transformation [Lamb (1932, p. 14)] can be found in Lamb (1932, p. 204). Instead of Weber's equations, a judicious use of Kelvin's circulation theorem can also lead to Cauchy's equations. But in either case Lagrangian coordinates are used, and the derivation is not simple. We propose to use Kelvin's circulation theorem, or (4), to prove Helmholtz' theorem directly.

First, we assume the vorticity vector to be continuous. Then no vortex line can intersect a surface if along *any* closed circuit lying therein the circulation is zero. For otherwise the ξ_n in (1.56) would be different from zero at the point of intersection and in its neighborhood and the circulation Γ along a circuit in that neighborhood would be different from zero. Now, on any point A on a vortex line, draw two lines BAC and DAE intersecting it and draw vortex lines passing through BAC and DAE. The surfaces thus formed will be called S_1 and S_2, which intersect in the vortex line under consideration. According to (1.56), the circulation along any closed circuit on S_1 or S_2 is zero. It will continue to be zero as we follow the fluid particles as they move, according to Kelvin's circulation theorem (4). Thus the circulation along any closed circuit situated on the material surface S_1 or S_2 remains zero. Hence these remain vorticity surfaces consisting of vortex lines. Their intersection, which is the material line which coincides with the original vortex line, will continue to be a vortex line, for otherwise at some point on it a vortex line would intersect at least one of the surfaces S_1 and S_2, contradicting the fact that the circulation along any circuit in S_1 or S_2 is zero for all times. Note that BAC and DAE must not be drawn on the same vorticity surface.

There is a connection between the theorem just arrived at and the rate of change of the magnitude ξ of the vorticity as given by (11). Since the vortex lines are "frozen" into the fluid, so to speak, the mass contained in the volume of a vortex tube with cross section dA and length ds is $\rho \, ds \, dA$ and is constant as the fluid flows. Since the vorticity is perpendicular to dA, the circulation around the bounding curve of dA is simply $\xi \, dA$, and is conserved, i.e., remains constant. Thus $\xi/(\rho \, ds)$ must be constant as the fluid particles move about, as stated by (11). As the vortex tube becomes thinner and thinner (or fatter and fatter), ξ becomes greater and greater (or smaller and smaller) in inverse proportion to dA and direct proportion to $\rho \, ds$. If ρ is constant, ξ is directly proportional to the length ds.

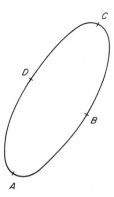

FIGURE 1. THE INDEPENDENCE OF THE
INTEGRAL $\int_A^C u_i \, dx_i$ OF PATH IF THE FLOW
IS IRROTATIONAL.

7. VELOCITY POTENTIAL FOR IRROTATIONAL FLOWS

If the vorticity is everywhere zero, so that the flow is irrotational, the circulation around the closed circuit (Fig. 1) $ABCDA$ is zero. This means that the circulation along the path ABC is exactly equal to that along the path ADC. In other words, the value of

$$\int_A^C u_i \, dx_i$$

for a fixed point A depends only on the coordinates of C. Therefore $u_i \, dx_i$ is an exact differential, which can be denoted by $d\phi$. Then

$$u_i = \frac{\partial \phi}{\partial x_i}, \tag{18}$$

in which ϕ is called the velocity potential. Its existence depends solely on irrotationality and not at all on the steadiness of the motion or the incompressibility of the fluid. From the foregoing discussion, it is evident that the value of the velocity potential ϕ at any point is exactly equal to the circulation along a curve starting at the point where ϕ is zero and ending at the point under consideration.

8. THE BERNOULLI EQUATION FOR STEADY FLOWS

For steady flows of an incompressible fluid the density is constant along a streamline, that is,

$$u_\alpha \frac{\partial \rho}{\partial x_\alpha} = 0. \tag{19}$$

For steady flows of a compressible fluid, the entropy along a streamline is constant if heat conduction is negligible and the process of change from point to point on the streamline can be considered as thermodynamically slow. In that case, ρ is a function of the pressure p alone along the streamline, as indicated by (A1.45), in which the entropy S is constant along the streamline. Let it be denoted by $f(p)$. Then

$$u_\alpha \frac{\partial \rho}{\partial x_\alpha} = u_\alpha \frac{\partial}{\partial x_\alpha} f(p),$$

or
$$u_\alpha \frac{\partial}{\partial x_\alpha} \frac{f(p)}{\rho} = 0. \tag{20}$$

The Euler equations are, for *steady* flows,

$$u_\alpha \frac{\partial u_i}{\partial x_\alpha} = -\frac{1}{\rho} \frac{\partial p}{\partial x_i} - \frac{\partial \Omega}{\partial x_i}, \tag{21}$$

Hence, for a liquid,

$$u_\alpha \frac{\partial}{\partial x_\alpha} \frac{u_i u_i}{2} = -u_i \left(\frac{1}{\rho} \frac{\partial p}{\partial x_i} + \frac{\partial \Omega}{\partial x_i} \right)$$
$$= -u_i \frac{\partial}{\partial x_i} \left(\frac{p}{\rho} + \Omega \right) = -u_\alpha \frac{\partial}{\partial x_\alpha} \left(\frac{p}{\rho} + \Omega \right), \tag{22}$$

in which ρ has been moved behind the differentiation sign by virtue of (19). Hence

$$\frac{D}{Dt} \left(\frac{p}{\rho} + \Omega + \frac{u_i u_i}{2} \right) = 0, \tag{23}$$

and
$$\frac{p}{\rho} + \Omega + \frac{u_i u_i}{2} = \text{const along a streamline,} \tag{24}$$

in which ρ is constant along the streamline. For a compressible fluid with isentropy along a streamline, $\rho = f(p)$ along each streamline, so that $\int dp/\rho$ has a meaning if it is understood to be evaluated along a streamline and

$$u_\alpha \frac{\partial}{\partial x_\alpha} \frac{u_i u_i}{2} = -u_\alpha \frac{\partial}{\partial x_\alpha} \left(\int \frac{dp}{\rho} + \Omega \right). \tag{25}$$

Hence
$$\frac{D}{Dt} \left(\int \frac{dp}{\rho} + \Omega + \frac{u_i u_i}{2} \right) = 0,$$

or
$$\int \frac{dp}{\rho} + \Omega + \frac{u_i u_i}{2} = \text{const along a streamline.} \tag{26}$$

Actually, (24) can also be written in the form of (26), which is then the general form of the Bernoulli equation for steady flows and is valid along a

streamline, provided the fluid is incompressible or the flow is isentropic.[1]
The "constant" in general varies from streamline to streamline.

9. THE BERNOULLI EQUATION FOR IRROTATIONAL FLOWS

For an irrotational flow, the Bernoulli equation has a more general form
and is valid over the entire space occupied by the fluid.   Since the flow is
irrotational, it is consistent to consider the fluid to be either a homo-
genous liquid or a homentropic gas.   In either case the Euler equations
can be written as

$$\frac{\partial u_i}{\partial t} + u_\alpha \frac{\partial u_i}{\partial x_\alpha} = - \frac{\partial}{\partial x_i}\left(\int \frac{dp}{\rho} + \Omega \right).$$

Since the flow is irrotational,

$$u_i = \frac{\partial \phi}{\partial x_i} \quad \text{and} \quad \frac{\partial u_i}{\partial x_\alpha} = \frac{\partial u_\alpha}{\partial x_i}, \tag{27}$$

so that the Euler equations can be written as $(i = 1,2,3)$

$$\frac{\partial}{\partial x_i}\left(\frac{\partial \phi}{\partial t} + \int \frac{dp}{\rho} + \Omega + \frac{u_\alpha u_\alpha}{2} \right) = 0. \tag{28}$$

Integration of these three equations produces

$$\frac{\partial \phi}{\partial t} + \int \frac{dp}{\rho} + \Omega + \frac{u_\alpha u_\alpha}{2} = F(t). \tag{29}$$

But the arbitrary function $F(t)$ can be absorbed into $\partial \phi / \partial t$ without
affecting the flow in any way, since the velocity components are space
derivatives of ϕ.   Consequently,

$$\frac{\partial \phi}{\partial t} + \int \frac{dp}{\rho} + \Omega + \frac{u_\alpha u_\alpha}{2} = \text{const}, \tag{30}$$

in which the constant can be assigned the value zero without loss of
generality.   It is well to remember, however, that $F(t)$ and the constant in
(30) depend on the boundary condition on the pressure when ϕ is given
or known.   Equation (30) is the Bernoulli equation for irrotational flow
of a homogeneous liquid or a homentropic gas, valid over the entire
field of flow.

If ρ is constant and all the boundaries are fixed, so that there is no
free surface on which p is zero and no interfaces, all the boundary conditions are kinematic and linear.   In this case, although the exact solution
of the Laplace equation is often still difficult, its approximate solution is

---

[1] Remember that isentropy, in the usage of this book, demands only equal entropy
along a pathline, not necessarily homentropy.

easy because it is linear and the boundary conditions are linear. After the velocity components are determined by solving the Laplace equation, the pressure can be calculated from (30). Thus the nonlinearity of the Euler equation is reflected only in the nonlinearity of (30), and there it presents no difficulty at all because the nonlinear terms are already determined and only the pressure is to be evaluated. However, if there is a free surface, the ghost of nonlinearity will reappear in the boundary condition through the very Bernoulli formula which previously kept it away (until the straightforward step of evaluating the pressure)—this time to stay and make considerable trouble except in a few special cases in which it can be appeased by inverse and hodograph methods.

10. THE BERNOULLI EQUATION FOR STEADY TWO-DIMENSIONAL FLOWS WITH CONSTANT VORTICITY

In Sec. 8 the Bernoulli equation (26) for steady flows was derived. It was shown that the Bernoulli quantity is constant along a streamline. If the flow is two-dimensional and steady and possesses *constant* vorticity ζ, then there is a Bernoulli equation [Kuo (1943)] for the entire field of flow, provided the density of the fluid is constant and the viscosity of the fluid can be neglected. Inviscidness is assumed for convenience, but it should be remembered that if the Reynolds number is high, viscous effects are important only within boundary layers in the neighborhood of solid boundaries, where the flow with constant vorticity does not satisfy the nonslip condition. Outside these layers, the viscous terms in the Navier-Stokes equations can be neglected, although the existence of a region of constant vorticity may be due precisely to a small viscosity, whose indirect effect on the flow is therefore taken into account.

The equations of motion are given by (2a), which for the flow under consideration can be written as

$$-v\zeta = -\frac{\partial \chi'}{\partial x}, \qquad u\zeta = -\frac{\partial \chi'}{\partial y}, \tag{31}$$

in which

$$\chi' = \frac{p}{\rho} + \Omega + \frac{q^2}{2}, \tag{32}$$

q being the speed. Aside from the slightly more general definition of χ', Eqs. (31) are the same as (2.17) if the temporal-acceleration and viscous terms are dropped in the latter. In (31) Cartesian coordinates x and y are used, and u and v are the corresponding velocity components. The equation of continuity allows the use of the stream function ψ. Substituting into (31) and integrating, we obtain

$$\frac{p}{\rho} + \Omega + \frac{q^2}{2} + \zeta\psi = \text{const.} \tag{33}$$

This equation is the same as that found by Kuo (1943), except that he neglected the gravity term. Note that (33) is valid over the entire field of flow wherever the basic assumptions are not violated, not just along a streamline.

11. THE BERNOULLI EQUATION FOR UNSTEADY TWO-DIMENSIONAL FLOWS WITH CONSTANT VORTICITY

The assumptions regarding the density and viscosity made in Sec. 10 are retained here, but the flow, though still two-dimensional, is no longer assumed steady. We assume, however, that the boundary geometry does not change with time. Note that if a cylinder moves unsteadily in the fluid-filled space between two parallel plates in a direction parallel to these plates, the boundary geometry does not change with time if we adopt a frame of reference moving with the cylinder. Since the speed of the cylinder varies with time, the flow relative to the moving frame is still unsteady.

We can then separate the flow into two parts: one with constant vorticity ζ and the other irrotational. Thus

$$\psi = \psi_1(x,y) + \psi_2(x,y,t), \qquad \nabla^2\psi_1 = -\zeta, \qquad \nabla^2\psi_2 = 0 \qquad (34)$$

if the subscript 1 denotes the rotational flow and the subscript 2 irrotational flow. Both ψ_1 and ψ_2 must be constant along the (stationary) solid boundaries. The rotational flow is steady, since ζ is constant and the boundary geometry does not change with time. The irrotational flow is unsteady, with velocity components

$$u_2 = \frac{\partial\phi_2}{\partial x} = \frac{\partial\psi_2}{\partial y}, \qquad v_2 = \frac{\partial\phi_2}{\partial y} = -\frac{\partial\psi_2}{\partial x}. \qquad (35)$$

Since the flow represented by ψ_1 is steady but the flow (u_2,v_2) is unsteady, (31) must now be replaced by

$$\frac{\partial u_2}{\partial t} - v\zeta = -\frac{\partial\chi'}{\partial x}, \qquad \frac{\partial v_2}{\partial t} + u\zeta = -\frac{\partial\chi'}{\partial y}, \qquad (36)$$

in which $u = u_1 + u_2 = \dfrac{\partial\psi}{\partial y}, \qquad v = v_1 + v_2 = -\dfrac{\partial\psi}{\partial x}.$

Substituting (35) into (36) and integrating, we have

$$\frac{\partial\phi_2}{\partial t} + \frac{p}{\rho} + \frac{q^2}{2} + \Omega + \zeta\psi = \text{const}, \qquad (37)$$

in which $q^2 = u^2 + v^2$.

If the boundary geometry does change with time, we can choose a ψ_1 satisfying the second equation in (34) but no longer constant along the moving boundaries. We then choose a ψ_2 such that the third equation in (34) is satisfied and all the kinematic boundary conditions (such as equality of normal velocity of fluid and that of the solid) are satisfied by $\psi = \psi_1 + \psi_2$. Then although ψ_1 is still independent of time, ϕ_2 will not only be time-dependent but its time dependence will be more complicated than in the case of constant boundary geometry. Granting that complication, we still have (37). Because of the complicated time dependence of ϕ_2, however, (37) is much less useful if the boundary geometry changes.

12. SOME GENERAL RESULTS FOR ROTATING FLUIDS

The nature of the motion produced by a small perturbation in a rotating fluid can be very different from that in a nonrotating fluid. The equations of motion referred to rotating coordinates are most convenient for studying the motion of a fluid with a basic solid-body rotation. The reason for this is twofold. First, geophysical phenomena, which constitute much of the raison d'être for the study of rotating fluids, are studied, measured, and recorded with respect to the Earth, which is rotating. Second, one of the most important, interesting, and strange phenomena involving rotating fluids occurs when the fluid is in almost rigid rotation. With respect to a frame of reference rotating with the fluid, the residue motion is very weak, and this weakness presents advantages in analysis. But one must not infer that every time a rotating frame of reference is adopted, the fluid itself necessarily has a basic rotation.

In this section we shall study two closely related subjects, geostrophic motion and the Proudman-Taylor theorem. The fluid is assumed inviscid and incompressible and the density constant. The governing equations are therefore (2.59) to (2.64).

12.1. Geostrophy

If the acceleration *relative to an observer on the rotating frame of reference* is small compared with the Coriolis acceleration, which is then predominant, and if the Reynolds number $\omega L^2/\nu$ is large, the limiting form of the governing equation is

$$\mathbf{grad}\, P = -2\boldsymbol{\omega} \times \mathbf{v}, \tag{38}$$

valid outside of boundary layers and possibly internal shear layers. The motion governed by (38) is called *geostrophic*. In geophysical applications, the gradient of $\omega^2 r^2/2$ in P is small compared with g (equal to $-\mathbf{grad}\,\Omega$), and can be neglected. At any latitude α, the local rate of rotation about the local vertical coordinate z is $\omega \sin \alpha$, ω being the angular speed of rotation of the Earth. If a set of local Cartesian co-ordinates (the x axis pointing eastward) is used, the geostrophic equations of motion are

$$\frac{\partial p}{\partial x} = 2\rho\omega(v \sin \alpha - \omega \cos \alpha), \tag{39}$$

$$\frac{\partial p}{\partial y} = -2\rho\omega \sin \alpha\ u, \tag{40}$$

$$\frac{\partial p}{\partial z} = -\rho g + 2\rho\omega\ u \cos \alpha. \tag{41}$$

The vertical acceleration is assumed small to start with, and w must be zero on the ground if the ground is flat or if the scale of motion is so large that ground topography is negligible. Hence w must be negligible **throughout, and the last term in (39) can be neglected.** Furthermore the last term in (41) is numerically small compared with ρg. Hence, under the stated conditions, the above three equations can be written as

$$\frac{\partial p}{\partial x} = 2\rho\omega \sin \alpha\ v, \tag{42}$$

$$\frac{\partial p}{\partial y} = -2\rho\omega \sin \alpha\ u, \tag{43}$$

$$\frac{\partial p}{\partial z} = -\rho g. \tag{44}$$

These equations have far-reaching consequences. The horizontal part of the velocity is normal to the pressure gradient and therefore parallel to the isobars. In the northern hemisphere α is between 0 and 90°, and, viewed from above, the horizontal velocity vector (u,v) leads the pressure gradient by 90°. A region of closed streamlines, where the circulation is counterclockwise (viewed from above), is therefore a low-pressure region, whereas a region enclosed by streamlines along which the circulation is clockwise is a high-pressure region. Exactly the opposite is true of the southern hemisphere.

12.2. The Proudman-Taylor Theorem

If a rigid body moves slowly with constant velocity in a rotating fluid, so that the acceleration relative to a coordinate system I moving with it is small compared with the Coriolis acceleration, the motion, astonishingly enough, is two-dimensional. To show this result of Proudman (1916) and Taylor (1917), experimentally verified by Taylor (1921, 1923), we note first that the acceleration relative to I is exactly the same as that relative to the rotating (but not translating) frame II, since I is moving with constant velocity relative to II. Hence we can use (38), which is relative to II. Expanded, it has the following component equations:

$$\frac{\partial P}{\partial x} = 2\omega v, \qquad \frac{\partial P}{\partial y} = -2\omega u, \qquad \frac{\partial P}{\partial z} = 0. \tag{45}$$

By differentiating the first two with respect to z and utilizing the third, one readily obtains

$$\frac{\partial u}{\partial z} = 0, \qquad \frac{\partial v}{\partial z} = 0.$$

If we eliminate P between the first two equations in (45), we have

$$\frac{\partial u}{\partial x} + \frac{\partial v}{\partial y} = 0,$$

from which, by virtue of the equation of continuity, we have

$$\frac{\partial w}{\partial z} = 0. \tag{46}$$

Thus the velocity components are independent of z, and the motion is mathematically two-dimensional!

Consider a solid cylinder of finite height and horizontal ends moving steadily and horizontally in a fluid rotating about the vertical axis. The flow is two-dimensional, as we have just seen. For convenience the lateral surface of the solid cylinder will be denoted by S and its prolongation in the fluid by S_e. The flow outside of the surface $S + S_e$ is called the *outside flow*, and the flow inside S_e the *inside flow*. The outside flow is irrotational relative to the rotating frame. This can be seen by using system I, by noting that the velocity far from the cylinder is uniform, and by invoking the persistence of vorticity (relative in this case) in two-dimensional flows. The inside flow cannot be determined from (45). However, since the velocity at S_e is tangent to S_e, the P along S_e must be constant. This is clear from the discussion in Sec. 12.1. It is also clear from (45), for it is evident there that P is a stream function, which must be constant along a streamline. Hence the state of no flow (in system I) inside S_e is a possible one. Experiments by Taylor (1923) showed that this is in fact the state. The fluid column moving with the solid cylinder has been called the *Taylor column*. Near the surface of the Taylor column there is a boundary layer, where viscous effects must be taken into account.

For axisymmetric flows and in cylindrical coordinates, (46) still holds. The equation of continuity being

$$\frac{\partial(ru)}{\partial r} + \frac{\partial(rw)}{\partial z} = 0$$

and ru being zero at $r = 0$, (46) implies

$$u = 0, \tag{47}$$

so that the radial motion is completely inhibited. For a body moving steadily along the axis of rotation, (46) and (47) describe the motion. They have been verified in front of the body rather well [Long (1953)] but less well in the rear. The fluid column moving with the solid body according to (46) and (47) is also called the Taylor column.

The phenomena of blocking in steady weak two-dimensional motions of a stratified fluid [Yih (1959a)] and of blocking in steady weak motions of a conducting fluid in a strong magnetic field [Yih (1959b)] are analogous to the phenomenon of the Taylor column.

13. THE BERNOULLI EQUATION FOR STEADY FLOWS WITH RESPECT TO A ROTATING FRAME

In the case of an incompressible fluid, (19) is valid if we assume the flow to be steady and ignore diffusion. In the case of a compressible fluid we assume isentropy along a streamline in steady flows, so that (20) is valid. The validity of (19) and (20) remains for a rotating frame of reference, provided the flow is steady with respect to that frame. The fluid is assumed inviscid, as it is throughout this chapter.

We shall write (2.56) to (2.58) in the form

$$\frac{Du}{Dt} - 2\omega v = -\frac{1}{\rho}\frac{\partial p}{\partial x} - \frac{\partial}{\partial x}\left(\Omega - \frac{\omega^2 r^2}{2}\right), \tag{48}$$

$$\frac{Dv}{Dt} + 2\omega u = -\frac{1}{\rho}\frac{\partial p}{\partial y} - \frac{\partial}{\partial y}\left(\Omega - \frac{\omega^2 r^2}{2}\right), \tag{49}$$

$$\frac{Dw}{Dt} = -\frac{1}{\rho}\frac{\partial p}{\partial z} - \frac{\partial}{\partial z}\left(\Omega - \frac{\omega^2 r^2}{2}\right). \tag{50}$$

First we note that the Coriolis acceleration vector $(-2\omega v, 2\omega u, 0)$ is perpendicular to the velocity vector (u,v,w). Hence if s is measured along the streamlines, the Coriolis acceleration has no component along it. Since the flow is steady,

$$\frac{Du_s}{Dt} = q\frac{\partial q}{\partial s}, \tag{51}$$

in which q is the speed and is hence equal to the velocity u_s along the streamline. Realizing that along the streamline ρ is a function of p or is constant, we can write

$$\frac{1}{\rho}\frac{\partial p}{\partial s} = \frac{\partial}{\partial s}\int\frac{dp}{\rho}. \tag{52}$$

The equation of motion along a streamline is then

$$q\frac{\partial q}{\partial s} = -\frac{\partial}{\partial s}\left(\int\frac{dp}{\rho} + \Omega - \frac{\omega^2 r^2}{2}\right),$$

integrating of which yields

$$\int\frac{dp}{\rho} + \frac{q^2}{2} + \Omega - \frac{\omega^2 r^2}{2} = \text{const along a streamline,} \tag{53}$$

which is the Bernoulli equation we seek. Evidently the quantity $\Omega - \omega^2 r^2/2$ replaces Ω in the Bernoulli equation in a nonrotating frame. It is called the *geopotential*. In the absence of motion relative to the rotating frame, the density and the pressure are constant in a surface of constant geopotential. This can be shown from (48) to (50). The reader is urged to verify this statement. *Hint:* Remember to use the total differential.

14. THE BERNOULLI EQUATION FOR STEADY TWO-DIMENSIONAL FLOWS WITH CONSTANT VORTICITY RELATIVE TO A ROTATING FRAME

If we seek a Bernoulli equation valid in the entire field of flow, not merely along a streamline, both the flow and the fluid will have to be more specialized. The flow is now assumed two-dimensional, with a constant vorticity ζ relative to the rotating frame, and the fluid is assumed to be inviscid and incompressible and to have a constant density. Equations (48) and (49) can then be written as

$$-(\zeta + 2\omega)v = -\frac{\partial}{\partial x}\left(P + \frac{q^2}{2}\right), \tag{54}$$

$$(\zeta + 2\omega)u = -\frac{\partial}{\partial y}\left(P + \frac{q^2}{2}\right), \tag{55}$$

with P given in (2.65) and q indicating the speed. Since u and v are expressible in terms of the stream function ψ as in (1.31), using (1.31) in (54) and (55) and integrating, we have

$$\frac{p}{\rho} + \frac{q^2}{2} + \Omega - \frac{\omega^2 r^2}{2} + (\zeta + 2\omega)\psi = \text{const}, \tag{56}$$

which is the equation sought.

15. THE BERNOULLI EQUATION FOR UNSTEADY TWO-DIMENSIONAL FLOWS WITH CONSTANT VORTICITY RELATIVE TO A ROTATING FRAME

The fluid is assumed to be the same as in Sec. 14, and so is the flow, except that now it is unsteady. The derivation is similar to that in Sec. 11, and the result is

$$\frac{\partial \phi_2}{\partial t} + \frac{p}{\rho} + \frac{q^2}{2} + \Omega - \frac{\omega^2 r^2}{2} + (\zeta + 2\omega)\psi = \text{const}, \qquad (57)$$

in which ϕ_2 is the (time-dependent) velocity potential of the irrotational part of the flow, the other part being steady. Again, if the boundary geometry does not change with time, the time dependence of ϕ_2 is quite simple [see, for instance, (4.91) and (4.92)]. Otherwise it is complicated, and (57) is nearly useless.

We conclude this section by referring the reader to a Bernoulli equation in Lamb (1932, p. 20), for whatever usefulness it may have. It is for a flow which is irrotational with respect to a nonrotating frame, but it is an equation in terms of coordinates of a moving frame and velocity components relative to it.

16. THE BIOT-SAVART LAW

It is a well-known result that any vector field can be resolved uniquely into two parts, one part (u_1, v_1, w_1) being curl-free (irrotational) and the other part (u_2, v_2, w_2) divergence-free. The curl-free part in general has a nonzero θ defined by

$$\theta = \frac{\partial u}{\partial x} + \frac{\partial v}{\partial y} + \frac{\partial w}{\partial z},$$

in terms of Cartesian coordinates and Cartesian velocity components. This θ is the strength of a source, denoted by $4\pi m$ in Sec. 5.2 of Chap. 4. If the coordinates of a point P are denoted by x, y, and z and the coordinates of the point where the divergence is θ' are x', y', and z', (4.20) gives, upon integration,

$$\phi = -\frac{1}{4\pi} \int \frac{\theta'}{R} dV', \qquad (58)$$

in which $dV' = dx'\, dy'\, dz'$ and

$$R = [(x - x')^2 + (y - y')^2 + (z - z')^2]^{1/2}, \qquad (59)$$

provided the integral in (58) exists, i.e., provided θ' vanishes outside of some finite distance from the origin or diminishes sufficiently fast toward infinity. From (58) the components of the curl-free part of the vector (here represented by the velocity) can be found by taking the gradient of ϕ.

The divergence-free part of the velocity can be expressed in terms of three functions F, G, and H as

$$u_2 = \frac{\partial H}{\partial y} - \frac{\partial G}{\partial z}, \qquad v_2 = \frac{\partial F}{\partial z} - \frac{\partial H}{\partial x}, \qquad w_2 = \frac{\partial G}{\partial x} - \frac{\partial F}{\partial y}. \qquad (60)$$

To the functions F, G, and H may be added, respectively,

$$\frac{\partial \chi}{\partial x}, \qquad \frac{\partial \chi}{\partial y}, \qquad \text{and} \qquad \frac{\partial \chi}{\partial z}$$

without affecting u_2, v_2, and w_2. Hence we can assume χ to be so chosen that

$$\frac{\partial F}{\partial x} + \frac{\partial G}{\partial y} + \frac{\partial H}{\partial z} = 0. \qquad (61)$$

Then (60) gives

$$\nabla^2 F = -\xi, \qquad \nabla^2 G = -\eta, \qquad \nabla^2 H = -\zeta, \qquad (62)$$

ξ, η, and ζ being the vorticity components. Equations (62) are analogous to the equation

$$\nabla^2 \phi = \theta, \qquad (63)$$

in which θ is the divergence of the velocity, and which has its solution given by (58). Hence the solution of (62) is

$$F = \frac{1}{4\pi} \int \frac{\xi'}{R} dV', \qquad G = \frac{1}{4\pi} \int \frac{\eta'}{R} dV', \qquad H = \frac{1}{4\pi} \int \frac{\zeta'}{R} dV', \qquad (64)$$

with R defined by (59), V' being the region in which the vorticity does not vanish, and $dV' = dx'\, dy'\, dz'$. Equation (61) is satisfied because, the divergence of vorticity being identically zero,

$$\frac{\partial F}{\partial x} + \frac{\partial G}{\partial y} + \frac{\partial H}{\partial z} = -\frac{1}{4\pi} \int_{V'} \left(\xi' \frac{\partial}{\partial x'} \frac{1}{R} + \eta' \frac{\partial}{\partial y'} \frac{1}{R} + \zeta' \frac{\partial}{\partial z'} \frac{1}{R} \right) dV'$$

$$= \int_S \frac{1}{R} (l\xi + m\eta + n\zeta)\, dS = 0, \qquad (65)$$

by virtue of Green's theorem, where l, m, and n are the direction cosines of the normal at the surface S bounding V', drawn into V'. S can be taken to be the spherical surface centered at (x,y,z) and with a radius R large enough to contain V'. The surface integral vanishes for finite R, by virtue of Green's theorem or the vanishing of the "vorticity discharge" through a closed surface. If V is finite so is R. If V' is infinite the surface integral can be considered as the limit of surface integrals, all zero, for finite but increasing values of R, and is therefore also zero.

Consider now a vorticity filament of length ds', cross section σ', and vorticity $2\omega'$, with components

$$\xi' = 2\omega' \frac{\partial x'}{\partial s'}, \qquad \eta' = 2\omega' \frac{\partial y'}{\partial s'}, \qquad \zeta' = 2\omega' \frac{\partial z'}{\partial s'}. \qquad (66)$$

Since $2\sigma'\omega'$ is the circulation κ and $dV' = \sigma' \, ds'$, (64) becomes

$$F = \frac{\kappa}{4\pi} \int \frac{dx'}{R}, \qquad G = \frac{\kappa}{4\pi} \int \frac{dy'}{R}, \qquad H = \frac{\kappa}{4\pi} \int \frac{dz'}{R}. \tag{67}$$

Equations (60) and (67) then give

$$u_2 = \frac{\kappa}{4\pi} \int \left(\frac{\partial y'}{\partial s'} \frac{z - z'}{R} - \frac{\partial z'}{\partial s'} \frac{y - y'}{R} \right) \frac{ds'}{R^2},$$

$$v_2 = \frac{\kappa}{4\pi} \int \left(\frac{\partial z'}{\partial s'} \frac{x - x'}{R} - \frac{\partial x'}{\partial s'} \frac{z - z'}{R} \right) \frac{ds'}{R^2}, \tag{68}$$

$$w_2 = \frac{\kappa}{4\pi} \int \left(\frac{\partial x'}{\partial s'} \frac{y - y'}{R} - \frac{\partial y'}{\partial s'} \frac{x - x'}{R} \right) \frac{ds'}{R^2}.$$

If δu_2, δv_2, and δw_2 are the components of the velocity due to the vortex filament of length ds' and circulation κ, then

$$[(\delta u_2)^2 + (\delta v_2)^2 + (\delta w_2)^2]^{1/2} = \frac{\kappa}{4\pi} \frac{\sin \chi \, ds'}{R^2}, \tag{69}$$

where χ is the angle between the vortex filament and the radius vector R, and the direction of this differential-induced velocity is normal to the plane containing R and the vortex-line element, because the integrand of (68) is the vector product of the unit vector along the vortex filament and the unit vector along R. The law consisting of (69) and the statement about the direction of the velocity induced by a vortex filament is exactly the same as the law found experimentally by Biot and Savart (1820) concerning the magnetic field induced by an element of electric current and established analytically by Ampère (1826). It is called the *Biot-Savart law*.

The Biot-Savart law as applied to hydrodynamics is completely kinematic in nature. Its validity does not depend on inviscidness (or any other property) of the fluid. It is only for convenience that it has been included in this chapter.

PROBLEMS

1 For steady two-dimensional flows of an inviscid fluid of constant density ρ, show that

$$\frac{p}{\rho} + \frac{q^2}{2} + \Omega + F(\psi) = \text{const},$$

in which $F(\psi)$ is related to the vorticity ζ by

$$\zeta = F'(\psi),$$

ψ being the stream function, p the pressure, q the speed, and Ω the body-force potential.

2 For steady axisymmetric flows of an inviscid fluid of constant density ρ, show that

$$\frac{p}{\rho} + \frac{q^2}{2} + \Omega + F(\psi) = \text{const},$$

in which ψ is now the Stokes stream function and $F(\psi)$ is related to the vorticity η (in the φ direction) by

$$\eta = rF'(\psi).$$

3 The flow of an inviscid incompressible fluid was started from rest. If the fluid is nonhomogeneous, show that the component of vorticity normal to any constant-density surface is zero. In what surfaces will the vortex lines lie? *Hint:* Take a loop in a constant-density surface and apply Kelvin's theorem on circulation to it.

4 At any point of a streamline in a rotational flow there is a vortex line. The vortex lines passing through all points of the streamline form a surface called a *Lamb surface*. Consider an inviscid incompressible or inviscid barotropic fluid. Show that, if the flow started from rest, surfaces of constant density or entropy are Lamb surfaces, and that in a Lamb surface the Bernoulli equation is valid for such a fluid in steady flow, whether or not the fluid is homogeneous. (If not homogeneous, barotropy refers to any fluid particle, not all the fluid particles.) *Hint:* Convert the left-hand side of the Euler equation to

$$\rho[\textbf{grad } (q^2/2) - \textbf{v} \times \boldsymbol{\xi}],$$

in which q is the speed, $\textbf{v}$ the velocity vector, and $\boldsymbol{\xi}$ the vorticity vector. For an incompressible fluid ρ is constant in a Lamb surface. For a barotropic gas the ρ can be absorbed into the pressure term.

5 If a fluid is inviscid and incompressible but nonhomogeneous, and if gravity effects are neglected, show that

$$\rho u_\alpha \frac{\partial u_i}{\partial x_\alpha} = -\frac{\partial p}{\partial x_i} \quad \text{and} \quad \frac{\partial u_\alpha}{\partial x_\alpha} = 0$$

can be reduced to

$$\rho_0 u_\alpha' \frac{\partial u_i'}{\partial x_\alpha} = -\frac{\partial p'}{\partial x_i} \quad \text{and} \quad \frac{\partial u_\alpha'}{\partial x_\alpha} = 0$$

by using

$$u_\alpha \frac{\partial \rho}{\partial x_\alpha} = 0,$$

in which

$$u_i' = \left(\frac{\rho}{\rho_0}\right)^{1/2} u_i \quad \text{and} \quad p' = p,$$

ρ_0 being a constant reference density. What is the implication of this result?

Is there a corresponding result for steady motions of an inviscid ideal gas of variable entropy or even of an inviscid nonhomogeneous barotropic fluid (see Prob. 2.3)?

6 At speeds very much less than $(gL)^{1/2}$, L being a representative linear scale, steady motions of an inviscid incompressible fluid of variable density are given by

$$0 = -\frac{\partial p}{\partial x}, \qquad 0 = -\frac{\partial p}{\partial y}, \qquad 0 = -\frac{\partial p}{\partial z} - g\rho,$$

z being measured in the direction of the vertical. Accepting these equations as correct, show that $w = 0$ and that, for two-dimensional motions in the xz plane,

$$\frac{\partial u}{\partial x} = 0.$$

What is the implication of this result? (Consider slow steady flows past a cylinder with horizontal axis and infinite length.)

7 Show that the velocity components in a constant-density surface of an incompressible inviscid stratified fluid are the components of grad $\phi(x,t,\rho)$ in that surface, provided the motion has been started from rest.

8 Show that for the motion and the fluid considered in the preceding problem the Bernoulli equation

$$\frac{\partial \phi}{\partial t} + \frac{p}{\rho} + \frac{q^2}{2} + \Omega = F(\rho)$$

holds in a constant-density surface, Ω being the body-force potential.

9 Solve Probs. 3 and 7 for an inviscid compressible fluid of variable entropy. Assume isentropy for the change of state of each fluid particle but nonhomogeneous entropy. *Hint:* The density should now be replaced by entropy.

10 Show that for an inviscid compressible fluid of variable entropy set in motion from rest, the Bernoulli equation

$$\frac{\partial \phi}{\partial t} + \int \frac{dp}{\rho} + \frac{q^2}{2} + \Omega = F(S)$$

holds in a constant-entropy surface provided the entropy of each particle remains invariant though the entropy distribution is not homogeneous. The velocity components in a constant-S surface are the components of grad ϕ in that surface.

REFERENCES

AMPÈRE, A. M., 1826: "Théorie mathématique des phénomènes électro-dynamiques," Paris.

BIOT, J.-B., AND F. SAVART, 1820: *J. Phys. (Paris)*, **91**: 151.

BJERKNES, V., 1918: *Vid.-Selsk. Skrifter*, Kristiania.

CAUCHY, A., 1827: Mémoire sur la théorie des ondes, *Mém. Acad. Roy. Sci.*, **1**.

HANKEL, H., 1861: "Zur allgemeinen Theorie der Bewegung der Flüssigkeiten," Göttingen.

HELMHOLTZ, H. L. F. VON, 1858: Über Integrale der hydrodynamischen Gleichungen welche den Wirbelbewegungen entsprechen, *Crelle*, **55**.

KUO, Y. H., 1943: On the Force and Moment Acting on a Shear Flow, *Quart. Appl. Math.*, **1**: 273–275.

LAMB, H., 1932: "Hydrodynamics," The Macmillan Company, London; reprinted (6th ed.) by Dover Publications, Inc., New York, 1945 (with the same pagination).

LONG, R. R., 1953: Steady Motion around a Symmetrical Obstacle Moving along the Axis of a Rotating Liquid, *J. Meteor.*, **10**: 197.

PROUDMAN, J., 1916: On the Motion of Solids in a Liquid Possessing Vorticity, *Proc. Roy. Soc. London Ser. A*, **92**: 408–424. (The first nine lines in Sec. 9, p. 420, mention the astounding phenomenon of the two-dimensionality of weak steady motions of a rotating fluid relative to the rotating system.)

TAYLOR, G. I., 1917: Motion of Solids in Fluids when the Flow Is Not Irrotational, *Proc. Roy. Soc. London Ser. A*, **93**: 99–113. (See the first seven lines on p. 102.)

TAYLOR, G. I., 1921: Experiments with Rotating Fluids, *Proc. Roy. Soc. London Ser. A*, **100**: 114–121. (This mentions the two-dimensionality of slow steady motions in rotating fluids more explicitly and emphatically than the paper of 1917 and contains experimental evidence of dye screens indicating tendency toward two-dimensionality.)

TAYLOR, G. I., 1923: Experiments on the Motion of Solid Bodies in Rotating Fluids, *Proc. Roy. Soc. London Ser. A*, **104**: 213–218. (The Taylor column effect is discussed in detail here. On the third paragraph on p. 217 the "virtual 'solid cylinder' " is mentioned. Sir Geoffrey told the writer that he had experimented with small fish in a rotating tank. When they swam toward the virtual solid cylinder, they tried to avoid it as if sensing the presence of a solid obstacle. He did not publish his results on piscine behavior.)

THOMSON, W. (LORD KELVIN), 1869: On Vortex Motion, *Edinburgh Trans.*, **25**: 217–260; reprinted in "Mathematical and Physical Papers," Cambridge University Press, London.

TRUESDELL, C., 1954: "The Kinematics of Vorticity," The University of Indiana Press, Bloomington, Ind.

YIH, C.-S., 1959a: Effect of Density Variation on Fluid Flow, *J. Geophys. Res.*, **64**: 2219–2223.

YIH, C.-S., 1959b: Effects of Gravitational or Electromagnetic Fields on Fluid Motion, *Quart. Appl. Math.*, **16**: 409–415.

CHAPTER FOUR

IRROTATIONAL FLOWS
OF AN
INVISCID FLUID
OF
CONSTANT DENSITY

1. INTRODUCTION OF THE GOVERNING EQUATION

As shown in Sec. 2 of Chap. 3, irrotationality will persist if the fluid is an inviscid homogeneous liquid (ρ = const) or an inviscid homentropic gas. Since the state of rest is irrotational, flows of such fluids started from rest will be irrotational. Furthermore, the flow caused by a moving body in such fluids, otherwise at rest, can be made steady, if the body moves with uniform velocity, by using coordinates fixed to the body. In that case the upstream velocity is uniform, and the irrotationality far upstream will persist downstream. For a real fluid, even if the motion started from an irrotational state (of which the state of rest is a special but important case), or if the flow originates from an upstream section at which the flow is irrotational, vorticity will be generated at the boundary through the action of viscosity, because it is there that the nonslip condition to be satisfied by a viscous fluid and the irrotationality of the motion are incompatible. However, as mentioned before (Sec. 4 of Chap. 2 and Sec. 1 of Chap. 3), if the Reynolds number is large, the region in which the vorticity is appreciable is concentrated in a thin layer near the boundary before the flow separates from the boundary and regions of reverse flow (as well as a wake) are created. Furthermore, it is often possible to prevent separation from the boundary by suction, in which case the flow is irrotational except in the boundary layer. For these reasons it is useful to study irrotational flows, in spite of the fact that all real fluids (except helium II at very low temperatures) are viscous. This chapter is devoted to the study of irrotational flows of a fluid of constant density, and it is only for the convenience of exposition that we have assumed inviscidness of the fluid to begin with.

The equation of continuity (1.14), when combined with (3.27), which owes its existence to irrotationality, produces the Laplace equation

$$\nabla^2 \phi = 0, \tag{1}$$

in which ϕ is the velocity potential and ∇^2 is the Laplacian operator $\partial^2/(\partial x_\alpha \, \partial x_\alpha)$. This is the equation governing irrotational flows if the density of the fluid is constant. In terms of x, y, and z, (1) can be written as

$$\left(\frac{\partial^2}{\partial x^2} + \frac{\partial^2}{\partial y^2} + \frac{\partial^2}{\partial z^2}\right)\phi = 0. \tag{1a}$$

The forms of the Laplace equation in other coordinate systems can be found in Appendix 2 or, more directly, can be derived from the equations of continuity for constant density by expressing the velocity as the gradient of ϕ, all in the coordinates under consideration. In cylindrical coordinates (r,φ,z), the Laplace equation is

$$\left(\frac{\partial^2}{\partial r^2} + \frac{1}{r}\frac{\partial}{\partial r} + \frac{1}{r^2}\frac{\partial}{\partial \varphi^2} + \frac{\partial^2}{\partial z^2}\right)\phi = 0. \tag{2}$$

In spherical coordinates (R,θ,φ), it is

$$\left[\frac{1}{R^2}\frac{\partial}{\partial R}\left(R^2\frac{\partial}{\partial R}\right) + \frac{1}{R^2 \sin \theta}\frac{\partial}{\partial \theta}\left(\sin \theta \frac{\partial}{\partial \theta}\right) + \frac{1}{R^2 \sin^2 \theta}\frac{\partial^2}{\partial \varphi^2}\right]\phi = 0. \tag{3}$$

The equation governing the irrotational motion of a homentropic gas will be given in Chap. 6. It is nonlinear, but if the effects of compressibility are neglected, it can be reduced to (1).

2. EQUATIONS FOR AXISYMMETRIC IRROTATIONAL FLOWS IN TERMS OF STREAM FUNCTIONS

If all the variables in a flow are independent of φ in cylindrical or spherical coordinates, the flow is called *axisymmetric*. The velocity component in the direction of increasing φ (in either coordinate system) may not be zero in general but is zero if the flow is *everywhere* irrotational. If it is zero, the vorticity components in the directions of increasing r and z in cylindrical coordinates are zero, and the remaining component is (see Appendix 2)

$$\eta = \frac{\partial u}{\partial z} - \frac{\partial w}{\partial r},$$ (4)

with u, v, and w indicating the velocity components in the three coordinate directions. Substituting (1.33) into this expression for η and requiring it to vanish, we obtain

$$\left(\frac{\partial^2}{\partial r^2} - \frac{1}{r}\frac{\partial}{\partial r} + \frac{\partial^2}{\partial z^2}\right)\psi = 0,$$ (5)

in which ψ is Stokes' stream function.

In spherical coordinates the only vorticity component which is not a priori zero for axisymmetric flow without swirl is (see Appendix 2)

$$\zeta = \frac{1}{R}\left[\frac{\partial(Rv)}{\partial R} - \frac{\partial u}{\partial \theta}\right],$$ (6)

with the velocity components in the directions of increasing R, θ, and φ denoted by u, v, and w. Substituting (1.34) in ζ and requiring it to vanish, we obtain

$$\left[\frac{\partial^2}{\partial R^2} + \frac{\sin\theta}{R^2}\frac{\partial}{\partial \theta}\left(\frac{1}{\sin\theta}\frac{\partial}{\partial \theta}\right)\right]\psi = 0.$$ (7)

3. UNIQUENESS OF THE SOLUTION OF THE LAPLACE EQUATION

Whether or not there are more than one essentially different solutions of the Laplace equation with specified boundary conditions depends on the nature of the boundary conditions. If a problem involves the solution of the Laplace equation in a fluid domain on the boundary of which the dependent variable is specified, the problem is called a *Dirichlet problem*. If, instead, the derivative of the dependent variable in a direction normal to the boundary is specified on the boundary, the problem is called a *Neumann problem*. The solution of a Dirichlet-Neumann problem is unique provided the dependent variable is single-valued and differentiable at least twice with respect to the space coordinates.

A proof of the uniqueness of the solution of the Laplace equation for ϕ will be given under the assumption that ϕ is single-valued and twice differentiable with respect to x_i throughout the fluid domain V and that on part of the boundary (S_1)

$$\phi = f(x_1, x_2, x_3, t)$$ (8)

whereas in the rest of the boundary (S_2)

$$\frac{\partial \phi}{\partial n} = g(x_1, x_2, x_3, t). \tag{9}$$

The letter n in (9) indicates the distance normal to S_2, which, with S_1, constitutes the entire boundary of the fluid domain under consideration. If there are two different solutions ϕ_1 and ϕ_2, both satisfying the Laplace equation and the specified boundary conditions, their difference, denoted by ϕ for simplicity, satisfies the Laplace equation and the boundary conditions

$$\phi = 0 \text{ on } S_1 \quad \text{and} \quad \frac{\partial \phi}{\partial n} = 0 \text{ on } S_2. \tag{10}$$

If, in (1.9),

$$U_i = \phi \phi_{,i}, \tag{11}$$

in which

$$\phi_{,i} = \frac{\partial \phi}{\partial x_i},$$

application of (1.9) produces

$$-\int_S \phi \frac{\partial \phi}{\partial n} dS = \int_V \phi_{,i} \phi_{,i} dV + \int_V \phi \nabla^2 \phi \, dV = \int_V \phi_{,i} \phi_{,i} dV, \tag{12}$$

since[1] $\nabla^2 \phi = 0$. The boundary conditions (10) then demand that

$$\int_V \phi_{,i} \phi_{,i} dV = 0. \tag{13}$$

Since

$$\phi_{,i} \phi_{,i} = \left(\frac{\partial \phi}{\partial x_1}\right)^2 + \left(\frac{\partial \phi}{\partial x_2}\right)^2 + \left(\frac{\partial \phi}{\partial x_3}\right)^2, \tag{14}$$

it follows from (13) that all the spatial derivatives of ϕ vanish and consequently that ϕ can be a function of t only or, in particular, a constant. In either case the flow represented by ϕ is a state of no motion, or the flow represented by ϕ_1 is identical to that represented by ϕ_2.

If the boundary conditions are not as assumed, the solution of the differential system may not be unique. One example of the lack of uniqueness is furnished by the case of a layer of an inviscid liquid with a free surface flowing horizontally (in the x_1 direction) at infinity at a speed equal to the speed of propagation of a solitary wave. Two flows are possible, one with a solitary wave and the other without it. The boundary conditions at the bottom of the layer and at infinity (or $x_1 = \pm\infty$) are of the type (9), but the boundary condition on the free surface

---

[1] Note that if an isolated singularity exists in the fluid domain, $\nabla^2 \phi$ is infinite there and the second equality sign of (12) cannot be used. In this development, the fluid domain is assumed to be free from singularities.

is $p = 0$, or

$$\frac{\phi_{,\alpha}\phi_{,\alpha}}{2} + gx_3 = \text{const.} \tag{15}$$

(Note that the flow with the solitary wave is also steady, since the fluid at infinity moves with a velocity equal and opposite to the velocity of propagation of the solitary wave relative to the fluid at infinity.) This is not of the type (8) or (9). If no free surface or interface (between two fluids of different densities) is present, the boundary conditions are always of the types (8) and (9), and if there are no irregularities in the fluid domain, the solution is always unique.

4. THE MAXIMUM AND MINIMUM OF HARMONIC FUNCTIONS

A function that satisfies the Laplace equation is called a *harmonic function*. (Satisfaction of the Laplace equation implies the existence of second derivatives wherever the equation is satisfied and hence existence of continuous first derivatives.) A harmonic function may become infinite in many different ways at some points, including infinity; or it may not be single-valued around certain points. All these points are its *singular points*. One interesting and significant property of a harmonic function is that in a domain free from singular points its maximum or minimum can occur only on the boundary of that domain and never in its interior. Although a mathematical proof of this property is not at all difficult, it is sufficient to demonstrate it merely from a physical point of view. If a maximum of the harmonic function ϕ exists at a point P in the interior of the domain, the irrotational flow corresponding to it must be wholly outward at the surface of a sphere with center at P, and continuity cannot be maintained unless there are sources in the sphere; sources, however, are not allowable in a fluid in irrotational flow and without singular points in its domain. Consequently a maximum, and similarly a minimum, of ϕ cannot exist in the interior of the fluid.

If ϕ is a harmonic function, its derivatives are also harmonic functions, whose maxima and minima can occur only on the boundary of a singularity-free region. This fact makes it possible to show that the speed of a fluid in irrotational flow cannot have a maximum in the interior of a singularity-free region. Suppose there were such a maximum at P. With the direction of the velocity at that point taken to be the x direction and with subscripts indicating partial differentiation, ϕ_x cannot be a maximum at P. Hence on a small sphere surrounding P there is at least one point Q where ϕ_x is greater than the ϕ_x at P, except in the trivial case in which ϕ_x is constant, which does not invalidate the conclusion. Then, at Q the speed is

$$(\phi_x{}^2 + \phi_y{}^2 + \phi_z{}^2)_Q{}^{1/2},$$

which is a fortiori greater than the speed at P. Hence the speed cannot be a maximum at P. The proof does not work for a minimum. As a matter of fact, minimum speed can occur in the interior of a fluid, the most obvious example being an internal stagnation point.

The theorem just established for a fluid of constant density in irrotational motion is also true for a homentropic gas in irrotational flow, provided the speed is everywhere less than the local sound speed. [For a proof, see Yih (1958).]

5. UNIFORM FLOW AND THREE-DIMENSIONAL SINGULARITIES

5.1. Uniform Flow

It is easily seen that

$$\phi = Ux \tag{16}$$

represents uniform flow with constant velocity U in the direction of increasing x. The stream function associated with the potential given by (16) is

$$\psi = Uy \tag{17}$$

if the flow is considered to be a two-dimensional flow or

$$\psi = \frac{Ur^2}{2} \quad \text{or} \quad \psi = \tfrac{1}{2}UR^2 \sin^2 \theta \tag{18}$$

if the flow is considered as axially symmetric with respect to the x axis. In (18), ψ is the Stokes stream function, r is the radial distance from the x axis, and θ is measured from the direction of increasing x.

5.2. Point Sources and Sinks

The simplest singularity is a point source or sink. If the source is situated at the origin and spherical coordinates are used, the purely radial velocity u at any point will be inversely proportional to the area of the spherical surface which passes through that point, in order to maintain continuity. Thus, with $4\pi m$ denoting the volume flux per unit time from the source,

$$u = \frac{m}{R^2}. \tag{19}$$

It is obvious that the flow represented by (19) is not rotational. The potential ϕ can be found by integration of (19) with respect to R, and is

$$\phi = -\frac{m}{R}, \tag{20}$$

with the constant of integration suppressed without loss of generality. It can be verified directly that the ϕ given by (20) satisfies the Laplace equation everywhere except at the origin. The flow is in fact everywhere irrotational, and ϕ does not satisfy the Laplace equation at the origin only because the equation of continuity is not valid there (where the source is situated). The Stokes stream function (in spherical coordinates) for the flow represented by (20) can be found by the use of (1.34). Thus, in the notation of (1.34)

$$u = \frac{m}{R^2} = \frac{1}{R^2 \sin \theta} \psi_\theta, \qquad v = 0 = -\frac{1}{R \sin \theta} \psi_R,$$

integration of which produces

$$\psi = -m \cos \theta. \tag{21}$$

If the source is situated at some point $P(x_0, y_0, z_0)$ other than the origin, R should be replaced by R_0, defined by

$$R_0^2 = (x - x_0)^2 + (y - y_0)^2 + (z - z_0)^2,$$

and θ should be replaced by θ_0, which is the angle measured from a polar axis through P.

5.3. Doublets

If a source of strength m is situated at $(a,0,0)$ and a sink of the same strength at $(-a,0,0)$, in Cartesian coordinates,

$$\phi = m\{[(x + a)^2 + y^2 + z^2]^{-\frac{1}{2}} - [(x - a)^2 + y^2 + z^2]^{-\frac{1}{2}}\}.$$

If a is infinitesimal, it follows from first principles in differential calculus that, to the first power in a,

$$\phi = m \frac{\partial}{\partial x} \frac{1}{R} 2a = -2ma \frac{x}{R^3}.$$

As a approaches zero, ϕ does also, and when $a = 0$, ϕ is also zero. The source and the sink have then annihilated each other. But if, as $a \to 0$, m increases in such a fashion as to maintain $2ma$ equal to a constant μ, then in the limit

$$\phi = -\frac{\mu x}{R^3} \tag{22}$$

not only to the first power but exactly. Since $-x/R^3$ is the derivative of $1/R$, it must also be a harmonic function. Since (22) is obtained from the potential for a source and that for a sink by a limiting process, its right-hand side represents the potential of a singularity which may be called a *doublet*. The axis of the doublet is the line drawn from the

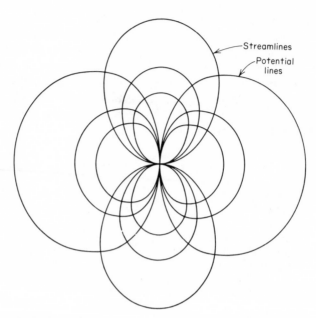

FIGURE 1.   STREAMLINES AND POTENTIAL LINES FOR A THREE-DIMENSIONAL DOUBLET. $\phi = -x/R^3$, $\psi = r^2/R^3$, $r^2 = R^2 - x^2$.

generic sink to the generic source, and in this case is in the positive x direction.   The flow pattern corresponding to (22) is shown in Fig. 1.

Similarly,

$$\phi = -\frac{\mu y}{R^3} \quad \text{and} \quad \phi = -\frac{\mu z}{R^3} \tag{23}$$

represent potentials for doublets of strength μ, with axes in the directions of increasing y and z, respectively.   If the axis of a doublet has direction cosines l, m, and n and its strength is μ,

$$\phi = -\frac{\mu}{R^3}\,(lx + my + nz). \tag{24}$$

Successive differentiation of (22), (23), or (24) produces singularities of higher orders.   Since the singularity represented by (24) is called a doublet, it is logical to call the higher-order singularities quadruplets, octuplets, and so on, but these terms are rarely used.

The stream function for a doublet can be obtained by use of (1.34) and by integration.   If θ is measured from the positive direction of the axis of the doublet, the Stokes stream function is

$$\psi = \frac{\mu \sin^2 \theta}{R}, \tag{25}$$

which is useful only if the flow is symmetric about that axis, i.e., only if other singularities do not destroy the symmetry of the flow due to the doublet.

6. DISTRIBUTED THREE-DIMENSIONAL SINGULARITIES

If there are point sources and sinks distributed over the domain D, the resultant ϕ can be obtained by integration. Thus

$$\phi = -\int_D \frac{m}{R'} dV, \tag{26}$$

in which m is a function of x', y', and z', which are Cartesian coordinates of any point in D, and

$$R'^2 = (x - x')^2 + (y - y')^2 + (z - z')^2.$$

Equation (26) is valid even if $P(x,y,z)$ is within D, so that $R' = 0$ when $x' = x$, $y' = y$, and $z' = z$, for the integral is still convergent. This can be seen by taking P to be the origin. Then dV contains the factor R'^2.

The velocity potential due to a distribution of doublets can also be found by integration, but it will presently be shown that distributed doublets are equivalent to a distribution of sources and sinks in D plus a surface distribution of sources and sinks on the surface S of D. Suppose there is a doublet distribution in D, so that the velocity potential is given by

$$\phi = -\int_D \frac{\mu_1(x - x') + \mu_2(y - y') + \mu_3(z - z')}{R'^3} dV, \tag{27}$$

in which μ_1, μ_2, and μ_3 are functions of x', y', and z'. It is sufficient to treat only the first term in (27) because the treatment of the other terms is similar. Writing dV as $dx' \, dy' \, dz'$ and carrying out the integration by parts with respect to x', we obtain, for the first term in (27),

$$\phi_1 = \iint_A \left[\frac{\mu_1}{R'}\right]_a^b dy' \, dz' - \int_D \frac{m_1}{R'} dV, \tag{27a}$$

in which A is the projection of D on the yz plane,

$$m_1 = \frac{\partial \mu_1}{\partial x}, \tag{28}$$

and a and b are the values of x at the end points of a line parallel to the x axis and piercing through D. The first term in (27a) corresponds to a surface source distribution of strength $\mu_1 l_1$ on the surface S of D, if l_1 is the first direction cosine of the outward normal to S. The second term corresponds to a source-sink distribution over D. Similar conclusions can be drawn from the other two terms in (27). Thus a doublet distribution in space is equivalent to a space distribution and a surface distribution of sources and sinks.

Repeated use of the development just given will show that a space distribution of singularities of any order can be reduced to the sum of a space distribution of sources and sinks plus a surface distribution of singularities the highest order of which is less by 1 than the order of the original singularities distributed in space.

Thus, we need deal only with sources and sinks if a space distribution is considered. The development given in this section applies equally to two-dimensional singularities.

7. METHOD OF SUPERPOSITION FOR AXISYMMETRIC FLOWS

7.1. The Half-body

Since the Laplace equation is linear, sums of its solutions are also solutions. Thus it is possible to construct new potential flows by superposition. For instance, a parallel flow with uniform velocity U in the x direction can be superposed with the flow due to a source of strength m situated at the origin. The combined potential is

$$\phi = Ux - \frac{m}{R}. \tag{29}$$

Since the equations governing the Stokes stream function, (5) and (7), are also linear, the stream function for the combined flow can also be obtained by superposition. Hence

$$\psi = \tfrac{1}{2} U R^2 \sin^2 \theta - m \cos \theta. \tag{30}$$

The straight line $\theta = 0$ corresponds to $\psi = -m$ and is therefore a streamline. The straight line $\theta = \pi$ corresponds to $\psi = m$ and is also a streamline. But the streamline $\psi = m$ has another branch, given by

$$R = \sqrt{\frac{m}{U}} \csc \frac{\theta}{2}, \tag{31}$$

which traces out a boundary (Fig. 2). This boundary is symmetric with respect to the polar axis and at infinity has a circular cross section with a radius equal to $2(m/U)^{1/2}$. This radius r_0 can be obtained either from (31) by multiplying both sides of it by $\sin \theta$ and letting θ approach zero or directly from the fact that $\pi r_0^2 U$ must be equal to the discharge m from the source, by continuity, since the velocity is equal to U at infinity. The boundary can be considered as the boundary of a solid body, which is called a *half-body*. The source inside of it serves the purpose of determining the flow outside the body, which is free from singularities. It is because of their usefulness in this and similar instances that the flows due to singularities such as sources, sinks, and doublets, artificial though

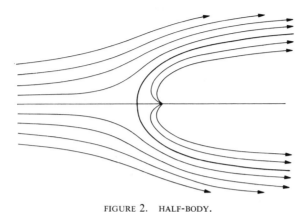

FIGURE 2.   HALF-BODY.

they may seem at first sight, are not mere mathematical exercises of no physical importance.

7.2.   Rankine Bodies

A closed boundary is generated by superposing a uniform flow on the flow due to a source and a sink.   The boundary can be considered to be that for an axisymmetric body, called a *Rankine body*.   Actually the term Rankine body is now used to denote a body generated by internal singularities.   Since any body can be generated by internal singularities, the term Rankine body indicates not a special class of bodies but emphasis upon the mode of its generation.

7.3.   Von Kármán's Method

Given an axisymmetric body, the flow past it can be approximated by superposing a uniform flow on the flow due to $n + 1$ sources or sinks of unknown strengths at given locations along the axis inside the body. The strengths can be determined by forcing the streamline occupying the axis of symmetry outside the body to pass through n specified points on the meridional trace of the body.   The resulting flow will not have the trace of the given boundary as a streamline but will have a streamline contiguous with the axis of symmetry and following nearly the trace of the boundary.   This method is due to Theodore von Kármán (1881–1963), who used distributed sources and sinks (1927).   The solution depends on nothing more than the solution of n linear simultaneous algebraic equations.   The total strength of the sources and sinks must be zero, for otherwise the body would not be closed.   Thus there are n unknown strengths of the sources and sinks.   These are the unknowns to be found from the n equations obtained by forcing the stream function to be

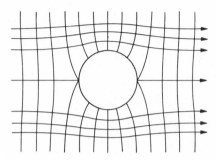

FIGURE 3.   FLOW PAST A SPHERE.

constant at the n points.   The positions of the sources and sinks are arbitrarily chosen.

7.4.   Flow Past a Sphere

If a uniform flow of velocity U in the x direction is superposed on that due to a doublet with an axis in the negative x direction, the flow past a sphere is obtained.   If the strength of the doublet is μ', the stream function is

$$\psi = \tfrac{1}{2}UR^2 \sin^2 \theta - \frac{\mu' \sin^2 \theta}{R} = \left(\frac{1}{2} UR^2 - \frac{\mu'}{R}\right) \sin^2 \theta. \tag{32}$$

The streamline (or stream surface) $\psi = 0$ consists of the lines $\theta = 0$ and $\theta = \pi$ and the sphere

$$R = \left(\frac{2\mu'}{U}\right)^{1/3}. \tag{33}$$

If the radius of the sphere is a,

$$\mu' = \tfrac{1}{2}Ua^3. \tag{34}$$

The flow pattern is shown in Fig. 3.

We have obtained (32) by superposition and have found that it is possible to determine μ' so that ψ is constant on the sphere.   We could have found the same ψ by solving (7) by the method of separation of variables.   With

$$\psi = \tfrac{1}{2}UR^2 \sin^2 \theta + \psi_1 \tag{35}$$

we seek to determine ψ_1 so that it satisfies (7) and does not disturb the condition at infinity and ψ is constant on the sphere $R = a$.   Since the term preceding ψ_1 in (35) satisfies (7), ψ also satisfies it if ψ_1 does.   Now (7) can be written as

$$\left[\frac{\partial^2}{\partial R^2} + \frac{1}{R^2} (1 - \mu^2) \frac{\partial^2}{\partial \mu^2}\right] \psi = 0, \tag{7a}$$

in which $\mu = \cos\theta$. In order that ψ can be independent of θ for $R = a$, the θ dependence of ψ_1 must be of the form $\sin^2\theta$ or $1 - \mu^2$. Substituting

$$\psi_1 = (1 - \mu^2)f(R) \qquad (36)$$

for ψ in (7a), we have

$$f'' - \frac{2}{R^2}f = 0, \qquad (37)$$

which has the solutions

$$f = AR^2 \quad \text{and} \quad f = \frac{B}{R}. \qquad (38)$$

The first solution corresponds to the term preceding ψ_1 in (35), and the condition at infinity demands that $A = 0$. The second solution gives

$$\psi_1 = \frac{B}{R}\sin^2\theta. \qquad (39)$$

The requirement that ψ be zero for $R = a$ determines B to be $-Ua^3/2$, so that the result is the same as (32).

To obtain the stream function for the motion due to the translation of the sphere, with quiescent fluid at infinity, it is necessary merely to remove the first part on the right-hand side of (32) and write

$$\psi = -\frac{Ua^3 \sin^2\theta}{2R}. \qquad (40)$$

The Laplace equation and the boundary condition on the sphere (that the velocity component normal to the boundary be the same for fluid as for solid) are still satisfied, but of course the sphere is moving to the left with speed U. The potential ϕ corresponding to (40) is given by (22), with μ equal to $-\mu'$, and μ' given by (34). The ϕ corresponding to (32) is given by (22) and (16).

8. A PROPERTY OF HARMONIC FUNCTIONS

One remarkable and useful property of harmonic functions is that if the absolute value of a harmonic function is bounded everywhere (and therefore also at infinity), it must be a constant. Note that the boundedness excludes isolated singularities and the satisfaction of the Laplace equation excludes distributed singularities. The theorem for the two-dimensional case is well known as *Liouville's theorem* [see Titchmarsh (1939, p. 85)]. The theorem for the three-dimensional case is curiously disguised in a theorem in Kellogg (1929, theorem XI, p. 227). Examination of Titchmarsh's presentation of Liouville's theorem shows how closely it is related to the theorem given by Kellogg.

The theorem will not be proved. Instead, its implications will be noted here. Consider the velocity component $u = \partial\phi/\partial x$. If ϕ is harmonic, so is u. Now if u has not singularities anywhere and is bounded at infinity, it must be constant. The same is true of v and w, which are given by

$$v = \frac{\partial\phi}{\partial y}, \qquad w = \frac{\partial\phi}{\partial z}. \tag{41}$$

Thus a flow with no singularity anywhere (for the velocity components) can only be a uniform flow, of which the state of rest is a special case.

If singularities are only in the finite part of space, the uniform flow at infinity is undisturbed, and the resulting flow may correspond to some flow caused by the motion or presence of solid bodies. Thus singularities are what make solid boundaries. When we solve a problem of irrotational flow, we explicitly or implicitly determine the distribution of singularities. They are, in a very real sense, the root of the matter.

9. KELVIN'S INVERSION THEOREM

If

$$x_i' = \frac{x_i}{R^2}, \qquad R^2 = x_\alpha x_\alpha, \tag{42}$$

and if $\phi(x_1,x_2,x_3)$ is a harmonic function, then

$$\frac{1}{R}\phi(x_1',x_2',x_3') \tag{43}$$

is also a harmonic function. This is *Kelvin's inversion theorem*. Since it will be needed in the next section, a direct proof will be given here. For convenience the symbol Δ will be used for ∇^2 and Δ' for the same operator in the coordinate x_i'.

We shall use spherical coordinates, and write

$$R' = \frac{1}{R}, \tag{44}$$

Thus, we are given

$$\Delta'\phi(R',\theta,\varphi) = 0, \tag{45}$$

and are required to show that

$$\Delta \frac{1}{R}\phi\left(\frac{1}{R},\theta,\varphi\right) = 0, \tag{46}$$

in which Δ is the Laplacian operator given in (3), and Δ' the Laplacian in R' instead of R. Substituting

$$\frac{1}{R}\phi\left(\frac{1}{R},\theta,\varphi\right) \tag{47}$$

for ϕ in the left-hand side of (3), and noting that

$$\frac{\partial}{\partial R} f\left(\frac{1}{R}, \theta, \varphi\right) = -\frac{1}{R^2} \frac{\partial}{\partial R'} f(R', \theta, \varphi), \qquad (48)$$

we obtain

$$\Delta \frac{1}{R} \phi\left(\frac{1}{R}, \theta, \varphi\right) = R'^5 \Delta' \phi(R', \theta, \varphi). \qquad (49)$$

By virtue of (45), (46) results from (49), and the Kelvin inversion theorem is proved.

10. SPHERE THEOREMS

The *image* of a singularity in a rigid boundary is a singularity or a distribution of singularities which, in combination with the original one, will give rise to a flow having streamlines along that boundary. Thus, if a source is situated at (a,b,c) and a, b, and c are all positive, its image in the yz plane is a source at $(-a,b,c)$ if that plane is occupied by a rigid boundary. An additional rigid boundary lying in the zx plane will call for two more sources, one situated at $(a,-b,c)$ and the other at $(-a,-b,c)$. If the xy plane is also occupied by a rigid boundary, four more sources, all with $z = -c$, are needed. Thus a source in the octant bounded by three rigid walls coinciding with the coordinate planes has seven direct or indirect images, all sources of the same strength.

If the boundary is not plane, the image of a singularity in that boundary is in general a complicated system. However, for the special case in which the boundary is a spherical surface, the image system can be found by the use of sphere theorems, a discussion of which follows.

10.1. Butler's Sphere Theorem

Consider an axisymmetric irrotational flow unbounded by any rigid boundary and given by the stream function $\psi_0 = \psi_0(R,\theta)$. If $\psi_0(0,\theta) = 0$ and there are no singularities in $R \leq a$, the stream function for the exterior flow after the sphere $R = a$ has been introduced is

$$\psi = \psi_0 + \psi_0^* = \psi_0(R,\theta) - \frac{R}{a} \psi_0\left(\frac{a^2}{R}, \theta\right). \qquad (50)$$

This is *Butler's sphere theorem*. It is implied in the statement of the theorem that the addition of ψ_0^* to ψ_0 will not introduce singularities exterior to the sphere or affect the condition of the flow at infinity. To prove that these implied requirement are met, it suffices to note that since ψ_0 has no singularities for $R \leq a$, ψ_0^* has none for $R \geq a$ and that since ψ_0 is of the order[1] R^2 near the origin, ψ_0^* behaves as $1/R$ at infinity and thus does not alter the condition there. Now ψ_0 must satisfy (7), since it describes an irrotational flow of a fluid of constant density. The reader can readily verify that ψ_0^* also satisfies (7) and hence ψ does too, as is required. It is evident that $\psi = 0$ for $R = a$. This completes the proof of Butler's sphere theorem.

10.2. Weiss' Sphere Theorem

Butler's sphere theorem is a special case of *Weiss' sphere theorem*, which is applicable not only to axisymmetric flow but to general three-dimensional flows as well.   Consider an irrotational flow of an inviscid fluid of constant density characterized by the velocity potential $\phi(x,y,z)$ with no singularities in $R \le a$. If the sphere $R = a$ is introduced into the fluid, the resulting potential is given by

$$\phi(x,y,z) + \phi_1, \tag{51}$$

in which

$$\phi_1 = \frac{a}{R} \phi\left(\frac{a^2x}{R^2}, \frac{a^2y}{R^2}, \frac{a^2z}{R^2}\right) - \frac{2}{aR} \int_0^a \lambda \phi\left(\frac{\lambda^2 x}{R^2}, \frac{\lambda^2 y}{R^2}, \frac{\lambda^2 z}{R^2}\right) d\lambda. \tag{52}$$

This is Weiss' sphere theorem.

To prove this theorem it is necessary to show that:

1 ϕ_1 is harmonic except at the singularities.

2 ϕ_1 has no singularities outside the sphere.

3 ϕ_1 does not disturb the condition at infinity.

4 $\dfrac{\partial \phi}{\partial R} = -\dfrac{\partial \phi_1}{\partial R}$ at $R = a$.

Condition 1 follows from Kelvin's theorem. Actually condition 3 is contained in condition 2, which is evident because all the singular points of ϕ are outside the sphere, so that the corresponding singularities of ϕ_1 are all inside it.   In condition 4 the partial differentiation implies that θ and φ are kept constant.   If x, y, and z are represented by x_i,

$$y_i = \frac{\lambda^2 x_i}{R^2} \quad \text{and} \quad x_i' = \frac{a^2 x_i}{R^2},$$

we have

$$-\frac{\lambda^2}{R^2} \frac{\partial y_i}{\partial \lambda} = \frac{2\lambda}{R} \frac{\partial y_i}{\partial R}, \qquad \frac{\partial}{\partial \lambda} = \frac{\partial y_\alpha}{\partial \lambda} \frac{\partial}{\partial y_\alpha}, \qquad \frac{\partial}{\partial R} = \frac{\partial y_\alpha}{\partial R} \frac{\partial}{\partial y_\alpha} = -\frac{y_\alpha}{R} \frac{\partial}{\partial y_\alpha}. \tag{53}$$

---

[1] Because $\psi_0(0,\theta) = 0$ and the velocity is finite at $R = 0$ [see (1.34)].

Thus
$$\frac{\partial}{\partial R}\left[\frac{\lambda^2}{R}\frac{\partial\phi(y_i)}{\partial\lambda}\right] = \frac{\partial}{\partial\lambda}\left[\frac{\lambda^2}{R}\frac{\partial\phi(y_i)}{\partial R}\right].$$
(54)

Since ϕ_1 can be written, on integration by parts, as

$$\phi_1 = \frac{1}{a}\int_0^a \frac{\lambda^2}{R}\frac{\partial\phi(y_i)}{\partial\lambda}\,d\lambda,$$

utilization of (54) gives

$$\frac{\partial\phi_1}{\partial R}\bigg|_{R=a} = \frac{\partial\phi(x_i')}{\partial R}\bigg|_{R=a} = -\frac{x_\alpha'}{R}\frac{\partial\phi(x_i')}{\partial x_\alpha'}\bigg|_{R=a} = -\frac{a^2}{R^2}\frac{x_\alpha'}{R'}\frac{\partial\phi(x_i')}{\partial x_\alpha'}\bigg|_{R=a}$$

$$= -\frac{\partial\phi(x_i')}{\partial R'}\bigg|_{R=a} = -\frac{\partial\phi(x_i)}{\partial R}\bigg|_{R=a},$$
(55)

in which
$$R'^2 = x_\alpha'x_\alpha' = a^4 R^{-2}.$$

Note that

$$\frac{\partial x_\alpha'}{\partial R'} = \frac{x_\alpha'}{R'},$$

which justifies the penultimate equality sign in (55), and that $R' = a$ when $R = a$, which, together with the forms of the quantities on the two sides of the last equality sign in (55), justifies that sign. The theorem is thus proved.

A sphere theorem similar to Weiss' theorem gives the potential after a spherical surface $R = a$ has been introduced into an original flow with no singularities *outside* that surface; it is due to Ludford, Martinek, and Yeh (1955). The result is the same as that given by (51) and (52) if the lower limit of the integral in (52) is changed to ∞. The proof is very much the same as that for Weiss' theorem. In the same paper the authors also extended Weiss' theorem and their own theorem to deal with a more general linear boundary condition on the spherical surface. A still more general theorem by Yeh, Martinek, and Ludford (1956) allows original (before the sphere is introduced) singularities both outside and inside the sphere, deals with general linear boundary conditions on its surface, and has applications to several fields governed by the Laplace equation.

11. ADDED MASSES

Consider a solid body of mass M immersed in a fluid. If the body is pushed with acceleration a, not only is the body accelerated but the fluid around it generally moves (in various directions) with greater and greater speed. Thus the force *pushing the body* has to do work to increase the kinetic energy not only of the body but also of the fluid. Hence the force F will be greater than Ma. If we set

$$F = (M + M')a,$$
(56)

M' is called the added mass and $M + M'$ the virtual mass. (In older literature, M' is sometimes called the virtual mass.)

But since the body not only can move in three directions but also can rotate, there are many added masses. As will be seen presently, the added masses, like the moments of inertia of a solid body, are components of a tensor.

First, for an irrotational flow characterized by the potential ϕ, the kinetic energy T in the fluid domain V is given by

$$T = \frac{\rho}{2} \int_V \phi_{,i} \phi_{,i} \, dV, \tag{57}$$

with i ranging over 1, 2, and 3. By virtue of (12), we can write

$$2T = -\rho \int_S \phi \frac{\partial \phi}{\partial n} \, dS. \tag{58}$$

Now let the translational velocity of the solid body have components U_1, U_2, and U_3 at the origin and the rotational (angular) velocity of the body have components U_4, U_5, and U_6. For each component α of motion of the body there is a corresponding velocity potential for the fluid. That potential, per unit of U_α, will be denoted by ϕ_α. Now the Laplace equation and the boundary conditions (in the absence of separation or cavitation, which create free surfaces) are linear, so that solutions can be superposed. Hence

$$\phi = U_\alpha \phi_\alpha, \qquad \alpha = 1, 2, \ldots, 6, \tag{59}$$

for the general case, in which the solid body is in any motion whatsoever. Substitution of (59) in (58) produces

$$2T = A_{\alpha\beta} U_\alpha U_\beta, \tag{60}$$

in which the Greek subscripts again range from 1 to 6 and

$$A_{\alpha\beta} = -\rho \int_S \phi_\alpha \frac{\partial \phi_\beta}{\partial n} \, dS. \tag{61}$$

We shall presently show, from (60), that the A's are components of the added-mass tensor. In the meantime, it will be shown that $A_{\alpha\beta}$ is symmetric, that is,

$$A_{\alpha\beta} = A_{\beta\alpha}. \tag{62}$$

Taking[1]

$$U_i = \rho \phi_\alpha \frac{\partial \phi_\beta}{\partial x_i} \tag{63}$$

[1] Note that U_i in (1.9) has nothing to do with the U_α in (59).

in (1.9), we have, since ρ is constant,

$$A_{\alpha\beta} = \rho \int_V \frac{\partial\phi_\alpha}{\partial x_i} \frac{\partial\phi_\beta}{\partial x_i} \, dV, \tag{64}$$

since ϕ_β is harmonic. Similarly, taking in (1.9)

$$U_i = \rho\phi_\beta \frac{\partial\phi_\alpha}{\partial x_i},$$

we obtain an equation identical with (64) except that $A_{\alpha\beta}$ is replaced by $A_{\beta\alpha}$. Hence (62) is true.

The components of $A_{\alpha\beta}$ form a tensor in the following sense. Consider a six-dimensional Cartesian space $(x_1, x_2, \ldots, x_6)$, with

$$x_4 = x_1, \qquad x_5 = x_2, \qquad x_6 = x_3.$$

Any transformation of coordinates from x_i to x_i' will be accomplished by the scheme

	x_1	x_2	x_3	x_4	x_5	x_6
x_1'	a_{11}	a_{12}	a_{13}	0	0	0
x_2'	a_{21}	a_{22}	a_{23}	0	0	0
x_3'	a_{31}	a_{32}	a_{33}	0	0	0
x_4'	0	0	0	a_{44}	a_{45}	a_{46}
x_5'	0	0	0	a_{54}	a_{55}	a_{56}
x_6'	0	0	0	a_{64}	a_{65}	a_{66}

in which

$$a_{\alpha\beta} = \frac{\partial x_\alpha'}{\partial x_\beta}$$

and

$$a_{(i+3)(j+3)} = a_{ij}, \qquad i, j = 1, 2, 3.$$

With this scheme

$$U_\beta' = a_{\beta\gamma} U_\gamma.$$

Since the left-hand side of (60) is a scalar,

$$T = T',$$

and

$$A_{\alpha\beta}' U_\alpha' U_\beta' = A_{\alpha\beta}' a_{\alpha\gamma} a_{\beta\delta} U_\gamma U_\delta = A_{\gamma\delta} U_\gamma U_\delta.$$

Since U_γ and U_δ are arbitrary,

$$A_{\gamma\delta} = A_{\alpha\beta}' a_{\alpha\gamma} a_{\beta\delta},$$

which has the same structure as (1.41). Hence the components $A_{\alpha\beta}$ form a (Cartesian) tensor.

To see why it is called the added-mass tensor, we compare (60) with the kinetic energy T_B of the displaced fluid considered as a solid, given by

$$2T_B = B_{\alpha\beta} U_\alpha U_\beta, \tag{65}$$

in which $B_{\alpha\beta}$ is given by comparison with (ρ_B = density of solid body)

$$2T_B = \int_V \rho_B[(U_1 + U_5x_3 - U_6x_2)^2 + (U_2 + U_6x_1 - U_4x_3)^2 \\ + (U_3 + U_4x_2 - U_5x_1)^2]\, dV.$$

Thus

$$B_{11} = B_{22} = B_{33} = M, \tag{66}$$

$$\begin{aligned} B_{44} &= I_{11}, & B_{55} &= I_{22}, & B_{66} &= I_{33}, \\ B_{56} &= -I_{23}, & B_{64} &= -I_{31}, & B_{45} &= -I_{12}, \end{aligned} \tag{67}$$

$$\begin{aligned} B_{15} &= M\bar{z}, & B_{16} &= -M\bar{y}, \\ B_{26} &= M\bar{x}, & B_{24} &= -M\bar{z}, \\ B_{34} &= M\bar{y}, & B_{35} &= -M\bar{x}, \end{aligned} \tag{68}$$

in which I_{ij} are the components of the moment-of-inertia tensor, $\bar{x}$, $\bar{y}$, and $\bar{z}$ are the coordinates of the center of mass, and M is the mass—all of the body. Furthermore $B_{\alpha\beta} = B_{\beta\alpha}$.

Comparing (65) and (60), we see that $A_{\alpha\beta}$ corresponds to $B_{\alpha\beta}$. When there is only translation, say in the x_1 direction, the total kinetic energy is

$$T + T_B = \tfrac{1}{2}(A_{11} + M)U_1^2. \tag{69}$$

The force F acting on the body to accelerate it is calculated from

$$FU_1 = \frac{d}{dt}(T + T_B) = (A_{11} + M)U_1 a,$$

so that

$$F = (A_{11} + M)a, \tag{70}$$

in which a is the acceleration. Comparing (70) with (56), we see that $A_{11} = M'$. Either (69) or (70) justifies calling A_{11} the added mass (for translation in the x_1 direction) and hence $A_{\alpha\beta}$ the added-mass tensor. Note that $A_{\alpha\beta}$ for any α and β does not depend on time or the velocity (linear or angular) or the acceleration but only on the shape of the body and the choice of the coordinates. When other boundaries are present, it of course also depends on the location and orientation of the body.

Note that by choosing the origin at the center of mass, the B's given by (68) vanish. Since $A_{\alpha\beta}$ is symmetric, there are only 21 distinct components of the added-mass tensor. Of these A_{12}, A_{23}, and A_{31} can be made to vanish by the proper choice of direction of the coordinates. The operation is much like the one used to find the principal stresses or rates of deformation, and its possibility is due to the symmetry of $A_{\alpha\beta}$. Another three, say A_{35}, A_{16}, and A_{24}, can be made to vanish by the proper choice of the origin.

$B_{\alpha\beta}$ is simpler than $A_{\alpha\beta}$ in that B_{12}, B_{23}, and B_{31} all are zero and that the proper choice of the origin makes six B's (instead of just three) vanish. This is because rigid-body motion is simpler than fluid motion: in the

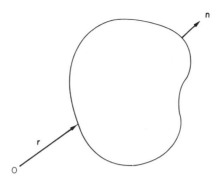

FIGURE 4. DEFINITION SKETCH.

first instance because there is no interaction between the translational
modes as far as forming the kinetic energy is concerned, and in the second
instance the interaction, though existent between translation and rotation,
is simpler and more regular in pattern.

Before proceeding further with the discussion of added masses, (61)
will be put in a simpler form. The kinematic boundary condition on the
surface of the body is

$$\frac{\partial \phi}{\partial n} = (\mathbf{U} + \mathbf{\omega} \times \mathbf{r}) \cdot \mathbf{n}, \tag{71}$$

in which $\mathbf{U}$ is the velocity vector, $\mathbf{\omega}$ the rotation vector, $\mathbf{r}$ the position
vector (from the origin to a point on the surface of the body), and $\mathbf{n}$ the
unit vector in the direction of the normal drawn outward from the solid
body (see Fig. 4).   Let n_4, n_5, and n_6 be defined by

$$(n_4, n_5, n_6) = \mathbf{r} \times \mathbf{n}. \tag{72}$$

Equations (72) and (59) enable us to write (71) as

$$\frac{\partial \phi}{\partial n} = U_\alpha \frac{\partial \phi_\alpha}{\partial n} = U_\alpha n_\alpha, \tag{73}$$

with α ranging from 1 to 6.   Since (73) is valid for *all* values of U_α,

$$\frac{\partial \phi_\alpha}{\partial n} = n_\alpha, \tag{74}$$

and we can write (61) as

$$A_{\alpha\beta} = -\rho \int_S \phi_\alpha n_\beta \, dS. \tag{75}$$

Since ϕ_α is the potential for $U_\alpha = 1$, (75) shows that the added masses are
dependent only on the shape and orientation of the body and the mode of

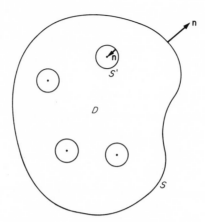

FIGURE 5. DEFINITION SKETCH. THE
NORMAL TO S' IS DRAWN OUTWARD FROM
D TOWARD THE SINGULARITIES. NOTE
THAT THE NORMAL n IS DRAWN AT S
INTO THE FLUID, CONSISTENT WITH THE
DIRECTION OF n ASSUMED FOR EQ. (1.9).

its motion, not on its linear or angular velocity or acceleration. This is
an important point to be borne in mind.

11.1. The Taylor Theorem

Of the 36 added-mass components, 9 are for translation, 9 for rotation,
and 18 signify the interaction between translation and rotation. Of
course, the symmetry of the added-mass tensor reduces the numbers of
distinct components to 6, 6, and 9, respectively, or 21 in all.

A theorem due to Taylor (1928) and extended by Birkhoff (1953)
and Landweber (1956) enables one to evaluate the added-mass components,
except the ones for pure rotation, in terms of the singularities generating
the flow. If the letter j ranges over 1, 2, and 3 and α ranges from 1 to 6,
the components $A_{j\alpha}$ or $A_{\alpha j}$ can be expressed in terms of the singularities.

Since

$$n_j = \frac{\partial x_j}{\partial n},$$

(75) can be written as

$$A_{\alpha j} = -\rho \int_S \phi_\alpha \frac{\partial x_j}{\partial n}\, dS. \tag{76}$$

As shown in Fig. 5, the surfaces enclosing the isolated singularities will
be denoted collectively by S', and, quite consistent with our definition
of $\mathbf{n}$ on S, the normal is now directed outward from the region D bounded

by S and S'. Taking the U_i in Eq. (1.9) to be

$$\phi_\alpha \frac{\partial x_j}{\partial x_i} \quad \text{and} \quad -x_j \frac{\partial \phi_\alpha}{\partial x_i}$$

in turn and adding the two equations so obtained, we have, after multiplication by ρ,

$$\rho \int_S \left(\phi_\alpha \frac{\partial x_j}{\partial n} - x_j \frac{\partial \phi_\alpha}{\partial n} \right) dS = \rho \int_D (-x_j \nabla^2 \phi_\alpha) \, dV$$

$$- \rho \int_{S'} \left(\phi_\alpha \frac{\partial x_j}{\partial n} - x_j \frac{\partial \phi_\alpha}{\partial n} \right) dS', \quad (77)$$

The sign of the left-hand side is changed because Green's theorem, or (1.9), has been applied to D, on the surface of which the normal is *outward* from D.

As explained before, the only distributed singularities which need be considered are distributed sources and sinks. Hence

$$\nabla^2 \phi_\alpha = 4\pi \sigma_\alpha, \quad (78)$$

in which σ_α is the strength of the distributed source per unit volume. Furthermore

$$\rho \int_S x_j \frac{\partial \phi_\alpha}{\partial n} \, dS = \rho \int x_j n_\alpha \, dS = B_{\alpha j}, \quad (79)$$

in which $B_{\alpha j}$ are some of the 36 components $B_{\alpha\beta}$ defined by (65).

The meanings of the first three terms in (77) are now clear. The last two terms involving isolated singularities will now be considered. First of all ϕ_α can be resolved into two parts, a part ϕ'_α, which is the potential at the singular point due to everything but the local singularity, and another part which is due to the local singularity. It is evident that the part ϕ'_α contributes nothing to the last integral in (77), for as the surfaces S' are shrunk to a point, the terms containing ϕ'_α approach zero in the limit. We shall, then, consider the ϕ_α in that integral to be due to the local singularity only, at each singular point in turn. For clarity we write, for any of the surfaces collectively denoted by S',

$$x_j = x_{js} - n_j R, \quad (80)$$

in which x_{js} is the value of x_j at the local singularity and R is the radius of the sphere whose center is at the singular point under consideration. For an isolated source of strength m_α (the α merely indicating the mode of motion),

$$I_2 \equiv \int_{S'} x_j \frac{\partial \phi_\alpha}{\partial n} \, dS' = -4\pi m_\alpha x_{js}. \quad (81)$$

For an isolated doublet,

$$\phi_\alpha = \frac{\mu_{\alpha i} n_i}{R^2}, \qquad n_i = -\frac{x_i}{R} \text{ on } S', \tag{82}$$

in which the summation convention is observed. Substituting (80) and (82) into I_2, we see that the term involving x_{js} is zero and the term involving n_j in I_2 is[1], with j unsummed,

$$-2 \int_{S'} \frac{\mu_{\alpha i} n_i n_j}{R^2} \, dS' = -2 \int_{S'} \frac{\mu_{\alpha j} n_j n_j}{R^2} \, dS' = -\frac{8\pi}{3} \mu_{\alpha j}. \tag{83}$$

It can be shown that for singularities of higher orders, I_2 is exactly zero. For the term containing x_{js}, antisymmetry makes its vanishing obvious. The vanishing of the integral in I_2 containing $n_j R$ for higher-order singularities is also due to antisymmetry, but it is necessary to show the antisymmetry of

$$n_j R \frac{\partial \phi}{\partial n}$$

on the spherical surface of radius R, a property that becomes evident only upon expressing n_j and ϕ in terms of the spherical coordinates R, θ, and φ. The reader is urged to satisfy himself on this point.

Now we turn to

$$I_1 \equiv \int_{S'} \phi_\alpha \frac{\partial x_j}{\partial n} \, dS' = \int_{S'} \phi_\alpha n_j \, dS'. \tag{84}$$

Again terms involving ϕ_α' vanish in the limit, so that ϕ_α can be considered to be the potential due to the local singularity. For an isolated source, I_1 vanishes on account of antisymmetry. For an isolated doublet,

$$I_1 = \int_{S'} \frac{\mu_{\alpha i} n_i n_j}{R^2} \, dS' = \frac{4\pi}{3} \mu_{\alpha j}. \tag{85}$$

For higher-order terms I_1 again vanishes, because of antisymmetry, just like I_2.

Substituting (76), (78), (79), (81), (83), and (85) into (77), we have, finally,

$$A_{\alpha j} + B_{\alpha j} = 4\pi\rho \left[\int_V x_j \sigma_\alpha \, dV + \sum (m_\alpha x_{js} + \mu_{\alpha j}) \right], \tag{86}$$

which is the extended Taylor theorem. The summation is over all the isolated singularities. It should be noted that vortex lines have been

---

[1] Be careful here. R is $-n$, and $\partial/\partial R$ implies that the other two spherical coordinates are kept constant and hence n_i is also kept constant, for $i = 1, 2, 3$. Note that the integral of $n_i n_j$ ($i \neq j$) over the spherical surface is zero, and that of $n_j n_j$ (j unsummed) is $\frac{1}{3}$.

excluded as singularities inside the body. So are singularities distributed on a line, which can be included by integration.

One may wonder why one needs to evaluate $A_{\alpha j}$ by the use of singularities and why not directly from (75) or (76), since, given a body, the singularities are known only if the flow problem is solved, and if it is solved, ϕ_α is known on S. The answer is that there are bodies generated by singularities, and for such bodies it is vastly more convenient to find $A_{\alpha j}$ by (86) than by (76).

Usually (86) is used only for a body moving in infinite fluid, although in the development other boundaries than that of the body itself are not excluded.

11.2. The Added Mass of a Sphere

For a sphere moving in infinite fluid, there is only one distinct component of the added-mass tensor, which is

$$A_{11} = A_{22} = A_{33}.$$

All other components are zero. Since $B_{11} = \rho V$, (86) gives

$$A_{11} = -\rho V + 4\pi\rho\mu_{11}.$$

But μ_{11} is the μ' in (34), with $U = 1$. Thus, if a is the radius of the sphere,

$$\mu_{11} = \tfrac{1}{2}a^3 \quad \text{and} \quad V = \frac{4\pi}{3}a^3.$$

Hence
$$A_{11} = \frac{2\pi}{3}\rho a^3 = \tfrac{1}{2}B_{11}, \tag{87}$$

or A_{11} is equal to one-half the displaced mass.

Consider now a sphere of density $\rho_B > \rho$ released at rest in a large body of an inviscid liquid. The acceleration of the sphere can be calculated by using the principle of conservation of energy. Since the rate of decrease of the potential energy must equal the rate of increase of kinetic energy,

$$-V(\rho_B - \rho)gU_3 = \frac{d}{dt}\left[\frac{1}{2}V(\rho_B + 0.5\rho)U_3{}^2\right], \tag{88}$$

in which U_3 is the velocity in the direction of the vertical. Thus the acceleration is, after U_3 is canceled,

$$a_3 = \frac{dU_3}{dt} = -\frac{2\rho_B - 2\rho}{2\rho_B + \rho}g, \tag{89}$$

which is true for any nonzero U_3 and therefore true in the limit as U_3 becomes zero. This is the initial acceleration of the sphere even if the liquid is viscous, because initially there is no velocity anywhere so that the nonslip condition is satisfied on the surface of the sphere. Although

ϕ is a constant everywhere in the fluid initially, its rate of change with time is not zero except at infinity, and $\partial\phi/\partial t$ can be calculated on the basis of irrotational-flow theory. This term is what makes the pressure unsymmetric with respect to z or x_3 when the pressure is calculated from Eq. (3.30).

We may write (88) in the form

$$-V\rho_B g + V\rho g - \tfrac{1}{2}V\rho a_3 = V\rho_B a_3. \tag{90}$$

Since the first term is the weight of the sphere, the second the buoyancy, and the last the mass of the sphere times its acceleration, the term

$$- \tfrac{1}{2}V\rho a_3$$

must be the resultant of the hydrodynamic force on the accelerating body. Indeed it is through this force that the kinetic energy of the fluid is increased by the moving body.

A discussion of the evaluation of the term $\partial\phi/\partial t$ in (3.30) may be instructive. For a sphere of radius a moving with velocity $W(t)$ in the direction of increasing z,

$$\phi = -\frac{W(t)a^3 z}{2R^3}.$$

Actually the coordinates used in this expression for ϕ are not fixed but are moving with the sphere. It is clearer to denote them by x', y', and z' and retain x, y, and z to denote fixed coordinates. Then

$$\phi = -\frac{W(t)a^3 z'}{2R'^3}, \qquad R'^2 = x'^2 + y'^2 + z'^2. \tag{91}$$

If the two coordinate systems coincide at $t = 0$,

$$x' = x, \qquad y' = y, \qquad z' = z - \int_0^t W(t)\,dt. \tag{92}$$

If the transformation (92) is made in (91), ϕ is then in terms of t, x, y, and z and the quantity $\partial\phi/\partial t$ is then evaluated by keeping x, y, and z constant. This can be done for any point on the sphere at any time, and in particular at $t = 0$.

12. FORCE AND MOMENT ON A BODY IN STEADY FLOWS

Since only inviscid fluids are considered in this chapter the forces acting on the surface of a submerged body arise from p alone and are normal to the surface. Now according to the Bernoulli equation for irrotational flows, (3.29), contributions to the pressure p are made by a temporal part $\rho\,\partial\phi/\partial t$ and a convective part $\rho u_i u_i/2$, aside from the hydrostatic part, which gives rise to the bouyancy. The calculation for the buoyant force or moment is well known and will not be discussed here.

The force and moment arising from unsteadiness can be further divided into two categories. In the first category the body accelerates in infinite fluid in the absence of other boundaries, and the total hydro-dynamic force arises purely from $\rho\, \partial\phi/\partial t$. Since we have shown how to calculate the nine components of the added-mass tensor A_{ij} ($i, j = 1, 2, 3$ for translation only) in terms of the singularities inside the body, this force can also be expressed in terms of the internal singularities. In fact, the relationship between the force[1] $\mathbf{F}$ (with components F_i) and A_{ij} is simply

$$-F_i U_i = \frac{d}{dt}\left(\frac{1}{2} A_{ij} U_i U_j\right), \tag{93}$$

which states that the rate of work done by the force $-\mathbf{F}$ exerted on the fluid by the body is equal to the rate of change of the kinetic energy of the fluid due to the motion of the body. As to the moment, we recall that the added-mass components for pure rotation, or $A_{\alpha\beta}$ with α and β ranging over 4, 5, and 6, are not expressible in terms of the internal singularities (see also Sec. 21). If they were, the moment would be expressible in terms of internal singularities through a formula similar to (93), but this is not the case.

In the second category the unsteadiness is due to the presence of other boundaries as the body moves past them, and thus the internal singularities are not merely changing in magnitude, as in the first category, but changing in location as well. Clearly, to express the hydrodynamic force and moment arising from this kind of unsteadiness in terms of internal singularities it would be necessary (and not sufficient in the case of moments) to know their time-dependent locations and magnitudes as the body moves from place to place. This would be a hopeless task in general. For this reason such an expression is not attempted here.

We turn now to the investigation of forces and moments on a body in steady flow. If the flow is unsteady, the force or moment to be given is that part of the force or moment corresponding to the convective part of the acceleration only.

Consider first the force $\mathbf{F}$ on the body (Fig. 5) with surface S and let the totality of the spherical surfaces surrounding the isolated singu-larities be denoted by S'. For the reason stated before in this chapter, the only three-dimensionally distributed singularities considered will be sources and sinks. Surface singularities are not considered, although a volume distribution of doublets and higher-order singularities, upon integrations by parts, may lead eventually to a source-sink distribution in the same volume plus a surface distribution of singularities. When

[1] Note that this is not the force pushing the body.

such a surface distribution is present, the additional terms it will call for will be clear from the results to be given for isolated singularities.

The components of the force F on S are given by

$$F_i = -\int pn_i \, dS, \qquad (94)$$

in which the direction of the normal n on either S or S' is as indicated in Fig. 5. Since steady flow is considered and buoyancy can be independently and easily evaluated, the hydrodynamic part of p is simply

$$p = -\frac{\rho}{2} u_i u_i,$$

the constant on the right-hand side being neglected because it contributes nothing to the net force. Thus (94) can be written as

$$F_i = \frac{\rho}{2} \int_S u_j u_j n_i \, dS = \frac{\rho}{2} \int_D \frac{\partial}{\partial x_i} u_j u_j \, dV - \frac{\rho}{2} \int_{S'} u_j u_j n_i \, dS', \qquad (95)$$

by Green's theorem. Now, because of irrotationality,

$$\frac{\partial u_i}{\partial x_j} = \frac{\partial u_j}{\partial x_i},$$

so that
$$\frac{1}{2} \frac{\partial}{\partial x_i} u_j u_j = u_j \frac{\partial u_j}{\partial x_i} = u_j \frac{\partial u_i}{\partial x_j} = \frac{\partial}{\partial x_j} u_i u_j - u_i \frac{\partial u_j}{\partial x_j}.$$

But
$$\frac{\partial u_j}{\partial x_j} = 4\pi\sigma,$$

in which σ is the strength of the source distribution. Hence (95) can be written, upon a second application of Green's theorem, as

$$F_i = -4\pi\rho \int_D \sigma u_i \, dV + \rho \int_{S'} u_i u_j n_j \, dS' - \frac{\rho}{2} \int_{S'} u_j u_j n_i \, dS'. \qquad (96)$$

The term

$$\int_S u_i u_j n_j \, dS = \int_S u_i u_n \, dS$$

is zero because u_n is zero on the surface S of the solid body. The task then is reduced to evaluating the last two integrals in (96). Before we evaluate them, it is instructive to notice that the second integral in (96) represents the momentum efflux at S' and the last integral (without the minus sign) the ith component of the force at S'. The last two terms could be written as

$$M_i' - F_i'.$$

The first integral is just the same quantity[1] integrated over the domain

---

[1] Or M_i' only, since F_i' is internal and therefore has an integral equal to zero.

of distributed sources and sinks, and (96) is therefore just a momentum equation. Without the term containing σ, it is the ordinary form of the momentum equation, with control surfaces S and S'. The term containing σ renders the control surface obscure and points to the advantage of using the Green's theorem.

The evaluation of the last two terms in (96) will be discussed in separate cases in the following subsections.

12.1. Force and Momentum Efflux at a Surface Surrounding a Point Source

Let a source of strength m be situated at the origin and be the center of a spherical surface of radius r. Let the components of the velocity of the flow exclusive of that due to the source be denoted by v_i. The velocity u_i is then given by

$$u_i = v_i + \frac{mx_i}{r^3} \tag{97}$$

over the spherical surface. Of course, v_i will vary over the spherical surface. If the values of v_i and its derivatives at the origin are indicated by the subscript zero,

$$v_i = v_{i0} + \left(\frac{\partial v_i}{\partial x_j}\right)_0 x_j + \frac{1}{2}\left(\frac{\partial^2 v_i}{\partial x_j \partial x_k}\right)_0 x_j x_k + \cdots. \tag{98}$$

The second integral in (96) has four parts: one arising entirely from the v_i, which is zero in the limit as r approaches zero; one due to the last term in (97) alone (or due to the source alone), which is zero because of the antisymmetry of the integrand; and two coupled terms, which are not zero. The terms are, after (97) is used and the terms with antisymmetry dropped,

$$-4\pi\rho v_{i0}m - \rho m \int_{S'} v_{i0}\, \frac{x_i{}^2}{r^4}\, dS' = -\frac{16\pi}{3}\,\rho m v_{i0},$$

in which the repeated indices i do not indicate summation. (Note the direction of n in Fig. 5.) This is the momentum efflux out of D through S'. Similarly, the last term in (96), representing the force on S', is

$$\frac{4\pi}{3}\,\rho m v_{i0}.$$

Thus, for a single source of strength m, the sum of the last two integrals in (96) is simply

$$-4\pi\rho m v_i, \tag{99}$$

which is the contribution of the source to the component F_i of the resultant force acting on the body. The subscript zero on v_i has been dropped for convenience. From now on v_i and its derivatives mean the ith component of the velocity and its derivatives at the location of a singularity, with the contribution of the singularity itself excluded.

The form of (99) is revealing.   Let there be two sources, of strengths m_1 and m_2 respectively, and for simplicity let the line joining them be parallel to the x axis.   Then at both sources v_2 and v_3 are zero.   If the first source is to the left of the second,

$$v_1 = -\frac{m_2}{r^2},$$

r being the distance between the sources.   At the second source,

$$v_2 = \frac{m_1}{r^2}.$$

Thus the quantity (99) evaluated at one source is equal in magnitude but opposite in sign to that evaluated at the other.   In general,

The contribution to the sum of the last two integrals at one singularity by another is equal and opposite to the contribution to that sum at the latter singularity by the former.

We have proved the case for sources (or sinks), and although their orientation with respect to the coordinate axes has been assumed in the proof for the sake of simplicity, the result obviously remains true whatever this orientation is.   For other kinds of singularities the same result can be similarly proved.   We shall omit the proof, which is straightforward but cumbersome.   Instead, we shall give a much simpler proof, which not only will serve our present purpose but will contribute to the economy of obtaining the final result sought in this section (Sec. 12).   This will be done in the next subsection.

12.2.   Reciprocity Between Singularities

The result obtained in Sec. 12.1 can be called the *reciprocity* of two sources.   (The sign of m may be negative for either source.   Therefore it applies to source-sink or sink-sink combinations too.   This will henceforth be understood.)   For higher singularities the same reciprocity exists, and the approach adopted for a proof is given in Fig. 6.   For simplicity, let us indicate the sum of the last two integrals in (96) by I. Let a sink of strength m be situated at A, a source of strength m be situated at B, and a source of strength m_1 situated at C.   The spherical surfaces surrounding the singularities at A, B, and C will be denoted by S_1', S_2', and S_3', and the larger sphere containing S_1' and S_2' will be denoted by S_4'.   To start with, the integral I_{12} at S_1' due to the singularities at A and B is always equal and opposite to the integral I_{21} at S_2' due to the same singularities, as just shown in Sec. 12.1.   We concentrate on I_{13} and I_{23}, which are the respective values of I at S_1' and S_2' due to the singularities at A and C in one case and at B and C in the other.   If the

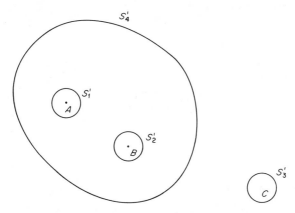

FIGURE 6. RECIPROCITY OF FORCES ON SINGULARITIES. THERE IS A SOURCE AT A AND A SINK OF EQUAL STRENGTH AT B. AS A APPROACHES B AND THE MOMENT OF THE STRENGTH IS KEPT CONSTANT, A DOUBLET IS FORMED. A SOURCE OF ANY STRENGTH IS SITUATED AT C. THE FORCE ON S'_3 IS EQUAL AND OPPOSITE TO THAT AT S'_4.

value of I on S'_4 is denoted by I_4, then (because $I_{12} = -I_{21}$ and because of steadiness)

$$I_4 = I_{13} + I_{23}, \tag{100}$$

which is merely the momentum equation applied to the control volume bounded by S'_1, S'_2, and S'_4. But, as proved in Sec. 12.1,

$$I_{13} = -I_{31} \quad \text{and} \quad I_{23} = -I_{32},$$

in which I_{31} and I_{32} are the values of I at S'_3 due to m_1 and $-m$ or m_1 and m, respectively. Thus

$$I_4 = -I_3 \tag{101}$$

if I_3 denotes

$$I_{31} + I_{32}.$$

Equation (101) shows the reciprocity between a source and a doublet, since as B approaches A and mAB is kept finite, a doublet will in the limit be obtained at A, with $I = I_4$ at S'_4.

12.3. Final Evaluation of the Force on the Body

With v_i denoting the velocity at A due to m alone, we have

$$I_{13} = 4\pi \rho m v_i \quad \text{and} \quad I_{23} = -4\pi \rho m \left(v_i + \frac{\partial v_i}{\partial x_j} a_j + \cdots \right) \tag{102}$$

if the components of the displacement AB are denoted by a_j. Thus with

$$m a_j = \mu_j,$$

as $a_j \to 0$, (100) and (102) give

$$I_4 = -4\pi\mu_j \frac{\partial v_i}{\partial x_j}, \tag{103}$$

with $\partial v_i/\partial x_j$ evaluated at A, where a doublet with axis in the direction AB is obtained in the limit. Equation (102) is valid in general, whether or not v_i is due to m alone. Thus the I for a doublet is obtained. By similar reasoning, for the doublet of a doublet, with

$$v_{jk} = \mu_j a_k,$$

we have
$$I = -4\pi v_{jk} \frac{\partial^2 v_i}{\partial x_j\, \partial x_k}.$$

Still higher singularities can be dealt with similarly. Thus, (96) becomes

$$F_i = -4\pi\rho\left[\int_V \sigma u_i\, dV + \sum\left(mv_i + \mu_j \frac{\partial v_i}{\partial x_j} + v_{jl}\frac{\partial^2 v_i}{\partial x_j\, \partial x_l} + \cdots\right)\right]. \tag{104}$$

A direct calculation for the terms in (104) corresponding to doublets and higher-order singularities could be made but would be tedious. Because of the principle of reciprocity, in the evaluation of v_i and its derivatives at each singularity, the influence of other internal singularities contained in S can be totally ignored. This applies in fact to u_i in the integral in (104). As mentioned before, distributed doublets or higher singularities over part of D can be reduced to a source distribution plus a singularity distribution on a surface. This surface singularity can be taken care of by integration, since the contributions at isolated singularities are given in (104). The only singularities not included in (104) are vortex lines. The contributions of these can be calculated, but we shall not dwell on the subject any longer. The force given by (104) is called the *Lagally force*, but Lagally (1922) did not treat singularities of orders higher than that of a doublet [see also Cummins (1953) and Landweber and Yih (1956)].

12.4. The Moment

From the forms of I given in (104) for isolated singularities, it is possible to write down the expression for the moment on the body by inspection, provided steady flow is still considered. Let

$$\epsilon_{ijk} = \begin{cases} 0 & i, j, k \text{ not all different,} \\ 1 & i, j, k \text{ in same cyclic order as } 1, 2, 3, \\ -1 & i, j, k \text{ in same cyclic order as } 2, 1, 3. \end{cases}$$

Then
$$M_i = -4\pi\rho\epsilon_{ijk}\left[\int \sigma x_j u_k\, dV + \sum\left(mx_j v_k + \mu_j v_k\right.\right.$$
$$\left.\left. + x_j\mu_\alpha\frac{\partial v_k}{\partial x_\alpha} + v_{j\alpha}\frac{\partial v_k}{\partial x_\alpha} + v_{\alpha\beta}x_j\frac{\partial^2 v_k}{\partial x_\alpha\, \partial x_\beta} + \cdots\right)\right]. \tag{105}$$

All the terms in (105) can be verified by inspection. The second and fourth terms after the summation sign should be compared with the first and third after that sign. When a source approaches a sink and the moment is kept constant, the moment of the source-sink pair produces a doublet. Hence the second term is to be looked upon as a limit of two terms of the type mx_jv_k, as the doublet is generated by a limiting process. A similar remark applies to the fourth term. The moment given by (105) is called the *Lagally moment*, although he did not treat singularities of orders higher than that of a doublet.

13. FLOW PAST A NEARLY SPHERICAL BODY

Most of the solutions in classical hydrodynamics for irrotational flows past a body have been obtained by the reverse method; i.e., they have been found by first specifying the solution and then finding what physical problem it corresponds to. Given a body or boundary of a specified shape, the exact solution is not easy to obtain. However, if the exact solution for the flow past a certain body or boundary is known, it should be possible to find the solution for the flow past a body having nearly the same shape. The method for finding such a solution is called the *perturbation method*.

The method is essentially one that utilizes a long ladder with a short one lashed to the top of it. The long ladder is the exact solution for a problem with nearly the same governing differential equations and boundary conditions, all of which do not have to differ from those of the actual problem at hand. The short ladder is the extra work one has to do in order to reach the solution desired. It is the perturbation, and it is able to function only because it rests on the long ladder which nearly reaches the goal. Sometimes it is not required that the perturbation part of the solution satisfy more boundary conditions than the primary part of the solution (or the exact solution of nearly the same problem). In that case the perturbation is regular. Often the opposite is true, in which case the perturbation is singular, and the short ladder, though short, is more elaborate than the long one on which it rests.

Singular perturbation will be discussed in Chap. 7, in connection with the boundary-layer theory. In this section we shall illustrate the method of regular perturbation by finding the solution for the flow past a nearly spherical body. In doing so, we incidentally indicate how spherical harmonics can be used for the solution of problems in the theory of potential flow. Although the scope of this book does not allow discussion of the use of other coordinates (such as ellipsoidal coordinates), the idea of the use of harmonic functions in any coordinate system will be demonstrated by the example to be given.

Since the body under consideration is nearly spherical, it is natural to use spherical coordinates.   The Laplace equation in spherical coordinates satisfied by the velocity potential ϕ is (3).   If we assume

$$\phi = R^n S_n, \tag{106}$$

S_n satisfies the equation

$$\frac{1}{\sin \theta} \frac{\partial}{\partial \theta}\left(\sin \theta \frac{\partial S_n}{\partial \theta}\right) + \frac{1}{\sin^2 \theta} \frac{\partial^2 S_n}{\partial \varphi^2} + n(n + 1)S_n = 0. \tag{107}$$

For axisymmetric flows, S_n is independent of φ, and (107) becomes, upon setting $\cos \theta = \mu$,

$$\frac{d}{d\mu}\left[(1 - \mu^2) \frac{dS_n}{d\mu}\right] + n(n + 1)S_n = 0, \tag{108}$$

the solution of which is

$$S_n = AF\left(-\frac{n}{2}, \frac{1 + n}{2}, \frac{1}{2}, \mu^2\right) + B\mu F\left(\frac{1 - n}{2}, \frac{2 + n}{2}, \frac{3}{2}, \mu^2\right), \tag{109}$$

with F indicating the hypergeometric functions defined by

$$F(\alpha,\beta,\gamma,x) = 1 + \frac{\alpha\beta}{1 \cdot \gamma} x + \frac{\alpha(\alpha + 1)\beta(\beta + 1)}{1 \cdot 2\gamma(\gamma + 1)} x^2$$

$$+ \frac{\alpha(\alpha + 1)(\alpha + 2)\beta(\beta + 1)(\beta + 2)}{1 \cdot 2 \cdot 3\gamma(\gamma + 1)(\gamma + 2)} x^3 + \cdots. \tag{110}$$

The two independent solutions given in (109) diverge at the poles, where $\theta = 0$ or π, as $\ln (1 - \mu^2)$ when they do not terminate; but the first solution terminates for even n and the second for odd n. For integral nonnegative values of n, whether even or odd, the terminating series, adjusted to make $P_n(\mu) = 1$ for $\mu = 1$, is

$$P_0(\mu) = 1 \text{ for } n= 0 \quad \text{and} \quad P_n(\mu) = \frac{1}{2^n n!} \frac{d^n}{d\mu^n} (\mu^2 - 1)^n \text{ for } n > 0,$$
$$\tag{111}$$

from which

$$P_0(\mu) = 1, \quad P_1(\mu) = \mu, \quad P_2(\mu) = \tfrac{1}{2}(3\mu^2 - 1),$$
$$P_3(\mu) = \tfrac{1}{2}(5\mu^3 - 3\mu), \quad \text{etc.} \tag{111a}$$

Thus, for integral nonnegative values of n

$$\phi = R^n P_n(\mu). \tag{112}$$

But, observing that the coefficient $n(n + 1)$ remains unchanged when n is changed to $-n - 1$, we see that

$$\phi = R^{-n-1} S_n \tag{113}$$

is also a solution of (3).   If n is a nonnegative integer,

$$\phi = R^{-n-1} P_n(\mu) \tag{114}$$

is a solution of (3), finite over a spherical surface of finite radius.
Since

$$u = \frac{\partial \phi}{\partial R} = \frac{1}{R^2 \sin \theta} \frac{\partial \psi}{\partial \theta}, \qquad v = \frac{1}{R} \frac{\partial \phi}{\partial \theta} = -\frac{1}{R \sin \theta} \frac{\partial \psi}{\partial R}, \qquad (115)$$

the two solutions for ψ corresponding to (112) and (114) are

$$\psi = \frac{1}{n+1} R^{n+1}(1 - \mu^2) \frac{dP_n}{d\mu} \qquad \text{and} \qquad \psi = -\frac{1}{n} R^{-n}(1 - \mu^2) \frac{dP_n}{d\mu},$$

$$(116)$$

respectively. It is a simple matter to show that (116) in fact satisfies (7).
Consider now a flow past a nearly spherical surface given by

$$R = a(1 - \epsilon \sin^2 \theta), \qquad \epsilon \ll 1. \qquad (117)$$

The flow is assumed to have a velocity uniform at infinity and parallel to
the axis of symmetry and of magnitude U. Since the stream function
for flow past a sphere is given by (32) and (34), we shall take

$$\psi = \psi_0 + \epsilon \psi_1, \qquad (118)$$

in which $\qquad\qquad \psi_0 = \tfrac{1}{2}U\left(R^2 - \frac{a^3}{R}\right) \sin^2 \theta. \qquad (119)$

The requirement that the surface of the body be a surface of streamlines
demands that[1]

$$\psi_0(a,\theta) - \epsilon a \sin^2 \theta \frac{\partial}{\partial R} \psi_0(R,\theta) + \epsilon \psi_1(a,\theta) = 0, \qquad (120)$$

the constant being fixed by the value (zero) of the streamline along the
axis of symmetry, since the perturbation does not affect the flow at
infinity. Since $\psi_0(a,\theta)$ is zero, (120) gives

$$\psi_1(a,\theta) = \frac{3Ua^2}{2} \sin^4 \theta. \qquad (121)$$

Now we have to choose the form for $\psi_1(R,\theta)$ so that it satisfies (7), is
finite for finite R, assumes the value given by (121) on $R = a$, and vanishes
for infinite R. The form of ψ given by the second equation in (116) is
appropriate if we take $n = 1$ and 3 and combine the results. Thus,
absorbing some constants in A and B, we have

$$\psi_1 = \frac{A}{R^3} \sin^2 \theta \, (15\mu^2 - 3) + \frac{B}{R} \sin^2 \theta.$$

---

[1] Note that in (120) the middle term represents the change of ψ_0 as R varies from a to
the actual surface along a *normal* to the surface $R = a$.

Equation (121) demands that

$$\frac{A}{a^3} (15\mu^2 - 3) + \frac{B}{a} = \frac{3Ua^2}{2} \sin^2 \theta = \frac{3Ua^2}{2} (1 - \mu^2),$$

or
$$A = - \frac{Ua^5}{10}, \qquad B = \frac{6Ua^3}{5}.$$

This determines ψ_1 and hence ψ in (118).

14. TWO-DIMENSIONAL IRROTATIONAL FLOWS, THE COMPLEX POTENTIAL

One could adopt the point of view that since point singularities can be integrated into line singularities, two-dimensional flows can be obtained from three-dimensional ones by integration. This is a valid but not very fruitful point of view, for the theory of functions of a complex variable is especially suited to the investigation of two-dimensional irrotational flows, and discussion of such flows should be conducted by means of that theory.

In Cartesian coordinates x and y, Eqs. (1.31) and (3.18) give

$$\frac{\partial \phi}{\partial x} = \frac{\partial \psi}{\partial y}, \qquad \frac{\partial \phi}{\partial y} = - \frac{\partial \psi}{\partial x}, \tag{122}$$

in which ψ is now the stream function for two-dimensional flows, since either side of the first equation is equal to u and either side of the second equation is equal to v, u and v being the velocity components in the directions of increasing x and y, respectively. From (122) it follows by cross differentiation that

$$\nabla^2 \phi = 0 \qquad \text{and} \qquad \nabla^2 \psi = 0, \tag{123}$$

in which ∇^2 is now the Laplacian operator in the two variables x and y

$$\nabla^2 = \frac{\partial^2}{\partial x^2} + \frac{\partial^2}{\partial y^2}.$$

It is evident that the first equation in (123) is a special form of (1a) when ϕ is independent of z, but the second equation in (123) is worthy of note. The stream function satisfies the Laplace equation only in two-dimensional flow. For axisymmetric flows the Stokes stream function satisfies (5) and (7), neither of which is the Laplace equation.

Equations (122) are called the *Cauchy-Riemann equations*. With the meaning assigned to ϕ and ψ, they provide the connection between two-dimensional irrotational flows and the theory of functions of a complex variable. To see this, consider the function

$$w = \phi(x,y) + i\psi(x,y). \tag{124}$$

In this equation w is no longer the z component of the velocity. Since the flow is two-dimensional, this meaning of the symbol, according to well-established usage, is not likely to cause confusion. We shall show that, because of (122), the w defined by (124) possesses in any region in which (122) is valid a unique derivative with respect to the complex variable

$$z = x + iy. \tag{125}$$

Again, since only two-dimensional flows are considered, the new meaning of z is not likely to cause confusion. Henceforth w and z are understood to represent the right-hand sides of (124) and (125), respectively. At any point z_0, where

$$z = z_0 = x_0 + iy_0,$$

w possesses a unique derivative with respect to z if $\Delta w / \Delta z$ possesses a unique limit as $|\Delta z| \to 0$, whatever the direction of the line representing Δz in the xy plane. Now[1]

$$\Delta w = \frac{\partial \phi}{\partial x} \Delta x + \frac{\partial \phi}{\partial y} \Delta y + i \left(\frac{\partial \psi}{\partial x} \Delta x + \frac{\partial \psi}{\partial y} \Delta y \right) + O[|\Delta z|^2], \tag{126}$$

provided the first derivatives of ϕ and ψ with respect to x and y exist in the neighborhood surrounding z_0, and they do if (122) is satisfied in that neighborhood, since the satisfaction of (122) would have no meaning if they did not exist. Now because of (122) we can write (126) as

$$\Delta w = \left(\frac{\partial \phi}{\partial x} + i \frac{\partial \psi}{\partial x} \right) \Delta z + O[|\Delta z|^2],$$

which means

$$\lim_{\Delta z \to 0} \frac{\Delta w}{\Delta z} = \frac{dw}{dz} = \frac{\partial}{\partial x} (\phi + i\psi), \tag{127}$$

whatever the direction along which z_0 is approached by z. If a unique derivative of w exists in a region, higher derivatives also exist. This is an important result in the theory of functions, but we shall not attempt to prove it here. The reader is referred to standard books on the theory of functions of a complex variable. Thus if a unique first derivative of w exists in a region, it is an analytic function, although in some other places, called *singularities*, it may cease to have a derivative.

The reverse theorem, that the Cauchy-Riemann equations are satisfied by the real and imaginary parts of an analytic function of z, is even easier to prove. Thus, if a unique dw/dz exists, it must be equal to

---

[1] The last term merely means a quantity which, after division by $(\Delta z)^2$, remains bounded as Δz approaches zero.

both

$$\frac{\partial \phi}{\partial x} + i \frac{\partial \psi}{\partial x} \quad \text{and} \quad \frac{1}{i}\left(\frac{\partial \phi}{\partial y} + i \frac{\partial \psi}{\partial y}\right).$$

Equating these, we have (122).

The practical significance of what has been shown in this section is as follows. If one writes down any analytic function of z, separates it into its real and imaginary parts, and calls them ϕ and ψ, then ϕ and ψ automatically qualify as the velocity potential and the stream function, respectively, of a two-dimensional irrotational flow of an incompressible homogeneous fluid. In the spirit of the inverse method of solution, one can already write down any analytic function w of z and see what problem of potential flow it solves. The function w is called the *complex potential*.

Since

$$u = \frac{\partial \phi}{\partial x} \quad \text{and} \quad v = -\frac{\partial \psi}{\partial x},$$

(127) can be written as

$$\frac{dw}{dz} = u - iv. \tag{128}$$

Thus $u - iv$, but not $u + iv$, is an analytic function of z, since the derivative of an analytic function is an analytic function.

15. THE FLOW NET

The lines of constant potential are given by

$$\phi(x,y) = \text{const}, \tag{129}$$

and the streamlines are given by

$$\psi(x,y) = \text{const}. \tag{130}$$

On a potential line

$$\frac{\partial \phi}{\partial x} dx + \frac{\partial \phi}{\partial y} dy = 0, \tag{131}$$

and on a streamline

$$\frac{\partial \psi}{\partial x} dx + \frac{\partial \psi}{\partial y} dy = 0. \tag{132}$$

From (131), (132), and (122) it follows that

$$\left.\frac{dy}{dx}\right|_{\phi} = -\left.\frac{dx}{dy}\right|_{\psi}, \tag{133}$$

with the subscripts denoting the variable that is kept constant in each case. Equation (133) states that the potential lines and the streamlines are orthogonal at all points in the flow.

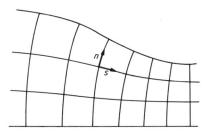

FIGURE 7.   FLOW NET.

If at any point we measure s along the streamlines and n along the potential lines, as shown in Fig. 7, and identify them as local Cartesian coordinates, (122) can be written as

$$\frac{\partial \phi}{\partial s} = \frac{\partial \psi}{\partial n}, \qquad \frac{\partial \phi}{\partial n} = -\frac{\partial \psi}{\partial s} = 0. \tag{134}$$

If we plot potential lines with uniform increment $\Delta\phi$ from one potential line to the next, plot streamlines with uniform increment $\Delta\psi$, and make $\Delta\phi$ and $\Delta\psi$ equal, the approximate form of the first equation in (134) becomes

$$\Delta s = \Delta n,$$

in which Δn and Δs correspond to $\Delta\psi$ and $\Delta\phi$, respectively. Thus the flow field can be described by a net of small squares, as shown in Fig. 7. Since $\Delta\phi$ and $\Delta\psi$ have been chosen according to the first equation in (134), the speed, equal to $\partial\phi/\partial s$, is inversely proportional to the side of the local square.

The flow net just described can be used to solve the Laplace equation in two dimensions graphically. One requires the fixed boundaries to be streamlines, determines the positions of streamlines and potential lines at stations far upstream or far downstream, and endeavors to fill the flow region with an orthogonal net of "curvilinear squares" making frequent use of erasers. The finished net provides an approximate determination of the velocity field. This procedure, useful two or three decades ago, has been superseded by methods of numerical computation made attractive by the availability of high-speed computers and perhaps even by the method of relaxation applied without an electronic computer.

16.   THE IDEA OF CONFORMAL MAPPING

The flow described by

$$w = f(z) \tag{135}$$

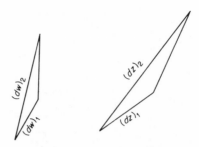

FIGURE 8.   CONFORMAL MAPPING.

represents a mapping between the xy plane, or z plane, and the $\phi\psi$ plane, or w plane. For, in differential form, (135) can be written as

$$dw = f'(z)\,dz.$$

At any point $z_0, f'(z_0)$ is a complex number. Hence, for two differentials $(dz)_1$ and $(dz)_2$, the corresponding differentials in w are related by

$$\frac{(dw)_1}{(dz)_1} = \frac{(dw)_2}{(dz)_2} = f'(z_0) = ae^{i\alpha}.$$

Thus the line segment represented by $(dw)_1$ is obtained from that represented by $(dz)_1$ by a stretching (with factor a) and a counterclockwise turning through an angle α. The same is true for $(dw)_2$ and $(dz)_2$. The triangle having $(dz)_1$ and $(dz)_2$ as two of its sides transforms to a triangle having $(dw)_1$ and $(dw)_2$ as its sides. Hence the triangle in the w plane is magnified (or shrunk) by a factor a and turned through an angle α (see Fig. 8, in which increments are used for differentials). The forms of the two triangles are similar. Since this is true for any infinitesimal triangle in the z plane and its counterpart in the w plane, the flow net in the z plane is mapped conformally into the rectangular grid made by the ϕ lines and ψ lines in the w plane. Hence the term *conformal mapping*. Any analytic relationship between two complex variables represents a conformal mapping.

17. ELEMENTARY TWO-DIMENSIONAL IRROTATIONAL FLOWS

Elementary two-dimensional irrotational flows are uniform flows and those due to sources, sinks, doublets, vortices, and higher-order singularities. They are also elemental in the sense that many flows can be constructed from them by superposition.

The flow due to a source of strength m is given by

$$w = m \ln z, \tag{136}$$

because
$$\phi = m \ln r, \qquad \psi = m\theta,$$

and the velocity is radial everywhere and equal to m/r. The source is situated at the origin, and the discharge is $2\pi m$ per unit length in the direction perpendicular to the xy plane. If the source is situated at $z = z_0$, (136) should be replaced by

$$w = m \ln (z - z_0). \tag{136a}$$

The transformation for a sink is identical to (136) or (136a), but m is negative for a sink.

The transformation

$$w = \frac{i\Gamma}{2\pi} \ln (z - z_0) \tag{137}$$

represents the flow due to a vortex situated at $z = z_0$. The circulation around it is Γ and is clockwise. Inspection of (136a) and (137) reveals that the potential lines for the flow due to a source are the streamlines for the flow due to a vortex, and vice versa.

Since the derivative of $\ln z$ is $1/z$, a discussion similar to that given in Sec. 5.3 will demonstrate that

$$w = -\frac{\mu}{z} \tag{138}$$

represents the flow due to a doublet of strength $|\mu|$ with its axis pointing in the direction of the complex number μ.

Successive differentiations of (136a) produce complex potentials of the form $(z - z_0)^{-n}$, $n = 2$, 3, etc., which correspond to higher singularities at z_0.

The transformation for uniform flow with velocity components U and V is simply

$$w = (U - iV)z. \tag{139}$$

The flows just discussed can be superposed to form new flows. For instance, flow past a two-dimensional half-body can be constructed by superposing the flow due to a source on a uniform flow, in the manner described in Sec. 7.1. Rankine bodies can be created in the manner of Sec. 7.2. Von Kármán's method can be applied equally well to two-dimensional flows. If there are plane walls, the flow caused by a singularity is obtained by superposing the flow due to it in the absence of the walls on the flows due to its images (direct or indirect) in these walls, as described briefly in Sec. 10. We shall not stop to present examples.

18. FLOW PAST A CIRCULAR CYLINDER

If a uniform flow with a velocity of magnitude U in the direction of increasing x is superposed on the flow due to a doublet, the resultant flow is described by

$$w = Uz - \frac{\mu}{z} = \left(U - \frac{\mu}{x^2 + y^2}\right)x + iy\left(U + \frac{\mu}{x^2 + y^2}\right). \quad (140)$$

Thus

$$\psi = y\left(U + \frac{\mu}{x^2 + y^2}\right), \quad (141)$$

and the streamline $\psi = 0$ consists of the branches

$$y = 0 \quad \text{and} \quad x^2 + y^2 = -\frac{\mu}{U}.$$

If

$$\mu = -Ua^2, \quad (142)$$

then the circle

$$x^2 + y^2 = a^2 \quad (143)$$

will be part of the streamline $\psi = 0$. The flow is uniform at infinity and there are no singularities outside the cylinder of cross section (143), Hence (140) and (142) give the flow past a circular cylinder (Fig. 9a).

If the flow due to the cylinder moving with speed U in the negative x direction is desired, the complex potential is simply

$$w = \frac{Ua^2}{z}. \quad (144)$$

The boundary condition at infinity (quiescence) is satisfied, and that at the surface of the cylinder is satisfied because (144) differs from (140) only in a uniform flow of speed U in the negative x direction for *both* fluid and solid, and since (140) already satisfies the boundary condition on the surface (that the normal component of the fluid and that of the solid on the surface must be equal), this difference does not affect that satisfaction at all. The boundary condition on the surface is, for the case of the moving cylinder,

$$\frac{\partial \phi}{\partial r} = -U \cos \theta = -\frac{xU}{a}. \quad (145)$$

The reader is urged to convince himself that (144) satisfies (145).

If there is a *clockwise* circulation Γ around the cylinder, (140) should be replaced by

$$w = U\left(z + \frac{a^2}{z}\right) + \frac{i\Gamma}{2\pi} \ln z. \quad (146)$$

Obviously the last term, which by itself would have circular streamlines, does not disqualify the circle (143) as a streamline. The flow patterns are shown in Fig. 9b, c, and d.

19. THE CIRCLE THEOREM

To illustrate the image of a singularity in a curved surface we shall present the circle theorem. If $f(z)$ is the complex potential in the absence of rigid

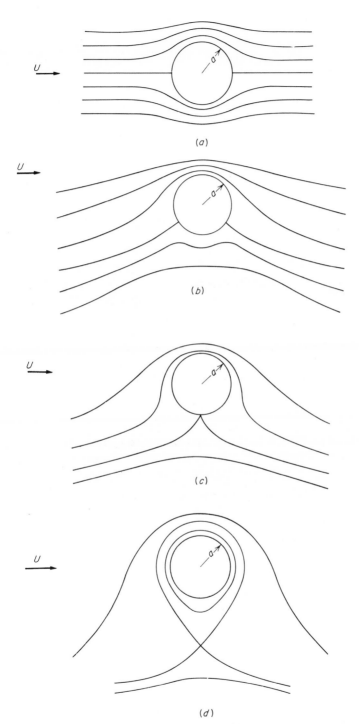

FIGURE 9. FLOW PAST A CIRCULAR CYLINDER. (*a*) WITHOUT CIRCULATION; (*b*) WITH CLOCKWISE CIRCULATION LESS THAN $4\pi aU$; (*c*) WITH CIRCULATION EQUAL TO $4\pi aU$; (*d*) WITH CIRCULATION MORE THAN $4\pi aU$.

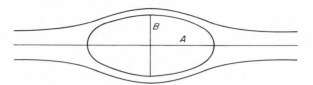

FIGURE 10. FLOW PAST AN ELLIPTIC CYLINDER AT ZERO INCIDENCE.

boundaries and $f(z)$ has no singularities within the circle $|z| = a$, then after that circle is introduced, the complex potential is

$$w = f(z) + f^*\left(\frac{a^2}{z}\right),$$ (147)

in which f^* is the complex conjugate form of f. (This means that if their arguments are identical and real but otherwise arbitrary, the two functions are complex conjugates.)

To prove the theorem, we note first that w given by (147) has no singularity outside the circle other than those of $f(z)$, which are of course allowable. This is because $f(z)$ has no singularities within the circle, and hence $f^*(a^2/z)$ has none without. Second, on the circle

$$w = f(z) + f^*(z^*) = \text{real},$$

and hence $\psi = 0$. Therefore the circle is a streamline, as required.

Note that if $f(z)$ is Uz, w is given by (141) and (142).

20.   SUCCESSIVE TRANSFORMATIONS

20.1.   Flow past an Elliptic Cylinder

Often the flow past a given body or the flow with prescribed boundaries is found by two or several transformations in succession. Consider the flow with velocity U at infinity (Fig. 10) past an elliptic cylinder with the major axis of its cross section coinciding with the x axis. The semimajor axis is A, and the semiminor axis is B.

We start with

$$w = U\left(z' + \frac{a^2}{z'}\right).$$ (148)

If we succeed in transforming the circle $|z'| = a$ into the ellipse without disturbing the condition at infinity, the solution is obtained. Now we know that

$$z = z' + \frac{a^2}{z'}$$

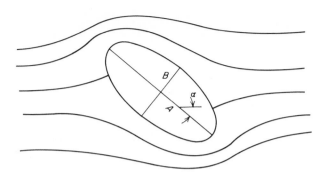

FIGURE 11. FLOW PAST AN ELLIPTIC CYLINDER AT AN ANGLE OF INCIDENCE α.

transforms the circle into a slit $y = 0$, $|x| \leq 2a$, which is a kind of ellipse but too flat for our purpose. Perhaps if we transform a smaller circle $|z'| = b$ into a slit, the larger circle $|z'| = a$ might become an ellipse. Try then

$$z = z' + \frac{b^2}{z'}. \qquad (149)$$

On the circle $|z'| = a$, $z' = ae^{i\theta}$. Hence

$$z = \left(a + \frac{b^2}{a}\right) \cos\theta + i\left(a - \frac{b^2}{a}\right) \sin\theta,$$

which means

$$\frac{x^2}{(a + b^2/a)^2} + \frac{y^2}{(a - b^2/a)^2} = 1. \qquad (150)$$

Thus the circle is indeed transformed into an ellipse, and

$$A = a + \frac{b^2}{a}, \qquad B = a - \frac{b^2}{a}, \qquad (151)$$

which allows one to solve for a and b in terms of A and B. Equations (148) and (149) represent successive transformations and together give the flow required.

If the ellipse has its major axis inclined as shown in Fig. 11, (149) should be replaced by a mapping that transforms the circle $|z'| = b$ into a line inclined at an angle $-\alpha$. The replacement is

$$z'' = z''' + \frac{b^2}{z'''}, \qquad (152)$$

in which

$$z'' = ze^{i\alpha}, \qquad z''' = z'e^{i\alpha}. \qquad (153)$$

The transformations (152) and (153) simply amount to the transformation (149) performed for coordinate axes rotated from their original positions

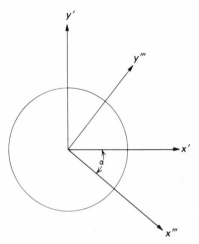

FIGURE 12.   TRANSFORMATION BETWEEN z' AND z'''.

through an angle $-\alpha$.   Combination of (152) and (153) produces

$$z = z' + \frac{b^2 e^{-i2\alpha}}{z'}.\tag{154}$$

The axes for x'', y'', x''', and y''' are shown in Figs. 12 and 13.

It is important to realize that for large $|z'|$ the transformation (149) or (154) does not alter the uniform velocity U.   Had the coefficient of z' in these equations been other than unity, that uniform velocity would have been altered.

If there is circulation around the elliptic cylinder, we merely have to start from

$$w = U\left(z' + \frac{a^2}{z'}\right) + \frac{i\Gamma}{2\pi}\ln z'\tag{155}$$

instead of (148).   It will be shown that the lift on the cylinder is equal to $\rho U\Gamma$.

For every value of z in (149) there are two of z'.   What does this signify?   It is immediately clear that not only has the point $|z'| = \infty$ transformed to the point $|z| = \infty$ but the point $|z'| = 0$ has done the same. Since the circle $|z'| = b$ has gone over to the slit $y = 0$, $|x| \leq 2b$, it is not difficult to see that the inside of that circle has also been mapped onto the whole z plane.   The flow just discussed corresponds to the outside of the circle $|z'| = b$.   The flow corresponding to the inside is a rather unrealistic flow with no closed body in it.   It would be instructive for the reader to sketch this unrealistic flow.   What has been said for the ellipse (150) applies also to the inclined ellipse.

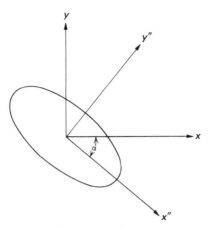

FIGURE 13. TRANSFORMATION BETWEEN z AND z''.

There is a noteworthy point about these successive transformations. Although the flows in the successive places (z' plane and z plane in the examples just given) seem different, some quantities remain unchanged at corresponding points or along corresponding circuits. For instance, ϕ and ψ at corresponding points in the z' plane and the z plane (and any intermediate planes there might be) are the same. If ϕ is multivalued at a point in the z' plane, which indicates that there is circulation around that point, the circulation around the corresponding point in the z plane will be the same [see the several lines following (3.18)]. Similarly, if there is a source of strength m in the z' plane, ψ will increase by the amount $2\pi m$ when the same point is reached after a circuit around the source has been traveled in the counterclockwise direction. Thus if there is a source in the z' plane, there is one of the same strength at the corresponding point in the z plane. These facts might have been in the Jeffreys' mind when they began the chapter on conformal mapping in their book [Jeffreys and Jeffreys (1956)] with a quotation from *Les Guêpes* by Alphonse Karr: "Plus ça change, plus c'est la même chose."

There are, however, apparent exceptions to the situation just described. For instance, for flow past the ellipse (150), the line segment

$$y = 0, \qquad 0 \geq x \geq -2b,$$

is a line source of variable strength, and the line segment

$$y = 0, \qquad 0 \leq x \leq 2b,$$

is a line sink antisymmetric with the line source. But in the z plane these line segments have as their counterpart the circle $|z'| = b$, on which there are no sources or sinks. The apparent contradiction is removed if we realize that as the circle $|z'| = b$ collapses into the line

$$y = 0, \qquad |x| \leq 2b, \tag{156}$$

the flow through the circle $|z'| = b$ must flow out of (or into) the slit (156) in the z plane. The slit (156) is the branch cut where the two Riemann sheets associated with z are joined. Where the flow is out of the slit in one sheet, it is into the slit in the other sheet, and vice versa. Thus if we consider the two sheets *together*, there are no sources or sinks along the slit. Indeed, if a small loop is described around a point $z' = be^{i\alpha}$, the corresponding loop in the z plane is double, with one loop in each sheet. If we start at a point near the corresponding point $z = c$ ($|c| < 2b$) and come back to the starting point after following the double loop around $z = c$, we shall indeed find ϕ and ψ unchanged. The apparent contradiction is therefore only superficial.

20.2.   Flow past an Inclined Plate and the Cisotti Paradox

If $b = a$, (154) and (155) together describe a potential flow past a flat plate of length $4a$ inclined at an angle $-\alpha$ with the x axis. We shall show in a later section that the lift is $\rho U\Gamma$, as for the circular cylinder, and that the drag is zero. Since the pressure on either side of the plate acts in a direction normal to the plate, it seems that the existence of a lift would entrain the existence of a drag. But there is a lift and no drag. This is called the *Cisotti paradox* by Birkhoff (1953), after the person who was first perplexed about the situation. Actually it is not much of paradox, since it is quite evident that there can be a finite force acting at the leading edge (but not at the trailing edge if we specify a Γ which is just sufficient to make the velocity tangent to the plate at the trailing edge). In fact, this force is along the plate pointing upstream and just sufficient to make the drag zero and at the same time, together with the finite pressure distribution, to give the correct lift. We are, of course, speaking of the idealized situation. The plate must be infinitely thin, there must be no separation, and the fluid is supposed to be inviscid and capable of sustaining infinite tension. But the Cisotti paradox is considered a paradox under these conditions. It is resolved under the same conditions.

20.3.   Flow past a Circular Arc

From the transformation

$$z = \zeta + \frac{b^2}{\zeta}, \tag{157}$$

in which now (Fig. 14)

$$m = a \cos \beta, \qquad b = a \sin \beta, \qquad \zeta = z' + mi, \tag{158}$$

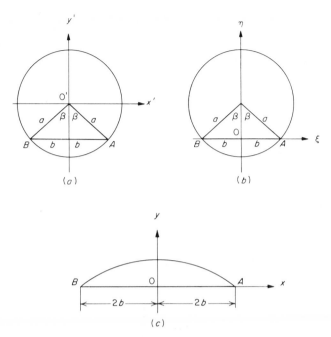

FIGURE 14.   TRANSFORMATION OF A CIRCLE INTO A CIRCULAR ARC.
(a) z' PLANE; (b) ζ PLANE, $\zeta = \xi + i\eta$; (c) z PLANE.

one easily obtains

$$(z - 2b)\zeta = (\zeta - b)^2, \qquad (z + 2b)\zeta = (\zeta + b)^2,$$

and hence
$$\frac{z - 2b}{z + 2b} = \left(\frac{\zeta - b}{\zeta + b}\right)^2. \tag{159}$$

By writing $\zeta - b = r_1 e^{i\theta_1}$, etc., it can be shown that the circle $|z'| = a$ transforms into a circular arc of chord length $4b$, as shown in Fig. 14c. Both the upper arc AB and lower arc AB in the z' plane transform to the arc AB in the z plane.

The mappings that transform the circle $|z'| = a$ into an inclined arc can be obtained in the same way that (154) is obtained.   Combining (148) with (157) and (158), we obtain the necessary transformations for the flow past a circular arc.   The flow past an inclined arc can be obtained in a similar way.

20.4.   Joukowsky Airfoil

In the last subsection we have seen that (157) and (158) transform the circle with radius a to a circular arc of chord length $4b$.   Now suppose that

$$\zeta = z' + me^{i\delta}, \qquad \delta \neq \frac{\pi}{2}, \tag{160}$$

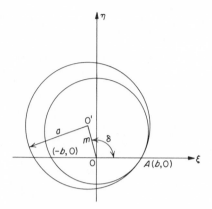

FIGURE 15. ζ PLANE FOR THE JOUKOWSKY
AIRFOIL.

so that O, the origin of ζ, is no longer directly under O', the origin of z'. Equation (157) will still transform the smaller circle in Fig. 15 into a circular arc of chord length $4b$, in spite of the fact that we no longer denote the radius of that circle by a. The circle with radius a is tangent to the smaller circle at A and contains it. Thus (157) must transform the circle with radius a into something which contains the circular arc (the transform of the smaller circle) and moreover is tangent to the arc at point A in the z plane, since tangency is a property preserved by conformal transformations. That "something" is called a *Joukowsky airfoil* (see Fig. 16). The reader can take appropriate values for m, δ, and b and obtain the exact form of the airfoil from (157), (160), and

$$|z'| = a.$$

The flow past a Joukowsky airfoil with zero angle of inclination is then given by (155), (157), and (160). If the angle of inclination is $-\alpha$ (or if the velocity of approach makes an angle α with the x'' axis), we first note that

$$z'' = \zeta + \frac{b^2}{\zeta} \quad \text{and} \quad \zeta = z''' + me^{i\delta} \tag{161}$$

transform the circle $|z'''| = a$ into a Joukowsky airfoil in the z'' plane, with the circular arc inside the airfoil having a chord along the x'' axis. Then it is easy to see that (155), (153), and (161) are the necessary transformations for the flow, parallel to the x axis at infinity, past an airfoil with an angle of inclination $-\alpha$.

Since the airfoil has a cusp at the trailing edge (denoted by A with little risk of confusion with the A in the plane of the intermediate variables),

FIGURE 16. THE JOUKOWSKY AIRFOIL.

the velocity there will be infinite in the framework of the theory of potential flow if the fluid flows around the corner. But with the right circulation the fluid does not have to flow around the corner and can have a
velocity tangent to the airfoil at the cusp. This circulation is found in
the following way.

The speed at the cusp is given by $|dw/dz|$, and

$$\frac{dw}{dz} = \frac{dw}{dz'} \frac{dz'}{d\zeta} \frac{d\zeta}{dz}.$$

Now $dz'/d\zeta$ is unity, and $d\zeta/dz$ is infinite at A, since

$$\frac{dz}{d\zeta} = 1 - \frac{b^2}{\zeta^2} = 0 \qquad \text{at } \zeta = \pm b.$$

Thus dw/dz can be finite only if dw/dz' is zero at A. But from (155)

$$\frac{dw}{dz'} = U\left(1 - \frac{a^2}{z'^2}\right) + \frac{i\Gamma}{2\pi z'}. \tag{162}$$

At A

$$z''' = -iae^{i\beta}, \tag{163}$$

and

$$z' = e^{-i\alpha}z''' = -iae^{i(\beta-\alpha)}.$$

The condition that dw/dz' vanish at A then becomes

$$U(1 + e^{i2(\alpha-\beta)}) - \frac{\Gamma}{2\pi a} e^{i(\alpha-\beta)} = 0,$$

which gives

$$\Gamma = 4\pi a U \cos(\alpha - \beta). \tag{164}$$

This then is the circulation (in the z' plane or in the physical plane
or z plane) needed. But how does the airfoil manage to enjoy this
particular circulation? If the airfoil started from rest and therefore had
no circulation around it, how could it manage to develop any circulation
at all? To answer these questions, we first remember that the fluid is
viscous—however small its viscosity or however large the Reynolds number
is compared with 1—and any slight viscosity will cause the fluid to separate
at the trailing edge so long as the velocity there is not tangent to the airfoil.
An eddy is then formed, as shown in Fig. 17a, with a circulation Γ
around it. The eddy soon detaches itself from the airfoil, leaving a
circulation $-\Gamma$ (or Γ in the clockwise direction) around the airfoil[1], since

[1] Just outside the boundary layer.

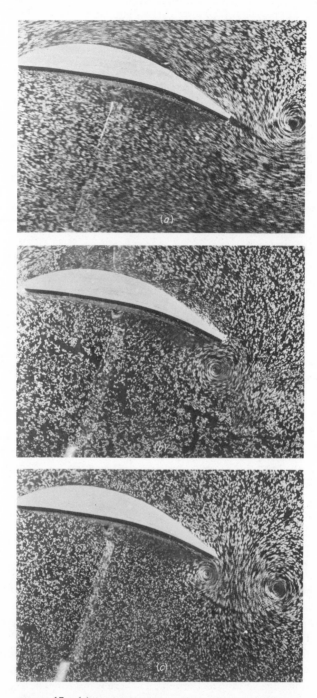

FIGURE 17. (*a*) FORMATION OF COUNTERCLOCKWISE EDDY AT TRAILING EDGE PRODUCING A CLOCKWISE CIRCULATION AROUND THE AIRFOIL AS IT STARTS TO MOVE. (*b*) FORMATION OF A CLOCKWISE EDDY AS THE AIRFOIL STOPS. (*c*) THE TWO EDDIES FORMED AS THE AIRFOIL STARTS AND THEN STOPS. (*Courtesy of Professor W. R. Debler.*)

the circulation along a large closed loop containing the airfoil and the detached eddy remains zero because the viscous effects are not felt so far away. This process stops only after the proper circulation has been developed and the velocity at the trailing edge is finite and tangent to the airfoil. Figure 17b shows the formation of a clockwise eddy as the airfoil stops, and Fig. 17c shows the two eddies, one formed as the airfoil starts and the other as it stops.

21. ADDED MASSES FOR TWO-DIMENSIONAL FLOWS

Since the Taylor theorem, as improved by Birkhoff (1953) and Landweber (1956), enables one to compute the added masses (except those for pure rotation) for three-dimensional bodies from the internal singularities, the corresponding theorem for two-dimensional flows can be obtained by integration. It is, however, much simpler to derive the two-dimensional counterpart of Taylor's theorem directly. If α denotes the subscripts, 1, 2, and 6, in the notation of Sec. 11.1, s denoting the distance along the boundary of the body, and $z = x_1 + ix_2$, (75) can be written, with dS changed to ds, as

$$A_{\alpha 1} + iA_{\alpha 2} = -\rho \oint \phi_\alpha n_1 \, ds - i\rho \oint \phi_\alpha n_2 \, ds = i\rho \oint \phi_\alpha \, dz$$

$$= i\rho \left(\oint w_\alpha \, dz - i \oint \psi_\alpha \, dz \right) = i\rho \oint w_\alpha \, dz - \rho \oint zn_\alpha \, ds,$$

because $\qquad n_1 \, ds + in_2 \, ds = dy - i \, dx = -i \, dz,$

$$\frac{\partial \psi_\alpha}{\partial s} = n_\alpha \quad \text{and} \quad \oint \psi_\alpha \, dz = -\oint z \frac{\partial \psi_\alpha}{\partial z} \, dz$$

$$= -\oint z \frac{\partial \psi_\alpha}{\partial s} \, ds = -\oint zn_\alpha \, ds.$$

But $\qquad \rho \oint zn_\alpha \, ds = B_{\alpha 1} + iB_{\alpha 2},$

in which $\qquad B_{\alpha 1} = \rho \oint x_1 n_\alpha \, ds, \qquad B_{\alpha 2} = \rho \oint x_2 n_\alpha \, ds,$

and in particular

$$B_{ij} = B\delta_{ij}, \qquad i, j = 1, 2,$$

$$B_{61} = -B\bar{x}_2, \qquad B_{62} = B\bar{x}_1,$$

with B as the displaced mass per unit length and $\bar{x}_1$ and $\bar{x}_2$ denoting the coordinates of the mass center of the cross section of the two-dimensional body. Thus

$$A_{\alpha 1} + B_{\alpha 1} + i(A_{\alpha 2} + B_{\alpha 2}) = i\rho \oint w_\alpha \, dz. \tag{165}$$

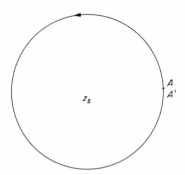

FIGURE 18. INTEGRATION AROUND
A SOURCE.

By the use of theorems in the theory of functions of a complex variable, the preceding equation can be written as

$$A_{\alpha 1} + B_{\alpha 1} + i(A_{\alpha 2} + B_{\alpha 2}) = 2\pi\rho \left[\int \sigma_\alpha z \, dA + \sum (m_\alpha z_s + \mu_\alpha) \right], \quad (166)$$

in which σ_α is the strength of sources distributed over an area A within the body, m_α is the strength of isolated sources, and μ_α is the strength of isolated doublets—all per unit length of the cylinder and for the αth motion.   Equation (165) is the two-dimensional counterpart of (86).   To demonstrate that (166) is equivalent to (165), we note that if there is an isolated doublet of strength μ_α situated at $z = z_d$ (μ_α is generally complex since the axis of the doublet may be tilted), then there is a term

$$-\frac{\mu_\alpha}{z - z_d}$$

in w_α and the contribution of this to the integral term in (165), by Cauchy's integral theorem, is

$$i\rho(-2\pi i\mu_\alpha) = 2\pi\rho\mu_\alpha.$$

If there is an isolated source of strength m_α situated at $z = z_s$ inside the body, there is a term

$$m_\alpha \ln (z - z_s)$$

in w_α, and (Fig. 18)

$$\oint m_\alpha \ln (z - z_s) \, dz = m_\alpha \{(z - z_s)[\ln (z - z_s) - 1]\}_A^{A'} = (z_{A'} - z_s)2\pi i m_\alpha,$$

since $\ln (z - z_s)$ at A' is

$$\ln |z_{A'} - z_s| + 2\pi i.$$

Since the body is closed, the algebraic sum of all sources and sinks for

α must be zero, or

$$\sum m_\alpha = 0$$

for any α, the summation being over all the sources and sinks for that particular α. Since $z_{A'}$ is a fixed number,

$$2\pi i \sum (z_{A'} - z_s)m_\alpha = -2\pi i \sum m_\alpha z_s.$$

(If the notation m_α seems confusing in connection with a summation sign, perhaps one should think of it as $m_{\alpha s}$. Then the summation is over s, or over all the sources for one α.) The contribution of isolated sources **and sinks to the last term in (165) is therefore**

$$2\pi\rho \sum m_\alpha z_s.$$

For distributed sources and sinks then the contributions to the integral in (165) are

$$2\pi\rho \int \sigma_\alpha z \, dA.$$

We recall that since distributed doublets can be reduced to source-and-sink distributions, they need not be considered. We have thus demonstrated the equivalence of (165) and (166).

An equation similar to (166) for the added mass of the body in rotation can also be developed, but such an equation involves a term

$$\oint w_6 z^* \, dz.$$

Since z^* is not an analytic function of z, such a formula is not very useful, and will not be given. If A_{66} is needed, a direct calculation by (75) can be performed.

For the case of translation of a circular cylinder, $\alpha = 1$, and the only singularity is a doublet with strength μ equal to a^2 since (for unit velocity in the x direction)

$$\mu = \mu_1 = a^2.$$

Thus $$A_{11} + B_{11} = A_{11} + B = A_{11} + \rho\pi a^2 = 2\pi\rho a^2,$$

and the added mass is

$$A_{11} = \rho\pi a^2 = B.$$

This result, of course, can be obtained by a direct calculation, without the use of (166).

22. FORCES AND MOMENT IN TWO-DIMENSIONAL FLOWS; BLASIUS THEOREMS

The forces and moment acting on a two-dimensional body can be obtained by integration from the generalized Lagally theorem presented in Sec. 12.

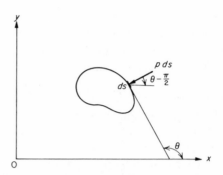

FIGURE 19.   AN ELEMENT OF FORCE EXERTED
ON A CLYINDER BY THE SURROUNDING
FLUID.

However, for steady two-dimensional flows the forces and moment can
be obtained most simply and directly by the Blasius theorems.

With reference to Fig. 19, the components of the force acting on
the surface element ds of the body are

$$dX = -p\,ds\,\cos\left(\theta - \frac{\pi}{2}\right),$$

$$dY = -p\,ds\,\sin\left(\theta - \frac{\pi}{2}\right),$$

in which p is the pressure.   Thus

$$dF = dX + i\,dY = ip\,ds\,e^{i\theta} = ip\,dz,$$

since $dz = ds\,e^{i\theta}$.   Then the complex conjugate of dF is

$$dF^* = dX - i\,dY = -ip\,ds\,e^{-i\theta} = -ip\,dz\,e^{-i2\theta}. \qquad (167)$$

The moment about the origin of the force acting on ds is

$$dM = -y\,dX + x\,dY = p\,ds\left[y\cos\left(\theta - \frac{\pi}{2}\right) - x\sin\left(\theta - \frac{\pi}{2}\right)\right]$$

$$= p\,ds(y\sin\theta + x\cos\theta),$$

which is the real part of

$$pz\,dz\,e^{-i2\theta} = iz\,dF^*.$$

Hence $\qquad\qquad dM + i\,dN = iz\,dF^* = pz\,dz\,e^{-i2\theta}, \qquad (168)$

in which $i\,dN$ is merely Im $(iz\,dF^*)$ and has no physical meaning.

Now, since the contribution of the hydrostatic pressure to forces
and moment can be separately and easily computed, p will be used to

denote only the dynamic pressure. Hence, for steady irrotational flows,

$$p = p_0 - \frac{\rho}{2} q^2,$$

in which q is the speed and p_0 a constant. The contribution of p_0 to forces and moment is of course zero, since the body is closed. Thus one can take

$$p = -\frac{\rho}{2} q^2.$$

Since

$$\frac{dw}{dz} = q e^{-i\theta},$$

the dynamic part of the pressure is

$$p = -\frac{\rho}{2} \left(\frac{dw}{dz}\right)^2 e^{i2\theta}. \tag{169}$$

Substituting (169) into (167) and (168) and integrating, one has

$$X - iY = \frac{i}{2} \rho \oint \left(\frac{dw}{dz}\right)^2 dz, \tag{170}$$

$$M + iN = -\frac{\rho}{2} \oint z \left(\frac{dw}{dz}\right)^2 dz. \tag{171}$$

These are the *Blasius theorems*.

22.1. Application of the Blasius Theorems

To facilitate the application of the Blasius theorems, several important theorems in the theory of functions of a complex variable will be recorded here, some of them without proof.

i THE CAUCHY INTEGRAL THEOREM **If a function $f(z)$ is analytic and single-valued inside and on a simple closed contour C, then**

$$\int_C f(z)\, dz = 0. \tag{172}$$

PROOF. *Let $f(z) = \xi + i\eta$. Then, under the conditions stated and by virtue of the Stokes theorem (1.56),*

$$\int_C f(z)\, dz = \int_C (\xi\, dx - \eta\, dy) + i \int_C (\eta\, dx + \xi\, dy)$$

$$= -\iint_D \left(\frac{\partial \eta}{\partial x} + \frac{\partial \xi}{\partial y}\right) dx\, dy + i \iint_D \left(\frac{\partial \xi}{\partial x} - \frac{\partial \eta}{\partial y}\right) dx\, dy,$$

in which D is the domain enclosed by C. Since f(z) is analytic, the Cauchy-Riemann equations

$$\frac{\partial \xi}{\partial x} = \frac{\partial \eta}{\partial y}, \qquad \frac{\partial \xi}{\partial y} = -\frac{\partial \eta}{\partial x}$$

are valid. Hence the theorem.

The converse of Cauchy's integral theorem is also true, and is known as Morera's theorem. The proof is omitted.

ii CAUCHY'S INTEGRAL   **Under the same conditions as stated for the Cauchy integral theorem, with z_0 inside C,**

$$f(z_0) = \frac{1}{2\pi i} \int_C \frac{f(z)}{z - z_0}\, dz. \tag{173}$$

This result can be generalized to

$$f^{(n)}(z_0) = \frac{n!}{2\pi i} \int_C \frac{f(z)}{(z - z_0)^{n+1}}\, dz, \tag{174}$$

in which $f^{(n)}(z)$ is the nth derivative of $f(z)$.

PROOF.   *For a proof of these results, we note first that, with*

$$z - z_0 = re^{i\theta},$$

on the unit circle C_1 surrounding the point z_0, $dz = ire^{i\theta}\, d\theta$ and

$$\int_{C_1} \frac{dz}{z - z_0} = i \int_{C_1} d\theta = 2\pi i, \tag{175}$$

$$\int_{C_1} \frac{dz}{(z - z_0)^n} = \frac{i}{r^{n-1}} \int_{C_1} e^{i(1-n)\theta}\, d\theta = 0, \qquad n \neq 1. \tag{175a}$$

These formulas are still valid if any closed circuit C containing z_0 inside is substituted for C_1, since there is no singularity between C and C_1. An application of (172) is implicit here. We can, for instance, consider the circuit C and C_1 in Fig. 20. If the doubly connected region between C and C_1 is cut, the sum of the integrals is

$$\int_C + \int_A^B + \int_{B'}^{A'} - \int_{C_1} = \int_C - \int_{C_1} = 0,$$

in which the integrands are omitted. Hence

$$\int_C = \int_{C_1}.$$

Now since $f(z)$ is analytic on and within C, it can be expanded in a power series of $z - z_0$ (see below). Then (173) and (174) follow from (172), (175), and (175a), with C_1 replaced by C in (175) and (175a).

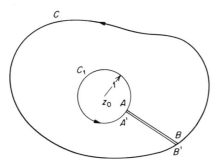

FIGURE 20.   EQUIVALENT CONTOURS.

iii TAYLOR SERIES   **If $f(z)$ is analytic on and inside a simple closed contour C, and if $z = z_0$ is a point inside C, then**

$$f(z) = f(z_0) + (z - z_0)f'(z_0) + \cdots + \frac{(z - z_0)^n}{n!} f^{(n)}(z_0) + \cdots,$$

convergent inside C.

iv LAURENT SERIES   **If $f(z)$ is analytic on and between two concentric circles C and C' with center at $z = z_0$, then $f(z)$ can be expanded in positive and negative powers of $z - z_0$, convergent at all points of the ring-shaped region.**

It may be mentioned that if $f(z)$ can be expanded in the form

$$f(z) = \sum_{n=-\infty}^{\infty} A_n(z - z_0)^n,$$

the integral

$$\int_C f(z)\, dz$$

will, by virtue of (172), (175), and (175a), be equal to $A_{-1} 2\pi i$. [The value can be different from zero because $f(z)$ is not necessarily analytic inside the smaller circle C.] The coefficient A_{-1} is called the *residue* of $f(z)$ at z_0.

We are now ready to give an example of the application of the Blasius theorem. For flow past a circular cylinder with circulation, the complex potential is given by (146), and

$$\left(\frac{dw}{dz}\right)^2 = \left[U\left(1 - \frac{a^2}{z^2}\right) + \frac{i\Gamma}{2\pi z} \right]^2,$$

$$z\left(\frac{dw}{dz}\right)^2 = z\left[U\left(1 - \frac{a^2}{z^2}\right) + \frac{i\Gamma}{2\pi z} \right]^2.$$

Thus the residue of $(dw/dz)^2$ is $i\Gamma U/\pi$, and (170) gives

$$X - iY = \frac{i}{2}\rho\,\frac{i\Gamma U}{\pi}\,2\pi i = -\rho\Gamma U i$$

or $$X = 0, \qquad Y = \rho\Gamma U, \tag{176}$$

which, as mentioned before, can also be obtained by a direct integration of the forces on the cylinder. The residue of $z(dw/dz)^2$ is real, so that the integral in (171), being $2\pi i$ times the residue, is purely imaginary. Thus (171) gives

$$M = 0.$$

This is hardly surprising, since the surface force due to the pressure is directed toward the center of the cylinder.

The second of (176), called the *Kutta-Joukowski theorem*, applies in general to any steady uniform flow (with velocity U at infinity) past a finite body with clockwise circulation Γ. A proof is as follows. Since the body is finite, the algebraic sum of the strengths of sources and sinks (isolated or distributed) is zero and contributes nothing to the residue. Doublets produce terms of the type A/z^2 in dw/dz, and since the flow is uniform at infinity, the term of the highest order in z in dw/dz is U. Thus doublets, too, contribute nothing to the residue of $(dw/dz)^2$. If Γ is the clockwise circulation around the body, it is then obvious that the residue of $(dw/dz)^2$ is $i\Gamma U/\pi$, and the theorem follows. Of course, singularities of higher (negative) orders contribute nothing to the residue, by virtue of (175a).

The Kutta-Joukowski theorem can also be proved by using the momentum principle applied to the fluid entering and leaving the control surface whose cross section is a very large circle C with the body, e.g., an airfoil, near its center. As stated before, the velocity field far from the body is that due to the uniform velocity U and the circulation, since the combined effect of the sources and sinks is that of a doublet, whose effect, like the effects of higher-order singularities, dies out so fast that at C it can be neglected. Let the radius of C be R and the *clockwise* circulation be denoted by Γ. Then at C the velocity components are (Fig. 21)

$$u = U + \frac{\Gamma}{2\pi R}\sin\theta, \qquad v = -\frac{\Gamma}{2\pi R}\cos\theta, \tag{177}$$

and the radial velocity component is

$$u_n = U\cos\theta,$$

which is positive as the fluid leaves and negative as it enters the circular area. Then the net flux of y momentum is

$$M_y = \int_0^{2\pi} \rho v u_n R\,d\theta = -\frac{\rho U\Gamma}{2}.$$

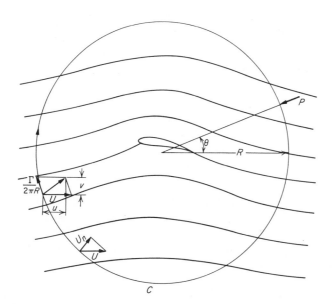

FIGURE 21.   CONTROL SURFACE FOR THE KUTTA-JOUKOWSKI
THEOREM.

The y component of the force acting on C by the fluid on the outside of
it is

$$(F_y)_C = -\int_0^{2\pi} (p \sin \theta)R \, d\theta. \tag{178}$$

But, with gravity neglected (since the buouyancy can be separately
calculated)

$$p = C' - \frac{\rho}{2} (u^2 + v^2), \tag{179}$$

and substitution of (177) into (179) and (179) into (178) yields, upon
neglecting any term in p not varying with θ because it makes no contribution whatever,

$$(F_y)_C = \frac{\rho U \Gamma}{2}.$$

If the lift on the body is denoted by L, the force acting on the fluid in C
by the body is $-L$, and the momentum principle states that

$$(F_y)_C - L = M_y,$$

or

$$L = \rho U \Gamma,$$

which is the Kutta-Joukowski theorem.

23. THE FREE-STREAMLINE THEORY

Problems involving two-dimensional potential flows with straight boundaries or with straight solid boundaries and free streamlines can be solved exactly if gravity effects are neglected for free-streamline flows. If no free boundaries are present, the flow region in the physical plane is a polygon. If free streamlines are present and gravity effects neglected, the flow region in some hodograph plane, which is the plane of

$$\Omega = -\ln \frac{dw}{dz},$$

is a polygon. Since the flow region in the w plane is obviously a special polygon with two angles, it is evident that if we can map a polygon unto the upper half of the plane of some parameter t (which is also a complex variable), then the relationship between z and w, or between dw/dz and w, is parametrically determined. There is a transformation, due to Schwarz and Christoffel, that will perform the task. In this section we shall concentrate on free-streamline flows only, since the method of solution for flows with straight boundaries then becomes obvious, but first we shall study the Schwarz-Christoffel transformation.

If the effect of gravity is to be taken into account, or if either the fluid at infinity or the body moving in the fluid is accelerating, the Schwarz-Christoffel transformation is not useful to deal with free-surface problems, because the domain in the logarithmic hodograph plane (see below) is not determined beforehand. For finite two-dimensional cavities when gravity or acceleration effects are present, see Yih (1960) and the references therein.

23.1. The Schwarz-Christoffel Transformation

The Schwarz-Christoffel transformation transforms the interior of a polygon into the entire upper (or lower) half plane of another variable. Since, as will soon be shown, this transformation is very useful for solving free-streamline problems in hydrodynamics, we shall present it first.

The polygon $A'B'C'D'E'A'$ (Fig. 22a) in the Ω plane (say) can be transformed into the upper half t plane by the Schwarz-Christoffel transformation

$$\Omega = M \int_0^t \frac{dt}{(t-a)^{\alpha/\pi}(t-b)^{\beta/\pi}(t-c)^{\gamma/\pi}(t-d)^{\delta/\pi}(t-e)^{\epsilon/\pi}} + N. \quad (180)$$

It can be seen from (180) that every time t crosses any of the values a, b, c, d, and e, the phase of $d\Omega/dt$, and hence of $d\Omega$ (since t is real except around the semicircles surrounding the points A, B, C, D, and E), will change by α, β, γ, δ, and ϵ. Of the seven constants (a, b, c, d, e, M, and N), three can be chosen arbitrarily and the remaining four determined

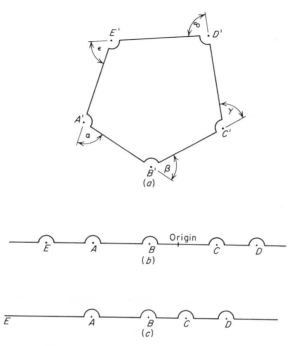

FIGURE 22. SCHWARZ-CHRISTOFFEL TRANSFORMATION. (*a*) POLYGON IN Ω-PLANE; (*b*) t PLANE, TO THE UPPER HALF OF WHICH THE POLYGON IS MAPPED; (*c*) t PLANE WITH E AT INFINITY.

from the positions of A', B', C', and D'—or any four of the five vertices. As long as the polygon is a closed one, so that

$$\alpha + \beta + \gamma + \delta + \epsilon = 2\pi,$$

there is no need to impose the condition that E' be where it should be, for if A', B', C', and D' are where they should be and the lines $E'A'$ and $D'E'$ have the correct inclinations, E' can only be at the place where it should be. The same argument applies to a polygon of n sides. Three (usually not M or N) of the $n + 2$ constants can always be chosen arbitrarily.

One point must be kept in mind. If any of the constants associated with the vertices in the t plane is infinite, the corresponding factor should be dropped. For instance, if e is infinite, the factor

$$(t - e)^{\epsilon/\pi}$$

should simply be dropped from Eq. (180). This can be readily seen from Fig. 22a and c. Let the angles of inclination of the straight lines in the

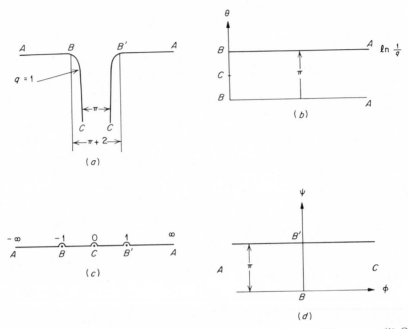

FIGURE 23.   TRANSFORMATION FOR THE JET ISSUING FROM A SLOT.   (a) z PLANE; (b) Ω
PLANE; (c) t PLANE; (d) w PLANE.

Ω plane be referred to $E'A'$, so that the angle of inclination of $E'A'$ is
zero.   According to the transformation

$$\Omega = M \int_0^t \frac{dt}{(t-a)^{\alpha/\pi}(t-b)^{\beta/\pi}(t-c)^{\gamma/\pi}(t-d)^{\delta/\pi}} + N,$$

after the points A, B, C, and D have been passed in the t plane (Fig. 22c),
the angle of inclination of $D'E'$ is

$$\alpha + \beta + \gamma + \delta = 2\pi - \epsilon.$$

Hence $D'E'$ and $E'A'$ must intersect at the angle ϵ, as shown in Fig. 22a.

23.2.   Kirchhoff's Jet

Two examples of free-streamline flow will be presented to illustrate the
use of the Schwarz-Christoffel transformation.   The first concerns a jet
issuing from a slot in an infinite flat plate (Fig. 23).   The flow is assumed
to be steady.   For convenience, the asymptotic width of the jet will be
taken to be π and the speed on the free streamlines unity.   The quantity
Ω is defined to be

$$\Omega = -\ln \frac{dw}{dz} = -\ln qe^{-i\theta} = \ln \frac{1}{q} + i\theta. \tag{181}$$

If gravity effects are neglected, q must be constant along the free stream-lines according to the Bernoulli equation, since the pressure is constant along these streamlines. The boundary $ABCCB'A$ in the physical plane then appears in the Ω plane (the logarithmic hodograph plane) as a poly-gon $ABCB'A$. In the t plane, the abscissas associated with B, C, and B' are chosen arbitrarily to be -1, 0, and 1. By symmetry the point A is mapped into the point at infinity in the t plane, so that the factor associated with the vertex A in the Ω plane is simply ignored in the Schwarz-Christoffel transformation:

$$\Omega = M \int \frac{dt}{\sqrt{t^2 - 1}} + N = M \cosh^{-1} t + N. \tag{182}$$

But
$$\Omega_B = 0 = M \cosh^{-1}(-1) + N = iM\pi + N,$$

$$\Omega_{B'} = -\pi i = M \cosh^{-1} 1 + N = N,$$

so that
$$N = -\pi i, \qquad M = 1, \tag{183}$$

and
$$\Omega = \cosh^{-1} t - i\pi. \tag{184}$$

In the w plane the streamlines are at a distance π apart, because in the physical plane the asymptotic speed of the jet is unity and the asymp-totic width π. The polygon in the w plane can be mapped into the upper half of the t plane by the transformation

$$w = M' \int \frac{dt}{t} + N' = M' \ln t + N'.$$

Since $w_{B'} = \pi i$ and $w_B = 0$ and $\ln t_B$ is πi according to the contour if $\ln t_{B'}$ is zero, one has

$$M' \ln 1 + N' = N' = \pi i,$$

$$M' \ln(-1) + N' = iM'\pi + \pi i = 0,$$

from which
$$M' = -1, \qquad N' = \pi i, \tag{185}$$

and
$$w = -\ln t + \pi i. \tag{186}$$

Actually, since the point C is the origin of the t plane and represents a sink of strength[1] 2π in the t plane, the term $-\ln t$ is to be expected. The term πi is just an additive constant resulting from the arbitrary choice of the origin in the w plane.

Now on the free streamline BC, $w = \phi$, since ψ is zero. If s is measured along BC from B,

$$\phi = s,$$

---

[1] The actual discharge in the jet is π. But for the entire t plane the strength of the sink at C must be 2π in order for the discharge from the upper half plane to be π.

since $q = 1$ on BC. Thus

$$ds = d\phi = dw = -\frac{dt}{t}.$$

But, on BC,

$$\Omega = i\theta, \qquad t = \cosh i(\pi + \theta) = \cos (\pi + \theta) = -\cos \theta,$$

so that

$$ds = \frac{\sin \theta}{\cos \theta} d\theta = \tan \theta \, d\theta.$$

The parametric differential equations for the curve BC are then

$$dx = \cos \theta \, ds = \sin \theta \, d\theta,$$

$$dy = \sin \theta \, ds = \frac{\sin^2 \theta}{\cos \theta} d\theta,$$

which can be integrated to be

$$x = 1 - \cos \theta,$$

$$y = -\sin \theta + \ln \tan \left(\frac{\theta}{2} + \frac{\pi}{4}\right), \tag{187}$$

the constants of integration being determined by taking B as the origin of the physical plane. These equations are the parametric equations for the free streamline BC. As C is approached, θ approaches $-\pi/2$ and x approaches 1. Taking into consideration the symmetry of the jet, the total width of the slot is $\pi + 2$, and the coefficient of contraction is then

$$C_c = \frac{\pi}{\pi + 2} = 0.611. \tag{188}$$

This is an early result of Kirchhoff (1869).

It should be noticed that the t plane serves as a parametric plane into which the Ω plane and the w plane can be mapped. Thus the relationship between w and dw/dz is parametrically determined and hence also the implicit relationship between w and z. On the free streamlines, where $ds = d\phi$, we have seen that dz is determined parametrically by θ and $d\theta$. The applicability of the Schwarz-Christoffel transformation depends crucially on the representation of the flow region into a polygon in the Ω plane, and this is possible only if all the solid boundaries are straight and if gravity effects are neglected on the free streamlines.

In order to prepare for the study of gas jets, we shall record here the expression of ψ in terms of θ and q. This is obtained by writing (184) as

$$t = \cosh (-\ln \zeta + i\pi) = -\frac{1}{2} \frac{\zeta^2 + 1}{\zeta},$$

in which

$$\zeta = \frac{dw}{dz} = qe^{-i\theta}.$$

Then (186) becomes

$$w = \ln \zeta - \ln (1 + \zeta^2) + \ln 2,\tag{189}$$

the imaginary part of which is

$$\psi = -\left[\theta + \sum_{n=1}^{\infty} \frac{(-1)^n}{n} q^{2n} \sin 2n\theta\right].\tag{190}$$

If we let $\theta' = \theta + \pi/2$, so that $\theta' = 0$ on the centerline of the jet and then drop the prime on θ' and the constant $\pi/2$, we have

$$\psi = -\left(\theta + \sum_{n=1}^{\infty} \frac{1}{n} q^{2n} \sin 2n\theta\right).\tag{191}$$

Nothing is changed except the line from which θ is measured, and this change is inconsequential. Equation (191) will be used for comparison with the stream function for a gas jet.

23.3. Deflection of Jets

As a second example we shall present the study by Siao and Hubbard (1953) of the deflection of jets by symmetrically placed V-shaped jets, as shown in Fig. 24a, where many quantities are defined. Since the flow is symmetric, we can consider the lower half only. Let the variables ζ and ζ' be defined by

$$\zeta = \frac{dw}{dz} \quad \text{and} \quad \zeta' = \zeta^{\pi/\alpha},$$

and let both the jet half-width far upstream and the speed on the free streamline be unity. Then the flow regions in the ζ plane and ζ' plane are shown in Fig. 24b and c. The t plane and w plane are shown in Fig. 24d and e. It is evident that the Schwarz-Christoffel transformation gives

$$t = -\tfrac{1}{2}\left(\zeta' + \frac{1}{\zeta'}\right)\tag{192}$$

and

$$w = \frac{1}{\pi}\left[\ln (t + 1) - \ln \left(t + \cos \frac{\pi\beta}{\alpha}\right)\right].\tag{193}$$

If

$$\alpha = \frac{n}{m} \pi,$$

we can introduce

$$\tau = \zeta^{1/n}.\tag{194}$$

Then

$$dz = \frac{dw}{\zeta} = \frac{1}{\zeta} \frac{dw}{dt} \frac{dt}{d\zeta'} \frac{d\zeta'}{d\zeta} \frac{d\zeta}{d\tau} d\tau,$$

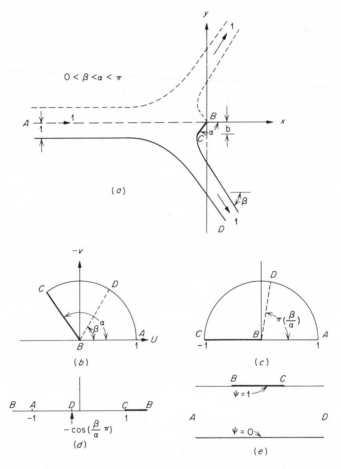

FIGURE 24.   DEFLECTION OF A JET BY A V-SHAPED BODY.   (a) z PLANE;
(b) ζ PLANE; (c) ζ′ PLANE; (d) t PLANE; (e) w PLANE.

and, with the several transformations given,

$$dz = -\frac{m}{\pi}\left(\frac{1}{\tau^m - e^{i\beta m/n}} + \frac{1}{\tau^m - e^{-i\beta m/n}} - \frac{1}{\tau^m - 1}\right)\tau^{m-n-1}\,d\tau. \quad (195)$$

By resolving the expression in parentheses into partial fractions, we have

$$dz = -\frac{1}{\pi}\sum_{r=0}^{m-1}\left[\frac{\exp\left[-i(2r\alpha + \beta)\right]}{\tau - \exp\left[i(2r\alpha + \beta)/n\right]}\right.$$
$$\left. + \frac{\exp\left[-i(2r\alpha - \beta)\right]}{\tau - \exp\left[i(2r\alpha - \beta)/n\right]} - \frac{2\exp\left(-i2r\alpha\right)}{\tau - \exp\left(ir\alpha/n\right)}\right]d\tau. \quad (196)$$

To integrate this, we note that

$$z_B = 0, \qquad \tau_B = 0, \qquad \text{and} \qquad \tau_C = \exp\frac{i\alpha}{n},$$

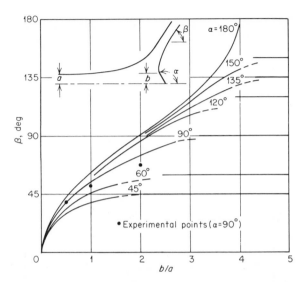

FIGURE 25. EFFECTS OF BOUNDARY WIDTH ON JET DEFLECTION.
THE HALF-WIDTH a IS NOW ARBITRARY. (*From T. T. Siao
and P. G. Hubbard, in J. S. McNown and C.-S. Yih (eds.),
"Free-streamline Analysis of Transition Flow and Jet Deflec-
tion," 1953, by permission of the Iowa Institute of Hydraulic
Research.*)

because $\zeta_C = \exp(i\alpha)$. Hence

$$
z_C = -\frac{1}{\pi} \sum_{r=0}^{m-1} \left\{ \exp\left[-i(2r\alpha + \beta)\right] \ln\left[1 - \exp\frac{i(\alpha - 2r\alpha - \beta)}{n}\right] \right.
$$

$$
+ \exp\left[-i(2r\alpha - \beta)\right] \ln\left[1 - \exp\frac{i(\alpha - 2r\alpha + \beta)}{n}\right]
$$

$$
\left. - 2\exp(-i2r\alpha) \ln\left[1 - \exp\frac{i(\alpha - 2r\alpha)}{n}\right] \right\}. \tag{197}
$$

If the absolute value of Im (z_C) is denoted by b,

$$
b =
$$

$$
\frac{1}{\pi} \sum_{r=1}^{m-1} \sin\left(\frac{2rn}{m}\pi\right) \left(-\cos\beta \ln\left|\cos\frac{2r-1}{m}\pi - \cos\frac{\beta}{n}\right| + 2\ln\sin\frac{2r-1}{2m}\pi\right)
$$

$$
+ \frac{1}{\pi} \sum_{r=0}^{m-1} \cos\left(\frac{2rn}{m}\pi\right) \left(\sin\beta \ln\left|\frac{\sin\frac{1}{2}\left(\frac{2r-1}{m}\pi - \frac{\beta}{n}\right)}{\sin\frac{1}{2}\left(\frac{2r-1}{m}\pi + \frac{\beta}{n}\right)}\right| + 1 - \cos\beta\right). \tag{198}
$$

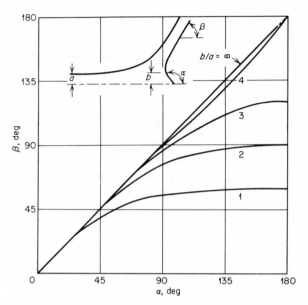

FIGURE 26.   EFFECTS OF BOUNDARY ANGLE ON JET DEFLECTION.
THE HALF-WIDTH a IS NOW ARBITRARY. (*From T. T. Saio
and P. G. Hubbard, in J. S. McNown and C.-S. Yih (eds.),
"Free-streamline Analysis of Transition Flow and Jet Deflec-
tion," 1953, by permission of the Iowa Institute of Hydraulic
Research.*)

The results are represented graphically in Figs. 25 and 26, where
only the upper half of the flow is shown for convenience.

24.   APPROXIMATE METHODS

For slender bodies placed in a stream with uniform velocity U at infinity,
a few approximate methods are available for determining the velocity
field.   In all these methods the longitudinal velocity is assumed to be
only very slightly different from U.   Since at the stagnation points this
is evidently not true, the determinations are useful only away from these
points.   In this section we shall present three approximate methods,
two for two-dimensional flows and one for axisymmetric flows.   Though
only approximate, these methods have the advantage that given the shape
of the body, the velocity fields are determined directly.

The relaxation method, which is approximate only in the sense that
it is a method of calculation using finite differences, will also be presented
in this section.

24.1. Source-Sink Method for Two-dimensional Symmetric Flow past a Slender Body

Consider a symmetric slender two-dimensional body symmetric with respect to the x axis and moving along that axis with velocity U in an otherwise quiescent fluid. The profile of the body is given by

$$y = \pm F(x), \qquad a \le x \le b.$$

If a continuous distribution of sources and sinks of strength $m(\xi)$ between a and b along the x axis is assumed, one has, with $f(\xi) = m(\xi)$,

$$\phi = \frac{1}{2\pi} \int_a^b f(\xi) \ln \sqrt{(x - \xi)^2 + y^2} \, d\xi, \tag{199}$$

in which $f(\xi)$ is the strength of the source (or sink) at $x = \xi$. This ϕ satisfies the Laplace equation, and can be identified with the velocity potential. The boundary condition on the surface of the body is, with higher-order terms neglected,

$$v = \pm U \frac{dy}{dx} = \pm U \frac{dF}{dx}, \tag{200}$$

in which the positive and negative signs are for the upper and lower halves of the body, respectively. The problem is to find the proper $f(\xi)$ so that the boundary conditions (200) are satisfied.

Differentiation of (199) with respect to x and y produces

$$u = \frac{1}{2\pi} \int_a^b \frac{f(\xi)(x - \xi) \, d\xi}{(x - \xi)^2 + y^2}, \qquad v = \frac{y}{2\pi} \int_a^b \frac{f(\xi) \, d\xi}{(x - \xi)^2 + y^2}. \tag{201}$$

The second of (201) can be written as

$$v = \frac{1}{2\pi} \int_{\eta_a}^{\eta_b} \frac{f(\xi) \, d\eta}{1 + \eta^2}, \tag{202}$$

in which

$$\eta = \frac{\xi - x}{y}.$$

For small positive y, (202) can be approximated by

$$v = -\frac{1}{2\pi} f(x) \int_{-\infty}^{\infty} \frac{d\eta}{1 + \eta^2} = \frac{f(x)}{2\pi} \tan^{-1} \eta \Big|_{-\infty}^{\infty} = \frac{f(x)}{2}. \tag{203}$$

For small negative y, the corresponding result is

$$v = -\frac{f(x)}{2}. \tag{204}$$

From (200) it then follows that

$$f(x) = 2U \frac{dF(x)}{dx}. \tag{205}$$

The component u can then be evaluated by the first of (201). For small y, u is the finite part (principal value) of

$$\frac{1}{2\pi} \int_a^b \frac{f(\xi)\, d\xi}{x - \xi},$$

or

$$u = \frac{1}{2\pi} \lim_{\epsilon \to 0} \left[\int_a^{x-\epsilon} \frac{f(\xi)\, d\xi}{x - \xi} - \int_{x+\epsilon}^b \frac{f(\xi)\, d\xi}{\xi - x} \right]. \tag{206}$$

Of course the v calculated from (205) is approximately correct only when u is nearly equal to 0, which is true over the major part of the body. However, at the stagnation point $u = U$, and (205) gives a v which is very wrong indeed. In spite of this, and with this in mind, (206) is still useful.

24.2. Munk's Vortex-sheet Theory

A sheet on which vorticity is concentrated, i.e., across which there is a discontinuity of the velocity component tangential to the line, is called a *vortex sheet*. In general there is a circulation around any segment of the sheet, which gives rise to a lift in the presence of a uniform stream.

If the upper surface of an (infinite) airfoil is given by $y = y_1(x)$ and the lower one by $y = y_2(x)$, the middle line of the foil is given by

$$y_m = \frac{y_1(x) + y_2(x)}{2}.$$

For very thin airfoils this middle line is assumed to be the airfoil. Furthermore, at small angles of attack this middle line can be assumed to coincide with the x axis, as far as the location of the vortex sheet representing the airfoil is concerned. Consider a potential of the form (with the uniform flow excluded)

$$\phi = \frac{1}{2\pi} \int_0^a f(\xi) \tan^{-1} \frac{y}{x - \xi}\, d\xi, \tag{207}$$

a being the chord of the airfoil. The tangential velocity on the airfoil is, approximately,

$$u = \frac{\partial \phi}{\partial x} = \frac{y}{2\pi} \int_0^a f(\xi) \frac{1}{(x - \xi)^2 + y^2}\, d\xi. \tag{208}$$

This is similar to (201), and results similar to (204) and (203) follow. For small y,

$$u_+ = \frac{f(x)}{2}, \qquad u_- = -\frac{f(x)}{2}, \tag{209}$$

and the discontinuity in u is

$$[u] = f(x). \tag{210}$$

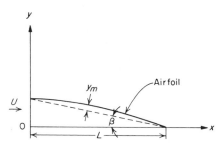

FIGURE 27. AIRFOIL AT SMALL ANGLE OF
ATTACK. L IS APPROXIMATELY EQUAL TO
CHORD LENGTH a FOR SMALL β.

The discontinuity in ϕ is therefore

$$[\phi] = \int_0^x f(\xi)\, d\xi, \tag{211}$$

with the potential higher on the upper surface if $f(\xi)$ is positive, or the vorticity is clockwise, and lower on the upper surface in the opposite case.

To find $f(\xi)$, we again consider the component v on the thin foil. With β as the small angle of attack (Fig. 27) the normal velocity component is approximately (y_m measured from chord if β is not zero)

$$v = U\left(\frac{dy_m}{dx} - \beta\right), \tag{212}$$

in which U is the speed of flight as shown in Fig. 27. Thus, from (207), for small y

$$v = \frac{\partial \phi}{\partial y} = \frac{1}{2\pi} \int_0^a \frac{f(\xi)}{x - \xi}\, d\xi, \tag{213}$$

with v given in (212). This integral equation is to be solved with the additional condition that

$$f(a) = 0, \tag{214}$$

which ensures that there is no infinite velocity at the trailing edge. The solution of (213) and (214) is given by Sedov (1939) and is

$$f(x) = \frac{2}{\pi} \sqrt{\frac{a - x}{x}} \int_0^a \sqrt{\frac{\xi}{a - \xi}} \frac{v(\xi)\, d\xi}{\xi - x}. \tag{215}$$

24.3. Linearized Theory for Axisymmetric Flow past Slender Bodies

The axisymmetric potential flow past a slender body can be approximated by superposing the potential of a line source of varying strength on that of a uniform flow.   The combined potential is

$$\phi = Uz - \frac{1}{4\pi} \int_0^a \frac{m(\xi)}{\sqrt{(z-\xi)^2 + r^2}} d\xi, \tag{216}$$

in which U is the uniform velocity at infinity in the z direction.   The slender body extends from $z = 0$ to $z = a$.   From (216) the velocity components can be obtained by differentiation:

$$w = \frac{\partial\phi}{\partial z} = U + \frac{1}{4\pi} \int_0^a \frac{m(\xi)(z-\xi)}{[(z-\xi)^2 + r^2]^{3/2}} d\xi, \tag{217}$$

$$u = \frac{\partial\phi}{\partial r} = \frac{1}{4\pi} \int_0^a \frac{m(\xi)r}{[(z-\xi)^2 + r^2]^{3/2}} d\xi. \tag{218}$$

Equation (218) can be written in the form

$$u = \frac{1}{4\pi r} \int_{-z/r}^{(a-z)/r} \frac{m(\xi)}{\{[(\xi-z)/r]^2 + 1\}^{3/2}} d\frac{\xi-z}{r} .$$

Since r is very small for a slender body, the term $[(\xi - z)/r]^2$ is very large except for values of ξ near z.   Hence $m(\xi)$ can be replaced by $m(z)$ and the limits by $-\infty$ and ∞, so that the last equation can be written as

$$u = \frac{1}{4\pi r} m(z) \int_{-\infty}^{\infty} \frac{d\eta}{(\eta^2 + 1)^{3/2}} = \frac{1}{2\pi r} m(z).$$

On the other hand, the velocity in the z direction is approximately U, so that to the first order of approximation

$$u = U \frac{dF(z)}{dz} \tag{219}$$

if the shape of the body is given by

$$r = F(z).$$

Thus, for the boundary, at which $r = F(z)$,

$$m(z) = 2\pi U F(z) \frac{dF(z)}{dz} . \tag{220}$$

The velocity component w is then evaluated from (217), by expanding $m(\xi)$ in powers of $\xi - z$:

$$m(\xi) = m(z) + m'(z)(\xi - z) + \tfrac{1}{2} m''(z)(\xi - z)^2 + \cdots ,$$

in which $m'(z) = \dfrac{dm(z)}{dz} ,$ $m''(z) = \dfrac{d^2m(z)}{dz^2} ,$ etc.

Near $z = 0$ and $z = a$, w is near zero, the difference between w and U is large, so that the linearization process is invalid.   However, the solution by linearization is valid elsewhere.   The method of solution

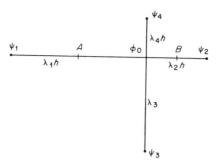

FIGURE 28.   THE IRREGULAR STAR.

can be modified to deal with subsonic flows of a compressible fluid past slender bodies.

24.4.   The Relaxation Method

A numerical method [Southwell (1946)] involving successive corrections can be used to solve the Laplace equation approximately, especially if only two dimensions are involved.   First, the Laplace equation will be expressed in terms of values of the unknown at finite distances apart. For instance, if the equation to be solved is

$$\frac{\partial^2 \psi}{\partial x^2} + \frac{\partial^2 \psi}{\partial y^2} = 0,$$

we shall consider a star as shown in Fig. 28.   Let A, B, C, and D be the midpoints of the first, second, third, and fourth rays, respectively.   Then, approximately,

$$\left(\frac{\partial \psi}{\partial x}\right)_A = \frac{\psi_0 - \psi_1}{\lambda_1 h}, \qquad \left(\frac{\partial \psi}{\partial x}\right)_B = \frac{\psi_2 - \psi_0}{\lambda_2 h}. \tag{221}$$

But

$$\left(\frac{\partial \psi}{\partial x}\right)_B = \left(\frac{\partial \psi}{\partial x}\right)_0 + \frac{\lambda_2 h}{2}\left(\frac{\partial^2 \psi}{\partial x^2}\right)_0, \tag{222}$$

$$\left(\frac{\partial \psi}{\partial x}\right)_A = \left(\frac{\partial \psi}{\partial x}\right)_0 - \frac{\lambda_1 h}{2}\left(\frac{\partial^2 \psi}{\partial x^2}\right)_0. \tag{223}$$

Subtraction of (223) from (222) produces, after Eqs. (221) are used,

$$\tfrac{1}{2}(\lambda_2 + \lambda_1)h\left(\frac{\partial^2 \psi}{\partial x^2}\right)_0 = \frac{\lambda_1 \psi_2 + \lambda_2 \psi_1 - (\lambda_1 + \lambda_2)\psi_0}{\lambda_1 \lambda_2 h}, \tag{224}$$

or

$$\left(\frac{\partial^2 \psi}{\partial x^2}\right)_0 = \frac{2[\lambda_1 \psi_2 + \lambda_2 \psi_1 - (\lambda_1 + \lambda_2)\psi_0]}{\lambda_1 \lambda_2 (\lambda_1 + \lambda_2)h^2}. \tag{225}$$

Similarly,

$$\left(\frac{\partial^2 \psi}{\partial y^2}\right)_0 = \frac{2[\lambda_3 \psi_4 + \lambda_4 \psi_3 - (\lambda_3 + \lambda_4)\psi_0]}{\lambda_3 \lambda_4 (\lambda_3 + \lambda_4)h^2}. \tag{226}$$

Thus the Laplace equation becomes, after multiplication by $\lambda_1\lambda_2\lambda_3\lambda_4(\lambda_1 + \lambda_2)(\lambda_3 + \lambda_4)h^2/2$,

$$\lambda_3\lambda_4(\lambda_3 + \lambda_4)[\lambda_1\psi_2 + \lambda_2\psi_1 - (\lambda_1 + \lambda_2)\psi_0]$$
$$+ \lambda_1\lambda_2(\lambda_1 + \lambda_2)[\lambda_3\psi_4 + \lambda_4\psi_3 - (\lambda_3 + \lambda_4)\psi_0] = 0, \quad (227)$$

or $$R = D_1\psi_1 + D_2\psi_2 + D_3\psi_3 + D_4\psi_4 - \psi_0 = 0, \quad (228)$$

in which

$$D_1 = \frac{\lambda_2\lambda_3\lambda_4}{(\lambda_1\lambda_2 + \lambda_3\lambda_4)(\lambda_1 + \lambda_2)}, \qquad D_2 = \frac{\lambda_1\lambda_3\lambda_4}{(\lambda_1\lambda_2 + \lambda_3\lambda_4)(\lambda_1 + \lambda_2)}, \qquad (229)$$
$$D_3 = \frac{\lambda_1\lambda_2\lambda_4}{(\lambda_1\lambda_2 + \lambda_3\lambda_4)(\lambda_3 + \lambda_4)}, \qquad D_4 = \frac{\lambda_1\lambda_2\lambda_3}{(\lambda_1\lambda_2 + \lambda_3\lambda_4)(\lambda_3 + \lambda_4)}.$$

The R in (228) is called the *residue*. The idea is to make R vanish by successive corrections at all the joints or centers of the stars. In practice, at least two of the λ's are equal to 1. At an internal point of the fluid, all four λ's are equal to 1. A star with four equal rays is called a *regular star*. At a regular star

$$D_1 = D_2 = D_3 = D_4 = \tfrac{1}{4}. \quad (230)$$

The procedure is as follows. First impose on this region a grid each joint of which is the center of a star—a regular star if the joint is not adjacent to the boundary. Compute the D's for each joint. Then make an intelligent guess, by the aid of a rough flow net for instance, of the values of ψ at alternate joints of the grid. The ψ values on the boundary of the domain of flow should of course be correctly prescribed. These will remain unchanged throughout the relaxation process. Then compute the ψ values at the other joints by requiring the residue there to be zero. Now the relaxation process begins, for at the joints where the ψ values were originally assigned, the residues will not be zero. The relaxation process consists of the following steps, to be repeated indefinitely at all the joints until the desired accuracy has been achieved:

1 At any joint where R is not zero, raise the value of ψ there by the amount R to make the new residue equal to zero.

2 Since that first step affects the neighboring points, calculate the residues at the neighboring points. For instance, if the ψ value at A (Fig. 29) has been raised by 8, and if

$$D_1 = 0.1, \qquad D_2 = 0.4, \qquad D_3 = 0.05, \qquad D_4 = 0.45, \qquad \text{at } C,$$

and B and D are regular stars, then the amount 0.8 is distributed

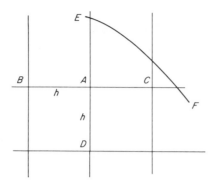

FIGURE 29.   *EF* IS PART OF A STREAMLINE.

to R_C (the residue at C), the amount 2 to R_B and R_D, and *nothing to* the joint E, where the ψ value is fixed if *EF* is part of a stream-line on which ψ is assigned. In interior joints, where the stars are regular, the distribution factors are all $\frac{1}{4}$, and any increase of ψ by the amount $\Delta\psi$ will increase each of the residues at the neighboring points by $\Delta\psi/4$.

3 As the residue at any point becomes large after the accumulation of increments, annihilate it as explained in step 1 and continue the relaxation procedure until the desired accuracy is achieved.

In carrying out the relaxation procedure, one must not allow one's attention to be hypnotically absorbed in one small region too long at the expense of the other regions. Great accuracy obtained in a region without regard to harmony at its borders or in the neighboring regions will be wasted, for residues will diffuse to the border from the neighboring regions and invade the region where residues have been annihilated assiduously, with the illusion of finality.

If the flow varies in speed and direction very fast in a certain region, a finer net may have to be used there. In that case a coarse net for the entire flow can be used first, and the finer net can be grafted unto the coarse net and the boundary ψ values considered as given by the coarse net.

The predominence of modern computing machinery does not render the relaxation method obsolete; it only eliminates the drudgery. A computation can be coded and fed into the machine. The speed with which the machine performs the job makes those who have computed the relaxation by hand look back to their early years more with a sigh of relief than of nostalgia. It must be added, however, that if only a rough estimate is needed, relaxation by hand computation can often give satisfactory results quickly wherever pencil and paper are available.

PROBLEMS

1 The half-body described by (31) is placed in a stream which has a velocity at infinity parallel to the axis of the body. Show that the total longitudinal force on the surface of the body is exactly the same as that acting on the circular area which is the longitudinal projection of the body far upstream.

2 Find the stream function for the irrotational flow of an inviscid fluid of constant density past a sphere of radius a by the use of Butler's theorem.

3 Show that if a body has a plane of symmetry, say $x_1 =$ const, then the added masses A_{12} and A_{13} vanish and that if it has an axis of symmetry, say $x_2 = 0 = x_3$, then A_{12}, A_{23}, and A_{31} all vanish. (Remember that $A_{\alpha\beta} = A_{\beta\alpha}$.)

4 Show that the added mass for a sphere (A_{11}) is equal to one-half the mass of the fluid displaced by it, by direct integration on the surface, using (22) and (75), the μ in (22) being $\frac{1}{2}Ua^3$.

5 A solid sphere is released in an inviscid fluid of constant density ρ. Show by a direct integration of the pressure that the net force acting on the surface of the sphere is vertical in direction and given by

$$F = -\tfrac{1}{2}\rho V \frac{dw}{dt} + g\rho V,$$

in which V is the volume of the sphere, w its velocity in the direction of the vertical, and F is positive if directed upward. *Hint:* In evaluating the term $\partial\phi/\partial t$ in the Bernoulli equation be sure to use fixed coordinates and express the coordinates moving with the sphere in terms of fixed coordinates.

6 A solid sphere of density ρ_s is released from rest in an incompressible fluid of density ρ. Find the initial acceleration. Is your answer valid for a viscous fluid?

7 A solid sphere of radius a and specific gravity 2 is released at rest at $t = 0$ in a large body of water. Find the initial pressure distribution on the sphere. Is the result valid initially if viscous effects are considered? Is the result valid subsequently if viscous effects are considered? *Hint:* Do not forget $\partial\phi/\partial t$ when evaluating p.

8 A source of strength m is situated at any point outside of the rigid sphere of radius a. There is nothing else causing the flow. Find the resultant *hydrodynamic* force acting on the surface of the sphere. Is the force the same if the source is replaced by a sink of the same strength? *Hint:* Find the image of the source in the sphere first.

9 The circle $|z| = a$ is a rigid surface submerged in a large body of an incompressible inviscid fluid of constant density. A source is introduced at any point outside the circle. Find the resulting flow. Solve the problem if the singularity is a doublet of any orientation instead of a source.

10 A source of strength m is situated at any point outside the rigid boundary $|z| = a$. There is nothing else causing the flow. Find the resultant hydrodynamic force acting on the surface of the boundary, i.e., the resultant force apart from buoyancy. Is the force the same if the source is replaced by a sink of the same strength? Give a qualitative explanation of why the force is always directed toward the source or sink.

11 Obtain (144) by the method of separation of variables for the flow it describes.

12 Use the circle theorem to verify that, for flow past a cylinder,

$$w = U\left(z + \frac{a^2}{z}\right),$$

which is (140) with μ given by (142).

13 An inviscid fluid of constant density fills the space between two concentric cylinders of radii r_1 and $r_2 > r_1$, respectively. If the plane AB starts to move from rest and rotates about the center with an angular speed ω, find the resulting flow. *Warning*. Solid rotation will not do. Why?

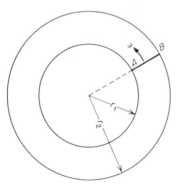

FIGURE P 13

14 An elliptic cylinder with a cross section described by

$$\frac{x^2}{a^2} + \frac{y^2}{b^2} = 1$$

rotates with angular speed ω about its axis. Outside the cylinder the space is filled with an inviscid fluid of constant density ρ. (x and y are coordinates moving with the cylinder.) Find the flow of the fluid if the cylinder started to move from rest.

15 Show that the added mass per unit length of an elliptic cylinder translating in the direction of the major axis of its cross section is the same as that of the inscribed circular cylinder with its diameter equal to the minor axis of the elliptic cross section.

16 Show that if the elliptic cylinder translates in the direction of its minor axis, the added mass per unit length is $\pi \rho a^2$, ρ being the density of the fluid and $2a$ the major axis of the elliptic cross section, or the diameter of the circumscribing circle.

17 Show by direct integration of the pressure on the cylindrical surface that the lift per unit length for the flow described by (146) is $\rho U \Gamma$.

18 Locate the stagnation points for the irrotational flow of an incompressible fluid past an ellipse of semiaxes a and b, its major axis being inclined to the direction of flow at infinity at an angle β. The circulation around the ellipse is Γ.

19 Find the flow past the nearly circular cylinder of a cross section described by

$$r = a(1 - \epsilon \sin^2 \theta), \qquad \epsilon \ll 1,$$

the velocity at infinity being U. The cylinder is stationary.

20 Find the coefficient of contraction π/c for the jet shown in the figure. Is it very much less trouble to find π/c for a particular c/b than to find π/c for a range of c/b?

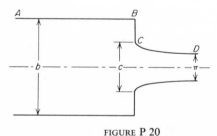

FIGURE P 20

21 An inviscid incompressible fluid issues from a two-dimensional reservoir bounded by the lines

$$x = 0, \ -\infty \le y \le -1, \qquad \text{and} \qquad y = 0, \ -\infty \le x \le -1.$$

Neglecting gravity effects, show that the asymptotic width of the jet is $0.747 \sqrt{2}$.

22 A hollow thin-walled cylinder of inner radius a and length $l \gg a$, open at both ends, is inserted into a plane wall at $x = 0$. The cylinder extends from $x = -l$ to $x = 0$. One side ($x < 0$) of the wall is filled with water, so that a circular jet issues from the pipe opening at

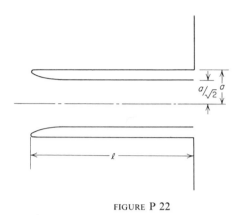

FIGURE P 22

$x = -l$. Assuming that the fluid is inviscid, neglecting gravity effects, and using a control volume and the momentum principle, show that the cross section of the jet decreases from πa^2 at $x = -l$ to $\pi a^2/2$ at $x = \infty$. The cylinder is the well-known Borda mouthpiece.

REFERENCES

BIRKHOFF, G., 1953: "Hydrodynamics," Princeton University Press, Princeton, N.J.

CUMMINS, W. E., 1953: The Forces and Moments on a Body Moving in an Arbitrary Potential Stream, *U.S. Navy David Taylor Model Basin Rept.* 708.

JEFFREYS, H., AND B. S. JEFFREYS, 1956: "Methods of Mathematical Physics," 3d ed., Cambridge University Press, New York.

KÁRMÁN, T. VON, 1927: Berechnung der Druckverteilung an Luftschiffkörpern, *Abhandl. Aerodynamischen Inst. Tech. Hoch. Aachen*, **6**: 3–17.

KELLOGG, O. D., 1929: "Foundations of Potential Theory," Murray Printing Co., New York; reprinted by Dover Publications, Inc., New York, 1953 (with the same pagination).

KIRCHHOFF, G., 1869: Zur Theorie freier Flüssigkeitsstrahlen, *Crelle*, **70**: 289–298.

LAGALLY, M., 1922: Berechnung der Kräfte und Momente, die strömende Flüssigkeiten auf ihre Begrenzung ausüben, *Z. Angew. Math. Mech.*, **2**: 409.

LANDWEBER, L., 1956: On a Generalization of Taylor's Virtual Mass Relation for Rankine Bodies, *Quart. Appl. Math.*, **14**: 51.

LANDWEBER, L., AND C.-S. YIH, 1956: Forces, Moments, and Added Masses for Rankine Bodies, *J. Fluid Mech.*, **1**: 319–336. [This contains an error in Eq. (61), where the term $16\pi^2 \rho \sigma \mu_j/3$ should not have appeared.]

LUDFORD, G. S. S., J. MARTINEK, AND G. C. K. YEH, 1955: The Sphere Theorem in Potential Theory, *Proc. Cambridge Phil. Soc.*, **51**: 389–393.

SEDOV, L. I., 1939: "Two-dimensional Motion of an Ideal Fluid," Oborongiz, Moscow.

SIAO, T. T., AND P. G. HUBBARD, 1953: Deflection of Jets, pt. I, Symmetrically Placed V-shaped Obstacle, J. S. McNown and C.-S. Yih (eds.) "Free-streamline Analysis of Transition Flow and Jet Deflection," State University of Iowa, Iowa City.

SOUTHWELL, R. V., 1946: "Relaxation Methods in Theoretical Physics," Oxford University Press, Fair Lawn, N.J. (The method was known in essence before 1946, but Southwell developed it and applied it fruitfully to many fields.)

TAYLOR, G. I., 1928: The Energy of a Body Moving in an Infinite Fluid, with an Application to Airships, *Proc. Roy. Soc. London Ser. A,* **120:** 13.

TITCHMARSH, E. C., 1939: "The Theory of Functions," 2d ed., Oxford University Press, Fair Lawn, N.J.

YEH, G. C. K., J. MARTINEK, AND G. S. S. LUDFORD, 1956: A General Sphere Theorem for Hydrodynamics, Heat, Magnetism, and Electrostatics, *Z. Angew. Math. Mech.,* **36:** 111–116.

YIH, C.-S., 1955: Maximum Acceleration in Two-Dimensional Steady Flows of an Ideal Fluid, *Quart. Appl. Math.,* **13:** 202.

YIH, C.-S., 1958: Maximum Speed in Steady Subsonic Flows, *Quart. Appl. Math.,* **16:** 178–180.

YIH, C.-S., 1960: Finite Two-dimensional Cavities, *Proc. Roy. Soc. London Ser. A,* **258:** 90–100.

ADDITIONAL READING

BIRKHOFF, G., AND E. H. ZARANTONELLO, 1957: "Jets, Wakes, and Cavities," Academic Press Inc., New York.

LAMB, H., 1932: "Hydrodynamics," 6th ed., The Macmillan Company, New York; reprinted by Dover Publications, Inc., New York, 1945 (with the same pagination).

MILNE-THOMSON, L. M., 1955: "Theoretical Hydrodynamics," 3d ed., The Macmillan Company, New York.

CHAPTER FIVE

WAVES
IN AN
INCOMPRESSIBLE
FLUID

1. INTRODUCTION

In a body of fluid, whether in motion or not, it often happens that when it is disturbed so that the fluid particles are displaced from their original positions or streamlines, restoring forces are called into play. These may be gravity, surface tension, centripetal force of a fluid in rotation, or the excess pressure caused by compression. The motion of the fluid brought about by these restoring forces will not cease when the fluid has reached its original position of static or dynamic equilibrium but will overshoot the mark and require the action of restoring forces again. Thus wave motion is generated and perpetuated until viscosity damps it out. The waves corresponding to the aforementioned restoring forces are called *gravity waves*, *capillary waves*, *inertial waves*, or *sound waves*, respectively. Although there are other kinds of waves, such as waves in an electrically conducting fluid in the presence of an electromagnetic field, only these four kinds of wave will be dealt with in this book. Gravity waves and capillary waves will be treated together as surface waves whenever possible. Unless stated otherwise, viscosity will be ignored. It is assumed that the

Reynolds number based on some representative length (such as the wave-length) and some representative speed (such as the wave speed) is large. The effect of viscosity will then be concentrated in the boundary layers along the solid boundary or the free surface. Compressibility waves will be treated in the next chapter.

2. SURFACE WAVES OF SMALL AMPLITUDE

If the density in the fluid or in each of the superposed fluid layers is constant, and if the wave motion is assumed to have started from rest relative to the undisturbed state of flow, which is itself assumed irrotational, then the wave motion will be irrotational, according to the results of Chap. 3. If, however, the density of a fluid varies continuously, the wave motion in that fluid will in general not be irrotational and we shall have to start from the equations of motion. Whenever a fluid is homogeneous or layerwise homogeneous, irrotationality of the wave motion is understood for the fluid or for each layer, although the interfaces between the layers will in general be vortex sheets of variable strength, because the discontinuity in the tangential velocity will vary from place to place along an interface.

2.1. The Linear Differential System Governing Surface Waves in a Homogeneous Liquid

Since spatially periodic waves are the blocks for building more general waves by the Fourier integral, we shall deal mainly with such waves. As will be seen in Sec. 5.2 of Chap. 9, the study of three-dimensional waves can be reduced to that of two-dimensional ones in the framework of a linear theory. We shall, then, discuss mainly two-dimensional waves, for which Cartesian coordinates x and y are used. The coordinate x is measured in the direction of wave propagation, and y is measured in a direction opposite to that of the gravitational acceleration g.

Since the wave motion is irrotational, there exists a potential ϕ satisfying

$$\frac{\partial^2 \phi}{\partial x^2} + \frac{\partial^2 \phi}{\partial y^2} = 0. \tag{1}$$

The boundary condition at any rigid stationary boundary is

$$\frac{\partial \phi}{\partial n} = 0, \tag{2}$$

in which n is the normal distance from the boundary.

At the free surface or an interface,

$$y = \eta(x,t)$$

if η is the displacement of that surface from its mean position (or equi-librium position). Since a fluid particle on that surface will remain there,

$$\frac{D}{Dt}(y - \eta) = 0, \tag{3}$$

in which

$$\frac{D}{Dt} = \frac{\partial}{\partial t} + u\frac{\partial}{\partial x} + v\frac{\partial}{\partial y}, \tag{4}$$

u and v being the total velocity components in the directions of increasing x and y, respectively, and t the time. Thus (3) becomes

$$v = \left(\frac{\partial}{\partial t} + u\frac{\partial}{\partial x}\right)\eta, \tag{5}$$

or, for irrotational flow,

$$\frac{\partial\phi}{\partial y} = \left(\frac{\partial}{\partial t} + u\frac{\partial}{\partial x}\right)\eta. \tag{5a}$$

For infinitesimal waves in a fluid without mean flow, the second-order term in (5a) can be neglected, giving

$$\frac{\partial\phi}{\partial y} = \frac{\partial\eta}{\partial t}. \tag{5b}$$

Equation (5) is the kinematic condition at the free surface or an interface.

The dynamic condition at the free surface is, for a single fluid with a free surface,

$$p = -T\left(\frac{1}{R_1} + \frac{1}{R_2}\right), \tag{6}$$

in which R_1 and R_2 are the two principal radii at the free surface, counted positive when the center of curvature is above the surface, or outside the liquid. For the coordinates adopted and for the two-dimensional motion being considered, the linearized form of (6) is

$$p = -T\frac{\partial^2\eta}{\partial x^2}. \tag{6a}$$

Both (6) and (6a) presume that the pressure above the free surface is zero. Otherwise p must be replaced by $p - p_a$, p_a being the atmospheric pressure. But we still need another equation for p. This is supplied by the Bernoulli equation

$$\frac{p}{\rho} = -\frac{\partial\phi}{\partial t} - gy - \frac{u^2 + v^2}{2} + F(t). \tag{7}$$

If the fluid is at rest except for the wave motion, u^2 and v^2 are of a higher order than the first two terms on the right-hand side of (7), and we can omit the term $(u^2 + v^2)/2$. But if there is a mean flow of constant velocity U in the x direction, then $u^2 + v^2$ must be replaced by $2Uu'$ if

$$u = U + u'.$$

If, further, periodicity in time is assumed for the wave motion, each term containing p, ϕ, and η will have the factor

$$e^{ik(x-ct)}, \tag{8}$$

in which k is the wave number equal to $2\pi/$wavelength and c is the wave velocity. If the pressure is assumed to be zero or constant at the free surface, where $y = \eta$, $F(t)$ in (7) can be put equal to zero, and (7) assumes the form

$$\frac{p}{\rho} = -\frac{\partial \phi}{\partial t} - gy, \qquad \text{no mean flow}, \tag{7a}$$

or $\qquad \dfrac{p}{\rho} = -\dfrac{\partial \phi}{\partial t} - gy - Uu', \qquad$ mean flow with x velocity U. $\tag{7b}$

For the case of a single fluid with a free surface and with no mean flow, (5b), (6a), and (7a) can be combined to form the single condition

$$\frac{\partial^2 \phi}{\partial t^2} = \left(\frac{T}{\rho}\frac{\partial^2}{\partial x^2} - g\right)\frac{\partial \phi}{\partial y} \qquad \text{at the free surface}, \tag{9}$$

or $\qquad \sigma^2 \phi = \left(g - \dfrac{T}{\rho}\dfrac{\partial^2}{\partial x^2}\right)\dfrac{\partial \phi}{\partial y} \qquad$ at the free surface $\tag{9a}$

if $\qquad \sigma = kc =$ angular frequency of the waves.

With the factor (8), (9a) can be further written as

$$k^2 c^2 \phi = \left(g + k^2\frac{T}{\rho}\right)\frac{\partial \phi}{\partial y} \qquad \text{at the free surface}. \tag{10}$$

A condition similar to (10) can be found for the interface of two liquids by using (5a) or (5b), (7a) or (7b), and

$$\Delta p = -T\left(\frac{1}{R_1} + \frac{1}{R_2}\right), \tag{11}$$

in which Δp is the difference between the pressure below the interface and that above.

2.2. Linear Surface Waves in a Semi-infinite Liquid

If the fluid fills the semi-infinite space

$$-\infty \leq y \leq 0, \qquad -\infty \leq x \leq \infty,$$

with a free surface at $y = 0$, and if k is the (positive) wave number of periodic waves, then in view of the form of (8), the solution of (1) by separation of variables is of the form

$$\phi = Ce^{ky} \cos k(x - ct), \tag{12}$$

which also satisfies the condition

$$\phi = 0 \quad \text{at} \quad y = -\infty.$$

The free-surface condition is (10), which, with $y = 0$ on the undisturbed free surface, has the form

$$c^2 = \frac{g}{k} + \frac{kT}{\rho}. \tag{13}$$

For very long waves, the effect of gravity predominates, and

$$c^2 = \frac{g}{k}. \tag{13a}$$

For such waves c^2 increases as the wavelength increases. For very short waves, capillary effects predominate, and

$$c^2 = \frac{kT}{\rho}. \tag{13b}$$

For such waves c^2 decreases as the wavelength increases. The c^2 given by (13) has the minimum value

$$c^2_{\text{min}} = 2\sqrt{\frac{Tg}{\rho}} \tag{14}$$

at $k = (\rho g/T)^{1/2} = k_{\text{cr}}$.

Equation (13) shows that the wave velocity is dependent on the wave number. Wave components with different wave numbers will then disperse as time goes on. For this reason (13) is called the *dispersion formula* or *dispersion equation*.

2.3. Gravity Waves in a Liquid Layer of Finite Depth

At the mean position of the free surface we shall again take y to be zero. Then $y = -h$ at the bottom of a liquid layer of uniform depth h. The appropriate solution of (1) for sinusoidal waves with wave number k is

$$\phi = C \cosh k(y + h) \cos k(x - ct) \tag{15}$$

since there is no flow other than that due to the waves. This satisfies not only (1) but also the boundary condition

$$\frac{\partial \phi}{\partial y} = 0 \quad \text{at} \quad y = -h. \tag{16}$$

The free-surface condition (10) is then

$$kc^2 \cosh kh = \left(g + \frac{k^2 T}{\rho}\right) \sinh kh,$$

or
$$c^2 = \left(g + \frac{k^2 T}{\rho}\right)\frac{\tanh kh}{k}. \tag{17}$$

For very long waves (or very small k), whatever the value of T,

$$c^2 = gh, \tag{17a}$$

and the waves are mainly gravity waves, since the curvature is so small that the effect of surface tension is not felt. For very short waves, k is very large, and

$$c^2 = \frac{kT}{\rho}, \tag{17b}$$

and the waves are mainly capillary waves. Comparison of (13b) with (17b) shows that for capillary waves the depth is quite unimportant.

From (15) we can obtain

$$u = \frac{\partial \phi}{\partial x} = -kC \cosh k(y + h) \sin k(x - ct), \tag{18}$$

$$v = \frac{\partial \phi}{\partial y} = kC \sinh k(y + h) \cos k(x - ct). \tag{19}$$

For very long waves, (18) shows that u is nearly independent of y in the domain of the liquid and that v is nearly zero throughout. Since the vertical acceleration is then negligible, the pressure distribution in the liquid is nearly hydrostatic for very long waves.

From (5b), applied at $y = 0$, it follows that

$$\eta = -a \sin k(x - ct), \qquad \text{with} \qquad a = \frac{C}{c} \sinh kh. \tag{20}$$

The locus of a material particle can be computed from (18) and (19). If $X(t)$ and $Y(t)$ are the components of the displacement of a material point from its mean position (x,y), and if the difference in the value of u or v as evaluated at (x,y) and $(x + X, y + Y)$ is neglected, we obtain from (18), (19), and

$$u = \frac{dX}{dt}, \qquad v = \frac{dY}{dt}$$

the result

$$X = -\frac{C}{c} \cosh k(y + h) \cos k(x - ct), \tag{21}$$

$$Y = -\frac{C}{c} \sinh k(y + h) \sin k(x - ct),$$

so that the locus is the ellipse

$$\frac{X^2}{A^2} + \frac{Y^2}{B^2} = 1,$$

in which $A = \dfrac{C}{c} \cosh k(y + h),$ $B = \dfrac{C}{c} \sinh k(y + h),$

the variable y serving only to indicate the mean elevation of the fluid particle under consideration. Note that at $y = -h$ the ellipse collapses into a horizontal line segment. For infinite depth the locus is a circle, as can be seen from

$$\lim_{h \to \infty} \frac{A}{B} = 1.$$

2.4. Standing Surface Waves and Stationary Surface Waves

Since (17) gives the value for c^2, there are two eigenvalues for c of the same magnitude but opposite signs. This is also physically obvious, since the waves can propagate in one direction as well as the other. If we superpose two wave trains propagating in opposite directions but having the same magnitude, (15) and an equation obtained from it on changing c (now assumed positive) to $-c$ give, upon superposition,

$$\phi = 2C \cosh k(y + h) \cos kx \cos kct. \tag{22}$$

This represents standing waves, because the positions of its nodal points and maximum displacement of the free surface are fixed as t varies. The origin of t can be fixed arbitrarily, so that we can replace t in (22) by $t + \epsilon$, as indeed we can in (15).

  If a frame of reference moves with the propagating waves represented by (15), the fluid will appear to have a general flow of speed c in the direction of decreasing x relative to the moving frame, with which the coordinate axes move. Thus not only must a term $-cx$ be added to the left-hand side of (15), but $x - ct$ must be changed to x, giving

$$\phi = -cx + C \cosh k(y + h) \cos kx. \tag{23}$$

This represents stationary waves, since the flow is now steady with respect to the moving frame. The harmonic conjugate of (23) is

$$\psi = -cy - C \sinh k(y + h) \sin kx, \tag{24}$$

because (23) and (24) can be combined to form

$$\phi + i\psi = -cz + C \cos k(z + ih), \tag{25}$$

where $z = x + iy$ is a complex variable. To find the surface streamline, we can set ψ equal to zero in (24), set y equal to η, and obtain as a first

FIGURE 1. SKETCH FOR EVALUATING THE
POTENTIAL ENERGY OF GRAVITY WAVES.

approximation

$$\eta = -\frac{C}{c} \sinh kh \sin kx.$$

This agrees with (20).

2.5. The Equipartition of Energy

The potential-energy increase due to the waviness of the free surface can best be seen in the following way. In Fig. 1, the water of the elemental volume $\eta \, dx$, which in the absence of the waves would occupy part of the trough, has been elevated above $y = 0$ to form part of the crest.   The rise in the center of gravity is η, and so the total gain in potential energy is

$$\text{PE} = g\rho \int_0^{\lambda/2} \eta^2 \, dx \qquad \text{or} \qquad \text{PE} = \frac{1}{2} g\rho \int_0^{\lambda} \eta^2 \, dx$$

per wavelength. For the waves discussed in Sec. 2.2,

$$\eta = -a \sin k(x - ct), \qquad a = \frac{C}{c}. \tag{26}$$

The corresponding ϕ, given by (12), is explicitly

$$\phi = Ce^{ky} \cos k(x - ct) = ace^{ky} \cos k(x - ct). \tag{27}$$

Thus, for unit length in the direction normal to the xy plane, the potential energy per unit wavelength is

$$\text{PE} = \frac{1}{4} g\rho a^2 \lambda,$$

and the kinetic energy per wavelength is

$$\text{KE} = \frac{1}{2} \rho \int_{-\infty}^{0} \int_0^{\lambda} (\phi_x^2 + \phi_y^2) \, dx \, dy.$$

By virtue of Eq. (4.12), the periodicity in x, and the vanishing of ϕ at $y = -\infty$, this can be written as

$$\text{KE} = \frac{1}{2} \rho \int_0^{\lambda} \left(\phi \frac{\partial \phi}{\partial y} \right)_{y=0} dx = \frac{1}{4} \rho a^2 c^2 k\lambda.$$

Since $c^2 = g/k$,

$$\text{PE} = \text{KE}. \tag{28}$$

Thus the energy is half potential, half kinetic.

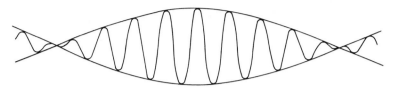

FIGURE 2.   GROUP OF WAVES (VERTICAL SCALE EXAGGERATED).

2.6.   Group Velocity

If two progressive wave trains with the same amplitude but slightly different wavelengths and therefore slightly different frequencies are super-posed, the result for the surface displacement η is (Fig. 2)

$$\eta = a \sin (kx - \sigma t) + a \sin (k'x - \sigma't)$$
$$= 2a \cos [\tfrac{1}{2}(k - k')x - \tfrac{1}{2}(\sigma - \sigma')t] \sin [\tfrac{1}{2}(k + k')x - \tfrac{1}{2}(\sigma + \sigma')t], \tag{29}$$

where $k' - k \ll k$ and $\sigma' - \sigma \ll \sigma$. Thus the result is a train of waves of wave number k and frequency σ but of heights varying with x at any instant, which are enveloped by another wave train of much smaller wave number $(k - k')/2$ and therefore much larger wavelength. The speed of propagation of the envelope is

$$c_g = \frac{d\sigma}{dk} = \frac{d(kc)}{dk} = c - \lambda \frac{dc}{d\lambda}, \tag{30}$$

called the *group velocity*, since it is the velocity with which waves within one wavelength of the envelope advance as a group. For deep-water waves,

$$c_g = \frac{1}{2}\left(gk + \frac{k^3 T}{\rho}\right)^{-1/2}\left(g + \frac{3k^2 T}{\rho}\right). \tag{31}$$

If k^2 is very small compared with $g\rho/T$,

$$c_g = \frac{1}{2}\left(\frac{g}{k}\right)^{1/2} = \frac{c}{2}.$$

If k^2 is large compared with $g\rho/T$, then approximately

$$c_g = \frac{3}{2}\sqrt{\frac{kT}{\rho}} = \frac{3c}{2},$$

c being now the wave velocity for surface-tension waves on deep water. Thus the group velocity for pure gravity waves is one-half that of the wave velocity, whereas the group velocity for pure capillary waves is three-halves of the wave velocity.

To understand the meaning of group velocity, consider pure gravity waves. Since c_g is less than c, the individual waves will advance in their envelope of much longer wavelength, and as they approach the nodal points of their wavy envelope, they will gradually die out, their places being taken by their successors, while at the same time waves will be "born" just ahead of the nodal points of the wavy envelope. Thus the waves are grouped, and these groups, each of length $2\pi/(k - k')$, are separated by bands of nearly smooth water in the neighborhood of the nodal points of the envelope (Fig. 2)

$$\eta_e^2 = 4a^2 \cos^2 \left[\tfrac{1}{2}(k - k')x - \tfrac{1}{2}(\sigma - \sigma')t\right].$$

Note that because it is η_e^2, not η_e, that is given for the envelope, the group length is $2\pi/(k - k')$, not $4\pi/(k - k')$, as might appear at first sight.

The group velocity for waves in water of finite depth can easily be computed from (17) and (30).

One important physical interpretation of the group velocity is that it is the speed with which the energy of the waves is propagated. If we consider deep-water waves and neglect surface-tension effects, at a fixed value of x, the rate of the work done by the fluid at the left on the fluid at the right is

$$\int_{-\infty}^0 p \frac{\partial \phi}{\partial x} \, dy,$$

in which the upper limit should be η but the upper limit zero introduces a negligible error only of a higher order. Since p, as computed from the linearized Bernoulli equation, has only the part $-\rho \, \partial\phi/\partial t$ which contributes to the final result, the rate of work, by virtue of (27) and (13a), is

$$\rho a^2 k^2 c^3 \sin^2 k(x - ct) \int_{-\infty}^0 e^{2ky} \, dy = \frac{1}{2} g\rho a^2 c \sin^2 k(x - ct),$$

the integral of which over a period $2\pi/\sigma$ is just $\tfrac{1}{4} g\rho a^2 \lambda$. This is just one-half of the total energy contained in one wavelength. Thus the supply of energy is sufficient only to advance an isolated group at a speed equal to half the velocity of individual waves.

The group velocity also has a kinematic significance [Lighthill and Whitham (1955)]. If we consider σ and k as functions[1] of x and t, the difference $\sigma(x) - \sigma(x + dx)$ at the two ends of the interval dx is at any instant

$$-\frac{\partial \sigma}{\partial x} \, dx.$$

---

[1] This implies taking Δt to be equal to at least one period of oscillation at any x and Δx to span over at least one wave at any time. The operators $\partial/\partial t$ and $\partial/\partial x$ are therefore idealizations only, but with this understanding what follows is not ambiguous.

This is equal to the rate of increase of waves in dx, or to

$$\frac{\partial k}{\partial t}\, dx,$$

if we remember that $k = 2\pi/\lambda$ and $\sigma = 2\pi f$, f being the frequency. Thus

$$\frac{\partial k}{\partial t} + \frac{\partial \sigma}{\partial x} = 0,$$

or

$$\frac{\partial k}{\partial t} + \frac{d\sigma}{dk}\frac{\partial k}{\partial x} = 0.$$

Thus the group velocity c_g $(= d\sigma/dk)$ has this significance: it is the speed with which k is convected. In other words, the wave number is conserved on a geometrical point moving with the group velocity.

2.6.1. *General Interpretation of the Group Velocity*

Consider the wave motion created by some arbitrary disturbance which has infinitely many components, each with a certain wave number **k**, of components k_x, k_y, and k_z. If the waves are dispersive, they will sort themselves out after sufficient time has elapsed. When that state has been reached, we can consider the wave number to be a function of time t and space x. Of course here we run into the same sort of difficulty as when we defined the density of a gas in continuum theory. Just as we cannot truly define the density of a gas at a point, because matter is not indefinitely divisible, so here we cannot truly define **k** at a point. However, the definition of density of a gas becomes meaningful if we take a cubic volume of sides 10^{-3} mm in length. In the same spirit, we can consider **k** as a function of t and x with the understanding that, to have meaning, we must allow a volume containing several waves in each direction and yet small when compared with the space in which the groups of waves propagate. In this volume the wave number **k** has a meaning, and when we say **k** is a function of t and x, we mean that, at any t and x, if we enclose the point in question by such a small volume, we shall have $1/2\pi$ times k_x, k_y, or k_z waves per unit length in the direction of x, y, or z, respectively. The following developments are therefore asymptotic in two senses. They are valid for large time and increasingly valid for short waves.

Under this assumption we may take a rectangular parallelepiped, of sides Δx, Δy, and Δz, each of which should, strictly speaking, contain several wavelengths in the respective direction. But while keeping the limitations of their smallness in mind, we shall now develop the continuum theory without further apology. If at x the frequency is σ, that at $x + \Delta x$ is $\sigma + (\partial\sigma/\partial x)\,\Delta x + O\,(\Delta x)^2$. The difference in $\sigma/2\pi$ at the two places must be equal to the time rate of change of the number of waves in Δx. Remembering that k_x is 2π times the number of waves per unit length in the x direction and neglecting higher orders in Δx, which vanishes in the limit, we have

$$\frac{\partial k_x}{\partial t} + \frac{\partial \sigma}{\partial x} = 0,$$

or, using the indicial notation,

$$\frac{\partial k_i}{\partial t} + \frac{\partial \sigma}{\partial x_i} = 0, \tag{32}$$

since the equations for the other two directions are identical in form. Equation (32) shows that

$$\frac{\partial}{\partial t} \, \text{curl } \mathbf{k} = 0 \qquad \text{or} \qquad \text{curl } \mathbf{k} = F(\mathbf{x}).$$

If $F(\mathbf{x})$ does not vanish, $\mathbf{k}$ will have a part independent of time. But we are studying waves originating in the finite part of the space and propagating in all directions toward infinity. Hence $\mathbf{k}$ cannot have a time-independent part. Thus $F(\mathbf{x}) = 0$, and the field of the vector $\mathbf{k}$ is irrotational, and there exists a function $\phi(\mathbf{x},t)$ such that

$$\mathbf{k} = \text{grad } \phi \tag{33a}$$

and

$$\sigma = -\frac{\partial \phi}{\partial t}. \tag{33b}$$

Note that (32) and (33a) give

$$\frac{\partial}{\partial x_i}\left(\frac{\partial \phi}{\partial t} + \sigma\right) = 0,$$

so that

$$\sigma = -\frac{\partial \phi}{\partial t} + F(t),$$

But $F(t)$ can be absorbed into ϕ without affecting $\mathbf{k}$; hence (33b). The function $\phi(\mathbf{x},t)$ in (33) is called the *phase function*.

Whitham (1960), motivated by the results for the one-dimensional case, postulated the existence of the phase function $\phi(\mathbf{x},t)$, such that the hypersurfaces

$$\phi(\mathbf{x},t) = \text{const}$$

in the $\mathbf{x}t$ space can be recognized as waves, and defined $\mathbf{k}$ and σ by (33). Then (32) follows. Aside from $\mathbf{k}$, σ may explicitly depend on location also. Let $\sigma = \sigma(\mathbf{k},\mathbf{x})$. Then (32) can be written as

$$\frac{\partial k_i}{\partial t} + (c_g)_j \frac{\partial k_j}{\partial x_i} + F_i = 0,$$

or, by virtue of the irrotationality of **k**,

$$\frac{\partial k_i}{\partial t} + (c_g)_j \frac{\partial k_i}{\partial x_j} + F_i = 0, \tag{34}$$

in which
$$(c_g)_i = \frac{\partial \sigma}{\partial k_i}, \qquad F_i = \frac{\partial \sigma}{\partial x_i}. \tag{35}$$

If σ depends only on **k**, (34) shows that k_i is conserved on a point moving with the velocity c_g, which we shall call the group velocity. If σ does depend on **x** as well as **k**, then multiplication of (32) by $(c_g)_i$ gives

$$\frac{\partial \sigma}{\partial t} + (c_g)_i \frac{\partial \sigma}{\partial x_i} = 0, \tag{36}$$

which says that the frequency is conserved on a point moving with the group velocity, whether or not σ depends on **x** in addition to **k**. The kinematic meaning of the group velocity is thus clear. This generalization and elucidation of the significance of group velocity is due to Whitham (1960), and part of the idea can also be found in Landau and Lifshitz (1959, p. 257).

Another way to see the dynamical significance of the group velocity is to consider a wave packet of mean wave number **κ**, represented by

$$A_0 = \exp(i\mathbf{\kappa} \cdot \mathbf{R}) f(\mathbf{R}),$$

where f is an amplitude function for the variable A_0 representing the waves and **R** is the radius vector. The variation in the wave number **k** in the wave packet is supposed to be small, say between $\mathbf{\kappa} - \mathbf{\epsilon}_1$ and $\mathbf{\kappa} + \mathbf{\epsilon}_2$. Then A_0 can be expressed by the Fourier integral as

$$A_0 = \int g(\Delta\mathbf{k}) \exp[i(\mathbf{\kappa} + \Delta\mathbf{k}) \cdot \mathbf{R}] \, d\Delta\mathbf{k},$$

the limits being $-\mathbf{\epsilon}_1$ and $\mathbf{\epsilon}_2$. At time t the function A becomes, if σ depends only on **k**,

$$A(t) = \int g(\Delta\mathbf{k}) \exp[i(\mathbf{\kappa} + \Delta\mathbf{k}) \cdot \mathbf{R} - i\sigma(\mathbf{\kappa} + \Delta\mathbf{k})t] \, d\Delta\mathbf{k}.$$

Upon expanding $\sigma(\mathbf{\kappa} + \Delta\mathbf{k})$ and neglecting second and higher orders in $\Delta\mathbf{k}$, we have

$$A(t) = \int g(\Delta\mathbf{k}) \exp\{i[\mathbf{\kappa} \cdot \mathbf{R} - \sigma(\mathbf{\kappa})t]\} \exp\left[i\Delta\mathbf{k} \cdot \left(\mathbf{R} - t\frac{\partial \sigma}{\partial \mathbf{k}}\right)\right] d\Delta\mathbf{k}$$

$$= \exp\{i[\mathbf{\kappa} \cdot \mathbf{R} - \sigma(\mathbf{\kappa})t]\} f\left(\mathbf{R} - t\frac{\partial \sigma}{\partial \mathbf{k}}\right).$$

This proves that the amplitude distribution has moved as a whole with the same group velocity as **k** (or **κ** here) and σ. We can then expect that the energy must move with the group velocity too.   This development is due to Landau and Lifshitz (1959).   The reader is urged also to read Broer (1950), who also discusses the dynamic significance of the group velocity.

2.7. Gravity Waves Caused by a Moving Body or a Moving Surface-pressure Distribution

It has been observed in experiments that when a body, floating or submerged, moves in a liquid with a free surface, the waves created by its motion are mainly behind it, although ripples due to surface tension are often observed in front of it also.   Unless the pertinent linear dimension of the body is small compared with $(T/\rho g)^{1/2}$, the waves behind the body, which are predominantly gravity waves, have much greater amplitude than the capillary ripples in front of the body.   In this subsection capillary effects will be ignored.

One point should be scrutinized immediately.   Upon any solution of the problem of creation of surface waves by a moving disturbance it is always possible to superpose the solution for waves of any amplitude (within the limits of the linear theory) and any wavelength propagating in any direction, since the differential system governing the flow is linear and homogeneous, provided the boundary conditions on the moving body (if any) are ignored.   If we demand that the flow relative to a constant disturbance moving with a constant velocity be steady, then only one of the solutions for waves can be superposed on the solution for the motion caused by the moving disturbance.   That solution is the one for waves propagating with the same velocity as the moving disturbance.   If we further demand that there be no waves ahead of the body, the solution is unique.   But while the demand of steadiness relative to the moving disturbance is readily acceptable, the demand of the absence of waves ahead of the disturbance requires some justification.

To provide such a justification, Rayleigh (1883) assumed a body force with the components

$$X = \mu(c - u), \qquad Y = -\mu v - g, \qquad Z = -\mu w,$$

where μ is a constant, eventually to be made to vanish, representing in a rough way the effects of a small viscosity (although of course it no longer denotes the viscosity of the fluid), u, v, and w are the components of the velocity of the fluid relative to the moving disturbance, and c is the speed of the disturbance moving in the $-x$ direction.   After the Rayleigh assumption is made, the solution obtained, and μ finally put to zero, it actually turns out that there are no waves ahead of the disturbance [Lamb (1932, pp. 399–409)].   Rayleigh's device has proved useful as well as ingenious and, although artificial, it has the great merit of simplicity.   Therefore, even if later writers (e.g., Engevik 1975) have shown that its use can be replaced by a development taking into consideration the establishment of waves from rest, we shall use it here.

With the body force components assumed by Rayleigh, the Euler equations become

$$\frac{Du}{Dt} = -\frac{1}{\rho}\frac{\partial p}{\partial x} + \mu\,(c - u), \tag{37}$$

$$\frac{Dv}{Dt} = -\frac{1}{\rho}\frac{\partial p}{\partial y} - \mu v - g, \tag{38}$$

$$\frac{Dw}{Dt} = -\frac{1}{\rho}\frac{\partial p}{\partial z} - \mu w. \tag{39}$$

With the circulation Γ defined by

$$\Gamma = \oint (u\,dx + v\,dy + w\,dz),$$

we obtain from the above equations, by the same procedure as that used on p. 61 (since ρ is constant), that

$$\frac{D\Gamma}{Dt} = -\mu\Gamma.$$

This gives

$$\Gamma = C\,e^{-\mu t}.$$

Thus, if the circulation is initially (for $t = 0$) zero, it will continue to be zero. We can then consider irrotational flows if we assume that they were stated from rest.

For a sinusoidal pressure distribution at the free surface the velocity potential, after the flow is made steady by adopting a coordinate system moving with the pressure distribution which travels with speed c to the left (relative to the frame of reference at rest), is

$$\phi = c(x - \beta e^{ky} \sin kx) \tag{40}$$

whose conjugate is

$$\psi = c(y - \beta e^{ky} \cos kx). \tag{41}$$

Upon setting ψ to zero and linearizing the Bernoulli equation applied to the free surface, we obtain the free-surface pressure:

$$p_0 = \rho\beta\,[(kc^2 - g)\cos kx + \mu c \sin kx] = \mathrm{Re}(Ce^{ikx}) \tag{42}$$

in which

$$C = \rho\beta(kc^2 - g - i\mu c). \tag{43}$$

In obtaining (42), the y in the term gy has been set equal to $\mu \cos kx$, thus making ψ in (41) equal to zero upon neglecting terms of higher order in the small quantity β. If we now consider p_0 to be the surface pressure, then $\psi = 0$ is the surface streamline, on which

$$\eta = \beta \cos kx. \tag{44}$$

Equation (44) can be written in terms of C as (real part meant)

$$g\rho\eta = \frac{k_e}{k - k_e - i\mu_1}\, Ce^{ikx}, \tag{45}$$

in which $$\mu_1 = \frac{\mu}{c}, \qquad k_e = \frac{g}{c^2}. \tag{46}$$

The subscript e on k_e means that k_e is the characteristic wave number pertaining to the speed c.

Although C as defined in (43) is complex so long as $\mu \neq 0$, μ can be made as small as we please and no difference is made in the final result if we regard C as real in (45). We shall do so, and taking real parts of (42) and (45), say that the surface pressure

$$p_0 = C \cos kx \tag{47}$$

produces the surface displacement

$$\eta = \frac{Ck_e}{\rho g} \frac{(k - k_e)\cos kx - \mu_1 \sin kx}{(k - k_e)^2 + \mu_1{}^2}. \tag{48}$$

Note that μ in (43) or μ_1 in (45) are not taken to be zero, for if we take C to be real, β is complex. The μ_1 in the denominator of (48) is, in fact, crucial.

The elemental solution just obtained can be used as building blocks of the solution for any surface pressure as k is allowed to take a continuous spectrum of values. Thus, if

$$p_0 = h(x) = \int_0^\infty g(k) \cos kx\, dk, \tag{49}$$

then $$\rho g\eta = k_e \int g(k) \frac{(k - k_e)\cos kx - \mu_1 \sin kx}{(k - k_e)^2 + \mu_1{}^2}\, dk. \tag{50}$$

In particular, if $h(x)/P$ is the Dirac delta function $\delta(x)$ defined by

$$\int_{-\epsilon}^{\epsilon} \delta(x)\, dx = 1, \qquad \delta(x) = 0 \text{ for } |x| > \epsilon, \tag{51}$$

where ϵ is a quantity however small, then from the Fourier integral (49) we obtain with the inverse transform that

$$g(k) = \frac{P}{\pi}, \tag{52}$$

and (50) becomes

$$\rho g\eta = \frac{k_e P}{\pi} \int_0^\infty \frac{(k - k_e)\cos kx - \mu_1 \sin kx}{(k - k_e)^2 + \mu_1{}^2}\, dk. \tag{53}$$

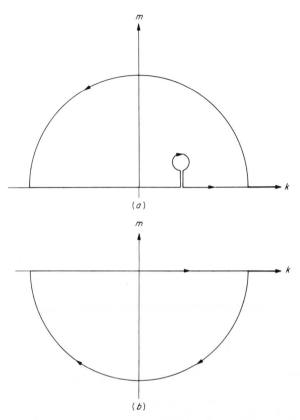

FIGURE 3. CONTOURS FOR EVALUATING THE INTEGRAL IN
EQ. (53). (a) FOR POSITIVE x, AND (b) FOR NEGATIVE x.

To evaluate (53), consider the contour integral

$$\int \frac{e^{ix\zeta}}{\zeta - a} d\zeta,$$

in which $\zeta = k + im$. In our case $a = k_e + i\mu_1$, k_e and μ_1 both being positive. The singularity of the integrand is at $\zeta = a$, which, since $\mu_1 > 0$, is located in the upper half of the complex ζ plane. For x positive, the contour is taken as shown in Fig. 3a, and the contribution of the small circle around the singularity to the integral is

$$-2\pi i e^{i(k_e + i\mu_1)x}.$$

The contribution of the large semicircle to the integral approaches zero as $|\zeta| \to \infty$, since x is positive and m is positive. Since the contour encloses

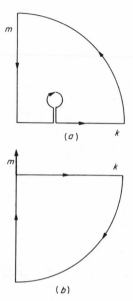

FIGURE 4. ALTERNATIVE
CONTOURS FOR EVALUAT-
ING THE INTEGRAL IN
EQ. (53). (*a*) FOR POSI-
TIVE *x* AND (*b*) FOR NEG-
ATIVE *x*.

no singularity, the integral is zero:

$$\int_{-\infty}^{0} \frac{e^{ikx}}{k - (k_e + i\mu_1)}\, dk + \int_{0}^{\infty} \frac{e^{ikx}}{k - (k_e + i\mu_1)}\, dk - 2\pi i e^{i(k_e + i\mu_1)x} = 0,$$

or
$$\int_{0}^{\infty} \frac{e^{ikx}}{k - (k_e + i\mu_1)}\, dk = 2\pi i e^{i(k_e + \mu_1)x} + \int_{0}^{\infty} \frac{e^{-ikx}}{k + (k_e + i\mu_1)}\, dk. \quad (54)$$

For *x* negative, the contour is as shown in Fig. 3*b*. It encloses no singularity; hence

$$\int_{0}^{\infty} \frac{e^{ikx}}{k - (k_e + i\mu_1)}\, dk = \int_{0}^{\infty} \frac{e^{-ikx}}{k + (k_e + i\mu_1)}\, dk. \quad (55)$$

Alternatively, for *x* positive, we can take a contour as in Fig. 4*a* and obtain

$$\int_{\infty}^{0} \frac{e^{-mx}}{im - (k_e + i\mu_1)}\, i\, dm + \int_{0}^{\infty} \frac{e^{ikx}}{k - (k_e + i\mu_1)}\, dk - 2\pi i e^{i(k_e + i\mu_1)x} = 0,$$

or
$$\int_{0}^{\infty} \frac{e^{ikx}}{k - (k_e + i\mu_1)}\, dk = 2\pi i e^{i(k_e + i\mu_1)x} + \int_{0}^{\infty} \frac{e^{-mx}}{m - \mu_1 + ik_e}\, dm. \quad (56)$$

Similarly, for x negative, integration along the contour shown in Fig. $4b$ produces

$$\int_0^\infty \frac{e^{ikx}}{k - (k_e + i\mu_1)} \, dk$$

$$= -\int_{-\infty}^0 \frac{e^{-mx}}{im - (k_e + i\mu_1)} \, i \, dm = \int_0^\infty \frac{e^{mx}}{m + \mu_1 - ik_e} \, dm. \quad (57)$$

It is evident that the integral in (53) is

$$\text{Re} \left[\int_0^\infty \frac{e^{ikx}}{k - (k_e + i\mu_1)} \, dk \right].$$

Substituting (54) and (56) into (53) and taking the real part, we have, for x positive,

$$\frac{\pi g \rho}{k_e P} \eta = -2\pi e^{-\mu_1 x} \sin k_e x + \text{Re} \left[\int_0^\infty \frac{e^{-ikx}}{k + (k_e + i\mu_1)} \, dx \right]$$

$$= -2\pi e^{-\mu_1 x} \sin k_e x + \int_0^m \frac{(m - \mu_1)e^{-mx}}{(m - \mu_1)^2 + k_e^2} \, dm,$$

which, as $\mu_1 \to 0$, becomes in the limit

$$\frac{\pi g \rho}{k_e P} \eta = -2\pi \sin k_e x + \int_0^\infty \frac{\cos kx}{k + k_e} \, dk$$

$$= -2\pi \sin k_e x + \int_0^\infty \frac{me^{-mx}}{m^2 + k_e^2} \, dm. \quad (58)$$

Since $\mu_1 \to 0$ as $t \to \infty$, the motion is established at $\mu_1 = 0$. For x negative, substitution of (55) and (57) into (53) produces, after the real part is taken, μ_1 put to zero, and k_e henceforth set equal to g/c^2,

$$\frac{\pi g \rho}{k_e P} \eta = \int_0^\infty \frac{\cos kx}{k + k_e} \, dx = \int_0^\infty \frac{me^{mx}}{m^2 + k_e^2} \, dm. \quad (59)$$

The integrals in (58) and (59) represent the "local" part of η and have the same value if the x in (59) has an absolute value equal to the x in (58). Hence it is sufficient to evaluate the integral in (58), for x positive. For convenience let it be denoted by I. By repeated integration by parts we have, with $n = mx$,

$$I = \frac{1}{k_e^2 x^2} - \int_0^\infty \frac{2n(n + 1)e^{-n}}{(n^2 + k_e^2 x^2)^2} \, dn$$

$$= \frac{1}{k_e^2 x^2} - \frac{3!}{k_e^4 x^4} + 8 \int_0^\infty \frac{n(n^2 + 3n + 3)e^{-n}}{(n^2 + k_e^2 x^2)^3} \, dn. \quad (60)$$

The asymptotic expansion is of the form

$$\frac{1}{k_e^2 x^2} - \frac{3!}{k_e^4 x^4} + \frac{5!}{k_e^6 x^6} - \cdots .$$

It is not convergent for any $k_e x$, but if finite sums like (60) are used, the question of convergence does not arise. Evidently for very large values of $k_e x$, the integrals in (60) are small and I is approximately equal to $(k_e x)^{-2}$. For values of $k_e x$ not so large, more terms in the series may give a closer approximation. But for small values of $k_e x$ the asymptotic representation is of course completely useless, and it is better to use the result

$$\int_0^\infty \frac{\cos kx}{k + k_e} \, dk = \int_{k_e x}^\infty \frac{\cos (k_e x - u)}{u} \, du$$

$$= -\text{Ci } k_e x \cos k_e x + \left(\frac{1}{2} \pi - \text{Si } k_e x \right) \sin k_e x, \quad (61)$$

in which
$$u = (k + k_e) x$$

and
$$\text{Ci } u = -\int_u^\infty \frac{\cos u}{u} \, du, \qquad \text{Si } u = \int_0^u \frac{\sin u}{u} \, du \qquad (62)$$

are tabulated functions.[1]

The important thing to note in (58) and (59) is that there are no upstream waves (x negative), only waves behind the disturbance (x positive). We shall return to this point in Sec. 2.9.

For a pressure distribution $p_0 = h(x)$ extending from $x = 0$ to $x = x_0$,

$$\pi g \rho \eta = 2\pi \int_0^{x_0} k_e h(\alpha) \sin k_e(x - \alpha) \, d\alpha + \int_0^{x_0} \int_0^\infty \frac{k_e h(\alpha) m e^{-m(x-a)}}{m^2 + k_e^2} \, dm \, d\alpha,$$

$$(63)$$

provided $x > x_0$. Similar results can be obtained for $x < 0$ and $0 < x < x_0$. Observe that the η given by (63) is bounded even as $c \to 0$. For certain forms of $h(x)$, such as

$$h(x) = \begin{cases} 0 & \text{for} \quad x < 0 \text{ and } x > x_0, \\ C & \text{for} \quad 0 < x < x_0, \end{cases}$$

or
$$h(x) = \frac{C}{x^2 + b^2},$$

the first integral in (63) can be evaluated exactly [see Lamb (1932, p. 406)], with limits changed to $\pm \infty$ for the second case.

A similar treatment with surface tension taken into account is given in Lamb (1932, pp. 464–468). In that case waves occur both ahead and

---

[1] See, for instance, Abramowitz and Stegun (1965, pp. 238–244).

behind the moving body, with the wavelength greater for the waves behind. The waves ahead of the body are mainly capillary waves, and those behind mainly gravity waves.

2.8. Generation of Gravity Waves by a Moving Submerged Cylinder

Consider a circular cylinder of radius b submerged at a depth f below the free surface, where $y = 0$. The cylinder is supposed to move horizontally in a direction normal to its axis, with a speed c, and it is desired to find the surface waves generated by this motion.

As we did in Sec. 2.7, we may consider the establishment of the motion, in order to render the solution unique, but this is tedious. After the demonstration in Sec. 2.7, we can simply accept the absence of upstream waves a priori and let that condition determine the solution uniquely. The motion relative to the cylinder can then be considered steady. A general consideration of the question of gravity waves upstream will be given in Sec. 2.9.

If the direction of motion of the moving cylinder is that of decreasing x, the velocity potential for the steady motion relative to the cylinder is

$$\phi = cx\left(1 + \frac{b^2}{r^2}\right) + \chi, \tag{64}$$

in which
$$r = [x^2 + (y + f)^2]^{1/2}. \tag{65}$$

The first part of ϕ given in (64) is the velocity potential if there is no free surface. If f is several times as large as b, χ is smaller than the first part on the right-hand side of (64) near the cylinder, and the violation of the boundary condition on the surface of the cylinder by χ can be ignored.

Since χ satisfies the Laplace equation and must approach zero as $y \to -\infty$, we can write

$$\chi = \int_0^\infty \alpha(k)e^{ky} \sin kx \, dk. \tag{66}$$

The corresponding η is

$$\eta = \int_0^\infty \beta(k) \cos kx \, dk, \tag{67}$$

where $\beta(k)$ is related to $\alpha(k)$ through the free-surface relation

$$c\frac{\partial \eta}{\partial x} = \frac{\partial \phi}{\partial y}. \tag{68}$$

The ϕ given by (64) can be written as

$$\phi = cx + cb^2 \int_0^\infty e^{-k(y+f)} \sin kx \, dk + \chi, \tag{69}$$

valid for positive values of $y + f$ only. Substitution of (69) into (68), which is for the free surface where $y + f$ is positive, produces

$$b^2 ce^{-kf} - \alpha(k) = c\beta(k). \tag{70}$$

Another relationship between $\alpha(k)$ and $\beta(k)$ is to be found from the dynamic condition that p is constant on the free surface if surface tension is ignored. The surface pressure is given by the Bernoulli equation

$$\frac{p}{\rho} = -gy - \frac{1}{2}\left(\frac{\partial \phi}{\partial x}\right)^2$$

$$= -gy - \frac{1}{2}c^2 - b^2 c^2 \int_0^\infty e^{-kf} \cos kx \, k \, dk - c \frac{\partial \chi}{\partial x}$$

$$= -gy - \frac{1}{2}c^2 - b^2 c^2 \int_0^\infty e^{-kf} \cos kx \, k \, dk - c \int_0^\infty \alpha(k) \cos kx \, kdk, \tag{71}$$

in which y has been set equal to zero and quadratic terms in χ have been neglected. If p is constant on the free surface, (67) and (71) give

$$g\beta(k) + kb^2 c^2 e^{-kf} + kc\alpha(k) = 0. \tag{72}$$

The solution of (70) and (72) is

$$\alpha(k) = -\frac{k + k_e}{k - k_e} b^2 ce^{-kf}, \qquad \beta(k) = \frac{2b^2 k e^{-kf}}{k - k_e}, \tag{73}$$

in which k_e is g/c^2. Hence (67) becomes

$$\eta = 2b^2 \int_0^\infty \frac{ke^{-kf} \cos kx}{k - k_e} \, dk$$

$$= \frac{2b^2 f}{x^2 + f^2} + 2k_e b^2 \int_0^\infty \frac{e^{-kf} \cos kx}{k - k_e} \, dk. \tag{74}$$

Since the integrand of the integral in (74) has a simple pole at $k = k_e$, the integral has the same value, when evaluated by integration along a contour, whether the contour has an indentation above or below the singular point $k = k_e$. Taking a contour as shown in Fig. 5, we obtain for the value of the last integral in (74) the real part of

$$i\pi e^{-k_e f + ik_e x} + i \int_0^\infty \frac{e^{-imf - mx}}{im - k_e} \, dm. \tag{75}$$

Thus (74) becomes

$$\eta = \frac{2b^2 f}{x^2 + f^2} - 2\pi k_e b^2 e^{-k_e f} \sin k_e x - 2k_e b^2 I(f, k_e, x), \tag{76}$$

in which $I(f, k_e, x) = \int_0^\infty \frac{(k_e \sin mf - m \cos mf)e^{-mx}}{m^2 + k_e^2} \, dm.$

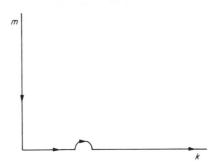

FIGURE 5.   CONTOUR FOR EVALUATING THE
INTEGRAL IN EQ. (74).

Since the η given by (74) is even in x, we know that for x negative

$$\eta = \frac{2b^2f}{x^2+f^2} + 2\pi k_e b^2 e^{-k_e f} \sin k_e x - 2k_e b^2 I(f,k_e,-x) \qquad (77)$$

satisfies all the conditions required of it at the free surface and corresponds
to a χ that satisfies the Laplace equation and the condition at $y = -\infty$.
Now since we wish to have no waves upstream, the second term in the
right-hand side of (77) must be made to vanish.   This can be done by
superposing a solution for η corresponding to stationary waves on the η
given by (76) and (77).   This solution is

$$\eta = -2\pi k_e b^2 e^{-k_e f} \sin k_e x,$$

which changes (76) and (77) into

$$\eta = \begin{cases} \dfrac{2b^2f}{x^2+f^2} - 4\pi k_e b^2 e^{-k_e f} \sin k_e x - 2k_e b^2 I(f,k_e,x) & \text{for } x > 0, \\[3mm] \dfrac{2b^2f}{x^2+f^2} - 2k_e b^2 I(f,k_e,-x) & \text{for } x < 0. \end{cases}$$

$$(78)$$

2.9. Location of Waves Produced by Moving Disturbances

The establishment of wave motion created by a moving surface pressure
discussed in Sec. 2.7 shows that the gravity waves created are behind the
moving disturbance.   A more general treatment of the location of waves
produced by a moving disturbance is given in Lamb (1932, pp. 413–414).
The development that follows is essentially the same as that of Lamb.
We assume that the waves produced by a (moving) concentrated
pressure at the free surface at a time t *antecedant* to the present

($t = 0$) to attenuate as $e^{-\mu t}$, integrate the waves produced in $0 \le t < \infty$, and finally let μ become zero. Referring x to coordinate axes moving with the disturbance with speed c in the negative x direction, the waves (whatever their kind, in fact) have either of the factors $e^{i\sigma t \pm ik(x-ct)}$. Thus one can write

$$\eta = \int_0^\infty \left\{ \int_0^\infty f(k)\, e^{i\sigma t + ik(x-ct)}\, dk + \int_0^\infty f(k)\, e^{i\sigma t - ik(x-ct)}\, dk \right\} e^{-\mu t}\, dt,$$

(79)

which, upon integration, becomes

$$\eta = \int_0^\infty \frac{f(k)e^{ikx}}{\mu - i(\sigma - kc)}\, dk + \int_0^\infty \frac{f(k)e^{-ikx}}{\mu - i(\sigma + kc)}\, dk, \quad (79a)$$

in which $f(k)$ is the Fourier coefficient of wave components. In (79) and (79a), σ is the circular frequency corresponding to k, and is a function of k only. Note that c is here the speed with which the disturbance moves, not the phase velocity of waves with wave number k. For convenience we shall write k_e for the solution of the equation $\sigma = kc$. Thus c is the phase velocity for waves with wave number k_e. We shall denote $d\sigma/dk$ evaluated at $k = k_e$ by c_g. Thus c_g is the group velocity of the same waves whose phase velocity is c.

Since μ can be considered small, the main part of (79a) comes from that range of k which is near k_e. Hence we can write

$$\sigma - kc = \left(\frac{d\sigma}{dk} - c \right)k' = (c_g - c)k', \quad (80)$$

in which $k' = k - k_e$ and c_g is the group velocity. With (80) substituted into it and c_g understood to be for $k = k_e$, (79a) becomes

$$\eta = \frac{f(k_e)e^{ik_e x}}{c - c_g} \int_{-\infty}^\infty \frac{e^{ik'x}}{\mu' + ik'}\, dk', \quad (81)$$

in which $\mu' = \mu/(c - c_g)$ and the limits of integration have been extended to $\pm\infty$ for convenience of the evaluation of the integral only. The error this extension introduces is not serious, since the main part of the integral comes from small values of k'. The integral in (81) can be evaluated by taking a contour in the plane of $k' + im'$, as shown in Fig. 6a for x positive or in Fig. 6b for x negative, and by application of Cauchy's integral theorems for analytic functions. Thus, for $c - c_g$ positive,

$$\int_{-\infty}^\infty \frac{e^{ik'x}}{\mu' + ik'}\, dk' = \begin{cases} 2\pi \exp\left(-\dfrac{\mu x}{c - c_g} \right) & \text{for } x > 0, \\[2mm] 0 & \text{for } x < 0. \end{cases}$$

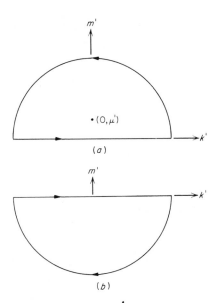

FIGURE 6. CONTOURS FOR EVALUATING
THE INTEGRAL IN EQ. (81). (*a*) FOR POSI-
TIVE *x* AND (*b*) FOR NEGATIVE *x*.

For $c - c_g$ negative,

$$\int_{-\infty}^{\infty} \frac{e^{ik'x}}{\mu' + ik'} \, dk' = \begin{cases} 0 & \text{for } x > 0, \\ -2\pi \exp \dfrac{\mu x}{c_g - c} & \text{for } x < 0. \end{cases}$$

On allowing μ to approach zero, we have, in the limit,

$$\eta = \begin{cases} \dfrac{2\pi f(k_e)e^{ik_e x}}{c - c_g} & \text{for } x > 0, \\ 0 & \text{for } x < 0 \end{cases}$$

if $c_g < c$, and

$$\eta = \begin{cases} 0 & \text{for } x > 0, \\ \dfrac{2\pi f(k_e)e^{ik_e x}}{c_g - c} & \text{for } x < 0 \end{cases}$$

if $c_g > c$. This explains the absence of upstream gravity waves and downstream capillary waves. If gravity and capillary effects are both taken into account, there will be both upstream and downstream waves, although the upstream waves, being mainly capillary in nature, will have a wavelength smaller than that of the downstream waves, which are mainly gravity waves.

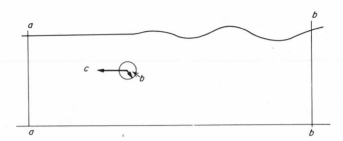

FIGURE 7.   SKETCH FOR CALCULATING WAVE RESISTANCE.

The use of (80) is valid only if higher-order terms in the Taylor expansion of $\sigma - kc$ can be neglected.   In particular, it is necessary that

$$\frac{1}{2}\frac{d^2\sigma}{dk^2}k'^2 \ll |c_g - c|\, k' \tag{84}$$

even when $k'x$ is a moderate multiple of 2π (beyond which the contribution to the integral can be neglected).   Condition (84) can be written as

$$\frac{d^2\sigma}{dk^2} \ll |c_g - c|\, x,$$

which can always be satisfied by sufficiently large values of x provided $c_g \neq c$.

2.10.   Wave Resistance

The resistance to the horizontal translation of a two-dimensional body in a fluid due to the gravity waves created behind can be computed by considering two fixed vertical planes normal to the direction of motion, one far ahead of the body and one far behind (see Fig. 7).   Nothing happens at the plane ahead, and the rate of work done on the fluid between the two planes at plane bb is $c_g E$ if E is the mean energy per unit area of the free surface, as shown in Sec. 2.6.   (Remember that the waves are progressing in the direction of decreasing x.)   On the other hand, the rate at which the wave energy is increasing between the planes is simply cE, since the distance between the body and the plane bb is increasing at the rate c.   If the resistance to the body is R, so that Rc is the rate at which work is being done by the body on the fluid between the planes, equating the rate of work done on the fluid between the planes with the rate of increase of its energy produces

$$Rc + Ec_g = cE \qquad \text{or} \qquad R = \frac{c - c_g}{c}\,E. \tag{85}$$

This is for gravity waves.   For capillary waves a similar consideration

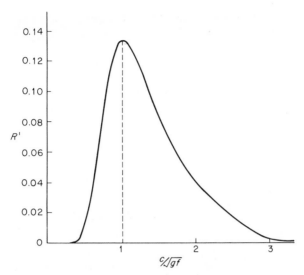

FIGURE 8. $R' = Rf^2/4\pi^2 g\rho b^4$ AS A FUNCTION OF $c/\sqrt{gf}$, R BEING THE RESISTANCE ON A DEEPLY SUBMERGED CYLINDER OF RADIUS b, c BEING ITS SPEED, AND f ITS DEPTH, OF SUBMERGENCE.

produces

$$Rc - Ec_g = -cE \quad \text{or} \quad R = \frac{c_g - c}{c}E. \tag{86}$$

According to (78), the amplitude of the waves behind the moving submerged cylinder of radius b is proportional to $k_e \exp(-k_e f)$. Since $c_g = c/2$ and

$$E = \frac{1}{2} g\rho [4\pi k_e b^2 \exp(-k_e f)]^2,$$

the resistance is

$$R = 4\pi^2 g\rho b^4 k_e^2 \exp(-2k_e f) = 4\pi^2 g^3 \rho b^4 c^{-4} \exp\left(-\frac{2gf}{c^2}\right), \tag{87}$$

which is depicted in Fig. 8. The maximum R occurs at $c = (gf)^{1/2}$.

2.11. Standing Waves in a Finite Mass of Fluid of Uniform Depth

Consider a mass of fluid contained in a cylindrical space of depth h. If the intersection of the lateral surface of the cylinder with a plane normal to the generatrix of that surface is denoted by C and the distance normal to the surface by n, then

$$\frac{\partial \phi}{\partial n} = 0 \quad \text{on the wall.} \tag{88}$$

Suppose that the origin is situated at the bottom and z is measured in the direction of the vertical; then

$$\frac{\partial \phi}{\partial z} = 0 \qquad \text{at } z = 0 \tag{89}$$

and the boundary condition at the free surface, corresponding to (9a), is

$$\sigma^2 \phi = \left(g - \frac{T}{\rho} \nabla_1^2 \right) \frac{\partial \phi}{\partial z} \qquad \text{at } z = h, \tag{90}$$

in which $$\nabla_1^2 = \frac{\partial^2}{\partial x^2} + \frac{\partial^2}{\partial y^2}$$

and σ is the angular frequency in the time factor $e^{\sigma t}$ contained in ϕ. The differential equation satisfied by ϕ is the Laplace equation

$$\left(\frac{\partial^2}{\partial x^2} + \frac{\partial^2}{\partial y^2} + \frac{\partial^2}{\partial z^2} \right) \phi = 0. \tag{91}$$

Wave motion in the fluid is governed by the system consisting of (91) and (88) to (90).

To solve the differential system just formulated, we seek a solution $S_n(x,y)$ satisfying the equations

$$(\nabla_1^2 + k_n^2) S_n(x,y) = 0, \qquad \frac{\partial S}{\partial n} = 0. \tag{92}$$

The k's are the eigenvalues and the S's the corresponding eigenfunctions. They all depend on the shape and size of the cross section of the cylinder, or on C. We then write

$$\phi_n = S_n(x,y) \cosh k_n z,$$

which satisfies (91), (88), and (89). Equation (90) demands that

$$\sigma^2 = \left(g + \frac{T}{\rho} k_n^2 \right) k_n \tanh k_n h. \tag{93}$$

Thus σ^2 becomes greater and greater as n increases, since k_n increases with n.

2.12. Edge Waves

The waves studied in Sec. 2.2 have the very striking property, if surface-tension effects are neglected, that the pressure along any streamline in a frame of reference moving with the waves is constant. This striking property has at least two consequences. One is the validity of the solution given in Sec. 2.2 (with $T = 0$) for a fluid of any stratification in density. Another is the existence of edge waves, also for any stratification.

To demonstrate the aforementioned property and its first consequence, the solution given in Sec. 2.2, with $T = 0$, will be given in coordinates moving with the waves. Thus (12) becomes

$$\phi = -cx + Ce^{ky} \cos kx, \tag{94}$$

and the corresponding stream function is

$$\psi = -cy - Ce^{ky} \sin kx. \tag{95}$$

The pressure corresponding to the ϕ and ψ given by (94) and (95) is given by the fundamental equations

$$-c \frac{\partial u'}{\partial x} = -\frac{1}{\rho} \frac{\partial p}{\partial x}, \qquad -c \frac{\partial v'}{\partial x} = -\frac{1}{\rho} \frac{\partial p}{\partial y} - g, \tag{96}$$

in which u' and v' are the components of the velocity perturbation pertaining to the waves. By multiplying the two equations in (96) respectively by dx and dy, we obtain

$$\frac{dp}{\rho} = -g \, dy + c \, du' = -g \, dy - Cck \, d(e^{ky} \sin kx), \tag{97}$$

since $\qquad u' = \dfrac{\partial}{\partial x} (Ce^{ky} \cos kx) \qquad$ and $\qquad \dfrac{\partial v'}{\partial x} = \dfrac{\partial u'}{\partial y}.$

By virtue of (95) and because $g = c^2 k$, (97) can be written as

$$\frac{dp}{\rho} = kc \, d\psi. \tag{98}$$

Now, since the flow is steady in the frame of reference adopted, ρ is a function of ψ only. Thus

$$p = kc \int \rho \, d\psi \tag{99}$$

is also a function of ψ only, being constant on every streamline, with any density variation from one streamline to another.

Now we adopt the coordinates shown in Fig. 9 and retain (94) as the solution for ϕ. We are certain that ϕ satisfies the Laplace equation and that there is no velocity component normal to the sloping shore $y' = -z' \tan \beta$, for there is simply no velocity component in the z direction. Since the component of the gravitational acceleration in the y direction is $-g \sin \beta$, we also know that if

$$c^2 = \frac{g}{k} \sin \beta, \tag{100}$$

the pressure on the streamline which is the intersection of the water surface and the sloping shore is constant. Since $y = y' \sin \beta - z' \cos \beta$, the waves die out away from the shore. All these are the results of the

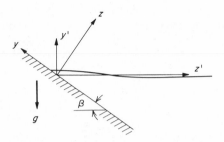

FIGURE 9. EDGE-WAVE PROFILE AND CO-
ORDINATES FOR EXPLAINING THE EDGE-WAVE
SOLUTION.

theory presented in Sec. 2.2. But the chief question remains: Is there a free surface near $y' = 0$ on which the pressure is constant?

We note that the pressure distribution is hydrostatic in the direction of z, according to the equation

$$g \cos \beta = -\frac{1}{\rho} \frac{\partial p}{\partial z}. \tag{101}$$

Because the coordinates are now tilted, Eqs. (96) have the form

$$-c \frac{\partial u'}{\partial x} = -\frac{1}{\rho} \frac{\partial p}{\partial x},$$
$$-c \frac{\partial v'}{\partial x} = -\frac{1}{\rho} \frac{\partial p}{\partial y} - g \sin \beta. \tag{102}$$

If these are multiplied respectively by dx and dy, (101) is multiplied by dz, and the results are added, we see that

$$\frac{dp}{\rho} = -g \sin \beta \, dy + c \, du' - g \cos \beta \, dz$$
$$= -g \sin \beta \, dy - Cck \, d(e^{ky} \sin kx) - g \cos \beta \, dz, \tag{103}$$

since

$$\frac{\partial v'}{\partial x} = \frac{\partial u'}{\partial y}.$$

Because of (100), and by virtue of (95), (103) can be written as

$$\frac{dp}{\rho} = d(kc\psi - gz \cos \beta).$$

This means that

$$p = F(\lambda), \qquad \rho = F'(\lambda), \qquad \lambda = kc\psi - gz \cos \beta, \tag{104}$$

which states that in any surface of constant density the pressure is constant.

We have demonstrated that edge waves are possible for any stratification, since the function $F(\lambda)$ is arbitrary. It is reassuring that the constant-λ surfaces are nearly horizontal. This can be seen from (95) and (104), which together give

$$\lambda = -gy' - Ccke^{ky} \sin kx = -gy' - Ccke^{k(y'\sin\beta - z'\cos\beta)} \sin kx,$$

by virtue of (100) and because

$$y' = y \sin \beta + z \cos \beta.$$

Since the term containing x corresponds to wave motion, the constant-λ surfaces are nearly constant-y' surfaces, or are nearly horizontal.

It was Stokes (1846) who discovered linear edge waves in a homogeneous fluid. The presentation given here not only extends the applicability of the solution to a fluid of any stratification but shows how edge waves owe their existence to a special property of gravity waves in a semi-infinite fluid, i.e., the property (99), which entrains (104). Every constant-density surface is constituted by streamlines, and the differences in pressure between these streamlines, which would exist if y were measured vertically, are annihilated by the pressure gradient in the z direction, so that on a surface of constant ρ the pressure is constant.

2.13. Wave Makers

In a very long channel filled with water of constant depth h, a wave maker at $x = 0$, and a perfect wave absorber at the far end, the waves made by the oscillating wave maker will propagate toward the far end and not be reflected. The solution can be constructed on the assumption of an infinitely long channel, in which waves are radiated away at $x = \infty$. The solution is simple.

If the frequency of the wave maker is given, then σ is fixed. For simplicity we shall ignore capillary effects and write (17) as

$$\sigma^2 = gk \tanh kh. \tag{105}$$

This determines all the allowable values for the k in (15). Solutions of the type of (15) will then be superposed to provide a solution. First, it is easy to see that if k satisfies (105), $-k$ does also. It is also easy to see that there is only one pair of real roots of (105), since the curves

$$Y = \frac{\sigma^2}{gk} \quad \text{and} \quad Y = \tanh kh$$

intersect only at two points, where $k = k_0$ and $k = -k_0$, say. The remaining roots of (105) are imaginary, and the substitution

$$k = ik'$$

transforms (105) to

$$\sigma^2 = -gk \tan kh \tag{106}$$

upon dropping the prime on k'. It is easy to see that there are infinitely many of these roots since they are determined by the intersections of

$$K = -\frac{\sigma^2}{gk} \quad \text{and} \quad K = \tan kh.$$

Let the positive roots of k satisfying (106) be denoted by k_1, k_2, ... Then the solution for ϕ is

$$\phi = A_0 \cosh k_0(y + h) \cos (k_0 x - \sigma t) + \sum_{n=1}^{\infty} A_n e^{-k_n x} \cos k_n(y + h) \sin \sigma t.$$

$$(107)$$

The satisfaction of the free-surface condition

$$\sigma^2 \phi = g \frac{\partial \phi}{\partial y},$$

which is simply (10) with $T = 0$ and σ replacing kc, is guaranteed by the satisfaction of (105). Remembering that the origin is at the undisturbed free surface, (107) obviously satisfies (16). Since at large values of x the solution represents waves propagating toward $x = \infty$, the radiation condition at $x = \infty$ is satisfied. It remains to satisfy the condition given at the wave maker, where

$$u = U(y) \sin \sigma t.$$

This is satisfied by demanding that

$$U(y) = A_0 k_0 \cosh k_0(y + h) - \sum_{n=1}^{\infty} A_n k_n \cos k_n(y + h), \qquad (108)$$

the right-hand side of which is $\partial \phi / \partial x$ with x set equal to zero after differentiation. The eigenfunctions $\cosh k_0(y + h)$ and $\cos k_n(y + h)$, satisfying a homogeneous linear differential system of the type

$$F'' + k^2 F = 0,$$

$$F'(-h) = 0, \qquad \sigma^2 F = gF',$$

are orthogonal. The orthogonality allows us to determine the A's by multiplying both sides of (108) by any of the eigenfunctions and integrating between $y = -h$ and $y = 0$.

If part of the wave energy is reflected at the far end, but only a small part, then the reflected waves will travel in the opposite direction and will be reflected again at the wave maker, which can be considered a perfect reflector. Successive reflections after that can be neglected, since a good part of these already reflected waves will be absorbed at the far end. Hence far from both ends

$$\phi = A_0 \cosh k_0(y + h) \cos (k_0 x - \sigma t)$$

$$+ 2\epsilon A_0 \cosh k_0(y + h) \cos k_0 x \cos \sigma t.$$

The corresponding surface form is

$$\eta = -\frac{1}{\sigma} A_0 k_0 \sinh k_0 h \sin (k_0 x - \sigma t) - \frac{2}{\sigma} \epsilon A_0 k_0 \sinh k_0 h \cos k_0 x \sin \sigma t,$$

or
$$\eta = A \cos \sigma t + B \sin \sigma t,$$

in which

$$A = -\frac{k_0 \sinh k_0 h}{\sigma} A_0 \sin k_0 x, \qquad B = \frac{1 - 2\epsilon}{\sigma} A_0 k_0 \sinh k_0 h \cos k_0 x.$$

If a slowly moving (in the x direction) instrument records η, the envelope of η has an amplitude whose square is

$$H = A^2 + B^2 = \left(\frac{A_0 k_0 \sinh k_0 h}{\sigma}\right)^2 [\sin^2 k_0 x + (1 - 2\epsilon)^2 \cos^2 k_0 x]$$

$$= \left(\frac{A_0 k_0 \sinh k_0 h}{\sigma}\right)^2 (1 - 2\epsilon - 2\epsilon \cos 2k_0 x)$$

upon neglecting a quantity of $O(\epsilon^2)$. The envelope has a wavelength of only one-half that of the waves made by the wave maker. If H_{max} is the maximum and H_{min} the minimum of H, then [Ursell, Dean, and Yu (1960)]

$$\frac{H_{max} - H_{min}}{H_{max} + H_{min}} = \frac{4\epsilon}{2 - 4\epsilon} = 2\epsilon \qquad \text{approx.} \qquad (109)$$

It is therefore possible to determine the coefficient of reflection ϵ by measurements of H.

3. NONLINEAR SURFACE WAVES

If the amplitude of the waves is not small enough for the linear theory presented so far in this chapter to be valid, not only will the squares of the velocity components of the fluid in wave motion have to be retained in the analysis, but the kinematic and dynamic conditions at the free surface must also be applicable at $y = \eta$, not at $y = 0$. These requirements render the analysis difficult. In this section, we shall first present an exact solution for waves of finite amplitude. These waves, found mathematically possible by Gerstner (1802), are rotational. Then Stokes' irrotational waves of finite amplitude, the solitary wave, and cnoidal waves will be briefly discussed. This section will be concluded by a presentation of shallow-water theory for long waves or for steady flows of a sheet of liquid with a depth much smaller than any relevant representative length in a horizontal direction.

3.1. Gerstner Waves

The solution for Gerstner waves is in Lagrangian coordinates. These coordinates were presented in Sec. 3 of Chap. 1. Although the coordinates c_1, c_2, and c_3 can be taken to be the initial Cartesian coordinates of the fluid particles at $t = 0$, such an assignment is not necessary. The Lagrangian coordinates of a particle constitute its name, which it retains for all values of the time t, but its name does not have to be determined by its initial position.

Consider a small parallelepiped along whose nonparallel edges the increments in Lagrangian coordinates are dc_1, dc_2, and dc_3. If A_1, A_2, and A_3 are the initial Cartesian coordinates of the fluid particles, the volume of the parallelepiped is, according to a theorem in analytic geometry,

$$\begin{vmatrix} \dfrac{\partial A_1}{\partial c_1} dc_1 & \dfrac{\partial A_2}{\partial c_1} dc_1 & \dfrac{\partial A_3}{\partial c_1} dc_1 \\[2mm] \dfrac{\partial A_1}{\partial c_2} dc_2 & \dfrac{\partial A_2}{\partial c_2} dc_2 & \dfrac{\partial A_3}{\partial c_2} dc_2 \\[2mm] \dfrac{\partial A_1}{\partial c_3} dc_3 & \dfrac{\partial A_2}{\partial c_3} dc_3 & \dfrac{\partial A_3}{\partial c_3} dc_3 \end{vmatrix} = \frac{\partial(A_1,A_2,A_3)}{\partial(c_1,c_2,c_3)} dc_1\, dc_2\, dc_3,$$

in which the determinant on the right-hand side is the Jacobian whose definition is given by the equation, and is a function of c_1, c_2, and c_3 only. Similarly, at any later time, the volume of the same material contained in the parallelepiped is

$$\frac{\partial(X_1,X_2,X_3)}{\partial(c_1,c_2,c_3)} dc_1\, dc_2\, dc_3.$$

Hence the conservation of mass demands that

$$\rho\, \frac{\partial(X_1,X_2,X_3)}{\partial(c_1,c_2,c_3)} = \rho_0\, \frac{\partial(A_1,A_2,A_3)}{\partial(c_1,c_2,c_3)} = \text{function of } (c_1,c_2,c_3). \qquad (110)$$

In the following, we shall use x_i for X_i, reserving X_i to denote the body force per unit mass, as we have done in all preceding chapters except where Lagrangian coordinates are considered. Thus x_1, x_2, and x_3 are functions of c_1, c_2, c_3, and t. This temporary inconsistency of notation is tolerable, since it is not likely that confusion will result. The equations of motion are

$$\frac{\partial^2 x_i}{\partial t^2} = X_i - \frac{1}{\rho} \frac{\partial p}{\partial x_i}, \qquad i = 1, 2, 3. \qquad (111)$$

If the i in (111) is changed to α and the result multiplied by $\partial x_\alpha/\partial c_i$ and summed over α, the equations of motion become

$$\left(\frac{\partial^2 x_\alpha}{\partial t^2} - X_\alpha \right) \frac{\partial x_\alpha}{\partial c_i} + \frac{1}{\rho} \frac{\partial p}{\partial c_i} = 0, \qquad (112)$$

to be solved with

$$\rho \frac{\partial(x_1, x_2, x_3)}{\partial(c_1, c_2, c_3)} = \text{function of } (c_1, c_2, c_3). \tag{110a}$$

Gerstner considered a two-dimensional flow of a liquid of homogeneous density independent of c_3. If the direction of gravitational acceleration is taken to be the direction of decreasing c_2,

$$X_1 = 0 = X_3 \quad \text{and} \quad X_2 = -g. \tag{113}$$

Denoting x_1 and x_2 by x and y and c_1 and c_2 by a and b, he found the solution

$$x = a + \frac{1}{k} e^{kb} \sin k(a - ct), \qquad y = b - \frac{1}{k} e^{kb} \cos k(a - ct), \tag{114}$$

in which c is the phase velocity of the waves. For Gerstner's solution

$$\frac{\partial(x,y)}{\partial(a,b)} = 1 - e^{2kb}, \tag{115}$$

so that $(88a)$ is satisfied. With (114), the equations of motion (112) become

$$\frac{1}{\rho} \frac{\partial p}{\partial a} + \frac{\partial}{\partial a} gy = kc^2 e^{kb} \sin k(a - ct), \tag{116}$$

$$\frac{1}{\rho} \frac{\partial p}{\partial b} + \frac{\partial}{\partial b} gy = -kc^2 e^{kb} \cos k(a - ct) + kc^2 e^{2kb}. \tag{117}$$

If the first of these equations is multiplied by da and the second by db and the results added, then at any instant t

$$\frac{dp}{\rho} = d\left(-gb + \frac{1}{2} c^2 e^{2kb}\right) + d\left[\left(\frac{g}{k} - c^2\right) e^{kb} \cos k(a - ct)\right],$$

or

$$\frac{dp}{\rho} = d\left(-gb + \frac{1}{2} c^2 e^{2kb}\right) \tag{118}$$

if

$$c^2 = \frac{g}{k}. \tag{119}$$

Thus if ρ is a function of b only, p too will be, and the solution (114) corresponds to a wave motion of finite amplitude in a liquid of any stratification, in which every surface of constant density is a surface of constant pressure. That ρ is indeed a function of b only can be seen in the following way.

With respect to a set of coordinate axes moving with the waves, (114) becomes

$$x = a - ct + \frac{1}{k} e^{kb} \sin k(a - ct), \qquad y = b - \frac{1}{k} e^{kb} \cos k(a - ct).$$

If we solve the first of these for $a - ct$ and substitute the result into the second, we obtain the equation for path lines and therefore streamlines for the *steady* flow relative to the moving axes. The only parameter is b, since k is constant for the entire flow. Thus to each constant b corresponds one streamline. Since ρ is constant along a streamline in steady flow, ρ must be a function of b only—not merely for the steady flow referred to moving coordinates but also for the unsteady flow referred to fixed coordinates.

Gerstner's waves are rotational, since a straightforward calculation shows that

$$u\,dx + v\,dy = \frac{\partial x}{\partial t}\left(\frac{\partial x}{\partial a}\,da + \frac{\partial x}{\partial b}\,db\right) + \frac{\partial y}{\partial t}\left(\frac{\partial y}{\partial a}\,da + \frac{\partial y}{\partial b}\,db\right)$$

$$= -\frac{c}{k}\,d[e^{kb}\sin k(a - ct)] - ce^{2kb}\,da,$$

which is not an exact differential. The circulation around the parallelogram with vertices

$$(a,b),\ (a + da,\ b),\ (a,\ b + db),\ (a + da,\ b + db)$$

is
$$\frac{\partial}{\partial b}\,(ce^{2kb}\,da)\,db.$$

But the area of the parallelogram is, by (115),

$$(1 - e^{2kb})\,da\,db.$$

Hence by virtue of the theorem of Stokes on circulation and vorticity,

$$\zeta = \frac{2kce^{2kb}}{1 - e^{2kb}}. \tag{120}$$

Gerstner's original work (1802) was for a homogeneous liquid. The extension to a stratified liquid was made by Dubreil-Jacotin (1932). Gerstner's solution was further shown by Yih (1966) to represent edge waves, with wave velocity c given by

$$c^2 = \frac{g\sin\beta}{k}, \tag{121}$$

in a liquid of any stratification in density, propagating along a sloping shore with an angle of inclination to the horizontal equal to β, and dying out toward sea.

The demonstration for edge waves is very similar to that given in Sec. 2.12 of this chapter. Thus, with reference to the coordinates in Fig. 9, we retain Gerstner's solution (114). Equations (116) and (117) remain valid, with g replaced by $g\sin\beta$. Since there is no motion in the direction of z, the third Lagrangian coordinate, c_3, can simply be taken to be z.

The equation of motion in the direction of z is

$$\frac{1}{\rho} \frac{\partial p}{\partial z} = -g \cos \beta. \tag{122}$$

If (116) and (117), with the g therein replaced by $g \sin \beta$, are multiplied by da and db, (122) is multiplied by dz, and the results are added, we have, by virtue of (114) and (121),

$$\frac{dp}{\rho} = d\left(-gb \sin \beta + \frac{1}{2} c^2 e^{2kb} - gz \cos \beta \right). \tag{123}$$

This means that

$$p = F(\lambda), \qquad \rho = F'(\lambda), \qquad \lambda = -g(b \sin \beta + z \cos \beta) + \frac{1}{2} c^2 e^{2kb}, \tag{124}$$

and the pressure is constant on any surface of constant density. Since apart from the wave motion, y is equal to b, the quantity

$$b \sin \beta + z \cos \beta$$

is equal to y' apart from the wave motion. Thus λ is equal to $-gy'$ apart from terms coming from the wave motion. Hence the constant-density and constant-pressure surfaces, being surfaces of constant λ, are largely horizontal, as is to be expected.

The success can again be traced to the special property of Gerstner waves that the pressure is constant along any streamline in the coordinate system moving with the waves, or that p is a function only of b, as shown in (118). Again the constant-density surfaces in the fluid in edge-wave motion consist of streamlines in the moving system, and the differences in pressure between these streamlines, which would exist if y were measured vertically, are again annihilated by the pressure gradient in the z direction, so that on a surface of constant density the pressure is also constant.

Explicitly, the edge-wave solution is still (114), except that the y in it is to be replaced by

$$y' \sin \beta - z' \cos \beta$$

if we wish to exhibit the solution explicitly. The wave speed c is given by (100). The constant-density surfaces become horizontal far from shore.

3.2. Stokesian Waves

As the amplitude of surface waves increases, not only does the surface form change but the speed of the waves increases. Stokes (1847) was the first to treat gravity waves of finite amplitude in a semi-infinite liquid. For this reason such waves will be called *Stokesian waves*.

If the amplitude (to be defined) is denoted by a and the wave length by λ, then if quantities of the order of a^4/λ^4 are neglected when compared with a^3/λ^3, the solution for Stokesian waves, with reference to coordinates moving with the waves, is contained in

$$\phi = c(x - \beta e^{ky} \sin kx), \qquad \psi = c(y - \beta e^{ky} \cos kx), \qquad (125)$$

in which c is the wave velocity. The flow represented by (125) is steady. These agree with (40) and (41) and differ from (94) and (95) only by a shift of the origin of x and the sign of c. Actually it is not at all obvious that to the order of accuracy stated it is unnecessary to have terms like

$$e^{2ky} \sin 2kx \qquad \text{and} \qquad e^{2ky} \cos 2kx$$

added in (125). But Stokes' calculations indeed showed that this is unnecessary. This is not to say that the waves will be simple-harmonic. Since y is a function of x on the free surface, the term e^{ky} in (125) will give rise to terms containing $\cos 2kx$, $\cos 3kx$, etc. It so happens that the term containing $\cos 2kx$ in the surface form is correctly determined by (125) if quantities of $O(k^4 a^4)$ are neglected. We shall return to this point later. Accepting this fact, pointed out by Rayleigh (1876), we shall proceed with (125).

The streamline $\psi = 0$ is given by

$$y = \beta e^{ky} \cos kx, \qquad (126)$$

and this, by successive approximations, can be written as

$$y = \frac{1}{2} k\beta^2 + \beta \left(1 + \frac{9}{8} k^2\beta^2\right) \cos kx$$
$$+ \frac{1}{2} k\beta^2 \cos 2kx + \frac{3}{8} k^2\beta^3 \cos 3kx + \cdots. \qquad (127)$$

If we set

$$\beta \left(1 + \frac{9}{8} k^2\beta^2\right) = a, \qquad (128)$$

Eq. (127) can be written as

$$y - \frac{1}{2} ka^2 = a \cos kx + \frac{1}{2} ka^2 \cos 2kx + \frac{3}{8} k^2a^3 \cos 3kx + \cdots.$$
$$(129)$$

The coefficient of $\cos kx$ is always defined to be the amplitude of the waves. To the degree of accuracy indicated in the preceding paragraph, a is given in terms of β by (128).

So far it has not been shown that (127) or (129) gives the form of the free surface. For the streamline $\psi = 0$ to qualify as the free-surface streamline, the pressure on it must be constant. Now the pressure is given by the Bernoulli equation [the speed being obtained from (125)]

$$\frac{p}{\rho} = \text{const} - gy - \frac{1}{2} c^2(1 - 2k\beta e^{ky} \cos kx + k^2\beta^2 e^{2ky}).$$

FIGURE 10. STOKESIAN WAVES.

On the streamline $\psi = 0$, (126) holds, and

$$\frac{p}{\rho} = \text{const} + (kc^2 - g)y - \frac{1}{2}k^2c^2\beta^2e^{2ky}$$

$$= \text{const} + (kc^2 - g - k^3c^2\beta^2)y + \cdots . \tag{130}$$

Thus the streamline $\psi = 0$ is the free-surface streamline if

$$c^2 = \frac{g}{k} + k^2c^2\beta^2 \quad \text{or} \quad c^2 = \frac{g}{k}(1 + k^2a^2). \tag{131}$$

Thus the wave velocity increases with the amplitude.

Further approximations by Stokes give, for the wave form,

$$y = \text{const} + a\cos kx$$

$$+ (\tfrac{1}{2}ka^2 + {}^{17}\!/_{24}k^3a^4)\cos 2kx + \tfrac{3}{8}k^2a^3\cos 3kx$$

$$- \tfrac{1}{3}k^3a^4\cos 4kx + \cdots \tag{132}$$

$$+ e$$

and, for the wave velocity,

$$c^2 = \frac{g}{k}(1 + k^2a^2 + \tfrac{5}{4}k^4a^4 + \cdots). \tag{133}$$

Comparison of (132) with (129) and (133) with (131) reveals that (129) and (131) are indeed accurate to terms of $O(k^3a^3)$, or if k^4a^4 is neglected compared with k^3a^3.

The waveform is more peaked at the crests and more stretched out at the troughs, as shown in Fig. 10 for $ka = \tfrac{1}{2}$. The maximum amplitude is the one that makes the highest points of the free surface the vertices of angles of 120°, as can be demonstrated by requiring the speed at these points to be zero.

The convergence of Stokes' series was questioned by at least one later investigator. This question of the existence of permanent waves of finite amplitudes was settled by Levi-Civita (1925), though even he did not determine how large the amplitude can be. It turns out that the question of maximum allowable amplitude of Stokes' solution is really of limited interest, since Benjamin (1967a) has shown that Stokesian waves are unstable for

$$0 < \delta \leq \sqrt{2}\,ka$$

when residual wave motions at adjacent sideband frequencies

$$\sigma(1 \pm \delta)$$

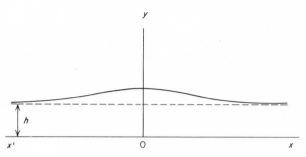

FIGURE 11.   THE SOLITARY WAVE.

are present, σ being the fundamental frequency of the Stokesian waves.

Finite-amplitude waves in a channel of finite depth h have been investigated by Struik (1926), but Benjamin (1967a) and Whitham (1967a and b), independently and by different approaches, showed that if $kh > 1.363$, Struik's waves are unstable [see also Benjamin and Feir (1967)].

3.3.   The Solitary Wave

If the length of a surface wave far exceeds the depth, it is often possible to take into account the amplitude of the wave in an analytical determination of the wave velocity.   Two solutions for long waves of permenent form are known: that for the solitary wave and that for cnoidal waves. We shall consider the solitary wave first.   The development follows closely that in Lamb (1932, p. 424).

Since irrotationality is still assumed, the complex potential is

$$\phi + i\psi = F(x + iy),$$

where F is assumed to be an analytic function of the complex variable $x + iy$ with real coefficients.   We may expand $F(x + iy)$ in the neighborhood of x, considering iy as the increment of the complex variable $x + iy$ over x.   Then, after separation of real and imaginary parts,

$$\phi = F - \frac{y^2}{2!} F'' + \frac{y^4}{4!} F^{\text{iv}} + \cdots, \tag{134}$$

$$\psi = yF' - \frac{y^3}{3!} F''' + \frac{y^5}{5!} F^{\text{v}} + \cdots, \tag{135}$$

in which F and its derivatives are evaluated at $y = 0$.

If x and y are coordinates in a frame of reference moving with the wave, the speed of the fluid at $x = \pm\infty$, where the depth is h, is simply equal to the wave velocity c.   With reference to Fig. 11, the Bernoulli equation written for the free surface is

$$u^2 + v^2 = c^2 - 2g(y - h), \tag{136}$$

or, by virtue of (134) or (135),

$$F'^2 - y^2 F' F''' + y^2 F''^2 + \cdots = c^2 - 2g(y - h). \tag{137}$$

According to (135), the bottom streamline has zero for the value of the stream function ψ. The value of ψ for the free surface is then ch. Thus, according to (135), the free surface is given by

$$yF' - \frac{y^3}{6} F''' + \cdots = ch. \tag{138}$$

Equations (137) and (138) implicitly determine y on the free surface as a function of x. To determine this function explicitly, it is necessary to eliminate F between (137) and (138) by successive substitutions, on the assumption that the derivatives $F'(x)$, $F''(x)$, ... are small enough for the terms in (134) or (135) to be diminishing in magnitude. Thus, from (138),

$$F' = \frac{ch}{y} + \frac{y^2}{6} F''' + \cdots = ch \left[\frac{1}{y} + \frac{1}{6} y^2 \left(\frac{1}{y} \right)'' + \cdots \right].$$

With this, (137) becomes, after some simplification,

$$\frac{1}{y^2} + \frac{2}{3} \frac{y''}{y} - \frac{1}{3} \frac{y'^2}{y^2} = \frac{1}{h^2} - \frac{2g(y - h)}{c^2 h^2},$$

which, after multiplication by y', can be integrated to produce

$$-\frac{1}{y} + \frac{1}{3} \frac{y'^2}{y} = -\frac{1}{h} + \frac{y - h}{h^2} - \frac{g(y - h)^2}{c^2 h^2}, \tag{139}$$

the constant of integration being determined by the condition (at infinity on the free surface)

$$y' = 0 \quad \text{at} \quad y = h.$$

Equation (139) can be simplified to

$$y'^2 = \frac{3(y - h)^2}{h^2} \left(1 - \frac{gy}{c^2} \right). \tag{140}$$

Thus the slope of the free surface vanishes at $y = h$ and at $y = c^2/g$. Since (140) implies that $y \leq c^2/g$, c^2/g must be the value of y at the crest. There is no other point at which y' vanishes except the point at infinity. Hence the wave is one of elevation only. If $a + h$ is the elevation of the crest above the bottom,

$$c^2 = g(h + a), \tag{141}$$

which shows that the wave velocity c increases with the amplitude of the wave. Equation (141) agrees exactly with the empirical formula adopted by Russell (1844).

Equation (140) can be written

$$\eta' = \pm \frac{\eta}{b}\left(1 - \frac{\eta}{a}\right)^{\!1/2},\tag{142}$$

in which $\eta = y - h$ and $b^2 = \dfrac{h^2(h + a)}{3a}$.

It can be readily verified that the integral of (142) is

$$\eta = a \operatorname{sech}^2 \frac{x}{2b} .\tag{143}$$

3.4. Cnoidal Waves

If, instead of the solitary wave, we inquire about the possibility of periodic long waves of finite amplitude, we can no longer write the Bernoulli equation in the form of (136), since it is no longer possible to say that $u^2 + v^2 = c^2$ where $y = h$, as in the case of the solitary wave.

In the linear theory of waves in an otherwise tranquil liquid, the mean position of any fluid particle is fixed, or, in other words, the time-mean velocity of any fluid particle is zero.   Therefore the meaning of the velocity of propagation of linear waves is not obscure: it is simply relative to the mean position of any fluid particle.   For nonlinear waves in infinitely deep water, the fluid is at rest at $y = -\infty$.   For the solitary wave just described, the fluid is at rest at $x = \pm\infty$.   In these cases the wave velocity is simply relative to the fluid particles at rest, wherever they happen to be.   For periodic waves of finite amplitude in water of finite depth, the speed of propagation is defined in the following way.   First, the mean depth of water over one wavelength is well defined, and will be denoted by h.   If a frame of reference moving with the waves is adopted, the flow is steady, and the ratio of the discharge through any vertical plane (per unit distance in the z direction or the horizontal direction normal to the direction of wave propagation) to h is denoted by c.   Hence if $\psi = 0$ at $y = 0$, $\psi = ch$ at the surface.   This c can be considered the wave velocity.

The Bernoulli equation for the free surface is

$$u^2 + v^2 = C - 2gy,\tag{144}$$

in which C is a constant to be determined from some characteristics (such as wavelength λ and wave height h_1 of the crests above the bottom) of the waves.   Then (137), (138), and (144) lead to

$$y'^2 = \frac{3g}{c^2h^2}\,(y - l)(h_1 - y)(y - h_2),\tag{145}$$

in the same way that (136) to (138) lead to (140). In (145), h_1 and h_2 are the upper and lower limits of y, and

$$l = \frac{c^2 h^2}{g h_1 h_2}. \tag{146}$$

Equation (145) implies that l cannot be greater than h_2. Of the two arbitrary constants h_1 and h_2, one may be considered to have arisen from the integration that produces (145), and the other may be considered to represent the unknown constant C in (144). Since Eq. (145) has the same form as that satisfied by Weierstrass' $\mathscr{P}$ function, we expect the solution to involve elliptic functions.

Using a parameter χ and writing

$$y = h_1 \cos^2 \chi + h_2 \sin^2 \chi, \tag{147}$$

we find the dependence of x on χ by differentiating (147) in χ, taking the square, and dividing (145) by the result. In so doing we obtain

$$\left(\frac{d\chi}{dx}\right)^2 = \frac{3g}{4c^2 h^2}(y - l),$$

$$\beta \frac{d\chi}{dx} = (1 - k^2 \sin^2 \chi)^{1/2}, \tag{148}$$

in which $\qquad \beta = \left[\frac{4 h_1 h_2 l}{3(h_1 - l)}\right]^{1/2}, \qquad k^2 = \frac{h_1 - h_2}{h_1 - l}, \tag{149}$

k^2 being smaller than unity. (Note that the k defined here is not the wave number.) Integration of (148) produces

$$x = \beta \int_0^\chi \frac{d\chi}{(1 - k^2 \sin^2 \chi)^{1/2}} = \beta F(\chi, k) \tag{150}$$

if the origin of x is taken at the crest, at which y attains the maximum h_1 and χ is zero. Now

$$y = h_2 + (h_1 - h_2) \cos^2 \chi, \qquad \cos \chi = cn \frac{x}{\beta}.$$

Hence $\qquad\qquad y = h_2 + (h_1 - h_2) cn^2 \frac{x}{\beta}, \tag{151}$

with the modulus k defined in (149). The wavelength is

$$\lambda = 2\beta \int_0^{\pi/2} \frac{d\chi}{(1 - k^2 \sin^2 \chi)^{1/2}} = 2\beta F_1(k), \tag{152}$$

and the volume of the liquid (per unit length in the z direction) in one wavelength is

$$h\lambda = \int_0^\lambda y \, dx$$

$$= 2\beta \int_0^{\pi/2} \frac{h_1 \cos^2 \chi + h_2 \sin^2 \chi}{(1 - k^2 \sin^2 \chi)^{\frac{1}{2}}} \, d\chi = 2\beta[lF_1(k) + (h_1 - l)E_1(k)].$$

Hence
$$(h - l)F_1(k) = (h_1 - l)E_1(k), \tag{153}$$

in which
$$F_1(k) = F\left(\frac{\pi}{2}, k\right) = \int_0^{\pi/2} \frac{d\chi}{\sqrt{1 - k^2 \sin^2 \chi}},$$

$$E_1(k) = E\left(\frac{\pi}{2}, k\right) = \int_0^{\pi/2} \sqrt{1 - k^2 \sin^2 \chi} \, d\chi.$$

Note that, with k^2 defined in (149),

$$h_2 - l = (h_1 - l)(1 - k^2),$$

and in deriving (153) we have used

$$-l + h_1 \cos^2 \chi + h_2 \sin^2 \chi = (h_1 - l) \cos^2 \chi + (h_2 - l) \sin^2 \chi$$

$$= (h_1 - l)(1 - k^2 \sin^2 \chi).$$

The six quantities h_1, h_2, l, k, λ, and β are connected by the four equations contained in (149), (152), and (153). If, for instance, λ and h_1 are given, the others can be found and c can be calculated from l by the use of (146), h being considered as given in any case. The solution is due to Korteweg and De Vries (1895).

If we put $l = h_2$, k is equal to unity and (152) gives $\lambda = \infty$. Then (153) shows that $l = h$, which means $h = h_2$, and (146) gives

$$c^2 = gh_1,$$

which is (141). Thus, the solitary wave is a special case of cnoidal waves. In fact (150), with $k = 1$ and $|\chi| < \pi/2$, becomes after integration

$$\frac{x}{\beta} = \ln \tan \left(\frac{\pi}{4} + \frac{\chi}{2}\right), \tag{154}$$

the constant of integration being determined by the condition that $x \to \infty$ as $\chi \to \pi/2$ and $x \to -\infty$ as $\chi \to -\pi/2$. Equation (154) can be simplified to

$$\cos \chi = \operatorname{sech} \frac{x}{\beta}.$$

Thus (151) can be written as

$$y - h_2 = (h_1 - h_2) \operatorname{sech}^2 \frac{x}{\beta}.$$

Since $y - h_2$ is η, $h_1 - h_2$ is a, and $\beta = 2b$, this agrees with (143) completely.

Struik's study (1926) of waves of finite amplitude in water of finite depth should have some connection with cnoidal waves. This connection has been brought out in a study by Laitone (1962).

3.5.   Shallow-water Theory: Propagation in One Dimension

We have seen in the treatment of surface waves propagating in water of finite depth in Sec. 2.3 that if the wavelength is large compared with the depth, the horizontal velocity at any section normal to the direction of propagation is very nearly constant and that the pressure distribution is nearly hydrostatic as a consequence of the negligibility of the vertical acceleration.   These facts can be made the basis for a theory of long waves with amplitudes not necessarily small.   The theory is called the *shallow-water theory*.   We shall treat one-dimensional propagation first.   (The wave motion itself is two-dimensional.)   It must be borne in mind that the depth d of the water (or any liquid) is "shallow" if the product kd is much less than unity, k being the wave number.   Thus tidal waves, with very long wavelengths, are shallow-water waves, in spite of the great depth of oceans in the ordinary sense.

With h denoting the mean depth of water and $h + \eta$ denoting the actual depth at any section, the uniformity of the horizontal-velocity component u enables us to write the equation of continuity in the form

$$\frac{\partial}{\partial x}[u(h + \eta)] = -\eta_t, \tag{155}$$

in which the subscript denotes partial differentiation.   The pressure distribution is given by the equation of hydrostatics:

$$p = g\rho(\eta - y), \tag{156}$$

the origin of y being taken at the level of the mean free surface.   Therefore the equation of motion assumes the form

$$u_t + uu_x = -g\eta_x, \tag{157}$$

since v is small compared with u, as can be deduced from the equation of continuity if kd is small, and since $\partial u/\partial y$ is small.   Equations (155) and (157) are the fundamental equations in the shallow-water theory for one-dimensional propagation.

3.5.1.   *Analogy with Gas Dynamics*

If we integrate ρ and p with respect to y from $-h$ to η and call the results $\bar{\rho}$ and $\bar{p}$ respectively,

$$\bar{\rho} = \rho(\eta + h), \tag{158}$$

$$\bar{p} = \frac{g\rho}{2}(\eta + h)^2 = \frac{g}{2\rho}\bar{\rho}^2. \tag{159}$$

If (158) is differentiated with respect to t, (155) is multiplied by ρ, and the results are added, we have

$$\bar{\rho}_t + (\bar{\rho}u)_x = 0, \tag{160}$$

If (157) is multiplied by $\bar{\rho}$,

$$\bar{\rho}(u_t + uu_x) = -\bar{p}_x + g\bar{\rho}h_x, \tag{161}$$

by virtue of (158). (Note that h is assumed to be a function of x only.) If h is constant, (161) becomes

$$\bar{\rho}(u_t + uu_x) = -\bar{p}_x, \tag{161a}$$

and Eqs. (160) and (161a) are the equations governing one-dimensional motion of a gas with the relationship between its pressure p and density $\bar{\rho}$ given by (159) for isentropic change of state, provided gravity effects are neglected. Thus there is an analogy between the dynamics of long waves and the dynamics of one-dimensional gas flows. Actually this analogy exists even if h is not constant but h_x is constant. In that case the direction of the analogous gas flow is inclined at an angle $\sin^{-1} h_x$ to the horizontal. The effect of the longitudinal component of gravity is therefore not neglected, although it is implicit in the assumption of a one-dimensional treatment that $\bar{\rho}$ is independent of y and z and this implies that the effect of the transverse component of gravity is neglected. Since in ordinary gas dynamics gravity effects are neglected, the treatment of (160) and (161a) in gas dynamics, with an equation for isentropic change of state like (159), can be used to deal with problems involving long water waves of constant h. The method of nonlinear characteristics was used by Riemann (1860) to investigate the propagation of plane gas waves of finite amplitude, and this is in essence the method to be presented in the next section for the solution of problems involving long waves of finite amplitude propagating in water. A more thorough and extensive treatment by Massau (1900) deals with the graphical integration of partial differential equations, with special emphasis on open-channel flow. Massau's investigations can be considered the first modern work on the use of nonlinear characteristics to solve problems involving water waves in open channels. A later work by Stoker (1948) develops further the ideas of Riemann and Massau.

As is to be shown in the next chapter, the speed c of sound waves, whose amplitude is infinitesimal, is given by

$$c^2 = \left(\frac{dp}{d\rho}\right)_S,$$

in which the subscript S indicates that the entropy is constant. As shown in the following chapter, the c so defined actually retains its significance as the local wave speed even for finite amplitude. This will also be clear by analogy after we have presented the method of characteristics for long water waves of finite amplitude. Using $\bar{p}$ and $\bar{\rho}$ of the fictitious gas, we see that, according to (159) and (158),

$$c^2 = \frac{d\bar{p}}{d\bar{\rho}} = g(\eta + h). \tag{162}$$

We may, for the time being, regard (162) as the definition of c. In the next section we shall show that the c so defined actually is the speed of propagation of long water waves of finite amplitude.

3.5.2. *Method of Characteristics for One-dimensional Propagation of Long Water Waves of Finite Amplitude*

With c defined in (162), (155) and (157) can be written as

$$2c_t + 2uc_x + cu_x = 0 \tag{163}$$

and

$$u_t + uu_x + 2cc_x - gh_x = 0. \tag{164}$$

Addition of (163) and (164) yields

$$\left[\frac{\partial}{\partial t} + (u + c)\frac{\partial}{\partial x}\right](u + 2c) = gh_x, \tag{165}$$

and subtraction yields

$$\left[\frac{\partial}{\partial t} + (u - c)\frac{\partial}{\partial x}\right](u - 2c) = gh_x. \tag{166}$$

If gh_x is constant, which will be denoted by m, (165) and (166) can be written as

$$\left[\frac{\partial}{\partial t} + (u + c)\frac{\partial}{\partial x}\right](u + 2c - mt) = 0, \tag{167}$$

$$\left[\frac{\partial}{\partial t} + (u - c)\frac{\partial}{\partial x}\right](u - 2c - mt) = 0. \tag{168}$$

In particular, for a channel of constant depth h, $m = 0$, and (167) and (168) become

$$\left[\frac{\partial}{\partial t} + (u + c)\frac{\partial}{\partial x}\right](u + 2c) = 0, \tag{169}$$

$$\left[\frac{\partial}{\partial t} + (u - c)\frac{\partial}{\partial x}\right](u - 2c) = 0. \tag{170}$$

We shall deal with (169) and (170) first; the way to deal with (165) to (168) will then be quite clear.

Curves in the xt plane described by the differential equation

$$\frac{dx}{dt} = u + c \tag{171}$$

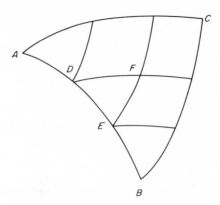

FIGURE 12.   THE NET OF TWO FAMILIES OF
CHARACTERISTICS.

constitute a family which will be denoted by C_1. Similarly, the family
of curves described by

$$\frac{dx}{dt} = u - c \tag{172}$$

will be denoted by C_2. The C_1 curves and C_2 curves are called the
characteristics of (169) and (170). It is evident that along a curve of the
family C_1 the quantity $u + 2c$ is constant, for, along such a curve

$$d(u + 2c) = \frac{\partial}{\partial t}(u + 2c)\, dt + \frac{\partial}{\partial x}(u + 2c)\, dx$$

$$= \left[\frac{\partial}{\partial t} + (u + c)\frac{\partial}{\partial x}\right](u + 2c)\, dt = 0, \tag{173}$$

according to (169). Similarly, along a curve of the family C_2 the quantity
$u - 2c$ is constant. Given u and c along any curve AB (Fig. 12), which is
itself not a characteristic, characteristics can be drawn through chosen
points on AB. At the point F, which is the intersection of a C_1 curve DF
and a C_2 curve EF,

$$(u + 2c)_F = (u + 2c)_D, \tag{174}$$

$$(u - 2c)_F = (u - 2c)_E, \tag{175}$$

the right-hand sides of which are known from the data along AB. Hence
the values of u and c at F can be computed from (174) and (175), and the
characteristics can be continued in the xt plane, producing "points of
knowledge" as the process is repeated. However, the process must stop
when the point C is reached, because data along AB (and not beyond A

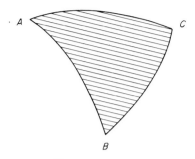

FIGURE 13. THE REGION OF DETER-
MINATION OF CURVE AB OR THE
REGION OF DEPENDENCE OF THE
POINT C.

and B) can determine u and c only in the shaded zone of Fig. 13, which is the region of determination of the curve AB or the region of dependence of the point C. The shaded region in Fig. 14 is called *domain of influence* of the point P by Stoker, because the value of u and c at P will influence the u and c in that region but not outside it.

If m is not zero but is constant, the solution of (167) and (168) is obtained in the same way. The additional term mt presents no difficulty, since at any point in the xt plane it is known. If h_x is not constant, (165) and (166) must be used. The characteristics are the same, but (173) must now be replaced by

$$d(u + 2c) = gh_x\, dt_1,$$

and the equation corresponding to (173) but for the C_2 curves must be replaced by

$$d(u - 2c) = gh_x\, dt_2.$$

Thus (174) and (175) become

$$(u + 2c)_F = (u + 2c)_D + gh_x\, dt_1, \tag{176}$$

$$(u - 2c)_F = (u - 2c)_E + gh_x\, dt_2, \tag{177}$$

in which dt_1 is the difference in t between F and D and dt_2 that between F and E. The procedure of determining u and c is only slightly changed.

Inspection of (174) reveals that the quantity $u + 2c$ is conserved on a curve on which (171) is satisfied, or, speaking graphically, at a geometrical point in the xt plane moving with velocity $u + c$. Since u is the fluid velocity in the x direction, this clearly means that the "message" $u + 2c$ is propagated *through* the fluid with speed c as well as *carried* with the speed u. Similarly, (175) means that $u - 2c$ is conserved on a geometrical point in the xt plane moving with speed $u - c$. This is to

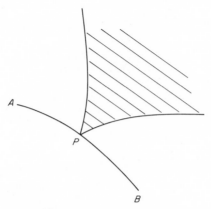

FIGURE 14.   THE DOMAIN OF INFLUENCE OF
POINT P.

say that the message $u - 2c$ is *carried* by the fluid with the velocity u but *propagated through the fluid* in the negative x direction with speed c. These considerations enable one to identify the c defined by (162) as the speed of propagation of nonlinear long waves.   By analogy, $(dp/d\rho)_S$ gives the square of the speed of long compression and expansion waves of finite amplitude.

3.5.3.   *Simple Waves*

The notion of simple waves initiated by Stoker (1948) is at once interesting and important.   If at $t = 0$ the fluid is in uniform motion with constant velocity u_0 and constant wave velocity c_0, and if a disturbance is initiated at $x = 0$, then in the xt plane there is a characteristic

$$\frac{dx}{dt} = u_0 + c_0 \tag{178}$$

belonging to the C_1 family.   Below this straight line is the zone of quiet, in which the C_1 characteristics are all straight lines parallel to the line described by (178).   In the zone of quiet $u = u_0$ and $c = c_0$.

Above the line (178) the C_1 characteristics are still straight lines, as shown by Stoker (1948), but are no longer parallel to the line (178).   In Fig. 15a C_1^0 is the C_1 characteristic described by (178).   Hence all along it $u = u_0$ and $c = c_0$.   At A,

$$u_A - 2c_A = u_0 - 2c_0, \tag{179}$$

since A and A_0 are on the same C_2 characteristic, and at A_0, $u = u_0$ and $c = c_0$.   At any arbitrary point B on the C_1 line through A,

$$u_B - 2c_B = u_0 - 2c_0, \tag{180}$$

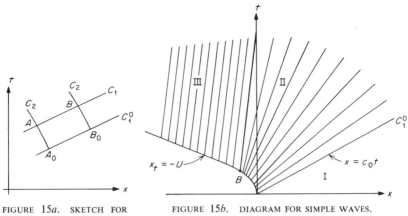

FIGURE 15a. SKETCH FOR
PROVING THAT AB IS A
STRAIGHT LINE.

FIGURE 15b. DIAGRAM FOR SIMPLE WAVES.

because B and B_0 are on the same C_2 characteristic, and at B_0, the values
of u and c are u_0 and c_0, respectively. If AB is a C_1 characteristic, then

$$u_A + 2c_A = u_B + 2c_B. \tag{181}$$

The three equations (179) to (181) demand that

$$u_A = u_B, \qquad c_A = c_B. \tag{182}$$

Hence not only must AB be a straight line, but all the C_2 characteristics
must intersect it at the same angle, because both $u + c$ and $u - c$ are
constant along AB, as a consequence of (182). Since AB is any C_1
characteristic, the conclusion reached for AB applies to all the C_1
characteristics for simple waves.

   The equation for the C_1 characteristics is (171). For simple waves,
u and c are constant on these, as indicated by (182). Equation (179)
gives (dropping the subscript A)

$$c = \tfrac{1}{2}(u - u_0) + c_0, \tag{183}$$

or

$$u + c = \tfrac{3}{2}u - \frac{u_0}{2} + c_0. \tag{184}$$

Thus, the C_1 characteristics can be constructed since u is known, say at
the piston creating the disturbance. In Fig. 15b, zone I is the zone of
quiet, and zone II is the zone of transition between the zone of quiet and
zone III of new uniform depth. The C_1 characteristic dividing zones II
and III starts from point B, to the left of which the receding piston
assumes a uniform velocity $-U$.

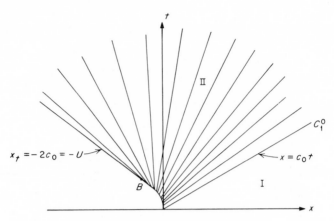

FIGURE 15c.   ZONE III SHRINKS TO A LINE.

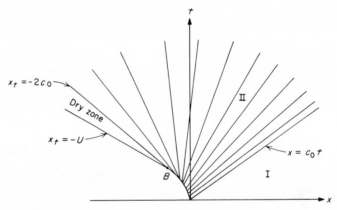

FIGURE 15d.   DRY REGION APPEARS BETWEEN B AND PISTON.

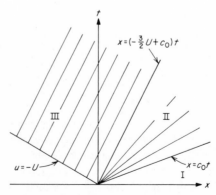

FIGURE 15e.   PISTON RECEDES WITH CONSTANT VELOCITY $-U > -2c_0$.

If the liquid in the zone of quiet is still, then $u_0 = 0$, and (183) indicates that if u on the piston is negative and equal to $-2c_0$, then $c = 0$ and the water depth is zero. The corresponding $u + c$ is then $-2c_0$, which is the algebraically least slope the characteristics in zone III can attain. If this is equal to $-U$, zone III shrinks to a line—the line describing the uniform receding motion of the piston. See Fig. 15c. If this slope is algebraically smaller than $-U$, zone III disappears, and a dry zone will start from B, occupying the space between the piston line (on which $dx/dt = -U$) and the last C_1 characteristic of zone II, passing through B and satisfying the equation (Fig. 15d)

$$\frac{dx}{dt} = -2c_0. \tag{185}$$

As has been shown, the u and c along any straight characteristic belonging to the C_1 family are constant. The u is determined by the value of dx/dt on the piston curve. Since any point on the piston curve is connected by a C_2 characteristic with the zone of quiet,

$$u - 2c = -2c_0 \tag{186}$$

at any point on the piston curve if u_0 is still assumed to be zero. With u known, c is also known. Hence the u and c along any straight characteristic are known. The free-surface profile or the distribution of u with x at any time t can be obtained by drawing a line in the xt plane parallel to the x axis and plotting the values of $h + \eta$ or of u obtained at the intersections of this line with the straight characteristics, the values of $h + \eta$ being equal to c^2/g.

The case in which the piston curve is a straight line passing through the origin of the xt plane is a degenerate case needing a special treatment, since all the straight characteristics in zone II radiate from that origin. For these (Fig. 15e),

$$\frac{dx}{dt} = \frac{x}{t}. \tag{187}$$

But since these are C_1 characteristics,

$$\frac{dx}{dt} = \frac{3}{2}u + c_0, \tag{188}$$

according to (184), if again u_0 is assumed to be zero. Thus

$$u = \frac{2}{3}\left(\frac{x}{t} - c_0\right). \tag{189}$$

Since (186) holds because every point in zone II is connected with the

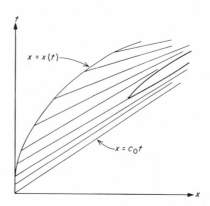

FIGURE 16. CONVERGING CHARACTER-
ISTICS, WITH AN ENVELOPE (CAUSTIC).

zone of quiet, it follows that

$$c = \frac{u}{2} + c_0 = \frac{1}{3}\left(\frac{x}{t} + 2c_0\right).$$ (190)

Equations (189) and (190) give u and c at any point in zone II. Zone II is separated from the zone of quiet by the line $x = c_0 t$, and from zone III by the line

$$x = (-\tfrac{3}{2}U + c_0)t,$$

$-U$ being the constant velocity of the piston.

If the piston curve is concave downward, as shown in Fig. 16, the straight characteristics will converge to form a *caustic*, which is their envelope. Such an envelope indicates the presence of a surge at which there is abrupt change in u and in c with x and t. The method of characteristics described here can be used only at a point which can be reached by a C_1 curve and a C_2 curve *not* meeting the envelope on the way to the point in question.

The region where characteristics converge corresponds to the front part of an elevation of the free surface. The convergence of the characteristics indicates that this part constantly steepens, until finally the wave breaks there. Evidence of a breaking wave can easily be produced in the laboratory by a moving piston, but this feature of the shallow-water theory must be squared with the permanence of the solitary wave and of the cnoidal waves, neither of which breaks or steepens in the front part of the elevation or elevations. Reconciliation resides in the fact that dispersive effects (or, equivalently, effects of velocity variation with height), however small, have been accounted for in the nonlinear theory of solitary waves and cnoidal waves whereas in the shallow-water theory they have been totally ignored.

3.5.4. *Method of Characteristics for Steady Supercritical Flow of Shallow Water in Two Horizontal Dimensions*

Cartesian coordinates x, y, and z are used, with the corresponding velocity components denoted by u, v, and w. The direction of increasing z is the direction of the vertical. If the representative horizontal length scale is much greater than the depth $h(x,y)$ at any point in the xy plane, we can again assume w to be small, the pressure distribution hydrostatic, and the velocity components u and v independent of z. Since irrotationality is assumed,

$$v_x - u_y = 0, \tag{191}$$

and as a consequence of steadiness and irrotationality the Bernoulli equation holds:

$$u^2 + v^2 + 2gh = \text{const}, \tag{192}$$

in which h corresponds to $h + \eta$ in the preceding section. The equation of continuity, on account of the independence of u and v with z, can be written in the integral form

$$(hu)_x + (hv)_y = 0. \tag{193}$$

Substitution of h obtained from (192) into (193) produces

$$(c^2 - u^2)u_x - uv(v_x + u_y) + (c^2 - v^2)v_y = 0. \tag{194}$$

Since (191) permits the use of the potential ϕ, in terms of which

$$u = \phi_x, \qquad v = \phi_y,$$

(194) can be written as

$$(c^2 - \phi_x{}^2)\phi_{xx} - 2\phi_x\phi_y\phi_{xy} + (c^2 - \phi_y{}^2)\phi_{yy} = 0. \tag{195}$$

This is seen to have exactly the same form as the equation governing two-dimensional steady irrotational flows of a gas. The c^2 for the gas flows is $(dp/d\rho)_S$, the subscript indicating isentropy.

The method of characteristics as derived directly from (195) is the same as that to be given in the treatment of that equation for gas flow in Chap. 6. We can also use the more natural characteristics in the hodograph plane, i.e., the epicycloids which are similar to those for gas flows. For demonstration, we shall consider a supercritical flow in an expanding channel, one boundary of which is shown in plan view in Fig. 17*b*, near

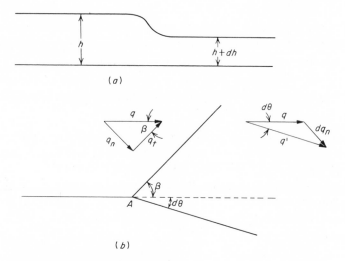

FIGURE 17. SUPERCRITICAL FLOW IN A CHANNEL EXPANSION.
(a) CROSS SECTION ALONG DOTTED LINE AND ITS EXTENSION IN
PLAN VIEW (b); AND (b) PLAN VIEW.

the point of expansion A, with the corresponding cross section shown in
Fig. 17a. At point A, where expansion of the boundary starts, a characteristic extends into the fluid with a constant angle of inclination β before
it meets other characteristics originating from the other boundary of the
channel. The upstream h and velocity q and the change of angle $d\theta$ of the
boundary are prescribed. The problem is to determine the other quantities after the expansion.

The first helpful relationship is the continuity equation

$$hq_n = (h + dh)(q_n + dq_n) = \text{const}, \qquad (196)$$

or, neglecting the product of differentials,

$$q_n \, dh + h \, dq_n = 0 \qquad (197)$$

along a characteristic line, along which changes in depth and velocity take
place. From now on we shall call this characteristic a *Mach line*, in
accordance with well-established usage in gas dynamics. The component
q_n is normal to the Mach line and the component q_t tangent to it. Since
the change in depth along the Mach line does not affect the tangential
momentum of a fluid particle as it passes through that line,

$$\rho h q_n q_t = \rho(h + dh)(q_n + dq_n)q'_t,$$

or, upon using (196),

$$q_t = q'_t. \qquad (198)$$

This means that the vector difference between $\mathbf{q}'$ and $\mathbf{q}$ is simply dq_n, as shown in Fig. 17b. The momentum equation for the direction normal to the Mach line is

$$d\frac{g\rho h^2}{2} = -d(\rho q_n^2 h),$$

or

$$gh\,dh = -q_n h\,dq_n,$$

since $q_n h$ is invariant across the Mach line. Thus

$$q_n\,dq_n = -g\,dh. \tag{199}$$

Combination of (197) and (199) gives

$$q_n^2 = gh = c^2. \tag{200}$$

This is significant. It states that the normal velocity is exactly equal to the propagation speed of long waves. The angle β is given by

$$\sin\beta = \frac{q_n}{q} = \frac{c}{q} = \frac{1}{F}, \tag{201}$$

in which F is the Froude number and β will be called the *Mach angle*. If we denote the numerical difference $q' - q$ by dq, then (Fig. 17b)

$$dq = dq_n \sin\beta, \tag{202}$$

and, for the case of positive $d\theta$ shown in Fig. 17b,

$$d\theta = \frac{dq_n \cos\beta}{q}, \tag{203a}$$

$$\frac{dq}{q} = \tan\beta\,d\theta = \frac{1}{\sqrt{F^2 - 1}}\,d\theta. \tag{203b}$$

Now the Bernoulli equation is

$$\tfrac{1}{2}q^2 + gh = c_0^2 = gH = \tfrac{1}{2}q_{\text{max}}^2, \tag{204}$$

in which H is the maximum depth attainable by the stream and attained only when $q = 0$, c_0^2 is defined to be gH, and q_{max} is the maximum q, attained only when $h = 0$. At the particular place where $q^2 = c^2$, we shall define the local c to be c^*, whose value depends only on the upstream condition. Thus

$$c_0^2 = \tfrac{3}{2}c^{*2}, \tag{205}$$

and from

$$\tfrac{1}{2}q^2 + c^2 = \tfrac{3}{2}c^{*2} \tag{204a}$$

we have

$$F^{*2} \equiv \left(\frac{q}{c^*}\right)^2 = \frac{3}{1 + (2/F^2)}, \tag{206}$$

or

$$F^2 = \frac{2F^{*2}}{3 - F^{*2}}, \tag{207}$$

Thus (203b) can be written as

$$d\theta = \pm\sqrt{\frac{F^{*2} - 1}{1 - (F^{*2}/3)}} \frac{dF^*}{F^*} , \tag{208}$$

where the negative sign is added to take care of the case of negative $d\theta$. If $\theta = \theta_0$ at $q = c^*$, integration of (208) gives

$$\pm(\theta - \theta_0) = \frac{\sqrt{3}}{2}\left[\frac{\pi}{2} - \sin^{-1}(2 - F^{*2})\right] - \frac{1}{2}\left[\frac{\pi}{2} + \sin^{-1}\left(2 - \frac{3}{F^{*2}}\right)\right],$$

which can be written as

$$\pm(\theta - \theta_0) = \sqrt{3}\cos^{-1}\frac{F^*}{F} - \cos^{-1}\frac{1}{F} ,$$

or finally $$\pm(\theta - \theta_0) = \sqrt{3}\tan^{-1}\sqrt{\frac{F^2 - 1}{3}} - \tan^{-1}\sqrt{F^2 - 1}. \tag{209}$$

These are the equations for the characteristics in the $F\theta$ plane, or $q\theta$ plane, or the hodograph plane. The curves described by (209) are epicycloids generated by a point on the circle of radius $\sqrt{3} - 1$ rolling in either direction on another of radius 1, starting from a point $F = 1$ and $\theta = \theta_0$. In practice we let $\theta_0 = 0$, $\pm\Delta\theta$, $\pm 2\Delta\theta$, etc., using a convenient increment $\Delta\theta$. The characteristics are shown in Fig. 20b.

To prove that the characteristics in the hodograph plane (or $F\theta$ plane) are epicycloids, see Fig. 18. There the velocity vector divided by c^* is represented by the radius vector from the origin O of the hodographic plane, so that

$$OP = \frac{q}{c^*} = F^*$$

and the angle θ is as indicated. The radius of the inner circle OB_0 is unity, so that the Froude number F^* or F at B_0 is 1. The radius of the big circle is $F^*_{max} = \sqrt{3}$. Then the diameter CB of the small circle with center at M is

$$CB = \sqrt{3} - 1.$$

The similarity of the triangles OAB and CPB gives

$$AP = AB\sqrt{3},$$

which is to say that

$$F^{*2} - (OA)^2 = 3[1 - (OA)^2],$$

or $$(OA)^2 = \frac{1}{2}(3 - F^{*2}) = \left(\frac{c}{c^*}\right)^2,$$

upon utilization of (204a). This shows that the angle OPA is indeed the

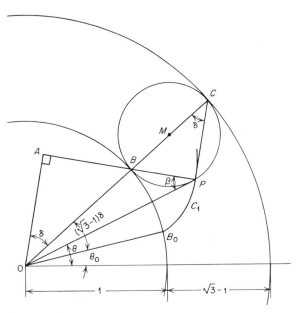

FIGURE 18.   SKETCH FOR THE PROOF THAT THE CHARACTERIS-
TICS ARE EPICYCLOIDS IN THE HODOGRAPH PLANE.

β defined by (201).   Let δ be defined by

$$\cot \delta = \sqrt{3} \tan \beta = \sqrt{3}(F^2 - 1)^{-\frac{1}{2}};$$

then since $AP = AB \sqrt{3}$, a look at the triangles OPA and OBA convinces us that the angle AOB is indeed δ.   Now the positive sign in (209) is taken for a C_1 curve.   If B_0P is a C_1 curve determined by (209), then

$$\theta = \theta_0 - \frac{\pi}{2} + \beta + \sqrt{3}\, \delta,$$

as demanded by (209).   On the other hand, a glance at the small circle in Fig. 18 convinces one that

$$\text{arc } BP = \frac{\sqrt{3} - 1}{2}\, 2\delta = (\sqrt{3} - 1)\, \delta.$$

Hence $\text{arc } B_0B = \theta - \theta_0 + \dfrac{\pi}{2} - \beta - \delta = (\sqrt{3} - 1)\, \delta = \text{arc } BP.$

This shows that the characteristic C_1 is an epicycloid obtained by rolling the small circle on the circle of unit radius.   The same is true of a C_2 curve.

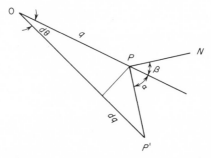

FIGURE 19. SIGNIFICANCE OF THE ANGLE β, PN BEING NORMAL TO THE CHARACTERISTIC PP' IN THE $q\theta$ PLANE. β IS THE MACH ANGLE, AND PN IS PARALLEL TO THE WAVEFRONT. (THE CHARACTERISTIC SHOWN HERE AND THAT SHOWN IN FIG. 18 BELONG TO DIFFERENT FAMILIES.)

The angle α shown in Fig. 19 is given by, upon using (203b),

$$\tan \alpha = \frac{q \, d\theta}{dq} = \sqrt{F^2 - 1}. \tag{210}$$

If NP is normal to the characteristic in the hodograph plane, i.e., to PP', then

$$\tan \beta = \frac{1}{\sqrt{F^2 - 1}} \qquad \text{or} \qquad \sin \beta = \frac{1}{F}. \tag{211}$$

In other words, β is exactly the Mach angle at the point P. Since OP is parallel to the velocity vector in the physical plane, PN is parallel to the Mach line (or the Mach wave) in the physical plane.

For instance, consider the flow of water with a free surface in an expanding channel (Fig. 20a). If the flow upstream is already supercritical, the expansion of the channel will make the Froude number even greater downstream. The gradual expansion of the channel can be approximated by straight line segments. We shall make the change in direction between consecutive lines equal to 2°. Before the initial point of expansion A is reached, the Froude number is the upstream Froude number, represented by the length 0-1 in the hodograph (Fig. 20b). At A the velocity changes its direction to that of AB. Since the hodograph gives the relationship between the direction and the speed of the velocity as it varies from point to point in the field of flow, the velocity in region 2, having an inclination of 2°, must lie along a characteristic in Fig. 20b passing through point 1, at the point where the velocity has the inclination 2°. It can be point 2 or 2″, but 2″ would correspond to a "compression," and calculation therefrom would lead us to physically unacceptable situations. Hence region 2 in Fig. 20*a* corresponds to point 2 in Fig. 20*b*. Similarly, region 2′ corresponds to the point 2′. The direction of *AQ* is given by the normal to the characteristic 1-2 at point 2 in the hodograph plane, as explained before. At the intersection *Q* of the Mach lines *AQ* and *A′Q*, a change of velocity occurs. The velocity in region 9 is along the centerline. Hence point 9 is the intersection of the characteristic 2-9 from point 2 and the characteristic 2′-9 from point 2′. Further calculation is now clear by inspection of Fig. 20*a* and *b*. With *F** known at any place, the *q* at the same place if found from (206) and then the *h* from (204). The stream-lines are shown in Fig. 20*c*.

After the slope at the boundary has reached a maximum, it will decrease again until the channel boundary has again become parallel to the center line. As the slope decreases, compression waves will be formed. If, as shown in Fig. 20*a*, the compression waves are made by design to neutralize the reflection of expansion waves by making the boundary agree in direction with the velocity in regions 15, 21, 26, etc., the flow in the expanded channel is eventually wave-free. The analysis presented here is entirely analogous to that for supersonic flow of gases in an expanding two-dimensional conduit [see Riabouchinsky (1932), Ippen and Knapp (1936), von Kármán (1938), and Preiswerk (1938)].

3.6. The Hydraulic Jump

The free surface of a layer of water flowing supercritically on a horizontal (or slightly inclined) bottom may jump to a greater height. The phenomenon is called a *hydraulic jump*, and the downstream and upstream depths are said to be *conjugate*.

Figure 21 shows the jump. The upstream depth and velocity are denoted by *h* and *u*, respectively, and the downstream depth and velocity by *h′* and *u′*. Using the sections *AA* and *BB*, the bottom, and the free surface as a control surface and equating the net horizontal force due to hydrostatic pressure at the sections *AA* and *BB* to the momentum efflux at *BB* minus the momentum influx at *AA*, we obtain

$$\frac{g\rho}{2}(h^2 - h'^2) = \rho q(u' - u), \tag{212}$$

in which ρ is the density and

$$q = uh = u'h' \tag{213}$$

is the discharge per unit length normal to the plane of flow. From (212) and (213) we obtain, after a factor $h - h'$ has been canceled on both sides and some further simplification,

$$\left(\frac{h'}{h}\right)^2 + \left(\frac{h'}{h}\right) - 2F^2 = 0$$

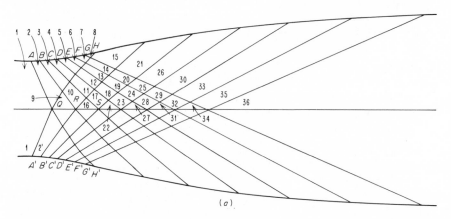

FIGURE 20. CONSTRUCTION OF THE FLOW IN A CHANNEL EXPANSION. (*a*) WAVE PATTERN
IN THE PLAN VIEW. (*Courtesy of Charles E. Hickox.*)

if the Froude number F is defined by

$$F^2 = \frac{u^2}{gh} = \frac{q^2}{gh^3}.$$

Hence $$\frac{h'}{h} = \frac{-1 + \sqrt{1 + 8F^2}}{2}.$$ (214)

It is evident that $h' = h$ if $F = 1$. It is also clear that if $F > 1$ (super-critical upstream flow), $h' > h$. At first sight it seems possible to have $h' < h$ and $F < 1$. But, as can easily be verified, this would imply an increase in available energy (or mechanical energy) after the (downward) jump, which would violate the second law of thermodynamics.

In a frame of reference moving with the upstream fluid, the hydraulic jump appears as a wave of finite amplitude propagating into tranquil fluid. Since its speed of propagation is then equal to u, which is greater than the speed of infinitesimal waves $\sqrt{gh}$, it furnishes an example of the increase of the speed of propagation of a wave with its amplitude.

Tidal bores observed in estuaries are hydraulic jumps propagating against flowing water. The analysis of a bore can be reduced to the analysis just presented if a frame of reference moving with the bore is adopted, with respect to which the flow is a steady flow with a stationary hydraulic jump. The upstream velocity relative to the jump is equal to

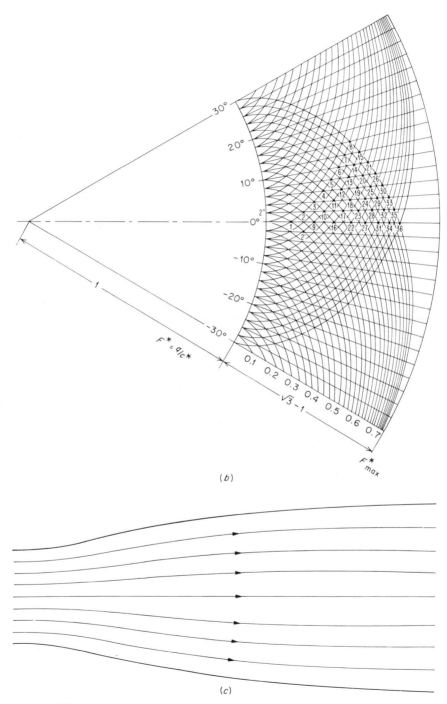

(b)

(c)

FIGURE 20.   CONSTRUCTION OF THE FLOW IN A CHANNEL EXPANSION. (b) THE CHARACTER-
ISTICS IN THE HODOGRAPH PLANE; (c) THE PATTERN OF THE STREAMLINES. (*Courtesy
of Charles E. Hickox.*)

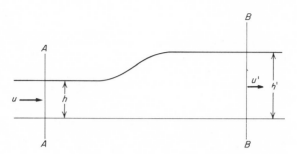

FIGURE 21.   THE HYDRAULIC JUMP.

the velocity of the river ahead of the bore augmented by the speed of propagation of the bore.[1]

If the Froude number is only slightly greater than 1, the jump is weak and often waves are observed after the jump. [For analysis of the undular jump for inviscid fluids see Meyer (1967) and his references; for the analysis of the undular jump of a viscous fluid, see Chester (1966) and his reference.] We must now hasten to a brief discussion of internal waves in a continuously stratified fluid and of inertial waves in a rotating fluid.

4. INTERNAL WAVES IN A CONTINUOUSLY STRATIFIED FLUID

In a stably stratified liquid, with the density decreasing upward, or in a stably stratified gas, with the entropy decreasing upward, a fluid particle displaced from its equilibrium position will, under the action of a restoring force, tend to return to its equilibrium position. But it will overshoot, tend to return to the equilibrium position again, and so on. Thus waves are possible in a stratified fluid in a gravitational field.

The limitation of space demands that our discussion of internal waves be brief. We shall restrict it to infinitesimal waves in a stratified liquid and a particularly simple class of waves of finite amplitude.

4.1. Infinitesimal Waves in a Stratified Liquid

If again u and v denote the velocity components in the directions of increasing x and y, respectively, with y measured in the direction of the

[1] The earliest graphical records of tidal bores of which the writer is aware are the exquisite paintings of two Chinese painters of the Sung period (A.D. 960–1260): one of a tidal bore on the Yangtze River, attributed to Hsü Tao-Ning, who lived in the early part of the eleventh century, and the other a tidal bore on the Chien Tang River (frontispiece), famous for tidal bores, by Li Sung (1166–1243). The former painting is in the Museum of Fine Arts, Boston, and the latter in the Old Palace Museum, Taipei, Taiwan, China.

vertical, the linearized equations of motion are, after quadratic and higher-order terms in the perturbation quantities (which include u and v) are neglected,

$$\bar{\rho}\frac{\partial u}{\partial t} = -\frac{\partial p'}{\partial x},$$ (215)

$$\bar{\rho}\frac{\partial v}{\partial t} = -\frac{\partial p'}{\partial y} - g\rho',$$ (216)

in which $\bar{\rho}$ is the mean density and a function of y alone, ρ' is the density perturbation, and of course t is the time and g the gravitational acceleration. The p' in (215) and (216) denotes the pressure perturbation. The missing term involving the mean pressure $\bar{p}$ is balanced by the term $-g\bar{\rho}$, which is also dropped. Inspection of (215) and (216) reveals that the viscous terms have been dropped. They can be retained; but if they are, the wave motion will be damped. In ignoring the viscous terms, we are implicitly studying slowly damping waves in a fluid of small viscosity. More precisely, we are studying the limiting (inviscid) case of waves in a fluid with large values of the Reynolds number $\sigma L^2/\nu$, in which σ is the frequency of the oscillations of the waves, L is a representative length, which is the depth or the wavelength, whichever is smaller, and ν is the kinematic viscosity. When we dropped the viscous terms in the study of waves, whatever the kind and whether the fluid is stratified or not, all this is what we have, or should have, in mind. As we shall see, for stratified fluids in wave motion there is another length scale, namely, the distance between one nodal plane and a neighboring one. When this is so small that the Reynolds number based on it (and σ and ν) is not large, viscous effects must be taken into account.

The equation of continuity is

$$\frac{\partial u}{\partial x} + \frac{\partial v}{\partial y} = 0.$$ (217)

The fluid is assumed to be incompressible and nondiffusive, so that

$$\frac{D}{Dt}(\bar{\rho} + \rho') = 0,$$ (218)

where D/Dt is the operator for substantial differentiation. If quadratic terms in the perturbation quantities are neglected, (218) has the linearized form

$$\frac{\partial \rho'}{\partial t} + v\frac{d\bar{\rho}}{dy} = 0.$$ (219)

Equation (217) allows the use of the stream function ψ, which we shall assume to be of the form

$$\psi = f(y)\exp[ik(x - ct)],$$

in which k is the wave number defined by

$$k = \frac{2\pi}{\lambda},$$

λ being the wavelength, and c is the phase velocity or wave velocity. The frequency of the motion, apart from a factor 2π, is

$$\sigma = kc. \tag{220}$$

The velocity components, expressed in terms of $f(y)$, are

$$u = \frac{\partial \psi}{\partial y} = f'(y) \exp [ik(x - ct)],$$

$$v = -\frac{\partial \psi}{\partial x} = -ikf(y) \exp [ik(x - ct)]. \tag{221}$$

If we assume ρ' to have the same exponential factor as ψ, (219) gives

$$ikc\rho' = v \frac{d\bar{\rho}}{dy}. \tag{222}$$

If now p is eliminated between (215) and (216), and (221) and (222) are substituted into the resulting equation, we obtain, after some simplifications,

$$(\bar{\rho}f')' - \left(k^2\bar{\rho} + \frac{g}{c^2}\frac{d\bar{\rho}}{dy}\right)f = 0, \tag{223}$$

in which the primes indicate differentiation with respect to y. Equation (223) can also be written as

$$(\bar{\rho}f')' - k^2\left(\bar{\rho} + \frac{g}{\sigma^2}\frac{d\bar{\rho}}{dy}\right)f = 0. \tag{223a}$$

The boundary condition at the bottom, where we shall assume $y = 0$, is simply

$$v = 0 \qquad \text{or} \qquad f(0) = 0.$$

If the depth is d and the fluid is bounded by a rigid upper plane, then

$$f(d) = 0.$$

If the upper boundary is free, the boundary condition is that $p = \bar{p} + p' = $ const on the free surface. But p has two parts. One part is just the perturbation pressure at $y = d$; for if η is the displacement of the free surface from its undisturbed position, the difference of p' evaluated at $y = d = \eta$ and that evaluated at $y = d$ is a quantity of a higher order. But there is another part, which comes not from p' but from $\bar{p}$, as a result of the displacement η. This is

$$\eta \frac{d\bar{p}}{dy} = -g\bar{\rho}\eta.$$

Hence the free-surface condition is

$$p' - g\bar\rho\eta = 0, \tag{224}$$

in which both p' and $\bar\rho$ are evaluated at $y = d$. The right-hand side is zero because the waves are sinusoidal in x, so that both p' and η have the same exponential factor as the stream function, and if the pressure is to be constant on the free surface, the left-hand side, containing this factor, can only be zero. Now the task is to express p' and η in terms of f. From (215) and (221) we have

$$p' = \bar\rho c f'(y) \exp\,[ik(x - ct)]. \tag{225}$$

Since there is no mean velocity,

$$v = \frac{\partial\eta}{\partial t}.$$

Hence, from (221),

$$\eta = \frac{1}{c} f(y) \exp\,[ik(x - ct)]. \tag{226}$$

Substitution of (225) and (226) into (224) produces

$$f'(d) - \frac{g}{c^2} f(d) = 0. \tag{227}$$

Thus, the boundary conditions are

$$f(0) = 0 \quad\text{and}\quad f(d) = 0 \tag{228}$$

for rigid boundaries, and

$$f(0) = 0, \quad f'(d) - \frac{g}{c^2} f(d) = 0 \tag{229}$$

if the upper surface is free.

The differential system (223) and (228) or (223) and (229) specifies an eigenvalue problem. Since the differential system is homogeneous, one obvious solution is $f(y) = 0$ identically. This is an unexciting solution because it says nothing happens. In order to have a nontrivial solution, c^2 must have certain values once k is specified along with $\bar\rho$ and d, or k^2 must have certain values if c^2 or σ^2 is specified. The values are the eigenvalues of c^2 or k^2.

Unless $\bar\rho$ varies exponentially with y, (223) is a differential equation with variable coefficients, and the eigenvalues must be found either entirely numerically by the use of a high-speed computer or by expansion of the solution of $f(y)$ in power series of y followed by numerical computation after the boundary conditions are imposed. But we can discuss (223) or (223a) quite generally without solving it. The governing differential system (223) and (228) constitutes the well-known Sturm-Liouville system, and many of the properties of the solution are the substance of the Sturm-Liouville theory. Since (229) contains c, when the eigenvalues of c are to be determined, the system (223) and (229) does not constitute the standard Sturm-Liouville system. But the work of Bôcher (1917) has extended the theory to include a system in which the parameter whose eigenvalues are to be determined appears both in the differential equation and in the boundary conditions. We can say therefore that the general properties of $f(y)$ can be described by the Sturm-Liouville-Bôcher theory. For lack of space we shall not present that theory, but instead shall present the properties of $f(y)$ in the following subsections in as convincing a way as possible without invoking it.

4.1.1. *The Maximum Frequency for Internal Waves*

If the upper boundary is fixed, all waves in a stratified fluid are internal waves. If the upper boundary is free, we define internal waves to be those waves which have at least one nodal plane in the interval $0 < y < d$. (The bottom where $y = 0$ is always a nodal plane.) We shall now show that internal waves are impossible if

$$\sigma^2 > -\frac{g}{\bar{\rho}}\frac{d\bar{\rho}}{dy} \tag{230}$$

everywhere. First, note that if $f(y)$ is a solution, so is $-f(y)$. Hence we may, without loss of generality, assume $f'(0) > 0$. Since $f(0) = 0$, in the immediate neighborhood of $y = 0$ we have $f(y) > 0$. Now, if (230) holds, (223a) shows that $(\bar{\rho}f')'$ will be nonnegative in that neighborhood. Thus $\bar{\rho}f'$, which is greater than zero at $y = 0$, will be greater than zero, and hence f' also will be positive in that neighborhood and $f(y)$ continues to increase. The argument repeats itself, and hence $f(y)$ constantly increases and can never be zero in the entire interval. Hence internal waves are impossible. The maximum of the square root of the right-hand side of (230) is called the *Brunt-Väisälä frequency.*

4.1.2. *The c Spectrum*

If k^2 is positive and $d\bar{\rho}/dy$ bounded, there is a maximum c^2 for internal waves. This is obvious; for we have just seen that there is a maximum σ^2 for internal waves and since $\sigma^2 = k^2 c^2$ and k^2 is nonzero, c^2 must have a maximum. (However, if the depth is infinite, c^2 may be unbounded as $k^2 \to 0$, for the lowest mode.) There are infinitely many discrete eigenvalues for c^2 less than this maximum value, with zero as their limit point. For rigid boundaries this can be seen in a rough way from (223). Let the coefficient of f in (223), including the minus sign, be denoted by Q for convenience. We have $f(0) = 0$, and, without loss of generality, $f'(0) > 0$. According to (223), so long as c^2 is small enough to make Q positive, $(\bar{\rho}f')'$ is negative, so that $\bar{\rho}f'(y)$ will decrease with y. If c^2 is sufficiently small, the rate of decrease of this quantity is sufficient to make $f'(y)$ itself decrease, and decrease fast enough to make $f'(y)$ negative for sufficiently large y, which then can make $f(y)$ decrease to zero.   Evidently we can make c^2 small enough to make $f(d) = 0$ but $f(y) > 0$ in $0 < y < d$. A still smaller c^2 can make $f(y)$ vanish not only at $y = d$ but also at some intermediate value, and so on.   Thus there are infinitely many values of c^2, decreasing toward zero, that are eigenvalues for the system governing wave motion between rigid boundaries.[1]   When the upper surface is free, the argument is not so simple [see Yih (1965, Sec. 12.3, Chap. 2)].   But the same conclusion holds, except that aside from internal waves, a stratified fluid with a free surface can have free-surface waves, for which σ^2 does not have a maximum.   Actually any discontinuity in $\bar{\rho}$, whether or not at the free surface [at which $\bar{\rho}$ jumps from $\bar{\rho}(d)$ to zero], will make not only the maximum of σ^2 but also the maximum of c^2 nonexistent, provided in the case of c^2 the depth is infinite.   For finite d and $d\bar{\rho}/dy$, c^2 is bounded even if $k = 0$.   Summarizing, we state the following theorem.

THEOREM   **If the depth is finite, then for any given k^2, whether or not the upper surface is free, the eigenvalues of c^2 for internal waves of a continuously stratified fluid have a finite maximum and converge to zero as the number of nodal planes increases indefinitely.   Thus for each k^2 there are infinitely many modes of internal waves.**

We note in passing that if the depth is finite, all waves due to density discontinuities, including free-surface waves, have a finite maximum c^2.

4.1.3.   *The k Spectrum*

For any given c, inspection of (223) and arguments similar to those employed in the preceding subsection[1] show that there are infinitely many eigenvalues of k^2.   However, there are only a finite number of positive values of k^2 or no positive value at all for k^2 if c^2 is sufficiently large. The other eigenvalues for k^2 are negative, corresponding not to waves but to local disturbances that do not propagate.

If there are several positive eigenvalues of k^2 for a given c^2, the highest value for k^2 corresponds to the least value of Q, the coefficient of f in (223), and hence to the gravest mode, with the least number of nodal planes in the fluid.   As the k^2 values decrease, the number of nodal planes increases.   But, as just mentioned, as the k^2 values become negative, the eigenfunctions corresponding to them no longer represent waves.

4.1.4.   *The Variation of c^2 with k^2*

Consider for the moment that the boundaries are rigid.   Then it is evident that as the wave number k (considered without loss of generality

[1] See Sturm's first oscillation theorem, presented in Sec. 6.4 of Chap. 9.   We can identify a with 0 and b with d.   The theorem is directly applicable here.

to be positive) increases, the coefficient of f in (223) would decrease unless c^2 decreases, which would make $f(d)$ different from zero if the number of zeros below d remains the same.[1]   Hence we have:

THEOREM   **The phase velocity decreases as the wavelength decreases.**

This theorem, demonstrated for a continuously stratified fluid between rigid boundaries, remains true even if there are surfaces of density discontinuity, including the free surface.   It remains true also for infinite depth of the fluid, but we have not the space for a thorough discussion.

4.2.   Waves of Finite Amplitude in a Stratified Fluid Flowing Steadily

If the wave amplitude is not assumed infinitesimal, (215) and (216) are replaced by

$$\rho\left(u\,\frac{\partial u}{\partial x} + v\,\frac{\partial u}{\partial y}\right) = -\,\frac{\partial p}{\partial x}, \tag{231}$$

$$\rho\left(u\,\frac{\partial v}{\partial x} + v\,\frac{\partial v}{\partial y}\right) = -\,\frac{\partial p}{\partial y} - g\rho \tag{232}$$

provided the flow is steady and two-dimensional and the fluid inviscid. Equation (218) now has the form

$$u\,\frac{\partial \rho}{\partial x} + v\,\frac{\partial p}{\partial y} = 0, \tag{233}$$

in which, as in (231) and (232), ρ is the (total) density.   First, we make [Yih (1958)] the simplifying transformation (ρ_0 = a reference density)

$$u' = \sqrt{\frac{\rho}{\rho_0}}\,u, \qquad v' = \sqrt{\frac{\rho}{\rho_0}}\,v. \tag{234}$$

Because of (233), (231) and (232) can be written as

$$\rho_0\left(u'\,\frac{\partial u'}{\partial x} + v'\,\frac{\partial u'}{\partial y}\right) = -\,\frac{\partial p}{\partial x}, \tag{235}$$

$$\rho_0\left(u'\,\frac{\partial v'}{\partial x} + v'\,\frac{\partial v'}{\partial y}\right) = -\,\frac{\partial p}{\partial y} - g\rho, \tag{236}$$

and the equation of continuity (217) becomes

$$\frac{\partial u'}{\partial x} + \frac{\partial v'}{\partial y} = 0.$$

[1] Again see Sturm's first oscillation theorem (Sec. 6.4 of Chap. 9) and identify a with 0 and b with d.   The coefficient of f in (223) is just Q.

Defining the vorticity of the field of (u', v') to be

$$\eta' = \frac{\partial v'}{\partial x} - \frac{\partial u'}{\partial y} = -\nabla^2 \psi', \tag{237}$$

in which ψ' is the stream function in terms of which u' and v' are given by

$$u' = \frac{\partial \psi'}{\partial y}, \qquad v = -\frac{\partial \psi'}{\partial x}, \tag{238}$$

Eqs. (235) and (236) can be written as

$$-\rho_0 \eta' \frac{\partial \psi'}{\partial x} = \frac{\partial}{\partial x}\left[p + \frac{\rho_0(u'^2 + v'^2)}{2}\right], \tag{239}$$

$$-\rho_0 \eta' \frac{\partial \psi'}{\partial y} = \frac{\partial}{\partial y}\left[p + \frac{\rho_0(u'^2 + v'^2)}{2}\right] + g\rho. \tag{240}$$

Multiplying (239) by dx and (240) by dy and adding the results, we have

$$-\rho_0 \eta' \, d\psi' = d\left[p + \frac{\rho(u^2 + v^2)}{2}\right] + g\rho \, dy = dH - gy \, d\rho, \tag{241}$$

in which
$$H = p + \frac{\rho(u^2 + v^2)}{2} + g\rho y$$

is the Bernoulli quantity, which for steady flow is constant along a stream-line and hence, like ρ, is a function of ψ' alone. With η' given by (237), (241) can be written as [Yih (1958)]

$$\nabla^2 \psi' + \frac{gy}{\rho_0}\frac{d\rho}{d\psi'} = \frac{1}{\rho_0}\frac{dH}{d\psi'} = h(\psi'). \tag{242}$$

In terms of ψ, which is related to ψ' by

$$d\psi' = \sqrt{\frac{\rho}{\rho_0}}\, d\psi,$$

(242) has the form [Dubreil-Jacotin (1935) and Long (1953a)]

$$\nabla^2 \psi + \frac{1}{\rho}\frac{d\rho}{d\psi}\left(\frac{\psi_x^2 + \psi_y^2}{2} + gy\right) = f(\psi). \tag{243}$$

Equation (242) or (243) governs steady two-dimensional flows of an incompressible and nondiffusive fluid of variable density, whether or not waves are present in the fluid.

The functions $d\rho/d\psi'$ and $h(\psi')$ depend on the upstream conditions. If far upstream

$$u' = U' \qquad \text{so that} \quad \psi' = U'y \tag{244}$$

and
$$\rho = \rho_0 - \beta y, \tag{245}$$

then
$$\frac{d\rho}{d\psi'} = -\frac{\beta}{U'} \qquad \text{and} \qquad h(\psi') = -\frac{g\beta}{\rho_0 U'^2}\psi'.$$

We are using the density at the bottom ($y = 0$) as the reference density. Thus (242) becomes

$$\nabla^2 \psi' + \frac{g\beta}{\rho_0 U'^2} \psi' = \frac{g\beta}{\rho_0 U'} y, \tag{246}$$

which is linear. With d denoting the depth and with

$$\Psi = \frac{\psi'}{U'd}, \qquad \xi = \frac{x}{d}, \qquad \eta = \frac{y}{d},$$

(246) can be written as

$$\Psi_\xi + \Psi_{\eta\eta} + F^{-2}\Psi = F^{-2}\eta, \tag{247}$$

in which the (modified) Froude number is defined by

$$F^2 = \frac{\rho_0 U'^2}{g\beta d^2}.$$

The solution of (247) is

$$\Psi = \eta + \Psi_1,$$

in which Ψ_1 satisfies

$$\left(\frac{\partial^2}{\partial \xi^2} + \frac{\partial^2}{\partial \eta^2} \right) \Psi_1 + F^{-2}\Psi_1 = 0. \tag{248}$$

If the upper boundary is assumed rigid,

$$\Psi_1 = 0 \qquad \text{at } \eta = 0 \text{ and } \eta = 1$$

since both boundaries must be streamlines and since $\Psi = 0$ at $\eta = 0$ and $\Psi = 1$ at $\eta = 1$. Upon separation of variables, we find that Ψ_1 can have the components

$$f_n(\xi) \sin n\pi\eta, \qquad \text{with } n \text{ an integer,}$$

provided

$$f_n''(\xi) + (F^{-2} - n^2\pi^2)f_n = 0. \tag{249}$$

Hence wave motion with up to $n + 1$ nodal planes in the closed interval $0 \le \eta \le 1$ is possible if

$$(n + 1)\pi > F^{-1} > n\pi,$$

since then f_n can be sinusoidal in ξ. If $n < 1$, there can be no waves. If $n = N$, there are N possible wave motions. These waves can be created by a barrier placed in the stream described by (244) and (245). Other upstream conditions giving a linear governing equation like (246) exist. We have chosen a particularly simple set of upstream conditions to illustrate the existence of finite-amplitude waves in a stratified fluid in steady flow [see Long (1953a) and Yih (1960, 1965)].

5. INERTIAL WAVES

By inertial waves we mean waves due to the presence of circulation in a fluid and in particular in a fluid which, apart from the wave motion, is in solid-body rotation.

Before the details of inertia-wave motion are presented, it is helpful to examine the basic physics of inertia waves.   Take a liquid of constant density rotating with constant angular velocity ω.   We ignore viscosity for the time being and displace a circular ring of fluid outward to a new position.   The circulation $2\pi vr$ along that ring will remain the same, according to the Kelvin theorem.   Hence v and a fortiori v^2/r will be smaller at the new position.   Since the prevailing pressure gradient at the new position is just enough to balance the v^2/r of the fluid surrounding the ring, which is greater than the v^2/r of the ring, the ring will be pushed back toward its original position.   It will, however, overshoot, and a restoring force will then push it outward toward its original position again, thus perpetuating the oscillation on and on.   It is evident that although viscosity will eventually damp out the oscillations, it cannot altogether destroy the mechanism of the wave motion.   In the following subsections we shall briefly discuss a few important cases of inertial-wave motion.

5.1.   Linear Axisymmetric Waves

If viscous effects are neglected and the amplitude of wave motion is small enough for us to neglect the nonlinear convective terms in the expressions for the acceleration components, Eqs. (2.59) to (2.61) can be written

$$\frac{\partial u}{\partial t} = -\frac{\partial P}{\partial r} + 2\omega v, \qquad \frac{\partial v}{\partial t} = -2\omega u, \qquad \frac{\partial w}{\partial t} = -\frac{\partial P}{\partial z}, \qquad (250)$$

provided axial symmetry is assumed.   It is implicitly assumed that ω, the mean angular velocity of the fluid, is constant.   Since the equation of continuity allows the use of the Stokes stream function ψ, in terms of which

$$u = -\frac{1}{r}\frac{\partial \psi}{\partial z}, \qquad w = \frac{1}{r}\frac{\partial \psi}{\partial r},$$

elimination of P and v from the three equations of motion above produces

$$\frac{\partial^2}{\partial t^2}\left(\frac{\partial^2}{\partial r^2} - \frac{1}{r}\frac{\partial}{\partial r}\right)\psi + \left(4\omega^2 + \frac{\partial^2}{\partial t^2}\right)\frac{\partial^2\psi}{\partial z^2} = 0. \qquad (251)$$

If the dependence of ψ on t is in the form $e^{-i\sigma t}$, this can be written

$$\sigma^2\left(\frac{\partial^2}{\partial r^2} - \frac{1}{r}\frac{\partial}{\partial r}\right)\psi + (\sigma^2 - 4\omega^2)\frac{\partial^2\psi}{\partial z^2} = 0, \qquad (252)$$

or $\qquad 4\sigma^2\eta\dfrac{\partial^2\psi}{\partial\eta^2} + (\sigma^2 - 4\omega^2)\dfrac{\partial^2\psi}{\partial z^2} = 0, \qquad$ with $\eta = r^2.$ $\qquad (253)$

For sinusoidal waves propagating in the z direction,

$$\psi = f(\eta)e^{i(kz-\sigma t)}, \qquad (254)$$

and (253) becomes

$$f'' - \frac{k^2}{4\eta}\left(1 - \frac{4\omega^2}{\sigma^2}\right)f = 0. \tag{255}$$

Since u must vanish at the centerline and at the wall of the cylindrical container, the boundary conditions are

$$f(0) = 0 \quad\text{and}\quad f(a^2) = 0 \tag{256}$$

if a is the inner radius of the cylindrical container. Equations (255) and (256) constitute a Sturm-Liouville system, and the Sturm-Liouville theory can be brought to bear on the problem.

From (253) it is evident that if $\sigma^2 > 4\omega^2$, the equation is elliptic in type and no real characteristics exist. Wave motion is then impossible. If $\sigma^2 < 4\omega^2$, waves are possible. This can be seen also from (255). If $\sigma^2 > 4\omega^2$, the coefficient of f is negative and the two boundary conditions cannot both be satisfied. If $\sigma^2 < 4\omega^2$, oscillation of f with η is possible; i.e., the boundary conditions can be satisfied for some k^2.

Actually, if the mean motion of the fluid is not solid rotation but has concentric-circular streamlines along which the mean circumferential velocity is $\bar{v}$, which can be but is not necessarily equal to ωr, the equation corresponding to (255) is

$$f'' - \frac{k^2}{4\eta}\left(1 - \frac{F}{\sigma^2}\right)f = 0, \tag{257}$$

in which

$$F = \frac{1}{r^3}\frac{d}{dr}(\bar{v}r)^2.$$

The boundary conditions are still given by (256). The derivation of (257) is left as an exercise.

Whether we have the general equation (257) or the special one (255), the Sturm-Liouville theory can be applied to give results similar to those presented in Sec. 4.1. For instance, for a given k^2, there is a maximum σ^2, below which the infinitely many eigenvalues of σ^2 converge to zero. For a given σ^2, there may be several positive eigenvalues of k^2 (or none at all if $\sigma^2 > 4\omega^2$) but always infinitely many negative ones for k^2, decreasing indefinitely toward minus infinity.

5.2. Rossby Waves

If viscous effects are neglected and ρ is assumed constant, Eqs. (2.62) to (2.64) become

$$\frac{Du}{Dt} = -\frac{\partial P}{\partial x} + 2\omega v, \tag{258}$$

$$\frac{Dv}{Dt} = -\frac{\partial P}{\partial y} - 2\omega u, \tag{259}$$

$$\frac{Dw}{Dt} = -\frac{\partial P}{\partial z}, \tag{260}$$

in which ω is the constant angular velocity of the rotating frame and Cartesian coordinates are used. Now consider a solid sphere with a fluid layer of uniform depth attached to parts of its surface by the gravitational attraction of the sphere, the whole system rotating about an axis with angular speed ω. This simple model of the oceans on the earth can be used to study certain large-scale motion of the oceans. For that purpose local Cartesian coordinates are often used, with z measured in the direction of the vertical, x measured eastward, and y northward. If α indicates the latitude, the equations of motion in local coordinates on the surface of the earth are

$$\frac{Du}{Dt} = -\frac{\partial P}{\partial x} + 2\omega v \sin \alpha, \tag{261}$$

$$\frac{Dv}{Dt} = -\frac{\partial P}{\partial y} - 2\omega u \sin \alpha, \tag{262}$$

$$\frac{Dw}{Dt} = -\frac{\partial P}{\partial z} - g \tag{263}$$

if a term $-2\omega w \cos \alpha$ is omitted from (261) and a term $2\omega u \cos \alpha$ from (263) and if the term $\omega^2 r^2/2$ is omitted from P, as explained in Sec. 12.1 of Chap. 3.

It is very significant that the latitude α is a function of the local coordinate y. In fact at any place on the earth

$$y = R_0 \alpha + \text{const},$$

in which R_0 is the radius of the earth. The y dependence of the coefficients of the last term in (261) and the last term in (262) has far-reaching effects. It is the cause of a certain kind of waves called *Rossby waves*. To illustrate this, we shall consider the simple case of horizontal motion, i.e., a motion with no velocity component in the radial (or vertical) direction or z direction. The motion is then two-dimensional and governed by (261), (262), and the equation of continuity

$$\frac{\partial u}{\partial x} + \frac{\partial v}{\partial y} = 0.$$

Eliminating P between (261) and (262) and writing $f = 2\omega \sin \alpha$ for brevity, we have

$$\frac{D\zeta}{Dt} = -v\frac{df}{dy} = -\frac{Df}{Dt}, \tag{264}$$

$$\frac{D}{Dt}(\zeta + f) = 0, \tag{265}$$

where ζ is the vorticity defined by

$$\zeta = \frac{\partial v}{\partial x} - \frac{\partial u}{\partial y} = -\nabla^2 \psi, \tag{266}$$

ψ being the stream function. The last equality sign in (264) holds because f is a function of y alone. From the definition of f it is clear that it is the vertical component of the vorticity due to the solid-body rotation, and (265) states that the total z component of the vorticity with respect to the (nonrotating) inertial system, $\zeta + f$, is conserved when we follow a fluid particle. The connection of this result with the Kelvin theorem is not immediately clear, as might appear at first sight, because the coordinates used here are local in nature; but there is a connection. If we remember that u and v are assumed independent of z or of the radial spherical coordinate R, it can be shown by using spherical coordinates that for large-scale horizontal flows on earth the azimuthal and meridional components of the vorticity are about one one-hundredth as large as the radial component, and hence can be neglected. Using a closed circuit on the surface of the ocean and applying Kelvin's circulation theorem in an inertial system and Stokes' theorem connecting circulation with vorticity, we obtain (265).

Having seen how and why the relative vorticity ζ can change from latitude to latitude, we shall demonstrate that wave motion is possible and interpret it physically. Rossby (1949) considered a flow characterized by

$$u = \text{const}, \qquad v = v(x,t), \qquad w = 0. \tag{267}$$

Equation (264) then becomes

$$\frac{\partial^2 v}{\partial t\,\partial x} + u\,\frac{\partial^2 v}{\partial x^2} + \beta v = 0, \tag{268}$$

in which $\beta = df/dy$. The solution is

$$v = Ae^{ik(x-ct)}, \tag{269}$$

with

$$c = u - \frac{\beta}{k^2}. \tag{270}$$

The waves represented by this solution are called Rossby waves because it was Rossby who saw the possibility of these waves.

Now we know that in order to have wave motion we must have a restoring force when a fluid particle is displaced from its equilibrium position. What is the nature of the restoring mechanism in the case of Rossby waves? First, we note that in the absence of wave motion $v = 0$, and we can still have a constant u. Let the pressure for this case be denoted by $\bar{p}$. Then (261) and (262) give

$$\frac{\partial \bar{p}}{\partial x} = 0, \qquad \frac{\partial \bar{p}}{\partial y} = 2\rho\omega u \sin \alpha.$$

This immediately gives

$$\bar{p} = 2\rho\omega \int_C^y u \sin \alpha \, dy.$$

Let
$$p = \bar{p} + p',$$

where p' is the pressure due to wave motion; then (261) and (262) become, still on the assumption that $u = $ const.

$$\frac{\partial p'}{\partial x} = 2\rho\omega v \sin \alpha,$$

$$\rho \frac{Dv}{Dt} = -\frac{\partial p'}{\partial y}.$$

Assuming the factor exp $[ik(x - ct)]$ for all perturbation quantities, we have

$$p' = \frac{2}{ik} \rho\omega v \sin \alpha,$$

and hence, since v is assumed independent of y,

$$\rho \frac{Dv}{Dt} = \frac{i\rho}{k} \beta v. \tag{271}$$

Now if η is the displacement of any material line with horizontal mean position from the mean position,

$$\left(\frac{\partial}{\partial t} + u\frac{\partial}{\partial x}\right)\eta = v,$$

or, in view of (270),

$$\frac{i\beta}{k} \eta = v. \tag{272}$$

Substituting this into (271), we have

$$\rho \frac{Dv}{Dt} = -\frac{\rho\beta^2}{k^2} \eta, \tag{273}$$

which says that the acceleration is 180° out of phase with the displacement, a situation common to simple harmonic waves. The right-hand side of (273) is the restoring force per unit volume and, as we have seen, a consequence of the variability of f and therefore, by virtue of (265), a consequence of the variation of ζ. Indeed, it is the purpose of some existing intuitive arguments to show that the distribution of ζ produces a restoring force opposite to the displacement.

We now show a less well-known kind of Rossby wave. If no mean flow is present, and if the waves are of small amplitude, (265) can be written as

$$\frac{\partial}{\partial t} \nabla^2\psi + \beta \frac{\partial\psi}{\partial x} = 0.$$

Assuming ψ to have the time factor $\exp(-i\sigma t)$, we have

$$\nabla^2\psi + \frac{i\beta}{\sigma}\frac{\partial\psi}{\partial x} = 0.$$

For instance, in the zone between two latitudes where

$$0 \le x \le a \qquad \text{and} \qquad 0 \le y \le b,$$

with $x = 0$ and $x = a$ at the same location, ψ must be required to be periodic, or $\psi(0,y) = \psi(a,y)$. It is easy to show that in these circumstances

$$\sigma = \pm\frac{m\pi\beta/a}{(m\pi/a)^2 + (n\pi/b)^2},$$

m and n being integers.

These frequencies correspond respectively to the factor

$$\exp(\mp im\pi x/a)$$

in the solution. Hence the waves always propagate westward. On the flow so obtained the flow with constant velocity u can be superposed. The waves are then carried with the velocity u while propagating *through* the fluid with westward velocity

$$\frac{\sigma a}{m\pi} = \frac{\beta}{(m\pi/a)^2 + (n\pi/b)^2}.$$

5.3. Axisymmetric Inertial Waves of Finite Amplitude in Steady Flows

Consider a steady axisymmetric flow in a tube with swirl. Cylindrical coordinates will be used, and u, v, and w will denote the radial, azimuthal, and longitudinal velocity components, respectively. The flow being axisymmetric, the Stokes stream function ψ can be used in spite of the existence of v. Furthermore, since any material circle with its center on the axis of the tube and its plane normal to the axis will remain so, by the Kelvin theorem of conservation of circulation vr must be constant along such a material circle, which in steady flow means it must be a function of ψ. For convenience we write.

$$(vr)^2 = f(\psi). \tag{274}$$

The equations of motion, referred to an inertial system, are then

$$w\left(\frac{\partial u}{\partial z} - \frac{\partial w}{\partial r}\right) - \frac{f(\psi)}{r^3} = -\frac{\partial}{\partial r}\frac{P}{\rho}, \tag{275}$$

$$u\left(\frac{\partial w}{\partial r} - \frac{\partial u}{\partial z}\right) = -\frac{\partial}{\partial z}\frac{P}{\rho}, \tag{276}$$

in which
$$\frac{P}{\rho} = \frac{p}{\rho} + \frac{u^2 + w^2}{2} + \Omega,$$

Ω being the body-force potential. Elimination of P in (275) and (276) produces

$$\frac{D}{Dt}\frac{\zeta}{r} + \frac{1}{r^4}\frac{\partial}{\partial z}f(\psi) = 0,$$

or, since

$$u = -\frac{1}{r}\frac{\partial \psi}{\partial z} \quad \text{and} \quad \frac{D}{Dt} = u\frac{\partial}{\partial r} + w\frac{\partial}{\partial z},$$

$$\frac{D}{Dt}\frac{\zeta}{r} - u\frac{f'(\psi)}{r^3} = \frac{D}{Dt}\left[\frac{\zeta}{r} + \frac{f'(\psi)}{2r^2}\right] = 0, \tag{277}$$

in which

$$f'(\psi) = \frac{df(\psi)}{d\psi} \quad \text{and} \quad \zeta = \frac{\partial w}{\partial r} - \frac{\partial u}{\partial z} = \frac{1}{r}\left(\frac{\partial^2}{\partial r^2} - \frac{1}{r}\frac{\partial}{\partial r} + \frac{\partial^2}{\partial z^2}\right)\psi.$$

Hence

$$\left(\frac{\partial^2}{\partial r^2} - \frac{1}{r}\frac{\partial}{\partial r} + \frac{\partial^2}{\partial z^2}\right)\psi + \frac{f'(\psi)}{2} = r^2 h(\psi). \tag{278}$$

If $u = 0$, $w = -W$ and $v = \omega r$ at $z = \infty$, so that the fluid there is in solid-body rotation and translation,

$$f(\psi) = \frac{4\omega^2}{W^2}\psi^2, \qquad h(\psi) = -\frac{2\omega^2}{W}, \tag{279}$$

and (278) becomes the *linear* equation (Long 1953b)

$$\left(\frac{\partial^2}{\partial r^2} - \frac{1}{r}\frac{\partial}{\partial r} + \frac{\partial^2}{\partial z^2}\right)\psi + \frac{4\omega^2}{W^2}\psi = -\frac{2\omega^2}{W}r^2. \tag{280}$$

The general solution of this equation is of the form

$$\psi = -\frac{Wr^2}{2} + r\sum_{n=1}^{\infty} A_n \exp\left\{-[\lambda_n{}^2 - (Ro)^{-2}]^{1/2}\frac{z}{b}\right\}J_1\left(\lambda_n\frac{r}{b}\right), \tag{281}$$

in which

$$Ro = \frac{W}{2\omega b}$$

is the Rossby number, b being the radius of the tube, and λ_n is the nth zero of $J_1(\lambda)$. If

$$\lambda_{N+1} > (Ro)^{-1} > \lambda_N,$$

it is evident from (281) that there can be N wave components. These waves can be made by inserting an axisymmetric body into the fluid axisymmetrically, and appear in the wake of the body.

PROBLEMS

1 Two fluids have an interface at $y = 0$ and are confined between the boundaries $y = -h$ and $y = h'$. If the upper fluid has density ρ' and the lower fluid has density ρ, show that the wave velocity c for gravity

waves is given by

$$kc^2 = \frac{g(\rho - \rho')}{\rho \coth kh + \rho' \coth kh'}$$

if k is the wave number and if surface-tension effect is neglected.   Show that:

a The limit of c^2 as both h and h' approach infinity is

$$c^2 = \frac{g(\rho - \rho')}{k(\rho + \rho')}.$$

b If both h and h' are finite, for small k the expansion formula for c is

$$c = c_0\left[1 - \tfrac{1}{6}k^2 \frac{\rho h^2 h' + \rho' h'^2 h}{\rho h' + \rho' h} + O(k^4 h^2 h'^2)\right],$$

where

$$c_0{}^2 = \frac{g(\rho - \rho')}{\rho/h + \rho'/h'}.$$

c If $h' = \infty$ but h remains finite, for small k the expansion formula for c is

$$c = c_0\left[1 - \frac{1}{2}\frac{\rho'}{\rho} h |k| + O(k^2 h^2)\right].$$

[This result was pointed out by Benjamin (1968).]

2 Show that the solution for wave motion in very deep water following an initial displacement of the free surface

$$\zeta_0 = J_0(kr)$$

is $-\phi = g\,\dfrac{\sin \sigma t}{\sigma}\,e^{kz}J_0(kr),\qquad \zeta = \cos \sigma t J_0(kr),\qquad \sigma^2 = gk.$

3 Using the results of the preceding problem and the Fourier-Bessel integral

$$f(r) = \int_0^\infty J_0(kr)k\,dk \int_0^\infty f(\alpha)J_0(k\alpha)\alpha\,d\alpha,$$

show that the solution for wave motion in very deep water following the initial conditions

$$\zeta_0 = f(r) \qquad \text{and} \qquad \phi_0 = 0$$

is $-\phi = g\displaystyle\int_0^\infty \dfrac{\sin \sigma t}{\sigma}\,e^{kz}J_0(kr)k\,dk\int_0^\infty f(\alpha)J_0(k\alpha)\alpha\,d\alpha,$

$$\zeta = \int_0^\infty \cos \sigma t J_0(kr)k\,dk\int_0^\infty f(\alpha)J_0(k\alpha)\alpha\,d\alpha.$$

4 If the initial elevation is concentrated in the neighborhood of the origin and

$$\int_0^\infty f(\alpha) 2\pi\alpha \, d\alpha = 1, \tag{a}$$

i.e., if $f(\alpha)$ is the Dirac "function," then

$$-\phi = \frac{g}{2\pi} \int_0^\infty \frac{\sin \sigma t}{\sigma} e^{kz} J_0(kr) k \, dk.$$

[The Dirac function, or distribution, for ζ_0 of course renders the linear theory invalid for small t. Nevertheless the result is a good approximation for the case $f(\alpha) = 0$ for $\alpha > \epsilon$, $f(\alpha)$ continuous for $\alpha \le \epsilon$ and satisfying (a), at least for large enough t.]

5 A gravity-wave train in water of depth h is described by

$$\phi_0 = A \cos k(x - ct) \cosh k(z + h),$$

the origin of the (vertical) ordinate z being at the undisturbed free surface and

$$c^2 = \frac{g}{k} \tanh kh.$$

A vertical circular cylinder is placed in the water. Its cross section is described by

$$r^2 \equiv x^2 + y^2 = a^2, \tag{b}$$

and it extends throughout the entire depth. Find the solution for the complex potential ϕ. *Hint:* Take

$$\phi = \phi_0 + \phi'$$

and make $\partial\phi/\partial r$ equal to zero on the cylindrical surface. $\partial\phi_0/\partial r$ is $(\partial\phi_0/\partial x) \cos \varphi$, where (r,φ) are polar coordinates in the xy plane. The solution of $\nabla^2\phi'$ is of the form

$$\phi' = \sum_{n=0}^\infty \cosh k(z + h)\{(B_n \cos kct + C_n \sin kct) \cos n\varphi J_n(kr)$$
$$+ (D_n \cos kct + E_n \sin kct) \cos n\varphi [J_n(kr) + G_n N_n(kr)]\}, \tag{c}$$

in which G_n is determined by

$$\frac{dJ_n(kr)}{dr} + G_n \frac{dN_n(kr)}{dr} = 0 \qquad \text{at} \qquad r = a,$$

N_n being the Neumann function of order n, or Bessel function of the second kind. The B_n and C_n in (c) are determined by $\partial\phi/\partial r = 0$ at $r = a$ by the use of the Fourier series. The D_n and E_n are determined by the *radiation condition* of Sommerfeld, namely, that at infinity each term in ϕ' should be of the form

$$[P_n \cos k(r - ct - \frac{\epsilon}{k})$$
$$+ Q_n \sin k(r - ct - \frac{\epsilon}{k})] \cos n\varphi \cosh k(z + h),$$

in which P_n and Q_n depend on r, $\epsilon = \pi(n/2 - \frac{1}{4})$. Remember that for large r

$$J_n(kr) = \left(\frac{2}{\pi kr}\right)^{\frac{1}{2}} [\cos (kr - \epsilon) - \sin (kr - \epsilon)] + O(r^{-\frac{3}{2}}),$$

$$N_n(kr) = \left(\frac{2}{\pi kr}\right)^{\frac{1}{2}} [\sin (kr - \epsilon) + \cos (kr - \epsilon)] + O(r^{-\frac{3}{2}}).$$

Finish the analysis. The problem is typical of many problems concerning diffraction. For large k (short waves) there will be an appreciable "shadow" on the lee side, namely, the side further along the direction of propagation of the original waves.

6 A straight canal with a horizontal bottom and vertical side walls extends from $x = -\infty$ to $x = \infty$. There is quiescent water of depth h in it and a vertical end wall originally at $x = 0$. At $t = 0$ the end wall moves to the right with a velocity U.

   a Show that there is always a hydraulic jump in the water to the right of the end gate, the strength of the jump approaching zero as U approaches zero.

   b Find the flow to the right of the gate, for any U.

   c Find the flow to the left of the gate, for any U.

7 Design a channel expansion for supercritical flow satisfying the requirement that there be no waves downstream from the expansion. Take the ratio of downstream to upstream width to be $2:1$ or any other ratio greater than 1.5.

8 Show that the loss in specific energy $u^2/2g + h$ in a weak hydraulic jump is proportional to $(\Delta h)^3/4$, where Δh is the height of the jump.

9 A supercritical stream of water flows in the x direction. An inward bend of the channel wall starting at P (see Fig. P9) causes an oblique hydraulic jump. Find β, the ratio of the upstream to downstream depth, and the downstream Froude number F' in terms of θ and the upstream Froude number F. Show that for small θ two kinds of jumps are possible, one with β near $\pi/2$ and the other with β near $\sin^{-1}(1/F)$, the former value corresponding to a subcritical and the latter to a supercritical flow after the jump.

10 Show from the answer to the preceding problem that for small θ and a given F, the height of the jump is proportional to θ and the loss of specific energy is proportional to θ^3 for the weak jump.

11 In an isothermal atmosphere the density is given by

$$\bar{\rho} = \rho_0 e^{-2\beta z},$$

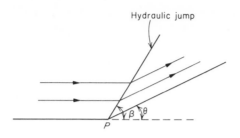

Plan view

FIGURE P9

in which z is measured in the direction of the vertical. If w denotes the vertical component of the velocity, show that infinitesimal wave motion of circular frequency σ is governed by the equation

$$\left[\left(1 - \frac{N^2}{\sigma^2}\right)\left(\frac{\partial^2}{\partial x^2} + \frac{\partial^2}{\partial y^2}\right) + \frac{\partial^2}{\partial z^2} - 2\beta \frac{\partial}{\partial z}\right] w = 0,$$

in which
$$N^2 = -\frac{g}{\bar{\rho}}\frac{d\bar{\rho}}{dz} = 2\beta g$$

and x, y, and z are Cartesian coordinates. Note that the type of equation depends on whether σ^2 is greater than, equal to, or less than N^2. Find the characteristics for two-dimensional waves when the equation is of the hyperbolic type, that is, $N^2 > \sigma^2$. See Sec. 18, Chap. 6.

12 Show that the solution of the equation in the preceding problem is

$$w = f(z) \exp\left[i(k_x x + k_y y - \sigma t)\right],$$

in which
$$f(z) = e^{\beta z}(A e^{ik_z z} + B e^{-ik_z z})$$

and
$$k_z = \left[\left(\frac{N^2}{\sigma^2} - 1\right)k_r^2 - \beta^2\right]^{1/2}, \qquad k_r^2 = k_x^2 + k_y^2.$$

The dispersion equation relating σ to the wave number k is then

$$\frac{\sigma}{N} = \begin{cases} \dfrac{k_r}{(k_r^2 + k_z^2 + \beta^2)^{1/2}} & \text{for } \sigma < \sigma_c = \dfrac{N}{\sqrt{1 + (\beta/k_r)^2}} \\[2mm] & k_z^2 > 0, \\[2mm] \dfrac{k_r}{(k_r^2 + \beta^2 - |k_z|^2)^{1/2}} & \text{for } \sigma > \sigma_c, \\[2mm] & k_z^2 < 0. \end{cases}$$

13 Show that for any oscillatory motion in x, y, and z with wave-number vector $\mathbf{k} = (k_x, k_y, k_z)$ the phase velocity $\mathbf{c}$ is given by

$$\mathbf{c} = \frac{\sigma}{k}\mathbf{n}, \qquad \mathbf{n} = \frac{\mathbf{k}}{k},$$

in which $k = |\mathbf{k}|$. Show that in the preceding problem if $k_z{}^2 > 0$,

$$c = |\mathbf{c}| = \frac{Nk_r}{k(k^2 + \beta^2)^{\frac{1}{2}}}$$

and that $c = A \sin \theta$, $\qquad A = \dfrac{N\sqrt{\epsilon}}{\beta}$, $\qquad \epsilon = \left(1 + \dfrac{k^2}{\beta^2}\right)^{-1}$

in terms of the polar coordinates (k, θ, φ) in the wave-number space defined by

$$k_x = k \sin \theta \cos \varphi, \qquad k_y = k \sin \theta \sin \varphi, \qquad k_z = k \cos \theta$$

(Wu, 1967).

14 The group velocity is

$$\mathbf{c}_g = \left(\frac{\partial \sigma}{\partial k_x}, \frac{\partial \sigma}{\partial k_y}, \frac{\partial \sigma}{\partial k_z}\right)$$

for real k_x, k_y, and k_z. Show that for Probs. 12 and 13

$$c_{gx} = A \cos \varphi \, \frac{k^2 \cos^2 \theta + \beta^2}{k^2 + \beta^2},$$

$$c_{g_y} = A \sin \varphi \, \frac{k^2 \cos^2 \theta + \beta^2}{k^2 + \beta^2},$$

$$-c_{gz} = \frac{Ak^2 \sin \theta \cos \theta}{k^2 + \beta^2}.$$

Show that c_{gz} is opposite in sign to c_z and $c_{gy}/c_{gx} = c_y/c_x$ and hence the plane containing $\mathbf{c}$ and $\mathbf{c}_g$ is vertical. (Wu, 1967).

15 For the wave motion treated in the Probs. 11 to 14, if $\sigma < N$, show that for a fixed σ

$$k_z{}^2 + \beta^2 = M^2 k_r{}^2, \qquad M^2 = \left(\frac{N}{\sigma}\right)^2 - 1,$$

or $\qquad \left(\dfrac{\beta}{k}\right)^2 = M^2 \sin^2 \theta - \cos^2 \theta,$

and that

$$c_{gr} = \frac{N}{k(\theta)} \frac{M^2 \csc \theta}{(1 + M^2)^{\frac{3}{2}}}, \qquad c_{gz} = -\frac{N}{k(\theta)} \frac{\cos \theta \csc^2 \theta}{(1 + M^2)^{\frac{3}{2}}}.$$

Hence θ is in the range

$$\theta_M \le \theta \le \pi - \theta_M, \qquad \text{with } \theta_M = \cot^{-1} M$$

(the equality signs valid only for $k = \infty$), and

$$\frac{c_{gz}}{c_{gr}} = -\frac{1}{M^2} \cot \theta.$$

Show then that waves created at the origin by oscillation are propagated along the straight line

$$\frac{z}{r} = -\frac{1}{M^2} \cot \theta$$

for each θ and that they exist only *outside* the vertical cone bounded by $r = \pm Mz$. (Wu, 1967).

16 Discuss (257). Is periodic wave motion possible if $F(r)$ is negative throughout? If not, what will happen? *Hint:* Remember that f must be zero at $r = 0$ and $r = a$, a being the radius of the tube. Can f be zero at $r = a$ if $F(r)$ is negative for all r between zero and a if σ^2 is to be positive? But then will a negative σ^2 do? And what does a negative σ^2 mean?

17 For a given $F(r)$ in (257), does $c = \sigma/k$ increase or decrease when k^2 increases?

REFERENCES

ABRAMOWITZ, M., and IRENE A. STEGUN, 1965: "Handbook of Mathematical Functions," Dover Publications, Inc., New York.

BENJAMIN, T. B., 1967*a*: The Instability of Periodic Wave-trains in Certain Nonlinear Systems, *Proc. Roy. Soc. London, Ser. A*, **299**: 59–75.

BENJAMIN, T. B., and J. E. FEIR, 1967: The Disintegration of Wave Trains on Deep Water, pt I, Theory, *J. Fluid Mech.*, **27**: 417–430.

BENJAMIN, T. B., 1968: Gravity Currents and Related Phenomenon, *J. Fluid Mech.*, **31**: 209–249.

BÔCHER, M., 1917: "Leçons sur les méthodes de Sturm dans la théorie des équations différentielles linéaires, et leur développements modernes," Gauthier-Villars, Paris.

BROER, L. J. F., 1950: On the Propagation of Energy in Linear Conservative Waves, *Appl. Sci. Res. Sect. A*, **2**: 329–344.

CHESTER, W., 1966: A Model of the Undular Bore on a Viscous Fluid, *J. Fluid Mech.*, **24**: 367–378.

DUBREIL-JACOTIN, M. L., 1932: Sur les ondes de type permanent dans les liquides héterogènes, *Atti Accad. Naz. Lincei, Rend. Classe Sci. Fis. Mat. Nat.*, (6)**15**: 814–819.

DUBREIL-JACOTIN, M. L., 1935: Complément à une note autérieure sur les ondes des types permanent dans les liquides héterogènes, *Atti Accad. Naz. Lincei, Rend. Classe Sci. Fis. Mat. Nat.*, (6)**21**: 344–346.

GERSTNER, F. J., 1802: Theorie der Wellen, *Abhandl. Kgl. Böhm. Ges. Wiss.*

IPPEN, A. T., and R. T. KNAPP, 1936: A Study of High Velocity Flow in Curved Channels of Rectangular Cross Section, *Trans. Am. Geophys. Union*, **17**: 516.

KÁRMÁN, T. VON, 1938: Eine praktische Anwendung der Analogie zwischen Überschallströmung in Gasen und überkritischer Strömung in offenen Gerinnen, *Z. Angew. Math. Mech.*, **18**: 49–56.

KORTEWEG, D. J., and G. DE VRIES, 1895: On the Change of Form of Long Waves Advancing in a Rectangular Canal, and on a New Type of Long Stationary Waves, *Phil. Mag.*, (5)**39**: 422.

LAITONE, E. V., 1962: Limiting Conditions for Cnoidal and Stokes Waves, *J. Geophys. Res.*, **67**: 1555–1564.

LAMB, H., 1932: "Hydrodynamics," The Macmillan Company, London; reprinted (6th ed.) by Dover Publications, Inc., New York, 1945 (with the same pagination).

LANDAU, L. D., and E. M. LIFSHITZ, 1959: "Fluid Mechanics," Pergamon Press, New York.

LEVI-CIVITA, T., 1925: Détermination rigoureuse des ondes permanentes d'ampleur finie, *Math. Ann.*, **92**: 264.

LIGHTHILL, M. J., and G. B. WHITHAM, 1955: On Kinematic Waves, pt. I, Flood Movement in Long Rivers, *Proc. Roy. Soc. London, Ser. A*, **229**: 281–316.

LONG, R. R., 1953a: Some Aspects of the Flow of Stratified Fluids, pt. I, A Theoretical Investigation, *Tellus*, **5**: 42–57.

LONG, R. R., 1953b: Steady Motion around a Symmetrical Obstacle Moving along the Axis of a Rotating Fluid, *J. Meteor.*, **10**: 197.

MASSAU, J., 1900: Mémoire sur l'intégration graphique des équations aux dérivées partielles, *Ann. Assoc. Ingr. Sortis Écoles Spéciales Gand*, **23**: 95–214.

MEYER, R. E., 1967: Note on the Undular Jump, *J. Fluid Mech.*, **28**: 209–222.

PREISWERK, E., 1938: Anwendung gasdynamischer Methoden auf Wasser-strömungen mit freier Oberfläche, *Mitt. Inst. Aerodynamik Eidgenoess. Tech. Hochschule, Zurich.*

RAYLEIGH, LORD, 1876: On Waves, *Phil. Mag.*, (5)**1**: 257.

RAYLEIGH, LORD, 1883: The Form of Standing Waves on the Surface of Running Water, *Proc. London Math. Soc.*, **15**: 69.

RIABOUCHINSKY, D., 1932: Sur l'analogie hydraulique des mouvements d'un fluide compressible, *Compt. Rend.*, **195**: 998.

RIEMANN, B., 1860: Über die Fortpflanzung ebener Luftwellen von endlicher Schwingungweite, *Abhandl. Kgl. Ges. Wiss.*, **8**: reprinted in "Collected Works of Bernhard Riemann," pp. 156–181, Dover Publications Inc., New York, 1953.

ROSSBY, C.-G., 1949: On the Dispersion of Planetary Waves in a Barotropic Atmosphere, *Tellus*, **1**: 1.

RUSSELL, S., 1844: Report on Waves, *Brit. Assoc. Rept.*

STOKER, J. J., 1948: The Formation of Breakers and Bores, *Comm. Pure Appl. Math.*, **1**: 1–88.

STOKES, G. G., 1846: Report on Recent Researches in Hydrodynamics, *Brit. Assoc. Rept.*

STOKES, G. G., 1847: On the Theory of Oscillatory Waves, *Cambridge Trans.*, **8**.

STRUIK, D. J., 1926: Détermination rigoureuse des ondes irrotationnelles périodiques dans un canal à profondeur finie, *Math. Ann.*, **95**: 595.

URSELL, F., R. G. DEAN, and Y. S. YU, 1960: Forced Small-amplitude Water Waves: A Comparison of Theory and Experiment, *J. Fluid Mech.*, **7**: 33–52.

WHITHAM, G. B., 1960: A Note on Group Velocity, *J. Fluid Mech.*, **9**: 347–352.

WHITHAM, G. B., 1967a: Nonlinear Dispersion of Water Waves, *J. Fluid Mech.*, **27**: 399–412.

WHITHAM, G. B., 1967b: Variational Methods and Applications to Water Waves, *Proc. Roy. Soc. London, Ser. A*, **299**: 6–25.

WU, T. Y., 1967: Radiation and Dispersion of Internal Waves, paper presented at International Conference on Stratified Fluids, held at the University of Michigan, Ann Arbor, April 11–14, 1967.

YIH, C.-S., 1958: On the Flow of a Stratified Fluid, *Proc. 3d U.S. Natl. Congr. Appl. Mech., Providence*, pp. 857–861.

YIH, C.-S., 1960: Exact Solutions for Steady Two-dimensional Flow of a Stratified Fluid, *J. Fluid Mech.*, **9**: 161–174.

YIH, C.-S., 1965: "Dynamics of Nonhomogeneous Fluids," The Macmillan Company, New York.

YIH, C.-S., 1966: Note on Edge Waves in a Stratified Fluid, *J. Fluid Mech.*, **24**: 765–767.

ADDITIONAL READING

BENJAMIN, T. B., 1966: Internal Waves of Finite Amplitude and Permanent Form, *J. Fluid Mech.*, **25**: 241–270.

BENJAMIN, T. B., 1967b: Internal Waves of Permanent Form in Fluids of Great Depth, *J. Fluid Mech.*, **29**: 559–592.

ECKART, C., 1960: "Hydrodynamics of Oceans and Atmospheres," Pergamon Press, New York.

ENGEVIK, L., 1975: On the indeterminacy of the problem of Stratified Fluid Flow over a Barrier and Related Problems, *Zeitsch. für ang. Math. u. Phy.*, **26**: 831–834.

PHILLIPS, N. A., 1965: Elementary Rossby Waves, *Tellus*, **17**: 295–301.

STOKER, J. J., 1957: "Water Waves," Interscience Publishers, Inc., New York.

CHAPTER SIX

THE DYNAMICS
OF
INVISCID
COMPRESSIBLE
FLUIDS

1. INTRODUCTION

The dynamics of compressible fluids is more complicated and more difficult than that of incompressible fluids for several reasons. First, since the density of a fluid particle is no longer constant but varies as it moves from place to place, and since the pressure is dependent on the density, the equation of state connecting the pressure p, the density ρ, and the (absolute) temperature T must be invoked. The appearance of T then necessitates the energy equation, which is just the equation representing the first law of thermodynamics, or the equation of heat budget. Second, the equation of continuity is nonlinear for a compressible fluid. This rules out the Laplace equation as the governing equation even if the flow is irrotational and a velocity potential exists. Third, as we shall see, under certain circumstances shock waves can appear in the flow, across which a fluid particle experiences an abrupt or at least a very steep change in entropy and corresponding changes in density, pressure, temperature, and velocity.

In this chapter we shall restrict our attention to ideal gases. Discussions of the effects of viscosity and diffusivity will be postponed to Chaps.

7 and 8, where the boundary-layer theory will be presented. Most of the results presented in this chapter represent the achievements of aerodynamicists in the first half of the twentieth century.

2. THE GOVERNING EQUATIONS

In any fluid flow the equation of continuity must of course be satisfied. This is Eq. (1.12), which we repeat for convenience:

$$\frac{\partial \rho}{\partial t} + \frac{\partial (\rho u_i)}{\partial x_i} = 0. \tag{1}$$

The equations of motion remain the Euler equations

$$\rho \frac{Du_i}{Dt} = -\frac{\partial p}{\partial x_i} + \rho X_i, \tag{2}$$

or Eq. (3.2). Since ρ is not constant and not known, there are five unknowns (ρ, p, u_1, u_2, and u_3) in the four equations contained in (1) and (2). For any fluid there is an equation of state. Since in this chapter we shall consider ideal gases exclusively, we shall present only the equation of state

$$\frac{p}{\rho} = RT \tag{3}$$

for ideal gases [see Eq. (A1.5)]. Although (3) provides one more equation, it also introduces one more variable, T. Still another equation is needed, which is provided by (2.79). With the diffusive and dissipation terms neglected, (2.79) becomes

$$\rho c_v \frac{DT}{Dt} + p\theta = 0, \tag{4}$$

in which θ is $\partial u_i / \partial x_i$. The neglect of the two terms on the right-hand side of (2.79) implies that the time rate of temperature change of a fluid particle is due predominantly to the compression or rarefaction of the gas, whose effect dominates the effect of heat diffusion or viscous dissipation. In thermal boundary layers, where the temperature gradient may be high, or in vorticity boundary layers, where the accompanying rate of dissipation happens to be large, the terms on the right-hand side of (2.79) must be retained. In this chapter we consider only inviscid and nondiffusive gases. Thus we shall use (2) and (4) and in so doing shall limit our attention to the flow outside of boundary layers. The six equations contained in (1) to (4) are the equations governing the flow of an ideal gas, assumed inviscid and nondiffusive. It is evident that these equations are nonlinear and that without further specialization there is little hope of finding solutions. We shall now see in what directions physically important specializations can be made.

Since (1) can be written as

$$\frac{1}{\rho}\frac{D\rho}{Dt} + \theta = 0, \tag{5}$$

(4) can be written as

$$\rho c_v \frac{DT}{Dt} - \frac{p}{\rho}\frac{D\rho}{Dt} = 0. \tag{6}$$

In (3) and (4) the R and c_v are in energy units, so that (see Appendix 1)

$$R = c_p - c_v \tag{7}$$

and

$$\frac{c_v}{R} = \frac{1}{\gamma - 1}, \qquad \gamma \equiv \frac{c_p}{c_v}. \tag{8}$$

By virtue of (3) and (8), (6) becomes

$$\rho \frac{D}{Dt}\frac{p}{\rho} - (\gamma - 1)\frac{p}{\rho}\frac{D\rho}{Dt} = 0,$$

which can be written as

$$\rho^\gamma \frac{D}{Dt}\frac{p}{\rho}\rho^{-(\gamma-1)} = \rho^\gamma \frac{D}{Dt}\frac{p}{\rho^\gamma} = 0. \tag{9}$$

Since (Appendix 1)

$$\frac{p}{\rho^\gamma} = \frac{p_0}{\rho_0{}^\gamma} e^{S/c_v}, \tag{10}$$

in which S is the entropy and p_0 and ρ_0 are the pressure and density, respectively, of the gas whose entropy is arbitrarily assigned the value zero, (9) can be written as

$$\frac{DS}{Dt} = 0, \tag{11}$$

which states that in an inviscid and nondiffusive gas with equation of state (3) the entropy of a fluid particle is preserved as it moves along its path. The same result would be obtained if the equation of state were other than (3), but in any case the very use of an equation of state implies that the rate of change of state is slow or, more precisely, that the time during which the state of the gas changes from state 1 to state 2 is large compared to the relaxation time of the fluid or, alternatively, the mean free path of the gas molecules divided by their mean speed. In aerodynamics, the only abrupt change of state is that suffered by a gas as it goes through a shock wave, across which the entropy is therefore not preserved even if heat conduction and viscous dissipation are neglected.

If the entropy of a gas is originally homogeneous, (11) ensures that S is constant subsequently. Hence

$$\frac{p}{\rho^\gamma} = \text{const} = C \tag{12}$$

provided that no shock waves are crossed. Then the governing equations are (1), (2), and (12). We have thus made a physically significant specialization. If the fluid indeed has homogeneous entropy, it or its flow is called *homentropic*.

3. THE BERNOULLI EQUATIONS

If the flow is steady, and if the change of state along a streamline is isentropic, then for an ideal gas (12) holds along it, and the Bernoulli equation (3.26) can be written as

$$\frac{\gamma}{\gamma - 1} \frac{p}{\rho} + \Omega + \frac{q^2}{2} = \text{const along a streamline,} \tag{13}$$

in which q is the speed and Ω the body-force potential, which is assumed to exist. Isentropy presumes that the change of state is not abrupt and that effects of heat conduction and viscous dissipation are negligible. The fluid does not have to be homentropic for (13) to hold.

If the entropy is homogeneous to start with and the change of state for any portion of the fluid is assumed isentropic, the entropy will remain homogeneous. For a homentropic fluid irrotationality persists, provided the effects of viscosity are negligible. Hence a motion started from rest will be irrotational. For an irrotational flow the Bernoulli equation is given by (3.29) or (3.30), which for an ideal gas can be written as

$$\frac{\partial \phi}{\partial t} + \frac{\gamma}{\gamma - 1} \frac{p}{\rho} + \Omega + \frac{q^2}{2} = F(t), \tag{14}$$

in which $F(t)$ can be equated to a constant or zero since an arbitrary function of time can be added to ϕ without affecting the velocity field. Of course, if ϕ, q, and p are given at some point, say infinity, then $F(t)$ must be determined from the data given there.

We now proceed to derive the Bernoulli equation for steady flows of a nonhomentropic gas which is valid even across a shock. The resulting equation is also called the *energy equation*. Figure 1 is a cross section of a stream tube, ABC and $A'B'C'$ are two of the streamlines bounding the tube, and BB' is a portion of a shock wave through which the gas flows. Since eventually we are going to obtain an equation valid along a streamline and the stream tube will be shrunk into a streamline, we shall assume the properties of the fluid to be homogeneous across AA' and also across CC'. Let the cross-sectional areas at AA' be a and that at CC' be a'. Then in time dt the fluid particles originally in section AA' will have moved a distance $q_A \, dt$ to section DD', and those at CC' will have moved a distance $q_C \, dt$ to EE'. The mass that has passed through either AA' or CC' is

$$\rho_A a q_A dt = \rho_C a' q_C \, dt.$$

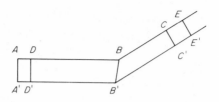

FIGURE 1. SKETCH FOR DERIVING THE
BERNOULLI EQUATION ALONG A STREAMLINE
IN STEADY FLOWS OF A GAS, VALID EVEN IF A
SHOCK WAVE IS PRESENT.

Applying the first law of thermodynamics,

$$p_A a q_A \, dt - p_C a' q_C \, dt$$

$$= \rho_A a q_A \, dt \left[\left(c_v T + \Omega + \frac{q^2}{2} \right)_C - \left(c_v T + \Omega + \frac{q^2}{2} \right)_A \right],$$

which can be written as

$$\left(\frac{p}{\rho} + c_v T + \Omega + \frac{q^2}{2} \right)_C = \left(\frac{p}{\rho} + c_v T + \Omega + \frac{q^2}{2} \right)_A,$$

or, since $c_v/R = 1/(\gamma - 1)$,

$$\frac{\gamma}{\gamma - 1} \frac{p}{\rho} + \Omega + \frac{q^2}{2} = \text{const along a streamline,} \qquad (15)$$

by virtue of (3). Equation (15) is identical to (13) except that isentropy
of change of state is *not* assumed. Thus (13) or (15) is valid across the
shock also. The first term in (15) is equal to $c_p T$, or the enthalpy per
unit mass.

In order to emphasize the assumption of isentropy along a streamline underlying (13) and the assumption of homentropy underlying (14),
we shall rewrite them as

$$\frac{\gamma}{\gamma - 1} \frac{p_0}{\rho_0{}^\gamma} \rho^{\gamma-1} + \Omega + \frac{q^2}{2} = \text{const along a streamline} \qquad (13a)$$

and

$$\frac{\partial \phi}{\partial t} + \frac{\gamma}{\gamma - 1} \frac{p_0}{\rho_0{}^\gamma} \rho^{\gamma-1} + \Omega + \frac{q^2}{2} = F(t) \qquad (14a)$$

over the entire field of flow. In (13a) and (14a), ρ_0 is a reference density
and p_0 the corresponding pressure at some point, so that

$$\frac{p}{\rho^\gamma} = \frac{p_0}{\rho_0{}^\gamma} \qquad (16)$$

along a streamline or everywhere, respectively, in regard to (13a) and

(14*a*). For steady irrotational flows of a homentropic fluid, (14*a*) becomes

$$\frac{\gamma}{\gamma - 1} \frac{p_0}{\rho_0{}^\gamma} \rho^{\gamma-1} + \Omega + \frac{q^2}{2} = \text{const.} \tag{17}$$

On earth the most important body force is the gravitational pull of the earth, so that Ω is simply gz, g being the gravitational acceleration and z the elevation measured from some datum. Thus all the Bernoulli equations, whether they are valid along a streamline only or valid everywhere in the fluid, can be looked upon as a means to evaluate the change in density (or in p/ρ at least, if a shock front has been crossed) due to the change in elevation or speed. In aerodynamics the body-force term is always neglected, for the airplane is too small and moves too fast for that term to be of any consequence compared with the term involving the speed. However, in meteorological problems great variations in height are encountered, and a change of elevation often has a much greater effect on the density than a change in speed.

It will be shown in a later section that the speed of sound c is given by

$$c^2 = \left(\frac{dp}{d\rho}\right)_S, \tag{18}$$

in which the subscript S signifies that the entropy is constant or, what amount to the same thing, that p and ρ are related as in (16). It follows that

$$c^2 = \frac{\gamma p}{\rho} = \gamma \frac{p_0}{\rho_0{}^\gamma} \rho^{\gamma-1}. \tag{18a}$$

Thus the terms involving p or ρ in (13) to (15), (13*a*), (14*a*), and (17) can be replaced by $c^2/(\gamma - 1)$. In particular, (17) can be written as

$$\frac{c^2}{\gamma - 1} + \Omega + \frac{q^2}{2} = \text{const} \tag{17a}$$

or, if gravity effects are negligible,

$$\frac{c^2}{\gamma - 1} + \frac{q^2}{2} = \text{const.} \tag{19}$$

Since $\gamma > 1$, it is evident from (18*a*) that $c = 0$ when $\rho = 0$. In that event q is a maximum. Equation (19) can therefore be written as

$$\frac{c^2}{\gamma - 1} + \frac{q^2}{2} = \frac{1}{2} q_{\max}^2. \tag{19a}$$

It should be noted that c as well as q varies from place to place. At the place where $c = q$ we shall denote the c by c^*. Then

$$\frac{\gamma + 1}{\gamma - 1} c^{*2} = q_{\max}^2. \tag{20}$$

Equations (19) and (20) are very useful in aerodynamics.

4. EQUATIONS GOVERNING STEADY IRROTATIONAL FLOWS OF A HOMENTROPIC GAS

As a consequence of Kelvin's theorem (Sec. 2 of Chap. 3), once irrotational, the flow in any portion of a homentropic gas will remain irrotational, provided the effects of viscosity and thermal conductivity are negligible. Thus the flow of a homentropic inviscid and nondiffusive gas, if started from rest or originating from a region where it is irrotational, is irrotational. For irrotational flows

$$u_i = \frac{\partial \phi}{\partial x_i}, \tag{21}$$

in which ϕ is the velocity potential. If the flow is unsteady, the equation in ϕ governing the flow is obtained by substituting (21) into (1) and evaluating ρ from (14a), in which $F(t)$ can be taken to be zero, as stated before, and

$$q^2 = |\text{grad } \phi|^2 = \frac{\partial \phi}{\partial x_\alpha} \frac{\partial \phi}{\partial x_\alpha}. \tag{22}$$

The result is lengthy, and will be given in Sec. 5.

If the flow is steady, the term $\partial \rho / \partial t$ in (1) is zero, and (1) and (21) together give

$$\nabla^2 \phi + \frac{\partial \phi}{\partial x_\beta} \frac{\partial}{\partial x_\beta} \ln \rho = 0, \tag{23}$$

upon changing the dummy indices i to β to avoid confusion later. The task is to eliminate ρ between (23) and (17). This is not difficult, for from (17) we obtain by direct differentiations that

$$c^2 \frac{\partial}{\partial x_i} \ln \rho = \frac{\gamma p}{\rho} \frac{\partial}{\partial x_i} \ln \rho = -g\delta_{i3} - \frac{\partial}{\partial x_i} \frac{q^2}{2}, \tag{24}$$

with q^2 given by (22) and δ_{i3} indicating the Kronecker delta. Thus (23) can be written, when x, y, and z are used for x_1, x_2, and x_3 and subscripts are used to indicate partial differentiation,

$$\left(1 - \frac{\phi_x^2}{c^2}\right)\phi_{xx} + \left(1 - \frac{\phi_y^2}{c^2}\right)\phi_{yy} + \left(1 - \frac{\phi_z^2}{c^2}\right)\phi_{zz}$$

$$- \frac{2\phi_y\phi_z}{c^2}\phi_{yz} - \frac{2\phi_z\phi_x}{c^2}\phi_{zx} - \frac{2\phi_x\phi_y}{c^2}\phi_{xy} - \frac{g}{c^2}\phi_z = 0, \tag{25}$$

with c^2 given by (17a) and Ω being gz in (17a). If gravity effects are negligible, the last term on the left-hand side of (25) can be neglected and (25) can be written in short as

$$\phi_{\alpha\alpha} - \frac{\phi_\beta\phi_\gamma}{c^2}\phi_{\beta\gamma} = 0, \tag{26}$$

the summation convention being strictly observed and c^2 being now given by (19), with q^2 defined by (22). Equation (25) or (26) is the equation governing steady irrotational flows of a homentropic gas.

If the flow is two-dimensional, all the terms involving y or x_2 can be dropped from (25) and (26). It is then customary to change the z in the resulting equations to y. Specifically, (26) becomes

$$\left(1 - \frac{\phi_x^2}{c^2}\right)\phi_{xx} - \frac{2\phi_x\phi_y}{c^2}\phi_{xy} + \left(1 - \frac{\phi_y^2}{c^2}\right)\phi_{yy} = 0. \tag{27}$$

The flow is then independent of z. For two-dimensional steady flows, one can obtain from (1.36), upon setting the g there equal to z, or directly from the equation of continuity

$$\frac{\partial(\rho u)}{\partial x} + \frac{\partial(\rho v)}{\partial y} = 0, \tag{28}$$

that
$$u = \frac{\rho_0}{\rho}\frac{\partial \psi}{\partial y}, \qquad v = -\frac{\rho_0}{\rho}\frac{\partial \psi}{\partial x}, \tag{29}$$

in which ψ is the stream function and ρ_0 a reference density. With Ω neglected, (17) can be written as

$$\frac{\gamma}{\gamma - 1}\frac{p_0}{\rho_0^\gamma}\rho^{\gamma-1} + \frac{\rho_0^2}{2\rho^2}(\psi_x^2 + \psi_y^2) = \tfrac{1}{2}q_{max}^2, \tag{30}$$

which enables one to express ρ in terms of ψ. Substituting (29) into

$$\frac{\partial v}{\partial x} - \frac{\partial u}{\partial y} = 0,$$

we obtain

$$\left[1 - \left(\frac{\rho_0}{\rho c}\right)^2\psi_y^2\right]\psi_{xx} + 2\left(\frac{\rho_0}{\rho c}\right)^2\psi_x\psi_y\psi_{xy} + \left[1 - \left(\frac{\rho_0}{\rho c}\right)^2\psi_x^2\right]\psi_{yy} = 0, \tag{31}$$

which is the companion of (27). In (31) ρc is given in terms of ψ by (30).

For steady axisymmetric irrotational flows of a homentropic gas, the equation for ϕ and the Stokes stream function ψ can be obtained in much the same manner. With Ω again neglected, and with r and z denoting two of the cylindrical coordinates, the equation of continuity is

$$\frac{\partial(r\rho u)}{\partial r} + \frac{\partial(r\rho w)}{\partial z} = 0, \tag{32}$$

in which u and w are the velocity components in the directions of increasing r and z, respectively, the z axis being the axis of symmetry. Then

$$u = \phi_r = -\frac{\rho_0}{r\rho}\psi_z, \qquad w = \phi_z = \frac{\rho_0}{r\rho}\psi_r, \tag{33}$$

in which ψ is the Stokes stream function. With (33) and (17) and omitting Ω in (17), we can obtain an equation in ϕ and one in ψ. They are, respectively,

$$\left(1 - \frac{\phi_z^2}{c^2}\right)\phi_{zz} + \left(1 - \frac{\phi_r^2}{c^2}\right)\phi_{rr} - \frac{2\phi_z\phi_r}{c^2}\phi_{zr} + \frac{\phi_r}{r} = 0 \qquad (34)$$

and

$$\left[1 - \left(\frac{\rho_0}{r\rho c}\right)^2 \psi_r^2\right]\psi_{zz} + \left[1 - \left(\frac{\rho_0}{r\rho c}\right)^2 \psi_z^2\right]\psi_{rr}$$

$$+ 2\left(\frac{\rho_0}{r\rho c}\right)^2 \psi_z\psi_r\psi_{zr} - \frac{\psi_r}{r} = 0. \qquad (35)$$

The lines of constant ϕ are orthogonal to the lines of constant ψ (streamlines) whether the flow is two-dimensional or axially symmetric, as can be readily demonstrated.

5. THE EQUATION GOVERNING UNSTEADY IRROTATIONAL FLOWS OF A HOMENTROPIC GAS

If the flow is unsteady, the term $\partial\rho/\partial t$ in (1) and the term $\partial u_i/\partial t$ contained in the left-hand side of (2) must be retained. If the fluid is homentropic and the motion started from rest, the subsequent flow is irrotational and (21) still holds. From (1) and (21) it then follows that

$$\frac{\partial}{\partial t}\ln\rho + \frac{\partial\phi}{\partial x_\beta}\frac{\partial}{\partial x_\beta}\ln\rho + \nabla^2\phi = 0, \qquad (36)$$

which replaces (23). To obtain a single equation for ϕ, the ρ in (36) needs to be eliminated. Differentiation of (14a) with respect to x_i and t produces, respectively,

$$\frac{\partial}{\partial t}\frac{\partial\phi}{\partial x_i} + c^2\frac{\partial}{\partial x_i}\ln\rho = -g\delta_{i3} - \frac{\partial}{\partial x_i}\frac{q^2}{2} \qquad (37)$$

and

$$\frac{\partial^2\phi}{\partial t^2} + c^2\frac{\partial}{\partial t}\ln\rho = -\frac{\partial}{\partial t}\frac{q^2}{2}. \qquad (38)$$

Multiplying (36) by c^2 and using (37) and (38), we arrive at

$$\frac{\partial^2\phi}{\partial t^2} = c^2\nabla^2\phi - \frac{\partial}{\partial t}q^2 - g\frac{\partial\phi}{\partial x_3} - \frac{\partial\phi}{\partial x_\beta}\frac{\partial}{\partial x_\beta}\frac{q^2}{2}, \qquad (39)$$

with q given by (22) and c^2 given by (18a) and (14) or by

$$\frac{\partial\phi}{\partial t} + \frac{c^2}{\gamma - 1} + \Omega + \frac{q^2}{2} = F(t). \qquad (40)$$

Hence (39) is an equation in ϕ alone and is the equation sought. The $F(t)$ in (40) can be absorbed into ϕ, as usual, so that it can be replaced by a constant or by zero.

6. THE SPEED OF SOUND

Let L represent a typical length scale, L_3 a typical length scale in the direction of the vertical, and τ a typical time scale for the unsteadiness (such as the time it takes to shut off a valve in a pipe containing flowing gas). Then if

$$\left(\frac{q}{c}\right)^2 \ll 1, \qquad \frac{qL}{c^2\tau} \ll 1, \qquad \frac{gL_3}{c^2} \ll 1, \tag{41}$$

the last three terms in (39) can be neglected with respect to the rest, and we have

$$\frac{\partial^2\phi}{\partial t^2} = c^2\, \nabla^2\phi. \tag{42}$$

This is the wave equation in mathematical physics. The simplest way to see that c is the speed of sound is by consideration of one-dimensional propagation, for which (42) assumes the simple form

$$\frac{\partial^2\phi}{\partial t^2} = c^2 \frac{\partial^2\phi}{\partial x^2}, \tag{42a}$$

which obviously has the general solution

$$\phi = f_1(x - ct) + f_2(x + ct). \tag{43}$$

The function $f_1(x - ct)$ can be considered as a signal which is propagated with speed c in the direction of increasing x; for so long as the change in x is equal to c times the change in t, the quantity $x - ct$ and hence $f_1(x - ct)$ are constant. Similarly, $f_2(x + ct)$ is a signal that propagates in the direction of decreasing x with speed c. Thus the c defined in (18) is indeed the speed of the infinitesimal waves of compression (or rarefaction), or the speed of sound.

It can be shown that

$$\phi = -\frac{f(t - c^{-1}R)}{R},$$

where R is the radial distance from the origin (the first of the spherical coordinates), satisfies (42). If c is infinite, the left-hand side of this equation is evidently the velocity potential for a point source in an incompressible fluid. The ϕ given here is called the retarded potential, for obvious reasons. From this potential other time-dependent solutions for distributions of sources and sinks can be obtained by superposition, and for doublets and higher singularities by differentiations with respect to x, y, or z. All these solutions satisfy the sound equation (42).

The solution

$$\phi = \tfrac{1}{2}[f(x - ct) + f(x + ct)] - \frac{1}{2c}[G(x - ct) - G(x + ct)], \quad (44)$$

being of the form of (43), satisfies (42a). As can be easily verified, it also satisfies the initial conditions

$$\phi = f(x) \quad \text{and} \quad \frac{\partial \phi}{\partial t} = g(x) \quad \text{at } t = 0$$

if

$$G(x) = \int_a^x g(\xi)\, d\xi,$$

the lower limit a being arbitrary.

It should be noted that under the assumptions (41) the right-hand sides of (37) and (38) can be neglected. Remembering that for sound waves the variations of ρ is very much smaller than ρ, (42), (37), and (38) then give

$$\frac{\partial^2 \rho}{\partial t^2} = c^2\, \nabla^2 \rho, \quad (45)$$

which is of the same form as (42a) but has a measurable quantity (ρ) as the dependent variable. Its solution is of the form (43) or (44). Taking the form (44), with $g(x) = 0$, we have

$$\rho = \tfrac{1}{2}[f(x - ct) + f(x + ct)], \quad (46)$$

which satisfies the initial conditions

$$\rho = f(x) \quad \text{and} \quad \frac{\partial \rho}{\partial t} = 0 \quad \text{at } t = 0.$$

If

$$f(x) = 0 \quad \text{except in } -b \le x \le b,$$

the solution (46) is for the case of motion due to a disturbance (compression or rarefaction from the mean state) in the range indicated, released *at rest* at $t = 0$. It states that half of the initial disturbance moves to the left with speed c and the other half to the right with the same speed, both keeping the same shape (variation of f with x at any instant) and moving farther and farther from each other.

7. POISSON'S SOLUTION OF THE WAVE EQUATION

We have seen the significance of c in (42) by consideration of one-dimensional propagation. It is more satisfying to demonstrate that c is the speed of sound in three-dimensional propagation. To do this, we need Poisson's solution of (42) or (45).

If $f(x,y,z)$ is an arbitrary function at least twice differentiable with respect to its arguments, if P has the coordinates x, y, and z, and if S is a spherical surface with its center at P and a radius $R = ct$, then

$$\phi \text{ (or } \rho) = tF(x,y,z,t) \tag{47}$$

is a solution of (42) [or (45)] if

$$F(x,y,z,t) = \frac{1}{4\pi R^2} \int_S f(x + \xi, y + \eta, z + \zeta) \, dS = \frac{1}{4\pi} \int f \, d\sigma, \tag{48}$$

in which $d\sigma$ is the solid angle subtended by dS, or $d\sigma = dS/R^2$. To distinguish from $R = ct$, we shall use r to denote the spherical distance from P, that is,

$$r^2 = \xi^2 + \eta^2 + \zeta^2.$$

In Eqs. (51) and (53) and the equation following (53), the arguments of df/dr are $x + \xi$, $y + \eta$, and $z + \zeta$.

Equation (47) is Poisson's solution. To show that it satisfies (42), we note first that

$$\nabla^2\phi = t \, \nabla^2 F = \frac{t}{4\pi R^2} \int \nabla^2 f \, dS = \frac{1}{4\pi c^2 t} \int \nabla^2 f \, dS \tag{49}$$

and

$$\frac{\partial^2\phi}{\partial t^2} = \frac{1}{t} \frac{\partial}{\partial t}\left(t^2 \frac{\partial F}{\partial t}\right). \tag{50}$$

To evaluate $\partial F/\partial t$, we note that the variation of F with t is due entirely to the variation of the radius r of S. Indeed for

$$dr = c \, dt,$$

the change in f at S is $(\partial f/\partial r) \, dr$, so that

$$\frac{\partial F}{\partial t} = \frac{c}{4\pi} \int \frac{\partial f}{\partial r} \, d\sigma. \tag{51}$$

Combining (50) and (51) and remembering that $R = ct$, we have

$$\frac{\partial^2\phi}{\partial t^2} = \frac{1}{4\pi ct} \frac{\partial}{\partial t} \int \frac{\partial f}{\partial r} \, dS. \tag{52}$$

But according to Green's theorem, with $n = r$,

$$\int \frac{\partial f}{\partial r} \, dS = \int \nabla^2 f \, dV. \tag{53}$$

Since $dV = c \, dt \, dS$ for an increment dt,

$$\frac{\partial}{\partial t} \int \frac{\partial f}{\partial r} \, dS = c \int \nabla^2 f \, dS$$

and (52) becomes

$$\frac{\partial^2 \phi}{\partial t^2} = \frac{1}{4\pi t} \int \nabla^2 f \, dS$$

or, upon the use of (49),

$$\frac{\partial^2 \phi}{\partial t^2} = c^2 \, \nabla^2 \phi.$$

Thus (42) is satisfied by (47).

The solution satisfying (42) and the initial conditions

$$\phi = f(x,y,z) \qquad \text{and} \qquad \frac{\partial \phi}{\partial t} = g(x,y,z) \qquad \text{at } t = 0 \qquad (54)$$

is

$$\phi(x,y,z,t) = \frac{\partial}{\partial t}(tF) + tG, \qquad (55)$$

where

$$G(x,y,z,t) = \frac{1}{4\pi} \int g \, d\sigma.$$

That (55) satisfies (42) is obvious, since tG is of the same form as (47) and obviously $\partial(tF)/\partial t$ satisfies (42) if tF does, and since superposition of the two solutions represented by the two terms in (55) is allowable because of the linearity and homogeneity of (42). To show that (55) satisfies (54), note that at $t = 0$ the radius R is zero and hence the mean value of f averaged over the solid angle 4π, which is F, is simply f. Furthermore $\phi = F$ at $t = 0$. Hence the first initial condition in (54) is satisfied. As to the second condition in (54), we note first that at $t = 0$

$$\frac{\partial \phi}{\partial t} = 2\frac{\partial F}{\partial t} + G.$$

But $\partial F/\partial t$ is given by (51), and as t approaches zero, the surface S shrinks to a point, so that $\partial f/\partial r$ has opposite signs at two diametrically opposite points on S as S reaches its limit. Hence in the limit $\partial F/\partial t$ is zero. As demonstrated for the case of F, the function G becomes g for $t = 0$. Hence the second condition in (54) is also satisfied.

Now, with reference to Fig. 2, let f and g vanish outside a domain D, so that F and G are zero if the spherical surface S centered at P and with radius $R = ct$ does not intersect D. Hence at P

$$\phi = 0 \qquad \text{for } t < t_1 \text{ or } t > t_2.$$

If the domain D shrinks to a point where the radius R from P equals R_1, then ϕ is different from zero at P only at $t = t_1$. All these facts indicate that c is indeed the speed of sound.

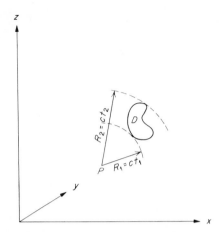

FIGURE 2. SKETCH FOR DEMONSTRATION
THAT c DEFINED BY (42) IS THE SPEED OF
SOUND.

8. TWO-DIMENSIONAL PROPAGATION

If the flow is independent of z,

$$f(x,y,z) = f(x,y).$$

Since the situation at any (x,y,z) is the same as at a point $(x,y,0)$, we shall consider the point $P(x,y,0)$. With S as defined in Sec. 7, its projection A on the plane $z = 0$ will be a circle of radius $R = ct$. The elemental areas dS and dA have the ratio

$$\frac{dS}{dA} = \frac{R}{\sqrt{R^2 - r^2}}, \qquad \text{where now } r^2 = \xi^2 + \eta^2. \tag{56}$$

We now use (48) and remember that the spherical surface S about P when projected on the plane $z = 0$ covers the circle $r = R$ *twice*. Hence

$$F(x,y,t) = \frac{1}{2\pi R} \int_A f(x + \xi, y + \eta) \frac{dA}{\sqrt{R^2 - r^2}}. \tag{57}$$

With F so defined and G similarly defined, (55) again provides the solution satisfying the two-dimensional forms of (42) and (54).

9. SUBSONIC AND SUPERSONIC SPEEDS

Consider a uniform stream flowing in the direction of increasing x and with a speed q and a weak disturbance created at the origin at $t = 0$. As time passes, the disturbance will propagate radially outward with

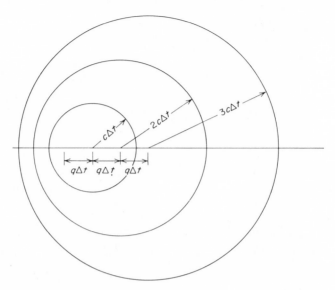

FIGURE 3.   PROPAGATION OF A SOUND WAVE IN A SUBSONIC STREAM.

speed c from a moving center located at $x = qt$. If $q < c$, the flow is called *subsonic*, and the disturbance can propagate upstream from the origin (see Fig. 3). If $q > c$, the flow is called *supersonic*, and the disturbance will be contained in a cone as time passes and can never propagate upstream from the origin, as shown in Fig. 4. The angle β defined by

$$\sin \beta = \frac{ct}{qt} = \frac{c}{q} \tag{58}$$

is called the *Mach angle*, and the cone, which is the envelope of the circles representing the disturbance at various times, the *Mach cone*. The case shown in Fig. 5 for $q = c$ divides the subsonic case from the supersonic case. The flow is sonic.

    If one imagines an idealized continuous weak source of disturbance situated at the origin, Figs. 3 to 5 depict the flows at subsonic, supersonic, and sonic speeds. In the supersonic case, the flow outside the Mach cone is undisturbed. In the sonic case the cone becomes a plane.

10. THE NORMAL SHOCK

We have seen that weak disturbances in a compressible fluid, which we call *sound*, propagate with the speed c defined by (18). It will now be shown that a disturbance of finite strength propagates with a speed

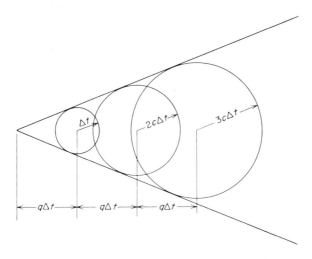

FIGURE 4.   PROPAGATION OF A SOUND WAVE IN A SUPERSONIC
STREAM.

greater than c.   For this demonstration we choose the particularly simple
case of the normal shock.   We first consider the shock wave to be
stationary, as shown in Fig. 6, in which the subscript 1 is used for upstream
quantities and the subscript 2 for downstream quantities and the velocity
is in the x direction and denoted by u.   One may consider the flow to
take place in a straight tube.

A control surface indicated by $ABCDA$ can be used in order to

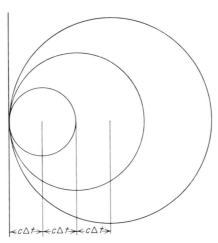

FIGURE 5.   PROPAGATION OF A SOUND WAVE
IN A SONIC STREAM.

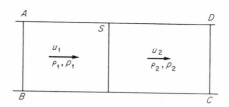

FIGURE 6. THE NORMAL SHOCK. S IS THE
SHOCK WAVE.

apply the momentum principle, the result of that application being

$$p_1 + \rho_1 u_1{}^2 = p_2 + \rho_2 u_2{}^2. \tag{59}$$

The equation of continuity is

$$\rho_1 u_1 = \rho_2 u_2, \tag{60}$$

and the energy equation (15) is, with the gravity term neglected,

$$\frac{\gamma + 1}{2(\gamma - 1)} c^{*2} = \frac{u_1{}^2}{2} + \frac{\gamma}{\gamma - 1} \frac{p_1}{\rho_1} = \frac{u_2{}^2}{2} + \frac{\gamma}{\gamma - 1} \frac{p_2}{\rho_2}, \tag{61}$$

in which c^* is defined by (19a) and (20). Equations (59) to (61) were
first used by Rankine (1870), then by Hugoniot (1887, 1889). Since the
upstream quantities are supposed to be known, Eqs. (59) to (61) can be
solved for the three unknowns p_2, ρ_2, and u_2. The simplest way to obtain
u_2 is to utilize (60) and write (59) in the form

$$u_1 - u_2 = \frac{p_2}{\rho_2 u_2} - \frac{p_1}{\rho_1 u_1}. \tag{62}$$

Substitution of (61) into (62) then produces

$$u_1 - u_2 = (u_1 - u_2) \left(\frac{\gamma + 1}{2\gamma} \frac{c^{*2}}{u_1 u_2} + \frac{\gamma - 1}{2\gamma} \right). \tag{63}$$

The equality $u_1 = u_2$ corresponds to the uninteresting case of no shock
wave. If there is a shock wave, u_1 is not equal to u_2, and (63) gives

$$u_1 u_2 = c^{*2}, \tag{64}$$

which determines u_2. We shall call this the *Rankine-Hugoniot condition*.
Equation (60) then determines ρ_2, and (59) determines p_2. Thus,

$$\frac{\rho_2}{\rho_1} = \frac{u_1}{u_2} = \left(\frac{u_1}{c^*} \right)^2, \tag{65}$$

$$p_2 - p_1 = \rho_1 u_1 (u_1 - u_2) = \rho_1 (u_1{}^2 - c^{*2}). \tag{66}$$

By the use of (61), (66) can also be written as

$$\frac{p_2}{p_1} = \frac{\dfrac{\gamma+1}{\gamma-1}\dfrac{u_1^2}{c^{*2}} - 1}{\dfrac{\gamma+1}{\gamma-1} - \dfrac{u_1^2}{c^{*2}}} \tag{67}$$

Equations (64) to (66) show that if $u_1 > c^*$, then $u_2 < c^*$, $\rho_2 > \rho_1$, and $p_2 > p_1$. The opposite case of $u_1 < c^*$, though dynamically possible, is not thermodynamically permissible, because the entropy is higher with the higher pressure and density and a shock wave with the downstream pressure lower than the upstream pressure would violate the second law of thermodynamics. Note that (10) gives

$$S_2 - S_1 = c_v \ln\left[\frac{p_2}{p_1}\left(\frac{\rho_1}{\rho_2}\right)^\gamma\right]. \tag{68}$$

It can be shown by substitution of (65) and (67) into (68) that $S_2 > S_1$ if $u_1 > c^*$ and $S_2 < S_1$ if $u_1 < c^*$. For example, letting

$$\left(\frac{u_1}{c^*}\right)^2 = 1 + \alpha, \tag{69}$$

we obtain, for fairly small α,

$$\frac{S_2 - S_1}{c_v} = \frac{\alpha^3}{12}\gamma(\gamma^2 - 1) + \text{terms of smaller magnitude.} \tag{70}$$

This shows that if α is positive, the entropy increases downstream but if α is negative, the entropy would decrease downstream if a shock wave were possible, in violation of the second law of thermodynamics.

It is important to see that if $u_1 > c^*$, then

$$u_1 > c_1 \quad \text{and} \quad u_2 < c_2, \tag{71}$$

so that the upstream flow is supersonic and the downstream flow subsonic. This is perhaps obvious, for c^*, which is the sound velocity when the flow speed equals the sound speed, is the same on both sides of the shock wave. If $u_1 > c^*$, we have seen that $u_2 < c^*$. Remembering that c^2 is given by $\gamma p/\rho$, we see from (61) that c_1 must be less than c^* and c_2 more than c^* and that the inequalities (71) are true a fortiori.

It remains to mention that in a frame of reference moving with speed u_1 in the direction of increasing x, the fluid upstream from the shock in the example just given will be stationary, and the shock front will propagate in the direction of decreasing x with speed u_1, which, as we have seen, is greater than c_1. Thus a disturbance of finite magnitude can travel with a speed greater than that of sound.

The effect of viscosity on the structure of the shock wave was investigated by Taylor (1910) in a paper which helped win him a Trinity fellowship when he was only twenty-four years old. In this paper he

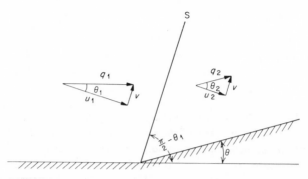

FIGURE 7. THE OBLIQUE SHOCK. S IS THE SHOCK WAVE.

considered the effect of the viscous part of the normal stress, namely, $2\mu\, du/dx$, and carried the calculation to a successful end to show the possibility of a shock wave with a continuous though sharp change in density.   In his solution the effect of heat conduction was not included. The complete problem, with the effects of both viscosity and thermal conductivity considered, was treated successfully only much later, by von Mises (1950).

11. OBLIQUE SHOCK WAVES

If we superimpose on the normal-shock flow a tangential velocity v parallel to the shock front, it is evident that the satisfaction of the equations of continuity, of normal and tangential (to the shock wave) momenta, and of energy will be unaffected.   The only thing affected is the total energy per unit mass of the stream, which will now be augmented by the amount $v^2/2$.   That is, with reference to Fig. 7, Eqs. (59) to (61) continue to stand, with the $u_1{}^2$ and $u_2{}^2$ in (61) respectively replaced by

$$q_1{}^2 = u_1{}^2 + v^2 \qquad \text{and} \qquad q_2{}^2 = u_2{}^2 + v^2. \tag{72}$$

With the same procedure used to obtain (63) and (64), we now have

$$u_1 - u_2 = (u_1 - u_2)\left(\frac{\gamma + 1}{2\gamma}\frac{c^{*2}}{u_1 u_2} - \frac{\gamma - 1}{2\gamma}\frac{v^2}{u_1 u_2} + \frac{\gamma - 1}{2\gamma}\right), \tag{73}$$

which (except in the uninteresting case of $u_1 = u_2$) gives

$$u_1 u_2 + \frac{\gamma - 1}{\gamma + 1}v^2 = c^{*2}. \tag{74}$$

Since, as shown in Fig. 7, the postshock velocity is parallel to the wall behind the shock,

$$\theta = \theta_2 - \theta_1. \tag{75}$$

Since $\qquad \tan \theta_1 = \dfrac{v}{u_1} \qquad$ and $\qquad \tan \theta_2 = \dfrac{v}{u_2},$

(74) and (75) give

$$\left(c^{*2} + \frac{2}{\gamma + 1} v^2 \right) \tan \theta = v(u_1 - u_2). \tag{76}$$

From (74) and the first equation in (72) we have

$$u_1(u_1 - u_2) + \frac{2}{\gamma + 1} v^2 = q_1^2 - c^{*2}, \tag{77}$$

which together with (76) give

$$u_1 \left(c^{*2} + \frac{2v^2}{\gamma + 1} \right) \tan \theta = (q_1^2 - c^{*2})v - \frac{2v^3}{\gamma + 1}. \tag{78}$$

On using the first equation in (72), this can be written, after squaring,

$$(q_1^2 - v^2) \left(c^{*2} + \frac{2}{\gamma + 1} v^2 \right)^2 \tan^2 \theta = v^2 \left(q_1^2 - c^{*2} - \frac{2}{\gamma + 1} v^2 \right)^2. \tag{79}$$

Given q_1 and θ, this is a cubic equation in v^2. With

$$Q_1 = \frac{q_1}{c^*}, \qquad V = \frac{v}{c^*},$$

(79) can be written as

$$(Q_1^2 - V^2) \left(1 + \frac{2}{\gamma + 1} V^2 \right) \tan^2 \theta = V^2 \left(Q_1^2 - 1 - \frac{2}{\gamma + 1} V^2 \right)^2. \tag{80}$$

Given Q_1, the variation of θ with V can be studied by plotting $\tan \theta$ as the ordinate against V as the abscissa. It is immediately clear that the curve is symmetric about the V axis, that it intersects that axis at the points

$$V = 0 \qquad \text{and} \qquad V = V_1 = \left[\frac{\gamma + 1}{2} (Q_1^2 - 1) \right]^{\frac{1}{2}}, \tag{81}$$

and that it has a vertical asymptote at

$$V = V_2 = Q_1. \tag{82}$$

Since, according to (20),

$$q_1^2 < q_{\max}^2 = \frac{\gamma + 1}{\gamma - 1} c^{*2},$$

a simple calculation will show that $V_1 < V_2$. Thus the curve is as shown in Fig. 8. Since the sign of u_1 is always positive, and since the sign of $\tan \theta$ is determined by (78), the sign of $\tan \theta$ is positive in the range $0 < V < V_1$ and negative in the range $V_1 < V < V_2$. That is, the solid line of the curve in Fig. 8 represents (78), from which (79) is obtained by

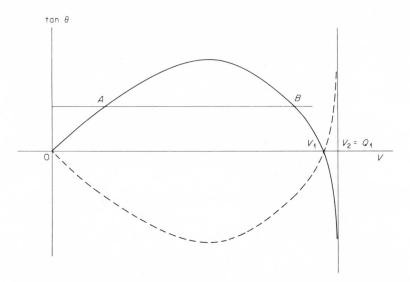

FIGURE 8. THE SHOCK GRAPH GIVING CONJUGATE STATES A AND B BEFORE AND BEHIND THE SHOCK WAVE.

squaring. If $V_1 < V$, $\tan \theta$ is negative, and from (76) it is seen that $u_1 < u_2$. This, as we have seen in the case of the normal shock, implies a decrease of entropy as the fluid crosses the shock wave and must be ruled out as physically impossible by virtue of the second law of thermo-dynamics. Thus, for a certain θ, there are two possibilities, as represented by points A and B in Fig. 8.

For $\theta = 0$ the origin O represents the normal shock, and the point $V = V_1$ represents an infinitesimal shock, or a Mach wave, whose front is inclined to the direction of flow. The latter fact becomes obvious if we remember that for $\theta = 0$ (76) gives $u_1 = u_2$ and (74) and (81) then give

$$u_1{}^2 = c^{*2} - \frac{\gamma - 1}{\gamma + 1} \frac{(\gamma + 1)}{2} (q_1{}^2 - c^{*2}) = \frac{\gamma + 1}{2} c^{*2} - \frac{\gamma - 1}{2} q_1{}^2. \quad (83)$$

But since the energy equation can be written as

$$\frac{q_1{}^2}{2} + \frac{1}{\gamma - 1} c_1{}^2 = \frac{\gamma + 1}{2(\gamma - 1)} c^{*2}, \quad (84)$$

it is evident that

$$u_1 = c_1, \quad (85)$$

which is what must be proved to show that the wave front is a sound-wave front, or a shock front of infinitesimal strength.

Thus the point A, nearer O, represents a stronger shock, and the point B, nearer $V = V_1$, represents a weaker shock. This can be demon-strated in any specific case, but we need not show that since the case of

FIGURE 9. THE CURVED SHOCK
WAVE BEFORE A BLUNT BODY.

$\theta = 0$ makes the conclusion fairly obvious.. Given Q_1 and θ, the solution is therefore not unique. Which V (V_A or V_B) will actually materialize depends on the condition further downstream.

Note also that there is a maximum value for tan θ, and therefore one for θ, between the origin and the point where $V = V_1$. A steady shock front becomes impossible when the actual θ exceeds the maximum allowable value.

When a blunt body is put in a supersonic stream, a detached shock wave is produced, as shown in Fig. 9. The curved shock can be considered to consist of a series of straight shocks of infinitesimal lengths. Since the slope of the shock point varies, the entropy jump across the shock also varies, and this variation makes the entropy behind the shock non-homogeneous. A solution for the flow of a compressible fluid with a curved shock has to satisfy the shock condition (as outlined in this section) everywhere along the shock front, the equation governing nonhomentropic flow behind the shock, and the boundary condition on the surface of the solid body producing the shock. The solution is very difficult to find, and the methods used for approximate solutions are quite outside of the scope of this book. To appreciate the difficulties, do Probs. 14 and 15.

12. STEADY ONE-DIMENSIONAL FLOWS: THE LAVAL NOZZLE

Consider the flow of a gas from a reservoir into a conduit (Fig. 10) with its cross section A gradually varying with its longitudinal distance x from the reservoir. We shall denote the pressure in the reservoir by p_0, the

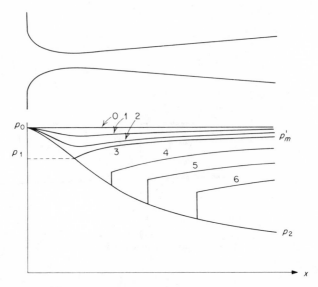

FIGURE 10.   THE LAVAL NOZZLE AND ITS OPERATION CURVES.

pressure at the throat by p_t, and the pressure at the mouth of the conduit by p_m.   If a one-dimensional approach is adopted, the continuity equation is

$$\rho A u = \text{const } C_1, \tag{86}$$

if u is the velocity and ρ the density.   If gravity effects are neglected, the equation of motion is

$$u \frac{du}{dx} = -\frac{1}{\rho} \frac{dp}{dx}, \tag{87}$$

or, in the form of differentials,

$$u \, du = -\frac{1}{\rho} dp. \tag{88}$$

Except where the fluid crosses a shock front, (88) can be written, by virtue of the isentropy of the flow, as

$$u \, du + c^2 \frac{d\rho}{\rho} = 0. \tag{89}$$

The differential form of (86) is

$$\frac{du}{u} + \frac{dA}{A} + \frac{d\rho}{\rho} = 0, \tag{90}$$

which combined with (89) gives

$$\frac{du}{u}\left(1 - \frac{u^2}{c^2}\right) = -\frac{dA}{A}.$$ (91)

Thus, if the sonic speed is ever reached by the flow, it must be at the place where the area is an extremum or, in the case of the Laval nozzle (convergent-divergent channel), where the area is a minimum. Equation (91) also states that at the throat the velocity must be an extremum (actually a maximum) if sonic speed is not reached there and that if sonic speed is reached there, u need not be an extremum. In fact, after the sonic speed has been reached at the throat, u increases as the channel expands after the throat, before a shock wave, if any, is reached.

Note, however, that there exists a singular case that for a particular exit pressure the value of $u = c = c_*$ at the throat is the greatest (or the smallest) in the nozzle with $u < c_*$ (or $u > c_*$) on both sides of the throat, across which $\partial u/\partial x$ is discontinuous, having the one-sided limits as

$$\frac{\partial u}{\partial x} = \pm c_* \left[\frac{1}{\gamma + 1}\frac{1}{A_*}\left(\frac{d^2A}{dx^2}\right)_*\right]^{1/2},$$

where the asterisk signifies the throat station. The above result can be deduced by applying the rule of L'Hospital to (91). This case is actually the limit of an entirely subsonic flow as a sonic throat condition is approached. Aside from this particular case, after subsonically reaching the sonic speed at the throat, u always increases as the channel expands after the throat, if the exit pressure is sufficiently low to prevent shock waves in the divergent part of the nozzle.

The situation can be similarly discussed if the flow upstream from the throat has already been made supersonic by the use of a previous Laval nozzle.

Figure 10 shows the trend of the pressure variation in the channel. If $p_m = p_0$, there is uniform pressure throughout and no flow, as indicated by the line labeled 0. As p_m is reduced, the pressure is indicated by curves 1, 2, etc. Curve 3 shows the pressure distribution when sonic speed is just reached, the pressure p_m for this case being denoted by p'_m. As p_m is further reduced, the pressure p_1 reached at the throat will remain invariant, and so will be the total mass discharge. But a shock wave will appear in the divergent part of the channel and will be "pushed" further and further downstream as the pressure p_m is further and further reduced, its location being exactly where the shock condition can be satisfied by the flow determined from the upstream data and the flow determined from the downstream data. When p_m reaches the pressure p_2, the flow in the divergent part of the channel is entirely supersonic and shock-free, and Schlieren pictures (pictures obtained by utilizing the variation of refractive index of the gas with density) show a diamond-shaped wave pattern in the gas as it expands after leaving the nozzle in the form of a jet.

13. UNSTEADY ONE-DIMENSIONAL FLOWS: RIEMANN'S METHOD OF CHARACTERISTICS

In a remarkable paper in a remarkably early year, Riemann (1860) showed how one-dimensional unsteady flows of a compressible fluid can be integrated by the method of characteristics. The velocity u, density ρ, and pressure p are now functions of x and t. Since the entropy[1] of the fluid is constant (in the absence of shocks, of course),

$$p = C\rho^\gamma \quad \text{and} \quad c^2 = \frac{\gamma p}{\rho} = \gamma C\rho^{\gamma-1}. \tag{92}$$

From these it can be shown directly that

$$\frac{1}{\rho}\frac{\partial p}{\partial x} = \frac{2}{\gamma - 1}\, cc_x, \tag{93}$$

subscripts now indicating partial differentiation. Then the equation of motion can be written as

$$u_t + uu_x + \frac{2}{\gamma - 1}\, cc_x = 0 \tag{94}$$

if gravity effects are neglected.

The equation of continuity is

$$\rho_t + u\rho_x + \rho u_x = 0, \tag{95}$$

which, by virtue of (92), can be written as

$$2c_t + 2uc_x + (\gamma - 1)cu_x = 0. \tag{96}$$

From (94) and (96) one obtains, by addition and subtraction,

$$\left[\frac{\partial}{\partial t} + (u + c)\frac{\partial}{\partial x}\right]\left(u + \frac{2c}{\gamma - 1}\right) = 0, \tag{97}$$

$$\left[\frac{\partial}{\partial t} + (u - c)\frac{\partial}{\partial x}\right]\left(u - \frac{2c}{\gamma - 1}\right) = 0, \tag{98}$$

which can be compared with (5.165) and (5.166). Again, along the

---

[1] It was in this regard that Riemann, assuming isothermal change of state, repeated Newton's mistake. In his short life (1826–1866), Georg Friedrich Bernhard Riemann did many truly great things in mathematics.

family of curves determined by

$$\frac{dx}{dt} = u + c \qquad (99)$$

the quantity $u + 2(\gamma - 1)^{-1}c$ is constant, and along the family of curves determined by

$$\frac{dx}{dt} = u - c \qquad (100)$$

the quantity $u - 2(\gamma - 1)^{-1}c$ is constant. The integration procedure is much as described in Secs. 3.5.2 and 3.5.3 of Chap. 5, where some examples were given. Envelopes of characteristics now indicate shock paths instead of wave breaks, as mentioned in Sec. 3.5.3 of Chap. 5. Incidentally the development above shows that the c defined in (18) retains its meaning as the speed of sound even if the disturbance is finite, so long as it is not discontinuous, such as a shock wave. (See the last paragraph of Sec. 3.5.2 of Chap. 5 for a corresponding discussion for shallow-water waves.)

14. STEADY TWO-DIMENSIONAL FLOWS: MOLENBROEK'S TRANSFORMATION

In this and the following five sections steady two-dimensional flows will be considered. In all but one of these sections the independent variables are the velocity components u and v or, equivalently, q and θ. The use of these variables often simplifies the governing equations and makes exact solutions of certain compressible-flow problems possible.

Equations (27) and (31) are highly nonlinear as they stand, but they can be made linear by either the *Molenbroek transformation* or the *Legendre transformation*. Molenbroek (1890) used, instead of x and y, the speed q and the angle of inclination θ of the velocity as the independent variables. These are related to u and v by

$$u = q \cos \theta, \qquad v = q \sin \theta. \qquad (101)$$

Solution of

$$d\phi = u \, dx + v \, dy \qquad \text{and} \qquad d\psi = \frac{\rho}{\rho_0} (-v \, dx + u \, dy) \qquad (102)$$

for dx and dy and use of (101) produce

$$dx = \frac{1}{q} \left(\cos \theta \, d\phi - \frac{\rho_0}{\rho} \sin \theta \, d\psi \right), \qquad (103)$$

$$dy = \frac{1}{q} \left(\sin \theta \, d\phi + \frac{\rho_0}{\rho} \cos \theta \, d\psi \right). \qquad (104)$$

The differentials dx, dy, $d\phi$, and $d\psi$ can all be expressed in terms of dq

and $d\theta$. Thus,

$$dx = x_q\, dq + x_\theta\, d\theta, \qquad d\phi = \phi_q\, dq + \phi_\theta\, d\theta, \qquad \text{etc.} \qquad (105)$$

When these differentials are substituted into (103) and (104) and the coefficients of the arbitrary differentials dq and $d\theta$ are equated on both sides of the equation in each case, we obtain

$$x_q = \frac{1}{q}\left(\cos\theta\,\phi_q - \frac{p_0}{\rho}\sin\theta\,\psi_q\right), \qquad x_\theta = \frac{1}{q}\left(\cos\theta\,\phi_\theta - \frac{p_0}{\rho}\sin\theta\,\psi_\theta\right)$$

$$(106)$$

$$y_q = \frac{1}{q}\left(\sin\theta\,\phi_q + \frac{p_0}{\rho}\cos\theta\,\psi_q\right), \qquad y_\theta = \frac{1}{q}\left(\sin\theta\,\phi_\theta + \frac{p_0}{\rho}\cos\theta\,\psi_\theta\right).$$

$$(107)$$

It follows from (17), upon neglecting the term representing gravity, that

$$\frac{d}{dq}\frac{p_0}{\rho} = \frac{p_0}{\rho}\frac{q}{c^2}, \qquad (108)$$

in which c^2 is given by (18a). With the aid of (108) and demanding that

$$x_{q\theta} = x_{\theta q}, \qquad y_{q\theta} = y_{\theta q}, \qquad (109)$$

we obtain from (106) and (107) the equations

$$-\sin\theta\left[\phi_q + \frac{p_0}{\rho q}\left(1 - \frac{q^2}{c^2}\right)\psi_\theta\right] - \cos\theta\left(\frac{p_0}{\rho}\psi_q - \frac{1}{q}\phi_\theta\right) = 0,$$

$$\cos\theta\left[\phi_q + \frac{p_0}{\rho q}\left(1 - \frac{q^2}{c^2}\right)\psi_\theta\right] - \sin\theta\left(\frac{p_0}{\rho}\psi_q - \frac{1}{q}\phi_\theta\right) = 0.$$

Thus $$\phi_q = -\frac{p_0}{\rho q}\left(1 - \frac{q^2}{c^2}\right)\psi_\theta, \qquad \phi_\theta = \frac{p_0 q}{\rho}\psi_q. \qquad (110)$$

The equation for ϕ is obtained by eliminating ψ from (110), and is

$$\left(1 - \frac{q^2}{c^2}\right)q^2\phi_{qq} + q\left(1 + \frac{q^4}{c^4}\right)\phi_q + \left(1 - \frac{q^2}{c^2}\right)^2\phi_{\theta\theta} = 0, \qquad (111)$$

in which c^2 is given by (19a) in terms of q^2. Similarly, elimination of ϕ from (100) produces

$$q^2\psi_{qq} + q\left(1 + \frac{q^2}{c^2}\right)\psi_q + \left(1 - \frac{q^2}{c^2}\right)\psi_{\theta\theta} = 0, \qquad (112)$$

with c^2 given by (19a). Equation (112) is called the equation of Chaplygin, in recognition of his remarkable work of 1904 on gas jets. The wonderful and outstanding feature of (111) and (112) is linearity! However, hodograph methods in gas dynamics often suffer from the difficulty of interpreting a solution in the physical plane, to which one must return. The mapping between the physical plane and the hodograph plane is not always one-to-one pointwise.

15. STEADY TWO-DIMENSIONAL FLOWS: LEGENDRE'S TRANSFORMATION

Equations (27) and (31) can also be made linear by the *contact transformation* of Legendre. For (27), the independent variables to be used are

$$u = \phi_x, \qquad v = \phi_y, \tag{113}$$

and the dependent variable is

$$\Phi = x\phi_x + y\phi_y - \phi. \tag{114}$$

The differential form of (114) is

$$d\Phi = u\,dx + v\,dy - d\phi + x\,du + y\,dv = x\,du + y\,dv.$$

Hence

$$x = \Phi_u, \qquad y = \Phi_v, \quad \text{and} \quad \phi = u\Phi_u + v\Phi_v - \Phi, \tag{115}$$

the last of which is by virtue of (114). Equations (113) to (115) show that there is a duality or a kind of reciprocity in the relationship between (u,v,Φ) and (x,y,ϕ). From the first two equations in (115) it follows that

$$dx = \Phi_{uu}\,du + \Phi_{uv}\,dv, \qquad dy = \Phi_{uv}\,du + \Phi_{vv}\,dv. \tag{116}$$

Solving these for du and dv, we obtain

$$du = \frac{1}{J}\,(\Phi_{vv}\,dx - \Phi_{uv}\,dy), \qquad dv = \frac{1}{J}\,(-\Phi_{uv}\,dx + \Phi_{uu}\,dy), \tag{117}$$

in which

$$J = \Phi_{uu}\Phi_{vv} - \Phi_{uv}^2. \tag{118}$$

On the other hand, it follows from (113) that

$$du = \phi_{xx}\,dx + \phi_{xy}\,dy, \qquad dv = \phi_{xy}\,dx + \phi_{yy}\,dy. \tag{119}$$

Comparison of (117) with (119) yields

$$\phi_{xx} = \frac{1}{J}\,\Phi_{vv}, \qquad \phi_{yy} = \frac{1}{J}\,\Phi_{uu}, \qquad \phi_{xy} = -\frac{1}{J}\,\Phi_{uv}. \tag{120}$$

Thus (27) can be written as

$$\left(1 - \frac{v^2}{c^2}\right)\Phi_{uu} + \frac{2uv}{c^2}\,\Phi_{uv} + \left(1 - \frac{u^2}{c^2}\right)\Phi_{vv} = 0, \tag{121a}$$

or, in terms of q and θ,

$$q^2\Phi_{qq} + q\left(1 - \frac{q^2}{c^2}\right)\Phi_q + \left(1 - \frac{q^2}{c^2}\right)\Phi_{\theta\theta} = 0, \tag{121b}$$

which follows from (121a) upon use of (101).

Similarly, with

$$\alpha = \psi_y, \qquad \beta = -\psi_x, \qquad \Psi = x\psi_x + y\psi_y - \psi \qquad (122a)$$

we obtain

$$x = -\Psi_\beta, \qquad y = \Psi_\alpha, \qquad \psi = \alpha\Psi_\alpha + \beta\Psi_\beta - \Psi \qquad (122b)$$

and

$$\left(1 - \frac{\rho_0^2}{\rho^2 c^2}\alpha^2\right)\Psi_{\alpha\alpha} - 2\frac{\rho_0^2}{\rho^2 c^2}\alpha\beta\Psi_{\alpha\beta} + \left(1 - \frac{\rho_0^2}{\rho^2 c^2}\beta^2\right)\Psi_{\beta\beta} = 0, \qquad (123)$$

in which ρ is a function of q according to (17) (with Ω neglected) and c^2 is a function of q according to (19a), and hence $(\rho_0/\rho c)^2$ can be expressed in terms of $(\rho q/\rho_0)^2$, which is just $\alpha^2 + \beta^2$, according to (122a) and (29).

Perhaps it is illuminating to explain the geometric implications of Legendre's transformation and why it is called a *contact* transformation. The dependent variable ϕ is a function of the independent variables x and y. In Cartesian coordinates x, y, and z,

$$z = \phi(x,y) \qquad (124)$$

is a surface. At any point on the surface its direction numbers are

$$\phi_x, \qquad \phi_y, \qquad -1,$$

and the equation of the tangent plane is

$$z = x\phi_x + y\phi_y - \Phi, \qquad (125)$$

where Φ is the (perpendicular) distance of the origin from the plane. The envelope of (125) is (124), and it is possible to describe (124) by the *plane coordinates* ϕ_x, ϕ_y, and Φ instead of the point coordinates x, y, and z. That there is a duality between the plane coordinates and the point coordinates can be seen clearly in (113) to (115). This duality is beautiful. As we have seen, if a partial differential equation of two independent variables is linear in its second-order derivatives, if these derivatives have coefficients which involve only the first derivatives (which are two of the plane coordinates), and if there are no terms not having a second-order derivative as a factor, this duality is also useful.

16. STEADY TWO-DIMENSIONAL SUBSONIC FLOWS: CHAPLYGIN'S GAS JETS

We have seen in Chap. 4 how the Schwarz-Christoffel transformation can be used to solve problems of potential flows of an incompressible fluid involving free streamlines provided all solid boundaries are straight. Chaplygin (1904) has shown how to obtain the solutions for gas jets from the corresponding solutions for liquid jets or, more generally, solutions involving free streamlines for a gas from the corresponding solutions for a liquid. We shall show in this section how Chaplygin obtained the solution for a gas jet corresponding to Kirchhoff's jet presented in Chap. 4 (see Fig. 11).

Given the properties of the gas far upstream in the reservoir, there is a maximum speed attainable by the gas. This has been denoted by q_{max}. It is convenient to use the new independent variable

$$\tau = \left(\frac{q}{q_{max}}\right)^2 \tag{126}$$

instead of q in (112). Then (19a) gives

$$c^2 = \frac{\gamma - 1}{2} q_{max}^2 (1 - \tau). \tag{127}$$

By the use of (126) and (127), (112) can be transformed to

$$P \frac{\partial}{\partial \tau}\left(Q \frac{\partial \psi}{\partial \tau}\right) - \frac{\partial^2 \psi}{\partial \theta^2} = 0, \tag{128}$$

in which $P = P(\tau) = \dfrac{2(\gamma - 1)\tau}{(\gamma + 1)\tau - (\tau - 1)}(1 - \tau)^{\gamma/(\gamma-1)}, \tag{129}$

$$Q = Q(\tau) = \frac{2\tau}{(1 - \tau)^{1/(\gamma-1)}}. \tag{130}$$

Equation (128) takes some effort to obtain, but the transformation is straightforward and presents no real difficulty.

Obviously (128) can be solved by separation of variables. If we assume

$$\psi = \tau^{n/2} e^{-in\theta} F(\tau) \tag{131}$$

and substitute this into (128), we obtain

$$\tau(1 - \tau)F'' + \left[n + 1 - \left(n - \frac{1}{\gamma - 1} + 1\right)\tau\right]F' + \frac{n(n + 1)}{2(\gamma - 1)} F = 0. \tag{132}$$

The right-hand side of (131) stands for either its real or its imaginary part. We shall understand it to stand for the imaginary part, i.e., the coefficient of i. The n in (131) is any constant, and the factor $\tau^{n/2}$ in (131) is merely to make the equation satisfied by $F(\tau)$ of the standard form (132), whose solution is the hypergeometric function

$$F(a,b; n + 1; \tau) = \sum_{r=1}^{\infty} \frac{ab(a + 1)(b + 1) \cdots (a + r - 1)(b + r - 1)}{(n + 1)(n + 2) \cdots (n + r)r!} \tau^r, \tag{133}$$

in which $a - b = n - (\gamma - 1)^{-1}, \qquad ab = -\dfrac{n(n + 1)}{2(\gamma - 1)}. \tag{134}$

To emphasize the dependence of a and b on n, we shall attach the subscript n to a and b and write the solution of (134) as

$$(a_n, b_n) = \tfrac{1}{2}\left[n - \frac{1}{\gamma - 1} \pm \sqrt{\frac{\gamma + 1}{\gamma - 1} n^2 + \frac{1}{(\gamma - 1)^2}} \right]. \qquad (135)$$

For convenience we shall also write

$$\psi_n(\tau) = \tau^{n/2} F(a_n, b_n; n + 1; \tau). \qquad (136)$$

Now for the Kirchhoff jet, we had

$$\psi = -\left(\theta + \sum_1^\infty \frac{q^{2n}}{n} \sin 2n\theta \right). \qquad (137)$$

For the gas jet, we have

$$\psi = -\left[\theta + \sum_1^\infty \frac{1}{n} \frac{\psi_{2n}(\tau)}{\psi_{2n}(\tau_1)} \sin 2n\theta \right], \qquad (138)$$

in which τ_1 is the constant value of τ on the boundary of the jet. (The speed and the density are constant on the jet boundary because the pressure is constant there.) Since $q = 1$ gives a constant value for ψ in (137), $\tau = \tau_1$ also gives a constant value for ψ in (138). Furthermore, it is obvious that for $\theta = \pm\pi/2$ (138) gives

$$\psi = \mp \frac{\pi}{2}. \qquad (139)$$

At infinity in the reservoir, $\tau = 0$ and $\psi_{2n}(\tau) = 0$, so that

$$\psi = -\theta,$$

as required. Hence (138) satisfies (128) and all the boundary conditions and is therefore the solution. We could have multiplied the right-hand side by a constant k; but since there is only one length scale l, which is the opening of the slit through which the jet issues, and since this can be (and is) left unspecified, we can just take k to be equal to 1. This does not affect the determination of the coefficient of contraction C_c defined by

$$C_c = \frac{l'}{l}, \qquad (140)$$

l' being the asymptotic width of the jet.

The determination of C_c will now be shown. The total flux being π according to (139),

$$\frac{\rho q l'}{\rho_0} = \pi, \qquad (141)$$

in which ρq is constant on the jet boundary and asymptotically constant throughout the jet and ρ_0 is the density far upstream in the reservoir.

Let us express q in terms of τ_1. By (18a) and (127),

$$\frac{\gamma p_0}{\rho_0{}^\gamma} \rho^{\gamma-1} = \frac{\gamma-1}{2} q_{max}^2(1-\tau_1) \tag{142}$$

asymptotically in the jet. Furthermore from (19a) we have

$$q_{max}^2 = \frac{2}{\gamma-1} c_0{}^2 = \frac{2\gamma}{\gamma-1} \frac{p_0}{\rho_0}, \tag{143}$$

where p_0 is the pressure far upstream in the reservoir and c_0 the sound speed at the same place. Equations (142) and (143) give

$$\rho = \rho_0(1-\tau)^{1/(\gamma-1)}.$$

Then asymptotically in the jet

$$\frac{\rho q}{\rho_0} = q_{max}\tau_1^{1/2}(1-\tau_1)^{1/(\gamma-1)}. \tag{144}$$

We have fixed the total flux to be π, but we have not specified l. Thus we are at liberty to assume an arbitrary value for q on the jet boundary. We shall take this to be 1. Then

$$q_{max}\tau_1^{1/2} = 1, \qquad \frac{\rho q}{\rho_0} = \frac{\rho}{\rho_0} = (1-\tau_1)^{1/(\gamma-1)}, \tag{145}$$

and

$$l'(1-\tau_1)^{1/(\gamma-1)} = \pi, \tag{146}$$

which gives l' as a function of τ_1. To find C_c, we need only calculate l as a function of τ_1. This is a little more complicated but not much more so.

Using the second equations in (107) and (110) and remembering $q = 1$ on the jet boundary, we have

$$\frac{\partial y}{\partial \theta} = \frac{\rho_0}{\rho}(\sin\theta\,\psi_q + \cos\theta\,\psi_\theta). \tag{147}$$

With ρ/ρ_0 given by (145) and ψ by (138), (147) can be used to evaluate l. Thus, in terms of τ and τ_1,

$$\frac{\partial y}{\partial \theta} = -\left(\frac{\tau_1}{\tau}\right)^{1/2} Q\left[\sum_1^\infty \frac{1}{n}\frac{\psi_{2n}'(\tau)}{\psi_{2n}(\tau_1)}\sin\theta\sin 2n\theta + \frac{1}{2\tau}\cos\theta \right.$$

$$\left. + \frac{1}{2\tau}\sum_1^\infty \frac{1}{n}\frac{\psi_{2n}(\tau)}{\psi_{2n}(\tau_1)} 2n\cos\theta\cos 2n\theta\right]. \tag{148}$$

If we put $\tau = \tau_1$ in (148), we see immediately that the second series does not converge. But for

$$\tau_1 < \frac{\gamma-1}{\gamma+1}, \tag{149}$$

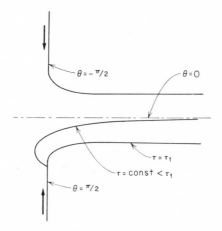

FIGURE 11.   CHAPLYGIN'S GAS JET.

which implies that the flow is subsonic, and for $\tau < \tau_1$ both series in (148) converge, and we can certainly integrate term by term from $\theta = 0$ to $\theta = \pi/2$. Note that the line $\tau = \text{const} < \tau_1$ extends from a place on the centerline of the jet to a place on the lower wall as θ varies from zero to $\pi/2$ (Fig. 11). Thus if (148) is integrated from zero to $\pi/2$ and then τ is allowed to approach τ_1, the result is $l/2$. [If τ is put equal to τ_1 to start with and $\partial y/\partial \theta$ in some valid form is integrated from $\theta = 0$ to $\theta = \pi/2$, the result is $(l - l')/2$.]   Hence

$$\frac{l}{2} = Q(\tau_1) \left\{ \frac{1}{2\tau_1} + \sum_1^\infty \frac{2(-1)^{n-1}}{4n^2 - 1} \left[\frac{\psi_{2n}'(\tau_1)}{\psi_{2n}(\tau_1)} + \frac{1}{2\tau_1} \right] \right\}, \qquad (150)$$

with Q defined in (130).   Thus (146) and (150) give

$$\frac{l}{l'} = \frac{2}{\pi} + \frac{4}{\pi} \sum_1^\infty \frac{(-1)^{n-1}}{4n^2 - 1} \left[1 + 2\tau_1 \frac{\psi_{2n}'(\tau_1)}{\psi_{2n}(\tau_1)} \right]. \qquad (151)$$

The values of l'/l for various values of τ_1, for $\gamma = 1.4$, as computed by Ferguson and Lighthill (1947), are listed in the following table:

τ_1	0	0.02	0.04	0.06	0.08	0.10	0.12	0.14	0.16	0.1667
l'/l	0.611	0.623	0.636	0.650	0.665	0.681	0.699	0.717	0.738	0.745

As mentioned before, the theory is valid only if the flow is entirely subsonic, i.e., if $\tau < \tau_1$ and (149) holds.   The discussion of convergence of (138) or (148) is beyond the scope of this book.[1]

From 1904 to the time of this writing it has been 63 years, but Chaplygin's elegant work remains a delight and an inspiration to workers

[1]The reader is referred to M. J. Lighthill's article in Howarth (1953) and the references given there.

in fluid mechanics and, like the songs piped by the youth on the Grecian Urn, somehow seems forever new.

17. THE CHAPLYGIN-KÁRMÁN-TSIEN APPROXIMATION FOR STEADY TWO-DIMENSIONAL SUBSONIC FLOWS

Even in two dimensions, the hodograph method does not furnish a solution for steady subsonic flow past a body of arbitrary shape. This is not surprising, since even for an incompressible fluid whose flow is governed by the relatively simple Laplace equation it is not possible to find an exact solution when the shape of the body is arbitrarily given. However the two equations in (110) vaguely resemble the Cauchy-Riemann equations, and one may ask what the relationship is between ρ and q that will make (110) reducible to the Cauchy-Riemann equations and how this relationship is to be interpreted. It seems that Chaplygin (1904) did implicitly pose such a question and suggest an approximation which can reduce (110) to the Cauchy-Riemann equations. He substituted the equation

$$p = -\frac{B}{\rho} + A \tag{152}$$

for the equation of homentropy

$$p = C\rho^\gamma \tag{153}$$

and evaluated A and B by requiring (152) to be a tangent to (153) in the plane of ρ^{-1} and p at the point corresponding to the stagnation point, i.e., at the point where

$$\rho = \rho_s, \qquad p = p_s, \tag{154}$$

the subscript indicating the point of stagnation. Von Kármán (1941) and Tsien (1939) evaluated A and B at infinity, where ρ is the undisturbed (by the body) density and p the undisturbed pressure, or

$$\rho = \rho_0, \qquad p = p_0, \tag{155}$$

the subscript indicating, for convenience, the point at *infinity*.

It is desirable to discuss the Kármán-Tsien approximation in some detail. The tangent (Fig. 12) to the curve represented by (153), at the point indicated by (155), is

$$p - p_0 = c_0^2 \rho_0^2 \left(\frac{1}{\rho_0} - \frac{1}{\rho} \right), \tag{156}$$

since

$$\left(\frac{dp}{d\rho^{-1}} \right)_0 = -\rho_0^2 \left(\frac{dp}{d\rho} \right)_0 = -\rho_0^2 c_0^2. \tag{157}$$

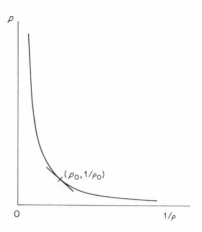

Equation (156) will be used instead of (153), as if any part of the gas when undergoing an isentropic change of state obeyed (156).   Then

$$c^2 = \frac{dp}{d\rho} = \frac{c_0{}^2 \rho_0{}^2}{\rho^2}. \tag{158}$$

The Bernoulli equation (3.26), upon neglecting Ω, is, for steady flows of a homentropic fluid,

$$\frac{q^2}{2} + \int \frac{dp}{\rho} = \text{const.} \tag{159}$$

Substitution of (156) into (159) gives

$$q^2 - c^2 = q_0{}^2 - c_0{}^2 = -a^2, \tag{160}$$

so that (158) can be written as

$$\left(\frac{\rho}{\rho_0}\right)^2 = 1 - \frac{q^2 - q_0{}^2}{c^2} \tag{161a}$$

or

$$\left(\frac{\rho_0}{\rho}\right)^2 = 1 + \frac{q^2 - q_0{}^2}{c_0{}^2}. \tag{161b}$$

(In Chaplygin's approximation, the subscript s is used instead of 0, and $q_s = 0$.)

By virtue of (160) and (158),

$$1 - \frac{q^2}{c^2} = \frac{a^2}{c^2} = \frac{a^2}{c_0{}^2} \frac{c_0{}^2}{c^2} = \left(\frac{a\rho}{c_0\rho_0}\right)^2. \tag{162}$$

Equations (110) can then be written as

$$\frac{\rho_0 q}{\rho} \phi_q = -\left(\frac{a}{c_0}\right)^2 \psi_\theta, \qquad \phi_\theta = \frac{\rho_0 q}{\rho} \psi_q. \tag{163}$$

The substitutions

$$\Phi = \frac{c_0}{a} \phi, \tag{164a}$$

$$\Psi = \psi, \tag{164b}$$

$$d\omega = \frac{a\rho}{c_0 \rho_0 q} dq \tag{164c}$$

then transform (163) to

$$\Phi_\omega = -\Psi_\theta, \qquad \Phi_\theta = \Psi_\omega. \tag{165}$$

These are the Cauchy-Riemann equations for the complex variables $\Phi + i\Psi$ and $\theta + i\omega$. If we define

$$U - iV = A' e^{\theta + i\omega} = A e^{\omega - i\theta}, \tag{166}$$

then $$\Phi + i\Psi = F(U - iV). \tag{167}$$

Note that, by virtue of (162) and (160) and with Q denoting the speed $\sqrt{U^2 + V^2}$, (164c) can be written as

$$\frac{dQ}{Q} = \sqrt{1 - \frac{q^2}{c^2}} \frac{dq}{q} = \frac{a}{q\sqrt{a^2 + q^2}} dq,$$

integration of which gives

$$Q = \frac{Cq}{a + \sqrt{a^2 + q^2}}.$$

Note that this Q has nothing to do with the Q in (128) or (130). If q is very small, the fluid can be considered incompressible and Q should be q exactly. Hence C must be $2a$, and we have

$$Q = \frac{2aq}{a + \sqrt{a^2 + q^2}}, \tag{168}$$

which connects Q and q. Now all the methods presented in Chap. 4 for incompressible fluids can then be brought to bear in seeking to determine the functional form of F in (167). In practice, one seeks to determine a complex potential $\Phi + i\Psi$ as an analytic function of the complex variable $X + iY$ in the physical plane or to determine the form of f in

$$\Phi + i\Psi = f(X + iY) \tag{169}$$

for a given profile of the body or the boundary. With f determined, F is also known implicitly, since

$$U - iV = \Phi_X + i\Psi_X. \tag{170}$$

From (167) and (166) we then know Φ and Ψ and therefore ϕ and ψ as functions of θ and Q or θ and q. To determine the form of a streamline in the physical plane, the $d\psi$ in (103) and (104) is put equal to zero and $d\phi$ is allowed to vary. Since q and θ are then functions only of ϕ, the shape of the streamline is determined parametrically by integration of (103) and (104). When obtained in this way, the streamline corresponding to the given body or boundary may differ from the given body or boundary, since q is different from Q for a compressible fluid. This difference can be reduced by further approximations. To maintain the scope of the book we cannot dwell on the Kármán-Tsien approximations any longer, except to say that if Chaplygin's approximation is used, the factor a/c_0 in (164a) and (164c) will not be necessary. The Kármán-Tsien approximation is better than the Chaplygin approximation because q and c of the flow are nearer q_0 and c_0 than q_s ($= 0$) and c_s over the major part of the flow.

18. SUPERSONIC TWO-DIMENSIONAL FLOWS: METHOD OF CHARACTERISTICS IN THE PHYSICAL PLANE

Steady two-dimensional flow of an inviscid homentropic fluid is governed by (27) or (28), whether or not the speed at any point is greater or less than the sound speed, provided there is no shock wave present. As we shall see, if the flow is steady and everywhere supersonic, the solution is rather straightforward in principle, by the use of the method of characteristics. In this section we shall discuss the method of characteristics in the physical plane.

Equation (27) or (28) is a special case of the general partial differential equation

$$a\phi_{xx} + b\phi_{xy} + c\phi_{yy} + f = 0, \tag{171}$$

in which a, b, c, and f are functions of x, y, ϕ, ϕ_x, and ϕ_y. For convenience of exposition we shall write

$$r = \phi_{xx}, \qquad s = \phi_{xy}, \qquad t = \phi_{yy}, \qquad p = \phi_x, \qquad q = \phi_y. \tag{172}$$

Thus, we have

$$ra + sb + tc + f = 0, \tag{173}$$

$$r\,dx + s\,dy - dp = 0, \tag{174}$$

$$s\,dx + t\,dy - dq = 0, \tag{175}$$

the first of which being (171) and the following two being obtained directly on taking the total differentials dp and dq.

Now consider the directions determined by

$$D \equiv \begin{vmatrix} a & b & c \\ dx & dy & 0 \\ 0 & dx & dy \end{vmatrix} = 0, \tag{176}$$

or
$$a(dy)^2 - b\,dx\,dy + c(dx)^2 = 0,$$

the solutions of which are

$$\frac{dy}{dx} = \frac{b \pm \sqrt{b^2 - 4ac}}{2a}. \tag{177}$$

Curves determined by (177) are called *characteristics*. If

$$b^2 > 4ac, \tag{178}$$

characteristics exist in the physical plane, because the slopes of the characteristics are real. These characteristics are essential for the solution of (171). To show this, we shall demonstrate first how the system of Eqs. (173) to (175) can be solved step by step for p and q.

Let the two families of characteristics determined by (177) be denoted by C_1 and C_2, respectively, and let ϕ and $\partial\phi/\partial n$ be specified along an arbitrary curve AB, which is itself not a characteristic, n being the distance normal to AB (see Fig. 13). Thus, p and q are both specified along AB. The curves AC, DI, EH, and FG belong to family C_1, and the curves DL, EK, FJ, and BC belong to family C_2. Along the characteristics (176) is satisfied, and for (173) to (175) to have a solution for r, s, and t it is necessary and sufficient that the rank of the matrix

$$\begin{bmatrix} a & b & c & f \\ dx & dy & 0 & -dp \\ 0 & dx & dy & -dq \end{bmatrix}$$

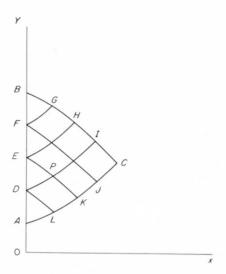

FIGURE 13.   TWO FAMILIES OF CHARACTER-
ISTICS, SHOWING ZONE OF DETERMINATION
OF THE LINE AB.

be less than 3.   That is, it is necessary and sufficient that (176) holds and

$$\begin{vmatrix} a & c & f \\ dx & 0 & -dp \\ 0 & dy & -dq \end{vmatrix} = 0. \tag{179}$$

If a finite number of points along AB (Fig. 13) are used to start with, calculation with finite differences is understood and the curves in Fig. 13 are made of a succession of straight lines.   Thus, along DP (179) has the form

$$(f\,\Delta x\,\Delta y + c\,\Delta x\,\Delta q + a\,\Delta y\,\Delta p)_1 = 0, \tag{180}$$

where f, c, and a are evaluated at D and

$$(\Delta x)_1 = x_P - x_D, \qquad (\Delta y)_1 = y_P - y_D,$$
$$(\Delta p)_1 = p_P - p_D, \qquad (\Delta q)_1 = q_P - q_D.$$

Similarly, along EP (179) has the form

$$(f\,\Delta x\,\Delta y + c\,\Delta x\,\Delta q + a\,\Delta y\,\Delta p)_2 = 0, \tag{181}$$

where f, c, and a are evaluated at E and

$$(\Delta x)_2 = x_P - x_E, \qquad (\Delta y)_2 = y_P - y_E,$$
$$(\Delta p)_2 = p_P - p_E, \qquad (\Delta q)_2 = q_P - q_E.$$

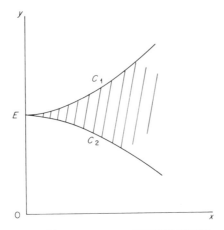

FIGURE 14.   DOMAIN OF INFLUENCE OF THE
POINT E.

The coordinates of P are known because P is the intersection of two lines DP and EP of known directions determined by (176).   Then (180) and (181) together determine p_P and q_P.   The process can be started from all chosen points along AB and repeated once the p and q at internal points such as P are known.   In this way p and q at a number of grid points on and within the area ABC can be determined. This number can be arbitrarily increased to achieve arbitrarily high accuracy.   From the values of p and q and the values of ϕ along AB the values of ϕ at the grid points are also known.

The area ABC in Fig. 13 is the zone of determination of the line AB, and the shaded area in Fig. 14 is the domain of influence of a point E. The terminology is obvious once the material in the previous paragraph is understood.

If the differential equation is (27), the characteristics determined by (176) are actually the Mach lines.   To see this remarkable fact, note that, for (27),

$$a = 1 - \frac{u^2}{c^2}, \qquad b = -\frac{2uv}{c^2}, \qquad c = 1 - \frac{v^2}{c^2}, \qquad (182)$$

which show that if $q > c$, real characteristics exist, since

$$\frac{b^2 - 4ac}{4} = \frac{u^2 + v^2}{c^2} - 1 = \frac{q^2}{c^2} - 1.$$

For a C_1 curve,

$$\frac{dy}{dx} = \frac{A}{B}, \qquad (183)$$

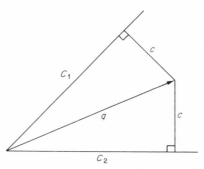

FIGURE 15. SKETCH SHOWING THAT THE
VELOCITY VECTOR BISECTS THE ANGLE
MADE BY THE CHARACTERISTICS, WHICH
RUN IN THE MACH DIRECTIONS.

in which $\qquad A = -\dfrac{uv}{c^2} + \left(\dfrac{q^2}{c^2} - 1\right)^{1/2}, \qquad B = 1 - \dfrac{u^2}{c^2}.$ $\qquad$ (184)

The velocity component normal to the C_1 curve is

$$\frac{-vB + uA}{(A^2 + B^2)^{1/2}},$$

and this, after some straightforward simplifications, can be shown to be exactly equal to c. Thus a C_1 curve is a Mach line (or curve). Similarly, the velocity component normal to a C_2 curve is also c, so that a C_2 curve is also a Mach line. It is evident then that the velocity vector must bisect the angle between two characteristic directions (or Mach directions), as shown in Fig. 15.

The method presented in this section can be applied to (5.195), which governs the flow of shallow water, so long as the flow is everywhere supercritical. Supersonic axisymmetric flows can also be treated by the method of characteristics.

19. SUPERSONIC TWO-DIMENSIONAL FLOWS: METHOD OF CHARACTERISTICS IN THE HODOGRAPH PLANE

The shape of the characteristics in the physical plane depends on the initial conditions given. The advantage of using hodographic variables is that for supersonic flows the shape of the characteristics is known a priori. The derivation of the equation for the characteristics in the hodograph plane is quite similar to the derivation of the corresponding equation for supercritical flows of shallow water, such as given in Sec. 3.5.4 of

Chap. 5. Now, however, the method presented in the preceding section is available, and we can take a shortcut.

Equation (111) is of the type (171) if we identify q with x and θ with y, since x and y in (171) are just any two independent variables, not necessarily spatial coordinates. Applying the analysis of the preceding section, we obtain

$$\frac{d\theta}{dq} = \pm \frac{1}{q}\left(\frac{q^2}{c^2} - 1\right)^{1/2} = \pm \frac{1}{q}(M^2 - 1)^{1/2}. \tag{185}$$

This is exactly the same as (5.203b), except that the Froude number is replaced by the Mach number M. The Bernoulli equation is now

$$\frac{q^2}{2} + \frac{c^2}{\gamma - 1} = \frac{\gamma + 1}{2(\gamma - 1)}c^{*2}, \tag{186}$$

which follows from (19a) and (20). Substituting the c^2 obtained from (186) into (185), and defining M^* by

$$M^* = \frac{q}{c^*},$$

we have

$$d\theta = \pm \left(\frac{M^{*2} - 1}{1 - h^{-2}M^{*2}}\right)^{1/2}\frac{dM^*}{M^*}, \tag{187}$$

in which

$$h^2 = \frac{\gamma + 1}{\gamma - 1}. \tag{188}$$

Equation (187) corresponds to (5.208), for which $\gamma = 2$. Integration of (187) yields

$$\pm(\theta - \theta_0) = \frac{h}{2}\left\{\frac{\pi}{2} - \sin^{-1}\left[\gamma - (\gamma - 1)M^{*2}\right]\right\}$$
$$- \frac{1}{2}\left\{\frac{\pi}{2} + \sin^{-1}\left[\gamma - (\gamma + 1)M^{*-2}\right]\right\}. \tag{189}$$

Since, as can be derived from (186),

$$M^2 = \frac{2M^{*2}}{\gamma + 1 - (\gamma - 1)M^{*2}},$$

(189) can be written in terms of M^2 as

$$\pm(\theta - \theta_0) = h\tan^{-1}\frac{(M^2 - 1)^{1/2}}{h} - \tan^{-1}(M^2 - 1)^{1/2}, \tag{190}$$

which corresponds to (5.209). In the hodograph plane of q and θ, (190) represents epicycloids. The proof is exactly the same as that for showing

that (209) are epicyloids[1].   Since the utilization of the hodographic characteristics to solve practical problems is similar to that described in detail
in Sec. 3.5.4 of Chap. 5, we may enjoy the economy sometimes possible
in a general treatment of fluid mechanics and leave the subject here.

20. ITERATION METHODS FOR TWO-DIMENSIONAL STEADY SUBSONIC FLOWS

We have seen that for supersonic flows exact solution is in principle
always possible and the accuracy of a calculation is limited only by the
necessity of using finite differences.   For steady subsonic flows no real
characteristics exist, and because of the compressibility of the fluid Eq.
(27) governing steady two-dimensional flows is much more complicated
than the Laplace equation for irrotational flows of incompressible fluids;
for this reason the theory of complex variables cannot be used to solve
problems of subsonic flows exactly, even if two-dimensional and steady.
True, we have seen how Chaplygin mangaged to give exact solutions for
two-dimensional steady subsonic flows, but these are only for straight
boundaries and for jets with boundaries of constant speed.   We have
also seen how the Kármán-Tsien approximation can solve the problem
approximately, by using hodograph variables.   In this section we shall
briefly mention how iteration methods can be used to provide approximate
solutions.

20.1.   The Rayleigh-Janzen Method

Equation (27) can be written as

$$\phi_{xx} + \phi_{yy} = \frac{1}{c^2} (\phi_{xx}\phi_x{}^2 + 2\phi_{xy}\phi_x\phi_y + \phi_{yy}\phi_y{}^2). \tag{191}$$

If we solve

$$\nabla^2\phi_0 = 0$$

for the given boundary conditions, substitute ϕ_0 into the right-hand side
of (191) to obtain a known function $F_0(x,y)$, using (19) to evaluate c^2,
and solve

$$\nabla^2\phi_1 = F_0(x,y)$$

with the appropriate boundary conditions (always making sure that the
actual boundary conditions are at no stage violated), then an approximate

---

[1] The use of the hodographic characteristics to solve problems of supersonic flow was
initiated by Prandtl and Busemann (1929).   However, the epicycloidal hodographs were
known to Thiessen, a student of Prandtl, who never published his work.   Busemann told
the author that his own contribution consists mainly in realizing that the Mach lines are
in general curved instead of straight, as some researchers then, misled by the nearly
straight Mach lines in Prandtl's photographs, were inclined to think they were.

solution of (191) is

$$\phi = \phi_0 + \phi_1.$$

The process can then be repeated as desired. This is the *Rayleigh-Janzen iteration*. As can be discerned from the form of (191), the convergence is good for small Mach numbers but becomes slower and slower as the Mach number of the oncoming stream increases toward unity. Before it reaches unity, supersonic pockets and shock waves will occur.

20.2. Prandtl's Iteration

If the Mach number of the oncoming stream, i.e., at infinity, is not very small, some advantage can be gained by writing (27) in the form

$$(1 - M_0^2)\phi_{xx} + \phi_{yy} = \frac{1}{c^2} [\phi_{xx}(\phi_x^2 - c^2 M_0^2) + 2\phi_{xy}\phi_x\phi_y + \phi_{yy}\phi_y^2]. \quad (192)$$

The iteration procedure is similar to that in Sec. 20.1. Here, as in Sec. 20.1, it is possible to write

$$\phi = Ux + \phi' \quad (193)$$

and to retain only terms up to a certain order in ϕ' on the right-hand side of (191) or (192) at each stage of iteration. The U in (193) is the velocity of the flow at infinity.

21. APPROXIMATE SOLUTION FOR STEADY FLOWS PAST SLENDER BODIES BY LINEARIZATION

For a uniform stream, of velocity U in the x direction at infinity and passing around slender bodies, the right-hand side of (192) can be neglected as compared with the left-hand side, and we can write, for two-dimensional flows,

$$(1 - M_0^2)\phi_{xx} + \phi_{yy} = 0, \qquad M_0 = \frac{U}{c_0}, \quad (194)$$

c_0 being the sound speed at infinity. In fact, the same reasoning allows us to write, for steady three-dimensional flows past slender bodies,

$$(1 - M_0^2)\phi_{xx} + \phi_{yy} + \phi_{zz} = 0. \quad (195)$$

We assume c_0 to be constant. Of course, in considering c_0 constant we implicitly neglect the effect of gravity.

If $M_0 > 1$, the method of characteristics can be used to solve (194) and (195). The solution for (194) is especially simple. If $M_0 < 1$, with the change of scale

$$x' = (1 - M_0^2)^{-\frac{1}{2}}x, \qquad y' = y, \qquad z' = z, \quad (196)$$

and dropping the primes *subsequently*, we can write

$$\nabla^2\phi = 0 \tag{197}$$

for (194) or (195), as the case may be. This equation was treated in detail in Chap. 4, and the approximate solutions for slender bodies, two-dimensional or axisymmetric, can be used to solve the problem for subsonic flows by considering irrotational motion of an incompressible fluid past a slender body which is more elongated than the actual body, according to (196). The transformation back to the original coordinates is straightforward. Of course, all the weaknesses of linearization, such as the improper behavior of the solution near stagnation points, remain as for incompressible fluids.

Since the flow of an incompressible fluid past a slender body has been discussed in detail in Chap. 4, we again take advantage of the economy possible in a general treatment of fluid mechanics to terminate the discussion here.

22. CONCLUDING REMARKS

In this chapter, which is necessarily a brief treatment of compressible fluids, we have not touched upon many subjects. For instance, the determinant

$$D = \frac{\partial(\phi,\psi)}{\partial(q,\theta)} ,$$

which, by virtue of (110), is equal to

$$-\frac{\rho_0}{\rho q} [q^2\psi_q{}^2 + (1 - M^2)\psi_\theta{}^2],$$

may vanish for supersonic flows. Lines in the physical plane which are images of lines in the hodograph plane where D vanishes are called *limit lines*. Along such lines the mapping between the physical plane and the hodograph plane becomes singular, and the flow near them needs further study. These lines and their significance in transonic flows are discussed in von Mises (1958).

PROBLEMS

1 Show that

$$\phi = -\frac{1}{R} F\left(t - \frac{R}{c}\right), \qquad \text{in which } R^2 = x^2 + y^2 + z^2,$$

satisfies the wave equation (42). This ϕ is called the *retarded potential*. Why? From this result can one interpret c as the speed of sound?

2 Plane sound waves propagating in the x direction are disturbed by a solid cylinder of cross section $x^2 + y^2 = a^2$. Solve the diffraction problem.

3 An ideal gas flows over a boundary with sinusoidal corrugations of an amplitude small compared with the wavelength. The mean position of the boundary is a plane. Treat the gas as inviscid and nonconductive of heat and show that when the flow is supersonic, the streamlines are in phase with the boundary along the Mach lines and when the flow is subsonic, the streamlines are in phase with the boundary along any line normal to the mean position of the boundary.

4 Show that for weak normal shocks the entropy jump varies as the cube of the pressure jump across the shock. *Suggestion:* Show from (67) that $p_2 - p_1$ is proportional to $u_1 - c^*$. Then from (69) show that α is proportional to $u_1 - c^*$.

5 For a weak oblique shock with a small θ (angle of inward boundary deflection), show that the entropy jump across the shock varies as θ^3.

6 In flows with oblique shocks, is the flow downstream from the shock necessarily subsonic? Discuss thoroughly.

7 For a small angle θ of inward deflection of the boundary, show that, to $O(\theta^2)$,

$$\frac{p_2 - p_1}{\frac{1}{2}\rho_1 U^2} = \frac{2}{\sqrt{M_1^2 - 1}}\,\theta + \frac{(\gamma + 1)M_1^4 - 4(M_1^2 - 1)}{2(M_1^2 - 1)^2}\,\theta^2,$$

where the subscript 1 refers to the upstream condition, U is the upstream speed, and the subscript 2 refers to the postshock condition. Show that the same formula is obtained, to $O(\theta^2)$, if one insists on using the method of Sec. 19 for isentropic change of state across the Mach line, considering θ to be infinitesimal. (This is not strictly justified if θ is finite but leads to only a small error if θ is small.) Does this have anything to do with the answer to Prob. 5?

8 Design a two-dimensional channel expansion for supersonic flow with no residual waves after the expansion. Use an expansion ratio of $1:1.5$.

9 Study in detail the analogy of steady shallow-water flow and two-dimensional steady flow of a gas. Include hydraulic jumps and shocks in your discussion. Enumerate all the analogous quantities.

10 An infinitely long pipe filled with a gas, initially at rest and in a uniform state, is partitioned by a plate normal to the pipe axis. The plate starts to move to the right with a speed U. Find the flow in front of and behind the plate.

11 Give the mathematical steps for solving Prob. 4.20 for a compressible fluid.

12 Find an approximate solution for two-dimensional steady irrotational flow past a circular cylinder with $M_\infty = \frac{1}{4}$.

13 Solve Prob. 12 for a Joukowsky airfoil. *Caution:* The governing equation, unlike the Laplace equation, is not invariant in form after a conformal mapping.

14 Find the equation governing steady two-dimensional flows of an inviscid nonhomentropic gas, neglecting the effects of heat conduction. *Hint:* So long as the streamlines do not cross a shock wave and the entropy along each streamline is constant, so that (13a) can be used, the constant on the right-hand side is a function of the stream function ψ. From (3.12) we have

$$\frac{\zeta}{\rho} = F(\psi),$$

and this can be put in terms of ψ upon elimination of ρ by the use of (13a).

15 Ignoring the effects of gravity and of heat conduction, find the equation governing steady axisymmetric flows of an inviscid nonhomentropic gas. *Hint:* Take the hint given for the preceding problem, except that the second equation in (3.13) should be used.

REFERENCES

CHAPLYGIN, S. A., 1904: On Gas Jets, *Sci. Mem. Moscow Univ. Math. Phys. Sec.*, **21**: 1–121; translation in *Natl. Advisory Comm. Aeron. Tech. Mem.* 1063, 1944.

FERGUSON, D. F., and M. J. LIGHTHILL, 1947: The Hodograph Transformation in Trans-sonic Flow, pt. IV, Tables, *Proc. Roy. Soc. London Ser. A*, **192**: 135–142.

HOWARTH, L. (ed.), 1953: "Modern Development in Fluid Dynamics: High Speed Flow," vols. I and II, Oxford University Press, London.

HUGONIOT, H., 1887: Mémoire sur la propagation du mouvement dans les corps, et spécialement dans les gaz parfaits, *J. École Polytech.*, (1)**57**: 1–97.

HUGONIOT, H., 1889: *idem, J. École Polytech.*, (1)**58**: 1–125.

KÁRMÁN, T. VON, 1941: Compressibility Effects in Aerodynamics, *J. Aeron. Sci.*, **8**: 337–356.

MISES, R. VON, 1950: On the Thickness of a Steady Shock Wave, *J. Aeron. Sci.*, **17**: 551–555.

MISES, R. VON, 1958: "Mathematical Theory of Compressible Fluid Flow," Academic Press Inc., New York.

MOLENBROEK, P., 1890: Über einiger Bewegungen eines Gases bei Annahme eines Geschwindigkeitspotentials, *Arch. Math. Phys.*, **9**: 157–195.

PRANDTL, L., and A. BUSEMANN, 1929: Näherungsverfahren zur zeichnerischen Ermittlung von ebenen Strömungen mit Überschallgeschwindigkeit, *Stodola Festschrift*: 85–122.

RANKINE, W. J. M., 1870: On the Thermodynamics of Waves of Finite Longitudinal Disturbance, *Phil. Trans.*, **160**: 277–286.

RIEMANN, B., 1860: Über die Fortpflanzung ebener Luftwellen von endlicher Schwingungweite, *Abhaudl. Kgl. Ges. Wiss.*, **8**: reprinted in "Collected Works of Bernhard Riemann," pp. 156–181, Dover Publications, Inc., New York, 1953.

TAYLOR, G. I., 1910: The Condition Necessary for Discontinuous Motion in Gases, *Proc. Roy. Soc. London Ser. A*, **84**: 371–377.

TSIEN, H. S., 1939: Two-dimensional Subsonic Flow of Compressible Fluids, *J. Aeron. Sci.*, **6**: 399–407.

ADDITIONAL READING

COURANT, R., and K. O. FRIEDRICHS, 1948: "Supersonic Flow and Shock Waves," Interscience Publishers, Inc., New York.

FRANKL, F. I., and E. A. KARPOVICH, 1953: "Gas Dynamics of Thin Bodies," Interscience Publishers, Inc., New York.

HAYES, W. D., and R. F. PROBSTEIN, 1959: "Hypersonic Flow Theory," Academic Press Inc., New York.

LIEPMANN, H. W., and A. E. PUCKETT, 1947: "Introduction to Aerodynamics of a Compressible Fluid," John Wiley & Sons, Inc., New York.

LIEPMANN, H. W., and A. ROSHKO, 1957: "Elements of Gasdynamics," John Wiley & Sons, Inc., New York.

MILES, J. W., 1959: "The Potential Theory of Unsteady Supersonic Flow," Cambridge University Press, New York.

OSWATITSCH, K., 1956: "Gas Dynamics," Academic Press Inc., New York.

SAUER, R., 1947: "Introduction to Theoretical Gas Dynamics," J. W. Edwards, Publisher, Incorporated, Ann Arbor, Mich.

SEDOV, L. I., 1959: "Similarity and Dimensional Methods in Mechanics," Academic Press Inc., New York.

WARD, G. N., 1955: "Linearized Theory of Steady High-speed Flow," Cambridge University Press, New York.

CHAPTER SEVEN

EFFECTS
OF
VISCOSITY

1. INTRODUCTION

In the discussion of fluid flow in the last three chapters the effects of viscosity have been neglected. The flows discussed in these chapters do not satisfy the nonslip condition at solid boundaries except when the boundaries are made to move in very special ways in order to satisfy that condition. Since the nonslip condition must be satisfied by real fluids (except for very rarefied gases with large mean free paths), we can inquire how realistic the solutions based on inviscidness are and how solutions satisfying the nonslip conditions can be obtained for a viscous fluid.

As will be made clear in this chapter, the flows based on inviscidness, although not satisfying the nonslip condition at solid boundaries, nevertheless are valid in regions outside of a layer attached to these boundaries and outside of wakes, provided the Reynolds number for the flow is high (see Chap. 2 for the definition of the Reynolds number). Except for unidirectional and circular flows and a few exact solutions, at high Reynolds numbers the solution of the Navier-Stokes equations in the majority of cases will be for boundary layers only. It must be noted, however, that the equations governing boundary-layer flow can also be used for jets and wakes, for which longitudinal diffusion of momentum or vorticity can often be neglected as compared with transverse diffusion of momentum or vorticity.

Before we concentrate on the boundary-layer theory, we shall first discuss the exact solutions of the Navier-Stokes equations. After the boundary-layer theory has been presented, flows at very low Reynolds numbers will be discussed.

Since most of the examples given are for an incompressible fluid of constant viscosity, we shall explicitly write the equations governing the motion of such a fluid. They are given by (2.7a), which written in full becomes

$$\rho \, \frac{Du}{Dt} = -\frac{\partial p}{\partial x} - \rho \, \frac{\partial \Omega}{\partial x} + \mu \, \nabla^2 u, \tag{1}$$

$$\rho \, \frac{Dv}{Dt} = -\frac{\partial p}{\partial y} - \rho \, \frac{\partial \Omega}{\partial y} + \mu \, \nabla^2 v, \tag{2}$$

$$\rho \, \frac{Dw}{Dt} = -\frac{\partial p}{\partial z} - \rho \, \frac{\partial \Omega}{\partial z} + \mu \, \nabla^2 w, \tag{3}$$

in which x, y, and z are Cartesian coordinates, u, v, and w are velocity components in the directions of the coordinate axes, Ω is the body-force potential, which is assumed to exist, and

$$\frac{D}{Dt} = \frac{\partial}{\partial t} + u \, \frac{\partial}{\partial x} + v \, \frac{\partial}{\partial y} + w \, \frac{\partial}{\partial z} \,, \qquad \nabla^2 = \frac{\partial^2}{\partial x^2} + \frac{\partial^2}{\partial y^2} + \frac{\partial^2}{\partial z^2} \cdot \tag{4}$$

The equation of continuity is

$$\frac{\partial u}{\partial x} + \frac{\partial v}{\partial y} + \frac{\partial w}{\partial z} = 0. \tag{5}$$

The Navier-Stokes equations in cylindrical coordinates are given in (A2.82). For those who prefer a separate, simple derivation of the Navier-Stokes equations in these coordinates, we present Synge's (1938) derivation.

To avoid confusion we shall denote the velocity components in cylindrical coordinates (r, φ, z) by v_r, v_φ, and v_z. The laws of transformation between Cartesian and cylindrical coordinates are

$$r^2 = x^2 + y^2, \qquad \varphi = \tan^{-1} \frac{y}{x} \,, \qquad z = z.$$

From these it follows that

$$\frac{\partial}{\partial x} = \frac{\partial r}{\partial x} \frac{\partial}{\partial r} + \frac{\partial \varphi}{\partial x} \frac{\partial}{\partial \varphi} = \cos \varphi \, \frac{\partial}{\partial r} - \frac{\sin \varphi}{r} \frac{\partial}{\partial \varphi} \cdot$$

Similarly,
$$\frac{\partial}{\partial y} = \sin \varphi \frac{\partial}{\partial r} + \frac{\cos \varphi}{r} \frac{\partial}{\partial \varphi},$$
so that
$$\frac{\partial}{\partial x} + i \frac{\partial}{\partial y} = e^{i\varphi}\left(\frac{\partial}{\partial r} + \frac{i}{r} \frac{\partial}{\partial \varphi}\right), \qquad \frac{\partial}{\partial x} - i \frac{\partial}{\partial y} = e^{-i\varphi}\left(\frac{\partial}{\partial r} - \frac{i}{r} \frac{\partial}{\partial \varphi}\right).$$
Hence
$$\nabla^2 = \frac{\partial^2}{\partial x^2} + \frac{\partial^2}{\partial y^2} + \frac{\partial^2}{\partial z^2} = \left(\frac{\partial}{\partial x} + i \frac{\partial}{\partial y}\right)\left(\frac{\partial}{\partial x} - i \frac{\partial}{\partial y}\right) + \frac{\partial^2}{\partial z^2}$$
$$= e^{i\varphi}\left(\frac{\partial}{\partial r} + \frac{i}{r} \frac{\partial}{\partial \varphi}\right)\left[e^{-i\varphi}\left(\frac{\partial}{\partial r} - \frac{i}{r} \frac{\partial}{\partial \varphi}\right)\right] + \frac{\partial^2}{\partial z^2}$$
$$= \frac{\partial^2}{\partial r^2} + \frac{1}{r} \frac{\partial}{\partial r} + \frac{1}{r^2} \frac{\partial^2}{\partial \varphi^2} + \frac{\partial^2}{\partial z^2}. \tag{6}$$

Furthermore, it can be readily verified that
$$u + iv = e^{i\varphi}(v_r + iv_\varphi), \qquad w = v_z, \tag{7}$$
and, for the body-force components,
$$X + iY = e^{i\varphi}(X_r + iX_\varphi), \qquad Z = Z, \tag{8}$$
in which the capital letters indicate body-force components in the directions designated by them or their subscripts. The Navier-Stokes equations for the directions of x and y can be written as
$$\rho \frac{D}{Dt}(u + iv) = -\left(\frac{\partial}{\partial x} + i \frac{\partial}{\partial y}\right)p + \rho(X + iY) + \mu \nabla^2(u + iv) \tag{9}$$
if μ is constant. Substituting (7) and (8) into (9), we have
$$\rho \frac{D}{Dt}[e^{i\varphi}(v_r + iv_\varphi)] = -e^{i\varphi}\left(\frac{\partial}{\partial r} + \frac{i}{r} \frac{\partial}{\partial \varphi}\right)p + \rho e^{i\varphi}(R + i\Phi)$$
$$+ \mu \nabla^2[e^{i\varphi}(v_r + iv_\varphi)].$$
When this is expanded, the real and imaginary parts separated, and v_r, v_φ, and v_z replaced by u, v, and w, we have the first two of the following equations:
$$\rho\left(\frac{Du}{Dt} - \frac{v^2}{r}\right) = -\frac{\partial p}{\partial r} - \rho \frac{\partial \Omega}{\partial r} + \mu\left(\nabla^2 u - \frac{u}{r^2} - \frac{2}{r^2} \frac{\partial v}{\partial \varphi}\right), \tag{10}$$
$$\rho\left(\frac{Dv}{Dt} + \frac{uv}{r}\right) = -\frac{1}{r} \frac{\partial p}{\partial \varphi} - \frac{\rho}{r} \frac{\partial \Omega}{\partial \varphi} + \mu\left(\nabla^2 v - \frac{v}{r^2} + \frac{2}{r^2} \frac{\partial u}{\partial \varphi}\right), \tag{11}$$
$$\rho \frac{Dw}{Dt} = -\frac{\partial p}{\partial z} - \rho \frac{\partial \Omega}{\partial z} + \mu \nabla^2 w, \tag{12}$$

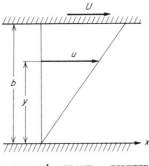

FIGURE   1.   PLANE   COUETTE
FLOW.

in which
$$\frac{D}{Dt} = \frac{\partial}{\partial t} + u\frac{\partial}{\partial r} + \frac{v}{r}\frac{\partial}{\partial \varphi} + w\frac{\partial}{\partial z},$$

∇^2 is given by (6), and Ω is the body-force potential.   Equation (12) follows from the third of the Navier-Stokes equations and is in fact identical in form with it, except that D/Dt and ∇^2 are now in cylindrical coordinates.   Equations (10) to (12) are identical with (A2.82).

The equation of continuity for an incompressible fluid in cylindrical coordinates can either be obtained by using (A2.67) or by the use of a curvilinear block; it is

$$\frac{1}{r}\frac{\partial}{\partial r}ru + \frac{1}{r}\frac{\partial v}{\partial \varphi} + \frac{\partial w}{\partial z} = 0. \tag{13}$$

2. STEADY FLOWS FOR WHICH THE NAVIER-STOKES EQUATIONS ARE LINEAR

Flows of an incompressible fluid having either unidirectional or circular streamlines are governed by linear Navier-Stokes equations, whether or not they are steady.   In this section steady flows will be discussed.   At the end we shall present a flow which is unidirectional at any elevation but whose velocity changes direction from elevation to elevation.

2.1. Plane Couette Flow

Consider the flow of a fluid between two parallel plates at a distance b apart due to the motion of the plate with velocity U in the x direction (Fig. 1).   For the moment we shall assume the plates to be horizontal, so that y is measured in the direction of the vertical and z is measured in a horizontal direction normal to that of x. The state of affairs does not vary with z. If, in addition, we assume that the flow is in the x direction, then

$$v = 0 = w. \tag{13a}$$

A glance at (5) then shows that u is independent of x and is a function of y only.

Because of the choice of the direction of increasing y,

$$\Omega = gy,$$

so that (3) states that

$$\frac{\partial p}{\partial z} = 0$$

and (2) states that

$$\frac{\partial p}{\partial y} = -\rho g; \tag{14}$$

i.e., the pressure variation with y is hydrostatic. The equation of incompressibility is

$$\frac{\partial \rho}{\partial t} + u\frac{\partial \rho}{\partial x} + v\frac{\partial \rho}{\partial y} + w\frac{\partial \rho}{\partial z} = 0, \tag{15}$$

which, for steady flows satisfying (13a), simply states that

$$\frac{\partial \rho}{\partial x} = 0.$$

Since ρ is also independent of z, ρ must be a function of y alone.

Now we can differentiate (14) with respect to x, and the result shows that

$$\frac{\partial^2 p}{\partial y\,\partial x} = 0,$$

which states that $\partial p/\partial x$ is independent of y. It is also independent of z, because the flow is two-dimensional. From (1), which now has the simple form

$$\mu\frac{d^2u}{dy^2} = \frac{\partial p}{\partial x}, \tag{16}$$

it is seen, however, that $\partial p/\partial x$ can only be a function of y. Thus it can only be a constant. If this constant is zero, i.e., if there is no pressure gradient in the x direction, integration of (16) quickly gives

$$u = \frac{Uy}{b} \tag{17}$$

if the lower plate is at rest. [If it is not, the modification of (17) is obvious.] The flow represented by (17) is called *plane Couette flow*.

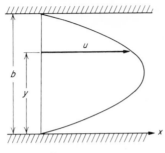

FIGURE 2. PLANE POISEUILLE FLOW.

2.2. Plane Poiseuille Flow

If the fluid between the parallel plates considered in Sec. 2.1 is set in motion not by the upper plate but by a pressure gradient in the x direction, all the arguments leading to (16) apply, and denoting $\partial p/\partial x$ by the constant K, we can write (16) as

$$\mu \frac{d^2u}{dy^2} = -K. \tag{18}$$

With the coordinates indicated in Fig. 2, the boundary conditions are

$$u = 0 \quad \text{at } y = 0 \text{ and } y = b.$$

Integration of (18), with the constants of integration determined by the boundary conditions, produces

$$u = \frac{K}{2\mu} y(b - y). \tag{19}$$

If the origin is taken midway between the plates, (19) can be written in the symmetric form

$$u = \frac{K}{2\mu}\left(\frac{b^2}{4} - y^2\right). \tag{20}$$

2.3. Plane Couette-Poiseuille Flow and Its Application

By virtue of the linearity of the governing differential system determining (17) and (19), the velocity profiles given by these equations can obviously be superposed. Thus, if the upper plate has the velocity U and the pressure gradient is denoted by $-K$, we have

$$u = \frac{Uy}{b} + \frac{K}{2\mu} y(b - y). \tag{21}$$

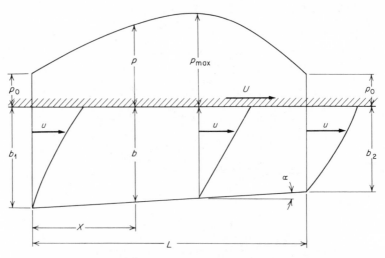

FIGURE 3.   FLOW IN A ROCKER BEARING.

The flow represented by (21) can be called *plane Couette-Poiseuille flow.*
The lubricating function of rocker bearings can be understood by approx-
imating the velocity distribution of the flow in such bearings by (21)
locally.

A rocker bearing usually consists of a plane upper surface sliding
over a lower surface, to which it is inclined.   The angle of inclination will
be denoted by α (Fig. 3).   The spacing of the upper and lower boundaries
is eventually to be determined by the geometry, the viscosity μ, the speed
of the upper boundary U, and above all by the vertical load on the bearing.
In the notation shown in Fig. 3, the discharge q, which must remain
constant for all x since the flow is now steady, is

$$q = \int_0^b u\, dy = \frac{Ub}{2} - \frac{\partial p}{\partial x}\frac{b^3}{12\mu}, \qquad (22)$$

in which, because of the slope of the lower boundary, $\partial p/\partial x$ is no longer
constant.   The spacing b is given by

$$b = b_1 - \alpha x$$

if α is so small that tan α can be replaced by α.   The q in (22) is so far
undetermined.   Solving (22) for $\partial p/\partial x$ and integrating with respect to x,
we have

$$p = \frac{6\mu U}{\alpha(b_1 - \alpha x)} - \frac{12\mu q}{2\alpha(b_1 - \alpha x)^2} + f(y).$$

The spacing of the boundaries is actually very small compared with L,
and the vertical scale in Fig. 3 has been exaggerated.   Thus the hydrostatic

variation of p can be ignored, $f(y)$ can be treated as a constant C, and q and C can be determined from the conditions

$$p = p_0 \quad \text{at } x = 0 \text{ and } x = L.$$

In this way we find

$$f(y) = C = p_0 - \frac{6\mu U}{\alpha b_1} + \frac{12\mu q}{2\alpha b_1{}^2}, \qquad q = \frac{Ub_1 b_2}{b_1 + b_2},$$

and

$$p = p_0 + \frac{6\mu Ux(L - x)(b_1 - b_2)}{Lb^2(b_1 + b_2)}. \tag{23}$$

The pressure distribution is shown in Fig. 3. Integration of (23) gives the total force per unit distance normal to the plane of flow

$$P = \frac{6\mu UL^2}{b_2{}^2(c - 1)^2}\left(\ln c - 2\,\frac{c - 1}{c + 1}\right), \qquad c = \frac{b_1}{b_2}. \tag{24}$$

For P to be positive, c must be greater than 1; that is, the upper boundary must move toward the narrow end of the spacing. The maximum value of P, as c varies, is

$$P_{\max} = \frac{0.4\mu UL^2}{b_m{}^2}, \qquad b_m = \frac{b_1 + b_2}{2}, \tag{25}$$

corresponding to $c = 2.2$. For a given P, a given surface geometry, and a certain fluid, U has a minimum below which the fluid is unable to support the load without contact of the solid boundaries.

2.4. Effect of Gravity; Parallel Flow Down an Inclined Plane

If the plates bounding the flow in Secs. 2.1 and 2.2 are parallel but not horizontal, the solution follows the line of development of these sections except that if the x direction is still the direction of flow, $\partial p/\partial x$ should be replaced by

$$\frac{\partial p}{\partial x} + \rho\,\frac{\partial \Omega}{\partial x} = \frac{\partial p}{\partial x} - \rho g \sin \beta \tag{26}$$

and (9) by

$$\frac{\partial p}{\partial y} = -\rho g \cos \beta, \tag{27}$$

in which β is the angle of inclination of the x axis (see Fig. 4).

If the density ρ is constant, the solution corresponding to (18) is

$$u = \frac{Uy}{b} + \frac{K + \rho g \sin \beta}{2\mu}\,y(b - y), \tag{28}$$

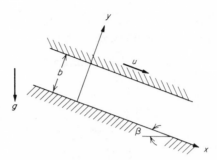

FIGURE 4. DEFINITION SKETCH FOR
INCLINED FLOW.

in which U is the velocity of the upper plate. It is convenient to write

$$p_d = p + \rho\Omega, \qquad (29)$$

where p_d is that part of the pressure p which is due to fluid motion alone, apart from a possible but inconsequential additive constant. In terms of p_d, (28) can be written simply as

$$u = \frac{Uy}{b} - \frac{1}{2\mu}\frac{dp_d}{dx}y(b - y). \qquad (30)$$

Note that (27) states that if ρ is constant, p_d is independent of y.

If ρ is not constant but a function of y, (12) is replaced by

$$\mu\frac{d^2u}{dy^2} = \frac{\partial p}{\partial x} - \rho g \sin \beta = -K - \rho g \sin \beta, \qquad (31)$$

integration of which yields

$$\mu u = -\int_0^y \int_0^{y'} \rho(y'')g \sin \beta \, dy'' \, dy' - \frac{K}{2}y^2 + C_1 y + C_2. \qquad (32)$$

The constant C_2 is zero if the lower plate is stationary, and the constant C_1 is determined by the condition at the upper boundary.

If ρ is constant but the upper surface is free, K is zero, and (Fig. 5)

$$\frac{du}{dy} = 0 \qquad \text{at } y = b,$$

which states that the shear stress at the free surface is zero. Integration of (31) for this case gives

$$u = \frac{\rho g \sin \beta}{2\mu} y(2b - y). \qquad (33)$$

If the origin of y is at the free surface, the velocity distribution is given by

$$u = \frac{\rho g \sin \beta}{2\mu} (b^2 - y^2). \qquad (34)$$

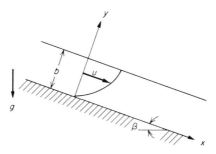

FIGURE 5.   FREE-SURFACE FLOW DOWN AN
INCLINED PLANE.

2.5.   Effect of Viscosity Variation

To illustrate the effect of viscosity variation, we consider a layer of fluid
of depth b flowing in the x direction along an infinite plate (Fig. 4).   The
upper surface may be rigid, as shown in Fig. 4, or free, as shown in Fig. 5.
Both the density ρ and the viscosity μ are assumed variable.

Since the viscosity is variable, (2.6) must be used instead of Eqs. (1)
to (3).   Following exactly the arguments in Sec. 2.1, we conclude that
both ρ and μ are functions of y alone and $\partial p/\partial x$ is a constant, which we
denote by $-K$.   The equation governing the flow is then

$$\frac{d}{dy}\left(\mu \frac{du}{dy}\right) = -K - \rho g \sin \beta, \tag{35}$$

which corresponds to (31) and is the first of the equations in (2.6).   Two
integrations of (35) give

$$u = -\int_0^y \left\{\frac{1}{\mu(y')} \int_0^{y'} [K + \rho(y'')g \sin \beta]\, dy''\right\} dy' + Cy. \tag{36}$$

The condition $u = 0$ at $y = 0$ is satisfied by (36), in which the constant C
can be determined to satisfy the upper boundary condition, whether the
upper boundary is rigid or free.

2.6.   Steady Unidirectional Flow through a
Rectangular Conduit

Steady flow through a rectangular conduit can be found by the method of
separation of variables.   The flow is again assumed to be in the direction
of increasing x, and the cross section of the conduit is shown in Fig. 6.
The angle of inclination between the x axis and the horizontal plane is
again denoted by β.   The governing equation, obtained from (1), is then

$$\mu\left(\frac{\partial^2}{\partial y^2} + \frac{\partial^2}{\partial z^2}\right)u = -K - \rho g \sin \beta, \tag{37}$$

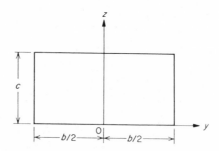

in which K is $-\partial p/\partial x$ and is constant but ρ may be a function of z. The solution of (37) can be written as

$$u = u_c + u_P, \tag{38}$$

where the particular solution

$$u_P(z) = -\frac{1}{\mu}\int_0^z\int_0^{z'}\rho(z'')g\sin\beta\,dz''\,dz' - \frac{K}{2\mu}z^2 + C_1 z + C_2 \tag{39}$$

satisfies

$$\mu\left(\frac{\partial^2}{\partial y^2} + \frac{\partial^2}{\partial z^2}\right)u_P = -K - \rho g\sin\beta \tag{40}$$

and the complementary solution u_c satisfies

$$\mu\left(\frac{\partial^2}{\partial y^2} + \frac{\partial^2}{\partial z^2}\right)u_c = 0. \tag{41}$$

The C_1 and C_2 in (39) can be determined to satisfy the conditions

$$u_P = 0 \qquad \text{at } z = 0 \text{ and } z = c.$$

Since u and u_P are both symmetric with respect to y, the solution of (41) by separation of the variables y and z is of the form

$$u_c = \sum_{n=1}^{\infty} A_n \cosh\frac{n\pi y}{c}\sin\frac{n\pi z}{c}. \tag{42}$$

This leaves the nonslip condition at $z = 0$ and c satisfied, since both u_P and u_c vanish at these two boundaries. But the nonslip conditions at $y = \pm b/2$ still has to be satisfied. Since u_c is symmetric with respect to y, these conditions reduce to the single one

$$\sum_{n=1}^{\infty} A_n \cosh\frac{n\pi b}{2c}\sin\frac{n\pi z}{c} = -u_P(z). \tag{43}$$

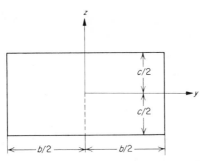

FIGURE 7. COORDINATES FOR A RECTAN-
GULAR CROSS SECTION, FLOW SYMMETRIC
WITH RESPECT TO BOTH THE y AXIS AND
THE z AXIS.

The left-hand side is a Fourier series, and the A's can be determined by the formula

$$A_n \cosh \frac{n\pi b}{2c} = -\frac{2}{c} \int_0^c u_P(z) \sin \frac{n\pi z}{c} \, dz, \tag{44}$$

because

$$\int_0^c \sin \frac{n\pi z}{c} \sin \frac{m\pi z}{c} \, dz = \frac{c}{2} \delta_{mn}, \tag{45}$$

where the Kronecker delta is

$$\delta_{mn} = \begin{cases} 0 & \text{if } m \neq n, \\ 1 & \text{if } m = n. \end{cases} \tag{46}$$

If ρ is constant, it is more convenient to use the coordinates shown in Fig. 7. Then

$$u_P(z) = \frac{1}{2\mu} (K + \rho g \sin \beta) \left(\frac{c^2}{4} - z^2 \right), \tag{47}$$

which is symmetric with respect to z, and

$$u_c(z) = \sum_{n=1}^{\infty} A_n \cosh \frac{(2n-1)\pi y}{2c} \cos \frac{(2n-1)\pi z}{2c}, \tag{48}$$

which is symmetric with respect to both y and z. The A's in (48) are determined by

$$A_n \cosh \frac{(2n-1)\pi b}{4c} = \frac{2c^2}{\mu \alpha_n^3} (K + \rho g \sin \beta) \left(\alpha_n \cos \frac{\alpha_n}{2} - 2 \sin \frac{\alpha_n}{2} \right), \tag{49}$$

with

$$\alpha_n = \frac{(2n-1)\pi}{2}.$$

The velocity contours are shown in Fig. 8 for $b = 2c$.

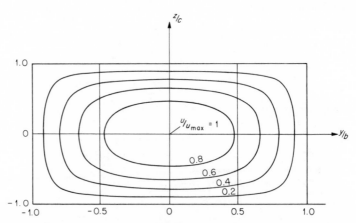

FIGURE 8.   VELOCITY DISTRIBUTION FOR LAMINAR UNIDIRECTIONAL FLOW
IN A RECTANGULAR CONDUIT WITH WIDTH-TO-HEIGHT RATIO EQUAL TO 2.
[*By permission from Hunter Rouse (ed.), "Advanced Fluid Mechanics,"
p. 220, John Wiley & Sons, Inc., New York, 1959.*]

The case of variable density and viscosity is reserved for the problems,
along with other cases of rectilinear flow with straightforward solutions.
Before going on to discuss flow in a circular pipe or an annulus, it is
appropriate to present a few analogies to laminar flow in conduits.

2.7.   The Vorticity-flow Analogy

Unidirectional flow in a conduit of any cross section is governed by (37),
which is Poisson's equation.   Therefore any other phenomenon governed
by the Poisson equation is analogous to unidirectional flow in a conduit.

The first analogy is the steady two-dimensional flow of an inviscid
fluid of constant density, within the same conduit and with constant
vorticity.   Such a flow is governed by (3.14a), except that the $f(\psi)$ there
should be replaced by $-\xi_0$, ξ_0 being the constant vorticity of the fluid;
i.e., it is governed by

$$\frac{\partial^2\psi}{\partial y^2} + \frac{\partial^2\psi}{\partial z^2} = -\xi_0 \qquad (50)$$

after we have replaced x and y by y and z.   Since (50) is the Poisson
equation, ψ is analogous to u in (37) and the velocity contours in Fig. 8
correspond to the streamlines of the flow with constant vorticity governed
by (50).

The vorticity flow just mentioned can be produced in the following
way.   Fill the conduit with an incompressible fluid, say water, and rotate
the conduit about its longitudinal axis.   If the fluid were really inviscid,
the motion of the fluid would always remain irrotational; but except for

helium at very low temperatures, all fluids have some viscosity. After a certain time the fluid will be in solid-body rotation, with angular speed $\omega_0 = \xi_0/2$. If the rotation of the conduit is then stopped, the fluid will eventually come to rest. But if the Reynolds number $\omega_0 L^2/\nu$ ($L = a$ transverse linear dimension of the conduit) is large, then aside from the region near the boundary the core flow will be governed by (50) until the core is destroyed by the viscous boundary layer growing toward the center. Initially the boundary layer is infinitely thin, and the boundary of the region of the flow governed by (50) is exactly the conduit boundary.

2.8. The Soap-bubble Analogy

Suppose that the solid walls of the conduit for the flow governed by (37) are infinitely thin, and suppose this conduit is cut at some value of x to produce a (hollow) cross section. If a soap film is spread over the cross section and the pressure inside the conduit is increased by an amount δp over the pressure outside, then the displacement ξ of the film from its original flat surface is governed by

$$\frac{\partial^2 \xi}{\partial y^2} + \frac{\partial^2 \xi}{\partial z^2} = -\frac{\delta p}{2T}, \tag{51}$$

in which T is the surface tension, provided the film is shallow, i.e., provided δp is not too large. The factor 2 in (51) is needed to account for the *two* surfaces of the soap film. Equation (51) is again the Poisson equation, so that ξ is analogous to u in (37). Thus, by measuring the height of a film at various locations we can determine the distribution of u in rectilinear flow of a viscous fluid in a conduit. This analogy was first noted by Prandtl.

2.9. The Torsion Analogy

For a long solid prism under torsion there exists a stress function satisfying the Poisson equation. (We shall refrain from deriving it here.) This stress function corresponds to u in (37), ψ in (50), and ξ in (51). The shear stress at any point in the cross section (not hollow) is given by the gradient of this stress function.

2.10. Poiseuille Flow

For unidirectional flow in a circular tube the governing equation is (12). The arguments concerning the constancy of $\partial p/\partial z$ are similar to those presented in Sec. 2.1. If the flow is steady and both ρ and μ are constant, the distribution of the longitudinal-velocity component w is governed by

$$\mu\left(\frac{d^2 w}{dr^2} + \frac{1}{r}\frac{dw}{dr}\right) = -K - g\rho \sin \beta \tag{52}$$

if the angle of inclination of the axis of the pipe to the horizontal is β and $\partial p/\partial z$ is denoted by $-K$. The only comment necessary concerning the integration of (52) is that r is the integrating factor needed to carry out the first integration, which gives

$$\mu r \frac{dw}{dr} = -(K + g\rho \sin \beta)\frac{r^2}{2} + C. \tag{53}$$

It is evident that C is zero since dw/dr is not infinite at $r = 0$. Dropping C in (53), dividing by r, and integrating again, we obtain

$$w = \frac{1}{4\mu}(K + g\rho \sin \beta)(r_0^2 - r^2), \tag{54}$$

in which r_0 is the inner radius of the pipe. The condition that w must vanish at $r = r_0$ is satisfied by (54). The flow described by (54) is called the *Poiseuille flow* or the *Hagen-Poiseuille flow*.

2.11.   Couette Flow

Consider a fluid with constant density and viscosity between two very long concentric circular cylinders of radius R_1 and R_2 ($>R_1$), set in motion by the outer cylinder's rotating about the common axis with angular velocity Ω_2 and the inner one's rotating with angular velocity Ω_1. For definiteness, the angular velocity is counted positive when the rotation is in the direction of increasing φ, which is one of the cylindrical coordinates r, φ, and z. The axes of the cylinders are assumed vertical for the sake of simplicity, although this assumption is not at all necessary.

Since the cylinders are very long, we shall ignore end effects and assume all flow variables to be independent of both φ and z. The appropriate Navier-Stokes equation to use is (11). Assuming no axial flow, so that $w = 0$, we can demonstrate from (13) that, since $v = v(r)$, u must also vanish. Equation (11) is then

$$\frac{d^2v}{dr^2} + \frac{1}{r}\frac{dv}{dr} - \frac{v}{r^2} = 0. \tag{55}$$

A glance at (55) shows that the solution is of the form r^n. The n is determined by (55) to be ± 1. Hence

$$v = Ar + \frac{B}{r}. \tag{56}$$

The boundary conditions are

$$v = \Omega_1 R_1 \text{ at } r = R_1 \qquad \text{and} \qquad v = \Omega_2 R_2 \text{ at } r = R_2, \tag{57}$$

which determine A and B to be

$$A = \frac{\Omega_2 R_2^2 - \Omega_1 R_1^2}{R_2^2 - R_1^2}, \qquad B = \frac{(\Omega_1 - \Omega_2)R_1^2 R_2^2}{R_2^2 - R_1^2}. \tag{58}$$

The flow described by (56) is called the *Couette flow*. If the circulations at the two cylinders are equal, $A = 0$, and (56) shows that the velocity distribution is that of a vortex concentrated at the axis. The flow is irrotational in the annulus. This illustrates the possibility of irrotational flow of a viscous fluid, provided the boundary obliges by moving with the required velocity. [Later, when we consider stability, we shall be forced to change the notation somewhat. Thus, in (9.83), V is written for v to emphasize that it is the velocity for the mean flow, and A_1 and B_1 will be written for A and B here.]

For Couette flow, (10) can be written as

$$\frac{v^2}{r} = \frac{1}{\rho}\frac{\partial p}{\partial r}, \tag{59}$$

which can be used to determine p after v is known.

Evidently the Poiseuille flow can be superposed on the Couette flow, since the governing equations are linear and uncoupled, so long as there is no dependence on z and φ.

2.12. Ekman Flow

We conclude the presentation of linear cases of steady flows with a discussion of a once curious but now commonly known flow, the *Ekman flow* [Ekman (1905)]. Ekman is said to have discovered this flow at sea on noting the discrepancy in movement between the sea current and floating pieces of ice.

The flow of a fluid of constant density and viscosity in a frame of reference rotating with angular velocity ω is governed by Eqs. (2.62) to (2.65) except that viscous terms should be added, as explained in Sec. 7 of Chap. 2. If the acceleration with respect to the rotating frame is zero, i.e., if only the Coriolis acceleration exists, the governing equations are, in the notation of Sec. 7, Chap. 2,

$$-2\omega v = -\frac{\partial P}{\partial x} + \nu\,\nabla^2 u, \tag{60}$$

$$2\omega u = -\frac{\partial P}{\partial y} + \nu\,\nabla^2 v, \tag{61}$$

$$0 = -\frac{\partial P}{\partial z} + \nu\,\nabla^2 w. \tag{62}$$

If w is zero, P is independent of z. If P is also independent of x and y, (60) and (61) become

$$-2\omega v = \nu\,\nabla^2 u, \tag{63}$$

$$2\omega u = \nu\,\nabla^2 v. \tag{64}$$

Ekman considered the case in which u and v are functions only of z. In that case the flow is governed by

$$-2\omega v = \nu \frac{d^2 u}{dz^2},$$ (65)

$$2\omega u = \nu \frac{d^2 v}{dz^2}.$$ (66)

The flow described by (65) and (66) is called Ekman flow.

Multiplication of (66) by i and addition of the result to (65) produce

$$2\omega i(u + iv) = \nu \frac{d^2}{dz^2}(u + iv),$$

the solution of which is

$$u + iv = (U + iV)\exp\left[\left(\frac{\omega}{\nu}\right)^{1/2}(1 + i)z\right]$$ (67)

if the velocity vanishes at $z = -\infty$ and $u = U$ and $v = V$ at $z = 0$. Equation (67) shows that for a given z the velocity is constant and that the velocity vector attentuates and rotates as z decreases from zero toward $-\infty$.

It should be noted that in the Ekman flow the inertial force is represented by the Coriolis force, and this force is balanced by purely viscous forces.

3. UNSTEADY FLOWS FOR WHICH THE NAVIER-STOKES EQUATIONS ARE LINEAR

Most of the flows discussed in Sec. 2 continue to possess straightforward solutions even if the cause of the flow is time-dependent. We shall discuss only a few representative cases. Others can be found in the problems, and one more will be given in Sec. 11.2 of Chap. 9. The approach can be applied to any unsteady flow so long as the Navier-Stokes equation governing that flow happens to be linear. In the following examples of unsteady flow both ρ and μ are assumed constant.

3.1. Impulsively Started Couette Flow

Unsteady Couette flow is governed by (11) or

$$\frac{\partial v}{\partial t} = \nu\left(\frac{\partial^2 v}{\partial r^2} + \frac{1}{r}\frac{\partial v}{\partial r} - \frac{v}{r^2}\right),$$ (68)

After v is determined, (59) can again be used to determine p. In terms of the new variables

$$\tau = \frac{t\nu}{R_1^2} \quad \text{and} \quad \xi = \frac{r}{R_1},$$

in which R_1 is the radius of the inner cylinder, (68) becomes

$$\frac{\partial v}{\partial \tau} = \left(\frac{\partial^2}{\partial \xi^2} + \frac{1}{\xi} \frac{\partial}{\partial \xi} - \frac{1}{\xi^2} \right) v. \tag{68a}$$

The velocity v can be resolved into a steady part v_s and an unsteady part v_u:

$$v = v_s + v_u, \tag{69}$$

in which v_s is given by (56). If the outer cylinder is stationary, the inner cylinder rotates with angular velocity ω for $t > 0$, and if R_2/R_1 is denoted by α,

$$v_s = -\frac{\omega R_1 \xi}{\alpha^2 - 1} + \frac{\omega R_1 \alpha^2}{\alpha^2 - 1} \frac{1}{\xi}. \tag{70}$$

The unsteady part v_u must satisfy (68a) and the boundary conditions

(i) $v_u \to 0$ as $\tau \to \infty$,

(ii) $v_u = -v_s$ at $\tau = 0$,

(iii) $v_u = 0$ at $\xi = 1$ and $\xi = \alpha$

if the motion is started from rest.

The solution of (68a), with v_u replacing v, by the method of separation of variables is

$$v_u = \sum_{n=1}^{\infty} A_n e^{-\lambda_n^2 \tau} Z_1(\lambda_n \xi), \tag{71}$$

in which $Z_1(\lambda_n \xi) = J_1(\lambda_n \xi) + \beta_n N_1(\lambda_n \xi), \tag{72}$

J_1 and N_1 being the Bessel function and Neumann function of the first order, respectively, and λ_n and β_n being the nth pair of the roots of

$$Z_1(\lambda) = 0 \quad \text{and} \quad Z_1(\alpha\lambda) = 0. \tag{73}$$

For $\alpha = 2$, the first six roots of λ are given below:

n	1	2	3	4	5	6
λ_n	3.1966	6.3124	9.4445	12.5812	15.7199	18.8595
β_n	−0.7100	−0.8380	−0.8884	−0.9147	−0.930	−0.96

It can be seen that the satisfaction of boundary condition (iii) is guaranteed by (73) and that of (i) guaranteed by the reality of the λ's. Only (ii) still needs to be satisfied. It demands that

$$\sum_{n=1}^{\infty} A_n Z_1(\lambda_n \xi) = -v_s. \tag{74}$$

Because of the orthogonality relation

$$\int_1^{\alpha} \xi Z_1(\lambda_n \xi) Z_1(\lambda_m \xi) \, d\xi = 0, \quad \text{for } m \neq n,$$

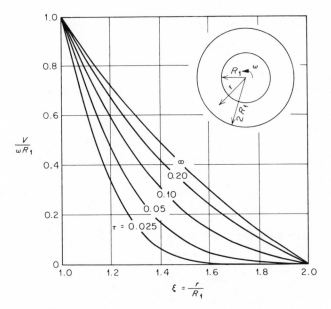

FIGURE 9.   ESTABLISHMENT OF FLOW IN AN ANNULAR CY-
LINDRICAL SPACE AFTER IMPULSIVE START OF THE ROTATION
OF THE INNER CYLINDER.

(74) gives

$$A_n = -\frac{\int_1^\alpha \xi v_s Z_1(\lambda_n \xi)\, d\xi}{\int_1^\alpha \xi Z_1{}^2(\lambda_n \xi)\, d\xi} = -\frac{2\omega R_1 Z_0(\lambda_n)}{\lambda_n[\alpha^2 Z_0{}^2(\alpha\lambda_n) - Z_0{}^2(\lambda_n)]}.$$

The velocity distribution given by (69) to (71) for various values of τ is
shown in Fig. 9.   The moments M_1 and M_2 on the inner and outer
cylinders are shown in Fig. 10.   Note in Fig. 9 that the fluid is essentially
at rest for small t except very near the moving cylinder and that the
motion is propagated by viscosity from the moving toward the stationary
cylinder.

3.2.   Unsteady Longitudinal Flow in the Space between Concentric Cylinders

Suppose that the fluid in the annulus region described in Secs. 2.11 and
3.1 is set in motion by a longitudinal pressure gradient $-K$ applied
abruptly at $t = 0$.   The governing equation is (12), which now assumes
the form

$$\frac{\partial w}{\partial t} = \frac{K}{\rho} + \nu\left(\frac{\partial^2 w}{\partial r^2} + \frac{1}{r}\frac{\partial w}{\partial r}\right). \tag{75}$$

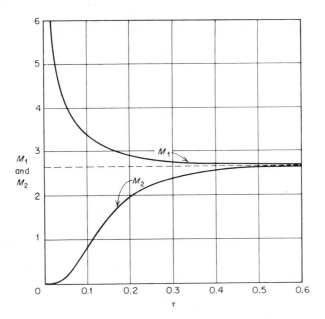

FIGURE 10. MOMENTS ON THE INNER CYLINDER (M_1) AND OUTER CYLINDER (M_2) AFTER IMPULSIVE START OF THE ROTATION OF THE INNER CYLINDER.

With τ and ξ defined as in Sec. 3.1 and

$$W = \frac{\mu w}{K R_1{}^2},$$

(75) can be written as

$$\frac{\partial W}{\partial \tau} = \frac{\partial^2 W}{\partial \xi^2} + \frac{1}{\xi} \frac{\partial W}{\partial \xi} + 1. \tag{76}$$

Employing the same approach as in Sec. 3.1, we obtain

$$W = W_s + \sum_{n=1}^{\infty} A_n Z_0(\lambda_n \xi) e^{-\gamma_n{}^2 \tau}, \tag{77}$$

in which W_s, the steady part of W, can be found by solving (76) with the left-hand side set equal to zero and is given by

$$W_s = \tfrac{1}{4}\left(-\xi^2 + \frac{\alpha^2 - 1}{\ln \alpha} \ln \xi + 1\right).$$

The A's in (77) are given by

$$A_n = \frac{2}{\lambda_n{}^3 [\alpha Z_1(\alpha \lambda_n) + Z_1(\lambda_n)]}, \tag{78}$$

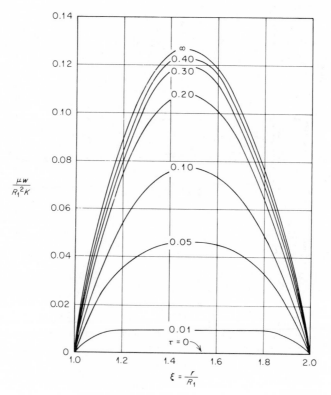

FIGURE 11. ESTABLISHMENT OF AXIAL FLOW IN AN ANNULAR
CYLINDRICAL SPACE AFTER A PRESSURE GRADIENT IS SUDDENLY
APPLIED AND THEN MAINTAINED.

in which $Z_1 = J_1 + \beta_n N_1$, and the Z_0 in (77) is defined by

$$Z_0(\lambda_n \xi) = J_0(\lambda_n \xi) + \beta_n N_0(\lambda_n \xi),$$

the λ_n and β_n being the nth pair of roots of

$$Z_0(\lambda_n) = 0 \quad \text{and} \quad Z_0(\alpha \lambda_n) = 0.$$

For $\alpha = 2$, the first six λ's and β's are given below:

n	1	2	3	4	5	6
λ_n	3.1230	6.2734	9.4182	12.5614	15.7040	18.8462
β_n	0.8922	0.9430	0.9610	0.9703	0.9764	1.0000

The distribution of W for $\alpha = 2$ and various values of τ is shown in Fig.
11. Note that for small τ there is an irrotational core away from the
boundaries. Vorticity is propagated into this core from the boundaries
by viscosity.

3.3.   Duhamel's Principle

In the problem treated in Sec. 3.1, if the angular velocity of the inner cylinder is not constant but equal to $\omega f(\tau)$, ω being a constant, then if the v given by (69) to (71) is denoted by v_1 for distinction from the actual v, we have

$$v = \int_0^\tau v_1(\tau - \tau_1) \frac{df(\tau_1)}{d\tau_1} \, d\tau_1 + f(0)v_1. \tag{79}$$

That (79) satisfies (68) and the boundary condition for v at $r = R_2$ is obvious, because v_1 satisfies them.   Since $v_1 = \omega R_1$ at $r = R_1$, from (79) we have

$$v = \omega R_1[f(\tau) - f(0)] + \omega R_1 f(0) = \omega f(\tau)R_1$$

at $r = R_1$, so that the boundary condition at the inner cylinder is satisfied.

It is revealing to derive (79) heuristically.   The area under the curve of the function $f(\tau)$ can be considered to be built of semi-infinite strips. To the first strip, of width $f(0)$, are added strips of width $df(\tau_1)$ at various values of τ_1.   The strip of width $f(0)$ corresponds to $f(0)v_1$ in (79), and the added strips correspond to the integral in (79).   Note that in the integrand the argument of v_1 is $\tau - \tau_1$, meaning that the elemental solution $v(\tau - \tau_1) \, df(\tau_1)$ corresponds to an impulsive start of the velocity $\omega R_1 \, df(\tau_1)$ at time τ_1.   These elemental solutions are added to form the integral in (79).   Note also that $df(\tau_1)$ can be negative at some τ_1.   The method of superposition represented by (79) is a contribution of Duhamel to the theory of heat conduction in solids [see Carslaw and Jaeger (1959, p. 31)].

Similarly, if the pressure gradient in Sec. 3.2 is $-Kf(\tau)$ instead of K, then denoting the W given by (77) by W_1 to distinguish it from the actual W, we have

$$W = \frac{\mu w}{R_1{}^2} = \int_0^\tau W_1(\tau - \tau_1) \frac{df(\tau_1)}{d\tau_1} \, d\tau_1 + f(0)W_1. \tag{80}$$

3.4.   Unsteady Flow of a Semi-infinite Fluid Caused by a Plate Moving in Its Own Plane

We shall consider two representative cases, impulsive start and periodic motion of the plate.   The coordinates are shown in Fig. 12, which also shows the velocity distribution for the latter case.   The origin of x is, of course, immaterial.   The equation of motion is

$$\frac{\partial u}{\partial t} = \nu \frac{\partial^2 u}{\partial y^2}, \tag{81}$$

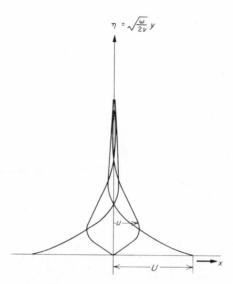

$$\eta = \sqrt{\frac{\omega}{2\nu}}\, y$$

FIGURE 12. VELOCITY DISTRIBUTION IN A FLUID SET IN MOTION BY AN OSCILLATING PLATE.

if the plate moves in the x direction. In the case of impulsive start,

$$u = \begin{cases} 0 & \text{for } t < 0 \\ U & \text{for } t \geq 0 \qquad \text{at } y = 0, \end{cases} \tag{82}$$

and $\qquad\qquad u = 0 \qquad$ for all finite t at $y = \infty$. $\tag{83}$

Examination of (81) to (83) shows that only the kinematic quantities u, U, t, ν, and y are involved. From these five variables three dimensionless parameters can be formed:

$$\frac{u}{U}, \qquad \frac{Uy}{\nu}, \qquad \text{and} \qquad \frac{U^2 t}{\nu}.$$

Thus $\qquad\qquad\qquad \dfrac{u}{U} = F\left(\dfrac{Uy}{\nu}, \dfrac{U^2 t}{\nu}\right). \tag{84}$

But this relationship is needlessly general for the problem at hand, because closer examination of (81) to (83) reveals that if y is magnified α times and t magnified α^2 times, (81) will still be satisfied, and so will (82) and (83). Thus clearly it is only on the ratio of $(Uy/\nu)^2$ and $U^2 t/\nu$ that u/U depends. That is to say,

$$\frac{u}{U} = F\left(\frac{y^2}{\nu t}\right),$$

or
$$\frac{u}{U} = f(\eta), \qquad \eta = \frac{y}{\sqrt{\nu t}}. \tag{85}$$

Once (85) is reached, the rest is simple. Equation (81) becomes

$$-\tfrac{1}{2}\eta f' = f'',$$

in which the primes indicate differentiation with respect to η. Two integrations yield

$$f = C_1 \int_0^\eta e^{-\eta^2/4} \, d\eta + C_2. \tag{86}$$

The boundary condition at $y = 0$ demands that $C_2 = 1$, and (83) demands that

$$C_1 \int_0^\infty e^{-\eta^2/4} \, d\eta + 1 = 0 \qquad \text{or} \qquad C_1 = -\frac{1}{\sqrt{\pi}}.$$

The flow described by (85) and (86) was studied by Stokes (1851), but the problem has been called *Rayleigh's problem* in the literature. The most important thing to observe in (85) and (86) is that for a fixed value of u/U, η is constant, or y varies as $\sqrt{\nu t}$. In this sense a signal propagates with the speed $\dfrac{1}{2}\sqrt{\dfrac{\nu}{t}}$, which is very small for an assigned finite t if ν is small.

If the motion of the plate is periodic, (81) and (83) stand, but (82) is replaced by

$$u = U \cos \omega t \qquad \text{at } y = 0. \tag{87}$$

The form of (87) suggests that we can assume

$$u = U e^{i\omega t} g(y), \tag{88}$$

it being understood that the real part of the right-hand side of (88) is to be used. Substituting (88) into (81), we have

$$i\omega g = \nu g'', \tag{89}$$

the primes denoting differentiations with respect to y. The solutions of (89) are

$$\exp\left\{ \pm \left[\sqrt{\frac{\omega}{2\nu}} (1 + i) y \right] \right\}. \tag{90}$$

Clearly (83) demands the negative sign, and (87) is satisfied by (88), with $g(\eta)$ given by (90), the negative sign being used. Hence

$$u = U \exp\left(-\sqrt{\frac{\omega}{2\nu}} y \right) \cos\left(\omega t - \sqrt{\frac{\omega}{2\nu}} y \right). \tag{91}$$

The velocity attentuates and oscillates with y, and for each y it oscillates with t. Figure 12 describes u in graphs. The depth of penetration of vorticity is $\delta \sim \sqrt{2\nu/\omega}$, which decreases as ω increases.

For convenience in this paragraph we shall denote the u given by (85) and (86) by u_1 and the u given by (91) by u_2. Given any specified motion $u = Uf(\tau)$ of the plate, we can always find the solution for u from u_1 by using the Duhamel principle. However, if $f(\tau)$ is periodic (not necessarily simple harmonic), after some time the initial condition does not matter, even if the motion has been started from rest. In that case building u from u_1 by the Duhamel principle, while correct, is cumbersome and inappropriate, since the procedure carries the undue and needless burden of history. The appropriate method is to resolve $f(\tau)$ into Fourier components, each corresponding to an elemental solution of the form of u_2, and to sum these solutions to give u. If $f(\tau)$ is not periodic but vanishes sufficiently fast for large τ, then u can be built either from u_1 by using the Duhamel principle or from u_2 by using a Fourier integral. If $f(\tau)$ does not vanish as $\tau \to \infty$ or does not vanish fast enough, use of u_1 in conjunction with the Duhamel principle is the only recourse.

3.5. Simple-harmonic Ekman Flow

In the example presented in Sec. 2.12, if the surface velocity is not constant $(U + iV)$ but equal to $U \cos \sigma t$, we can still assume P to be constant, and (60) and (61) become

$$\frac{\partial u}{\partial t} - 2\omega v = \nu \frac{\partial^2 u}{\partial z^2}, \tag{92}$$

$$\frac{\partial v}{\partial t} + 2\omega u = \nu \frac{\partial^2 v}{\partial z^2}, \tag{93}$$

in which, as before, ω is the angular velocity of the rotating fluid and frame of reference. Elimination of v between (92) and (93) produces

$$\nu^2 \frac{\partial^4 u}{\partial z^4} - 2\nu \frac{\partial^3 u}{\partial t \, \partial^2 z} + \frac{\partial^2 u}{\partial t^2} + 4\omega^2 u = 0. \tag{94}$$

The nature of the surface motion and of (94) allows u to be written

$$u = F(z)e^{i\sigma t}. \tag{95}$$

The right-hand side of (95) is understood to mean its real part only. Then (94) becomes

$$\nu^2 F^{iv} - i2\nu\sigma F'' - (\sigma^2 - 4\omega^2)F = 0. \tag{96}$$

With

$$\zeta = \left(\frac{\sigma}{\nu}\right)^{1/2}$$

(96) becomes

$$F^{iv} - i2F'' - \left(1 - \frac{4\omega^2}{\sigma^2}\right) = 0, \tag{97}$$

in which the primes indicate differentiation with respect to ζ. The boundary conditions are

(i) $F \to 0$ as $\zeta \to -\infty$,

(ii) $F(0) = U$,

(iii) $v = 0$,   at $\zeta = 0$.

The solution of (97) that satisfies (i) is

$$F = Ae^{\alpha\zeta} + Be^{\beta\zeta}, \tag{98}$$

where (the real part of α and β being positive)

$$(\alpha,\beta) = \left(i \pm \frac{i2\omega}{\sigma}\right)^{\frac{1}{2}}.$$

Boundary conditions (ii) and (iii) demand, respectively,

$$A + B = U \quad \text{and} \quad A - B = 0.$$

Hence $$A = B = \frac{U}{2}. \tag{99}$$

The solution for u is then given by (95), (98), and (99), and the solution for v can then be obtained easily from (92).

4. EXACT SOLUTIONS OF THE NAVIER-STOKES EQUATIONS FOR STEADY FLOWS: NONLINEAR CASES

To save space, the known exact solutions of the Navier-Stokes equations, when these are nonlinear, are described in Table 1. For all these cases the governing equations can be reduced to an ordinary differential equation, and the solutions, as will be explained later, are called *similarity solutions*.

4.1. The Round Laminar Jet Due to a Concentrated Force

In discussing exact solutions of the Navier-Stokes equations when they are nonlinear, the place of honor must be given to the beautiful solution of Squire (1951) for the round laminar jet[1] due to a point momentum source, or a concentrated force. Since the flow is symmetric with respect to the axis coinciding in direction with the force, the equation of continuity, as can be obtained on setting (A2.70) equal to zero, is in this case

$$\frac{1}{R^2} \frac{\partial}{\partial R} R^2 u + \frac{1}{R \sin\theta} \frac{\partial}{\partial\theta} (v \sin\theta) = 0. \tag{100}$$

[1] L. D. Landau "*Fluid Mechanics*," Pergamon Press, New York, 1959, mentioned that he had solved the problem in 1943—but gave no reference in detail.

The equations of motion are

$$u\frac{\partial u}{\partial R} + \frac{v}{R}\frac{\partial u}{\partial \theta} - \frac{v^2}{R} = -\frac{1}{\rho}\frac{\partial p}{\partial R} + \nu\left(\nabla^2 u - \frac{2u}{R^2} - \frac{2}{R^2}\frac{\partial v}{\partial \theta} - \frac{2v\cot\theta}{R^2}\right),$$

(101)

$$u\frac{\partial v}{\partial R} + \frac{v}{R}\frac{\partial v}{\partial \theta} + \frac{uv}{R} = -\frac{1}{\rho R}\frac{\partial p}{\partial \theta} + \nu\left(\nabla^2 v + \frac{2}{R^2}\frac{\partial u}{\partial \theta} - \frac{v}{R^2\sin^2\theta}\right),$$ (102)

in which $\quad \nabla^2 = \dfrac{1}{R^2}\dfrac{\partial}{\partial R}\left(R^2\dfrac{\partial}{\partial R}\right) + \dfrac{1}{R^2\sin\theta}\dfrac{\partial}{\partial \theta}\left(\sin\theta\dfrac{\partial}{\partial \theta}\right).$ (103)

Inspection of (100) to (102) reveals that if u and v are proportional to $1/R$ and, apart from an additive constant, p to $1/R^2$, it is possible to eliminate R from the governing equations. Assume that

$$\psi = \nu R f(\theta),$$ (104)

so that, according to (1.34) or directly from (100),

$$u = \frac{1}{R^2\sin\theta}\frac{\partial\psi}{\partial\theta} = \frac{\nu}{R\sin\theta}f'(\theta),$$

$$v = -\frac{1}{R\sin\theta}\frac{\partial\psi}{\partial R} = -\frac{\nu}{R\sin\theta}f(\theta).$$

(105)

We shall now endeavor to simplify (101) and (102). First,

$$\nabla^2 u = \frac{1}{R^2\sin\theta}\frac{\partial}{\partial\theta}\left(\sin\theta\frac{\partial u}{\partial\theta}\right),$$

$$\nabla^2 v = \frac{1}{R^2\sin\theta}\frac{\partial}{\partial\theta}\left(\sin\theta\frac{\partial v}{\partial\theta}\right) = \frac{1}{R^2}\left(\frac{\partial^2 v}{\partial\theta^2} + \frac{\partial v}{\partial\theta}\cot\theta\right),$$

$$\frac{\partial u}{\partial R} = -\frac{u}{R}, \qquad \frac{\partial v}{\partial R} = -\frac{v}{R}.$$

Equation (100) now has the form

$$u + \frac{\partial v}{\partial\theta} + v\cot\theta = 0,$$

which, after differentiation with respect to θ, becomes

$$\frac{\partial u}{\partial\theta} + \frac{\partial^2 v}{\partial\theta^2} + \frac{\partial v}{\partial\theta}\cot\theta - v\csc^2\theta = 0.$$

With all these relations (101) and (102) become

$$-\frac{u^2 + v^2}{R} + \frac{v}{R}\frac{\partial u}{\partial\theta} = -\frac{1}{\rho}\frac{\partial p}{\partial R} + \frac{\nu}{R^2\sin\theta}\frac{\partial}{\partial\theta}\left(\sin\theta\frac{\partial u}{\partial\theta}\right)$$ (106)

$$\frac{v}{R}\frac{\partial v}{\partial\theta} = -\frac{1}{\rho R}\frac{\partial p}{\partial\theta} + \frac{\nu}{R^2}\frac{\partial u}{\partial\theta}.$$ (107)

Integration of (107) with respect to θ produces

$$\frac{p - p_0}{\rho} = -\frac{v^2}{2} + \frac{vu}{R} + \frac{c_1}{R^2}, \tag{108}$$

where p_0 is the pressure at infinity, at which u and v vanish, and the form of c_1/R^2 is dictated by (106). Substitution of (105) into (108) and differentiation with respect to R give

$$\frac{1}{\rho}\frac{\partial p}{\partial R} = \frac{v^2}{R} - \frac{2vu}{R^2} - \frac{2c_1}{R^3}.$$

Thus (106) can be written as

$$-\frac{u^2}{R} + \frac{v}{R}\frac{\partial u}{\partial \theta} = \frac{v}{R^2}\left[2u + \frac{1}{\sin\theta}\frac{\partial}{\partial\theta}\left(\sin\theta\frac{\partial u}{\partial\theta}\right)\right] + \frac{2c_1}{R^3}. \tag{109}$$

If instead of θ we use $\mu = \cos\theta$ as the independent variable,

$$\frac{1}{\sin\theta}\frac{d}{d\theta} = -\frac{d}{d\mu}, \qquad u = -\frac{v}{R}f'(\mu), \qquad v = -\frac{v}{R}\frac{f(\mu)}{(1 - \mu^2)^{1/2}}, \tag{110}$$

and (109) becomes

$$[f'(\mu)]^2 + f(\mu)f''(\mu) = 2f'(\mu) + \frac{d}{d\mu}[(1 - \mu^2)f''(\mu)] - 2c_1. \tag{111}$$

Two integrations give, successively,

$$ff' = 2f + (1 - \mu^2)f'' - 2c_1\mu - c_2 \tag{112}$$

and $\qquad f^2 = 4\mu f + 2(1 - \mu^2)f' - 2(c_1\mu^2 + c_2\mu + c_3). \tag{113}$

Equation (113) is of the Riccati type. Its solution can be obtained by first finding a particular solution and then finding a more general one. Inspection of (113) shows that a linear function $A + B\mu$ can satisfy (113) exactly, with the c's being functions of A and B. This is another way of saying that to be a solution, $A + B\mu$ requires one relation between the three c's. Squire chose the form

$$f_1 = \alpha(1 + \mu) + \beta(1 - \mu) \tag{114}$$

for $A + B\mu$, which is assumed to satisfy (113) with the proper choice of the c's. If we write

$$f = f_1 + f_2 \tag{115}$$

and require (113) to be satisfied, we have

$$2f_1f_2 + f_2^2 = 4\mu f_2 + 2(1 - \mu^2)f_2',$$

or $\qquad 2(1 - \mu^2)f_2' = (2f_1 - 4\mu)f_2 + f_2^2,$

This equation of the Bernoulli type can be solved by the substitution

$$f_2 = \frac{1}{G}, \tag{116}$$

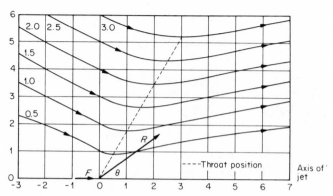

FIGURE 13. STREAMLINES OF A ROUND LAMINAR JET, $a = 1.0$, $F/\rho v^2 = 34.76$. [*From H. B. Squire, Quart. J. Mech. Appl. Math.,* **4** (1951), *by permission of the Quarterly Journal of Mechanics and Applied Mathematics.*]

which transforms (115) to

$$2(1 - \mu^2)G' + (2f_1 - 4\mu)G + 1 = 0. \tag{117}$$

The integrating factor of this equation is

$$\frac{(1 + \mu)^\beta}{(1 - \mu)^\alpha}.$$

After (117) is multiplied by this factor, it can be integrated, and the result is

$$\frac{(1 + \mu)^\beta}{(1 - \mu)^\alpha} (1 - \mu^2)G + \tfrac{1}{2}\left[a - \int_1^\mu \frac{(1 + \mu)^\beta}{(1 - \mu)^\alpha} d\mu\right] = 0, \tag{118}$$

in which a is a constant of integration. Substituting G from (118) into (116), then (116) and (114) into (115), we have

$$f = \alpha(1 + \mu) + \beta(1 - \mu) + \frac{2(1 - \mu^2)(1 - \mu)^{-\alpha}(1 + \mu)^\beta}{a - \displaystyle\int_1^\mu (1 - \mu)^{-\alpha}(1 + \mu)^\beta d\mu}.$$

Squire's jet corresponds to the case $\alpha = \beta = 0$, for which this solution becomes

$$f = \frac{2(1 - \mu^2)}{a + 1 - \mu} = \frac{2\sin^2\theta}{a + 1 - \cos\theta}. \tag{119}$$

After (119) is substituted into (104), the streamlines can be drawn. The flow patterns for $a = 1, 0.1$, and 0.01 are shown in Figs. 13 to 15. Clearly the solution corresponds to a round laminar jet.

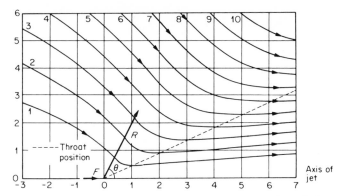

FIGURE 14.   STREAMLINES OF A ROUND LAMINAR JET, $a = 0.1$, $F/\rho v^2 =$
3.140×10^2.   [*From H. B. Squire, Quart. J. Mech. Appl. Math.*, **4**
(1951), *by permission of the Quarterly Journal of Mechanics and Applied
Mathematics.*]

The momentum flux M across the surface S of any sphere with center at the origin is

$$M = \int_0^\pi \rho u (u \cos \theta - v \sin \theta) 2\pi R^2 \sin \theta \, d\theta$$

$$= 2\pi \rho v^2 \left[\frac{32(a+1)}{3a(a+2)} - 16(a+1) + 8a(a+2) \ln \frac{a+2}{a} \right].$$

The force in the x direction exerted *by* the sphere on the surrounding fluid is

$$F_1 = -\int_0^\pi \left[\left(-p + 2\mu \frac{\partial u}{\partial R} \right) \cos \theta \right.$$

$$\left. - \mu \left(R \frac{\partial}{\partial R} \frac{v}{R} + \frac{1}{R} \frac{\partial u}{\partial \theta} \right) \sin \theta \right] 2\pi R^2 \sin \theta \, d\theta$$

$$= 2\pi \rho v^2 \left[24(a+1) - (12a^2 + 24a + 4) \ln \frac{a+2}{a} \right].$$

The sum of $M + F_1$ is given by

$$\frac{M + F_1}{2\pi \rho v^2} = \frac{32(a+1)}{3a(a+2)} + 8(a+1) - 4(a+1)^2 \ln \frac{a+2}{a},$$

which gives $(M + F_1)/\rho v^2 = 34.76$, 314.0, and $3,282$ for $a = 1$, 0.1, and 0.01, respectively. The quantity $M + F_1$ can be considered as the strength of a momentum source at the origin, since it is constant for all radii and only the origin is singular. This sum will be denoted by F, after Squire, to suggest a concentrated force acting on the fluid at the origin.

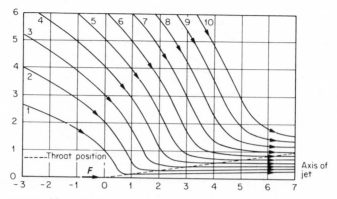

FIGURE 15. STREAMLINES OF A ROUND LAMINAR JET, $a = 0.01$, $F/\rho v^2 = 3.282 \times 10^3$. [*From H. B. Squire, Quart. J. Mech. Appl. Math.*, 4 (1951), *by permission of the Quarterly Journal of Mechanics and Applied Mathematics.*]

4.2. Two-dimensional Flow against an Infinite Plate

Consider the two-dimensional flow of a fluid of constant density and viscosity against a plate, as sketched in Table 1. The plate is infinite in extent, and the velocity of the fluid is normal to the wall at infinite y. Strictly speaking, the position of the origin is indeterminate, since the plate is infinite. We may consider the case to be the limiting case of flow against a plate of finite width, for which the origin of x is midway between the edges, and the solution is valid away from the edges.

It can easily be verified that if the fluid were inviscid and the flow irrotational, the complex potential would be $\beta(x + iy)^2$, and the Bernoulli equation gives

$$-\frac{1}{\rho}\frac{\partial p}{\partial x} - \frac{\partial \Omega}{\partial x} = \beta^2 x, \tag{120}$$

in which, as before, Ω is the body-force potential and β is a constant. Although the fluid is now considered viscous and the motion is no longer irrotational, (120) nevertheless leads to an exact solution of the Navier-Stokes equation and the boundary conditions, as we shall now show.

Admitting (120), we can write (1) as

$$u\frac{\partial u}{\partial x} + v\frac{\partial u}{\partial y} = \beta^2 x + v\left(\frac{\partial^2 u}{\partial x^2} + \frac{\partial^2 u}{\partial y^2}\right). \tag{121}$$

Since u and v are related through the equation of continuity, the only dependent variable is the stream function ψ. From ψ, β (of dimension T^{-1}), v, x, and y, we can obtain the three dimensionless variables and the

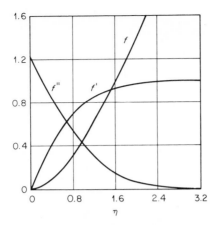

FIGURE 16. HOWARTH'S SOLUTION FOR
EQ. (125), WHICH GOVERNS TWO-DIMEN-
SIONAL LAMINAR FLOW AGAINST A PLATE.

relationship

$$\frac{\psi}{\nu} = F\left[\left(\frac{\beta}{\nu}\right)^{\frac{1}{2}}x, \quad \left(\frac{\beta}{\nu}\right)^{\frac{1}{2}}y\right].$$ (122)

Remembering that u is ψ_y and v is $-\psi_x$, we see that if (122) is specialized
in the form

$$\frac{\psi}{\nu} = \left(\frac{\beta}{\nu}\right)^{\frac{1}{2}}xf(\eta), \qquad \eta = \left(\frac{\beta}{\nu}\right)^{\frac{1}{2}}y,$$ (123)

(121) can be made free from x.   Using (123), we have

$$u = \beta xf'(\eta), \qquad v = -(\nu\beta)^{\frac{1}{2}}f(\eta)$$ (124)

and can write (121) as

$$f'^2 - ff'' = 1 + f'''.$$ (125)

The boundary conditions

$$u = 0 = v \text{ at } y = 0 \quad \text{and} \quad u = \beta x \text{ at } y = \infty$$ (126)

can be written in terms of f as

$$f'(0) = 0 = f(0) \quad \text{and} \quad f'(\infty) = 1.$$ (127)

The last condition in (106) demands that far away from the plate the flow
be potential.   Or, if we impose it and find a solution possible, it *indicates*
that far away from the plate the flow is potential.   The solution of (125)
with (127) was obtained numerically by Hiemenz (1911) and improved by
Howarth (1935).   Now it can be obtained, after the computation is
programmed, by the digital computer in a few seconds.   Howarth's
results are shown graphically in Fig. 16.

TABLE 1†

Flow	Flow pattern	Original coordinates	Transformation	Differential equation(s)	Boundary conditions
Two-dimensional flow with stagnation line		Cartesian	$\psi = (\nu\beta)^{1/2}xf(\eta)$, $\eta = \left(\dfrac{\beta}{\nu}\right)^{1/2}y$ $u = \beta x f'(\eta)$, $v = -(\nu\beta)^{1/2}f(\eta)$ (Hiemenz)	$f'^2 - ff'' = 1 + f'''$ (Numerical solution by Hiemenz and Howarth)	$f(0) = 0,$ $f'(0) = 0$ $f'(\infty) = 1$
Axisymmetric flow with stagnation point		Cylindrical, but r and z are denoted by x and y, respectively, and w is written v	$\psi = -(\beta_1\nu)^{1/2}x^2 f(\eta)$, $\eta = \left(\dfrac{\beta_1}{\nu}\right)^{1/2}y$ $u = \beta_1 x f'(\eta)$, $v = -2(\beta_1\nu)^{1/2}f(\eta)$ (Homann)	$f'^2 - 2ff'' = 1 + f'''$ (Numerical solution by Homann)	$f(0) = 0$ $f'(0) = 0$ $f'(\infty) = 1$
Axisymmetric flow due to rotation of a plate Ω = angular speed of plate		Cylindrical	$u = r\Omega F(\eta)$, $v = r\Omega G(\eta)$, $w = (\nu\Omega)^{1/2}H(\eta)$ $p = -\mu\Omega P(\eta)$ $\eta = \left(\dfrac{\Omega}{\nu}\right)^{1/2}z$ (von Kármán)	$F^2 - G^2 + F'H = F''$ $2FG + G'H = G''$ $HH' = P' + H''$ $2F + H' = 0$ (von Kármán) (Numerical solution by Cochran)	$F(0) = 0,$ $G(0) = 1$ $H(0) = 0$ $F(\infty) = 0$ $G(\infty) = 0$

† Note that ψ denotes Lagrange's stream function in two-dimensional flows and Stokes' stream function in axisymmetric flows. For two-dimensional flows, $u = \partial\psi/\partial y$, $v = -\partial\psi/\partial x$. For axisymmetric flows $u = -(1/r)(\partial\psi/\partial z)$, $w = (1/r)(\partial\psi/\partial r)$ in cylindrical coordinates, and $u = (1/R^2 \sin\theta)(\partial\psi/\partial\theta)$, $v = -(1/R \sin\theta)(\partial\psi/\partial R)$ in spherical coordinates. $u, v,$ and w are the velocity components in the directions of increasing $x, y,$ and z, or $r, \varphi,$ and z, or $R, \theta,$ and φ, as the case may be.

TABLE 1 (*Continued*)

Description	Diagram	Coordinate system	Notes / Equations	Boundary conditions
Axisymmetric flow due to rotation of two parallel plates rotating with angular speeds Ω_1 and Ω_2 and with spacing d		Cylindrical	Same as above, with $\Omega = \Omega_1$. / Same as above, definitive solution still lacking	$F(0) = 0,$ $G(0) = 1$ $H(0) = 0$ $F(\eta_1) = 0$ $G(\eta_1) = \dfrac{\Omega_2}{\Omega_1}$ $H(\eta_1) = 0$ $\eta_1 = \left(\dfrac{\Omega_1}{\nu}\right)^{\frac{1}{2}} d$
Two-dimensional flow in a converging channel of angle 2α		Polar coordinates r and φ	$u = \dfrac{f(\varphi)}{r}$ $v = 0$ $w = 0$ (Jeffrey, Hamel, and others) $\quad$ $f'' + 4f + \dfrac{f^2}{\nu} + k = 0$ $k = \text{const}$ Hence $f'^2 = \dfrac{2}{3\nu}(h - 3\nu k f)$ $\quad - 6\nu f^2 - f^3$ $h = \text{const}$ Soluble by elliptic function (Jeffrey, Hamel, and others)	$f(\alpha) = f(-\alpha)$ $= 0$ $\displaystyle\int_{-\alpha}^{\alpha} f(\varphi)\,d\varphi$ $= $ given discharge

TABLE 1 (*Continued*)

Flow	Flow pattern	Original coordinates	Transformation	Differential equation(s)	Boundary conditions
Two-dimensional spiral flow		Polar or Cartesian	$w = \alpha + i\beta$ $= (A + Bi)\ln z$ (complex potential for an irrotational flow) $\alpha = A \ln r - B\varphi$ $\beta = A\varphi + B\ln r$ α and β are the new independent variables; Lagrange's stream function ψ is assumed to depend on β only (Hamel)	$\psi^{iv} + (a + b\psi')\psi'' = 0,$ $\psi' = \dfrac{d\varphi}{d\beta}$ Hence $\psi''' + a\psi' + \dfrac{b}{2}\psi'^2 = C_1$ or, with $\psi' = h$, $h'' + ah + \dfrac{b}{2}h^2 = C_1$ Compare with above (Hamel)	$\psi'(0) = 0$ $\psi'(\beta_1) = 0$ $\psi(0) = 0$ $\psi(\beta_1) = \text{const } q$
Round laminar jet	See Figs. 13 to 15.	Spherical R, θ, φ	$\psi = \nu R f(\theta)$ $\mu = \cos\theta$ (Squire)	$ff' = 2f + 2(1 - \mu^2)f''$ $- (2c_1\mu + c_2)$ $\left(f' = \dfrac{df}{d\mu}, \quad f'' = \dfrac{d^2f}{d\mu^2}\right)$ Solution for round jet: $f = \dfrac{2(1 - \mu^2)}{a + 1 - \mu}$ (Squire)	$f = 0$ at $\theta = 0$ and $\theta = \pi$

After f and therefore u and v have been obtained, (2) permits the calculation for p. Thus, integration of (2) and (120) produces

$$\frac{p}{\rho} + \Omega = C - \tfrac{1}{2}(u_1^2 + v^2) + v\frac{\partial v}{\partial y}. \tag{128}$$

Hiemenz solved (125) as the boundary-layer equation near the stagnation point. That (125) represents the full Navier-Stokes equation seems to have been pointed out first by Tollmien (1931).

Note that the curve for f' in Fig. 16 is essentially a plot of u against η or y and that at about $\eta = 3$ the function $f'(\eta)$ is 0.99. Thus at

$$y = 3\left(\frac{v}{\beta}\right)^{\!\frac{1}{2}} = \delta \tag{129}$$

the velocity u is almost equal to the potential velocity. For a given β and a very small v, δ can be very small compared with x. That is, only in a very thin layer of depth δ are viscous effects important. Outside the layer the flow is essentially potential. The layer is called the boundary layer, and its thinness for small v is the justification for studying potential flows even of a viscous fluid. We shall now turn to the boundary-layer theory, trusting to Table 1 to provide the essentials concerning other exact solutions of the nonlinear Navier-Stokes equations.

5. EQUATIONS GOVERNING THE FLOW IN BOUNDARY LAYERS

We have seen in Sec. 4 of Chap. 2 that if ρ and μ are constant, irrotational flows are possible even for a viscous fluid, as far as the differential equations (the Navier-Stokes equations or the vorticity equations derived therefrom) are concerned. Irrotational flows of a viscous fluid of constant ρ and μ are ruled out only because (except in one or two special cases) the nonslip conditions at the boundaries are never satisfied. Now let us imagine an irrotational flow past a body whose surface obligingly creeps with just that velocity which is needed for the irrotational flow to satisfy the nonslip condition. If this creeping motion of the surface suddenly stops, at that moment the fluid right at the surface will stop moving also, because the nonslip condition must be satisfied. From the several examples given in Sec. 3, we expect that a layer of rotational flow, initially of zero thickness, will develop with time. Because the fluid outside of this layer is moving, however, the vorticity created at the boundary and propagated into the fluid by viscous diffusion will be convected downstream; and if the kinematic viscosity (which is a momentum diffusivity) is small, or, more properly, if the Reynolds number is large, convection will dominate diffusion except in a thin layer where both are important, and eventually a steady flow with a thin boundary layer will result.[1] Outside this layer the flow remains essentially irrotational.

[1] Except in the wake, after the boundary layer separates from the body.

We need not consider the establishment of steady flow to see the plausibility of a boundary layer. The irrotational flow mentioned in the preceding paragraph would be admissible but for its failure to satisfy the nonslip condition at the boundary. Satisfaction of this condition creates vorticity at the boundary which is diffused into the fluid. If the Reynolds number is large, the convection of vorticity downstream by the flow will dominate this diffusion except in a thin layer attached to the boundary, where both convection and diffusion are important. Outside this layer the flow will remain essentially irrotational. Since convection of vorticity is associated with the convective terms of the Navier-Stokes equations, which, being associated with the inertia of the fluid, are called the *inertial terms*, and since diffusion of vorticity is associated with the viscous terms of the Navier-Stokes equations, we can also say that outside the boundary layer inertial forces completely dominate viscous forces and inside it inertial forces and viscous forces are of the same order of magnitude. Actually very near the (stationary) body the velocity is very low, and there viscous forces dominate inertial forces. While true enough, this does not necessitate splitting the boundary layer into two regions, since all calculations for the velocity distribution in the boundary layer naturally account for the variation of the inertial terms in the layer.

The ideas and heuristic arguments presented in the preceding paragraph are clearly discernible in the brief but truly epoch-making paper (1904) of Ludwig Prandtl (1875–1953). It can be fairly stated that no other single idea in the history of modern fluid mechanics has been so fruitful or influential as Prandtl's idea of the boundary layer. The marvel is that this idea is so simple!

To put the boundary-layer theory on more formal grounds, we shall derive the equations governing the flow in the boundary layer. For convenience we shall split the pressure p into two parts, a hydrostatic part p_s and a part p_d that is due to the motion of the fluid. Since, in the absence of flow and for constant ρ,

$$p_s + \rho\Omega = \text{const} = C,$$

it is clear that $\qquad p_d = p - p_s = p + \rho\Omega - C.$ $\qquad\qquad$ (130)

Since the constant C vanishes after differentiation, the term $p + \rho\Omega$ in the Navier-Stokes equations can be replaced by p_d. We shall treat only the case of constant ρ and μ in this section. For constant ρ and μ, the Navier-Stokes equations can be written as

$$\frac{Du_i}{Dt} = -\frac{1}{\rho}\frac{\partial p_d}{\partial x_i} + \nu\,\nabla^2 u_i.$$ $\qquad\qquad$ (131)

If we set $\qquad u_i' = \dfrac{u_i}{U_0}, \qquad p_d' = \dfrac{p_d}{\rho U_0^2}, \qquad x_i' = \dfrac{x_i}{L},$ $\qquad\qquad$ (132)

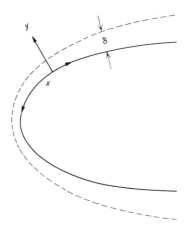

FIGURE 17.   COORDINATES FOR THE
BOUNDARY LAYER.

in which U_0 is a reference velocity and L a reference length, substitute
(132) into (131), and then drop the primes, we have

$$\frac{Du_i}{Dt} = -\frac{\partial p_d}{\partial x_i} + \frac{1}{R}\,\nabla^2 u_i, \tag{133}$$

in which
$$R = \frac{UL}{\nu} \tag{134}$$

is the Reynolds number and both D/Dt and ∇^2 are in dimensionless terms.
The essentials of the boundary-layer theory can be described by considering two-dimensional flows.   If the curvature of the solid boundary is
small compared with $1/\delta$, δ being the thickness of the boundary layer, we
can measure x along the boundary and y in a direction normal to it
(Fig. 17) and still use (133).   The equations for two-dimensional flows
are then

$$\frac{\partial u}{\partial t} + u\frac{\partial u}{\partial x} + v\frac{\partial u}{\partial y} = -\frac{\partial p_d}{\partial x} + \frac{1}{R}\left(\frac{\partial^2 u}{\partial x^2} + \frac{\partial^2 u}{\partial y^2}\right), \tag{135}$$

$$\frac{\partial v}{\partial t} + u\frac{\partial v}{\partial x} + v\frac{\partial v}{\partial y} = -\frac{\partial p_d}{\partial y} + \frac{1}{R}\left(\frac{\partial^2 v}{\partial x^2} + \frac{\partial^2 v}{\partial y^2}\right), \tag{136}$$

in which u is now the tangential and v the normal component of the
velocity.   In the same natural coordinates,

$$\frac{\partial u}{\partial x} + \frac{\partial v}{\partial y} = 0 \tag{137}$$

is the equation of continuity.

To start with, we assume that δ is small compared with L if R is large and verify it later. That means

$$\epsilon = \frac{\delta}{L} \ll 1,$$

and the range of y for the boundary layer is ϵ. Since u and x are all of the order of unity, (137) states that v is of the order of ϵ. Now the convective terms in (135) are all of $O(1)$. The term $\partial u/\partial t$ may or may not be of the same order as the convective terms. In either case it is the term or terms of prevailing magnitude that should be used. We shall assume $\partial u/\partial t$ to be of $O(1)$ for convenience of exposition. A glance at the viscous terms in (135) reveals that

$$\frac{\partial^2 u}{\partial x^2} \ll \frac{\partial^2 u}{\partial y^2},$$

so that the first can be neglected and the viscous terms in (135) can be replaced by $R^{-1}\,\partial^2 u/\partial y^2$. Since in the boundary layer the viscous terms are of the same order of magnitude as the inertial terms, $R^{-1}\,\partial^2 u/\partial y^2 = O(1)$; this shows that

$$\epsilon^2 = \frac{1}{R} \quad \text{or} \quad \delta = \frac{L}{\sqrt{R}}. \tag{138}$$

To see how p_d varies, we turn to (136). Again the term $\partial^2 v/\partial x^2$ can be neglected since it is *added* to a much larger term $\partial^2 v/\partial y^2$. Then all the terms involving v are of $O(\epsilon)$. Hence the pressure variation with respect to y in the boundary layer is of $O(\epsilon^2)$, and can be neglected. Thus we can take the pressure outside the boundary layer to be the pressure inside. But outside the boundary layer the Bernoulli equation applies, since the flow is irrotational there. The Bernoulli equation takes the dimensionless form

$$p_d = C - \tfrac{1}{2}U^2, \tag{139}$$

in which U is the dimensionless velocity just outside the boundary layer, measured in units of U_0, and is a function of x alone. Thus the flow in the boundary layer is governed by (137) and

$$\frac{\partial u}{\partial t} + u\frac{\partial u}{\partial x} + v\frac{\partial u}{\partial y} = U\frac{dU}{dx} + \frac{1}{R}\frac{\partial^2 u}{\partial y^2}, \tag{140}$$

which is the dimensionless boundary-layer equation for two-dimensional flows. In *dimensional* units, (140) has the form

$$\frac{\partial u}{\partial t} + u\frac{\partial u}{\partial x} + v\frac{\partial u}{\partial y} = U\frac{dU}{dx} + v\frac{\partial^2 u}{\partial y^2}. \tag{141}$$

We shall present the boundary-layer equation for axisymmetric flows later. Before proceeding to use (140) to solve problems, we note that (138) verifies the assumption that the boundary layer is thin if R is large.

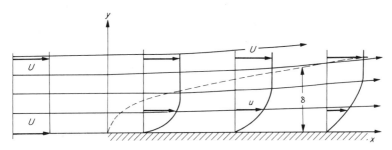

FIGURE 18.   BOUNDARY LAYER ON A PLATE AND THE STREAMLINES.

6. SOLUTIONS OF THE BOUNDARY-LAYER EQUATION FOR STEADY TWO-DIMENSIONAL FLOWS

We shall present a few typical solutions of (141) for steady flows.   They do not necessarily call for the presence of solid boundaries, since (141) can be used if longitudinal diffusion of momentum is negligible with respect to transverse diffusion of momentum and if the pressure distribution is either known or can be correctly assumed.   Thus in this section, aside from actual boundary layers we shall also treat laminar jets. The reader is reminded that (141) is based on the assumption of constant ρ and μ.

6.1.   Steady Flow along a Flat Plate

For the steady flow of a viscous fluid along a plate (Fig. 18), the flow outside the boundary layer has uniform velocity U.   Hence (141) becomes

$$u \frac{\partial u}{\partial x} + v \frac{\partial u}{\partial y} = \nu \frac{\partial^2 u}{\partial y^2}, \tag{142}$$

in which all quantities are in their dimensional forms.   The boundary conditions are

$$u = 0 = v \text{ at } y = 0 \qquad \text{and} \qquad u = U \text{ at } y = \infty. \tag{143}$$

The U in (143) is no longer dimensionless.   The equation of continuity allows the use of a stream function ψ.   Treating ψ as the dependent variable and x, y, U, and ν as independent variables, a dimensional analysis gives the relationship

$$\frac{\psi}{\nu} = F\left(\frac{Ux}{\nu}, \frac{Uy}{\nu}\right). \tag{144}$$

Since $u = \psi_y$ and $v = -\psi_x$, an examination of (142) and (143) brings out the interesting fact that if x, y, and ψ are changed to $\alpha^2 x$, αy, and $\alpha\psi$, respectively, neither (142) nor (143) is affected, which suggests that (144)

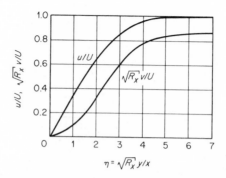

FIGURE 19. VELOCITY DISTRIBUTION IN
BLASIUS FLOW. $R_x = Ux/\nu$.

can be specialized to the form

$$\frac{\psi}{\nu}\left(\frac{\nu}{Ux}\right)^{1/2} = f(\eta),$$

or
$$\psi = (\nu Ux)^{1/2} f(\eta), \tag{145}$$

in which
$$\eta = \frac{Uy/\nu}{(Ux/\nu)^{1/2}} = \sqrt{\frac{U}{\nu x}}\, y. \tag{146}$$

It is a matter of straightforward calculation to show that with (145),

$$u = Uf'(\eta), \qquad v = \frac{1}{2}\left(\frac{U\nu}{x}\right)^{1/2}(\eta f' - f), \tag{147}$$

(142) becomes

$$2f''' + ff'' = 0, \tag{148}$$

and the boundary conditions become

$$f'(0) = 0 = f(0), \qquad f'(\infty) = 1. \tag{149}$$

Solution of the system consisting of (148) and (149) was given by Blasius (1908), who used a power series for f for small η and an asymptotic form for f at large η and matched the two expressions for f at some intermediate value of η to obtain a solution. Now a digital computer will perform the computation in a few seconds. The results are presented graphically in Fig. 19. Numerical results can be found in Schlichting (1968). The most important result is

$$f''(0) = 0.332. \tag{150}$$

Since $\partial u/\partial y$ at $y = 0$ is proportional to $f''(0)$, (150) enables us to compute the drag D. Thus, for *both* sides of the plate, the drag over a length l per unit width is, with the Reynolds number R based on l,

$$D = 2 \int_0^l \mu \frac{\partial u}{\partial y}\bigg|_0 dx = 2\mu U \sqrt{\frac{U}{\nu}} \int_0^l f''(0) x^{-\frac{1}{2}} dx$$

$$= 4f''(0)\rho U^2 l R^{-\frac{1}{2}}. \tag{151}$$

The drag coefficient defined by

$$C_D = \frac{D}{\rho U^2 l}$$

is then
$$C_D = 4f''(0) R^{-\frac{1}{2}} = 1.328 R^{-\frac{1}{2}}. \tag{152}$$

It is worth stating that the two-point boundary problem (148) and (149) can be changed to a one-point boundary problem by setting $f(\eta) = cF(c\eta)$, c being a constant, so that (148) and (149) are transformed into

$$2F''' + FF'' = 0, \qquad F(0) = F'(0) = 0, \qquad F'(\infty) = c^{-2}.$$

Since c is arbitrary, one may replace the last condition by $F''(0) = 1$, say. This one-point boundary problem is much simpler to solve numerically, giving a solution for all η, whose value $F'(\infty)$ determines c for the transformation. That this method works is a consequence of the invariance of (148) under the transformation $f \to cf$, $\eta \to \eta/c$.

From the definition of η in (146) and the expression for u in (147) it can be seen that along any parabola $x = Cy^2$, η, and therefore u, is constant. If u is plotted for each x against y, the velocity profiles are similar in the sense that a reduction of the y scale at each x station will make all the profiles into one single profile—in fact the curve of u vs. η. It is for these reasons that the solution is called a similarity equation. All solutions of the Navier-Stokes equations or of the boundary-layer equations obtained by first reducing these equations into an ordinary differential equation are similarity solutions, because the velocity distributions at various sections, whether they are sections at constant x, R, or z in the various coordinate systems, exhibit a similarity among themselves. Thus, all the solutions presented in Sec. 4 are similarity solutions, and so are the solutions presented in this section.

6.2. Two-dimensional Laminar Jets

Consider a slit in a wall from which a fluid issues into infinite space filled with the same fluid. The plane of flow is perpendicular to the slit. In this plane, the origin is the trace of the slit, and the line normal to the wall passing through the origin is the x axis. The y axis is the intersection of the plane of flow and the wall.

The mass discharge Q_m (per unit length in the z direction) at the wall, where u can be considered constant, is $b\rho u$, and the momentum flux is $b\rho u^2$, for the moment also taken at $x = 0$, b being the width of the slit.

If M is kept fixed and b is made to approach zero, u^2 will approach infinity as $M/b\rho$, so that Q_m will approach zero as a limit. In practice b can never be zero, of course, but the above consideration shows that it is reasonable to consider a momentum source which is a very weak mass source. We shall consider the steady flow caused by a two-dimensional momentum source.

Far from the origin the pressure can again be considered constant, although the situation is not as obvious as for the flow along a flat plate. Thus the governing equation is again (142). We shall first discover something about the momentum flux

$$M = \int_{-\infty}^{\infty} \rho u^2 \, dy \tag{153}$$

calculated at each value of x. Multiplying (142) by $\rho \, dy$ and integrating, by parts if necessary, we have

$$\frac{d}{dx} M + \rho v u \Big|_{-\infty}^{\infty} - \int_{-\infty}^{\infty} \rho u \frac{\partial v}{\partial y} \, dy = \mu \frac{\partial u}{\partial y} \Big|_{-\infty}^{\infty}. \tag{154}$$

At infinity $u = 0$. By virtue of the equation of continuity

$$\frac{\partial u}{\partial x} + \frac{\partial v}{\partial y} = 0, \tag{155}$$

(154) then becomes $\quad\quad \dfrac{d}{dx} M = 0. \tag{156}$

Thus M is a constant. This is to be expected, since p is assumed constant and the term $\mu \, \partial^2 u/\partial x^2$ representing the gradient of the viscous part of the normal stress has been neglected. Application of the momentum principle at a control surface consisting of two lines (or planes) $x = x_1$ and $x = x_2$ would have produced (156).

Thus we have ψ as the dependent variable and x, y, v, and M as the independent variables. It is more convenient to replace M by M/ρ, to make all variables kinematic in nature. A dimensional analysis produces the relationship

$$\frac{\psi}{\nu} = F\left(\frac{Mx}{\rho \nu^2}, \frac{My}{\rho \nu^2}\right). \tag{157}$$

The three variables in (157) are represented by ψ, x, and y. Supposing ψ to be multiplied by β, x by β, and y by γ and remembering that u is ψ_y and v is $-\psi_x$, we find that (142) will remain invariant if

$$\frac{\alpha^2}{\beta \gamma^2} = \frac{\alpha}{\gamma^3} \quad\quad \text{or} \quad\quad \beta = \alpha \gamma. \tag{158}$$

On the other hand, M is constant only if

$$\frac{\alpha^2}{\gamma} = 1 \quad\quad \text{or} \quad\quad \gamma = \alpha^2. \tag{159}$$

Equations (158) and (159) give $\beta = \alpha^3 = \gamma^{3/2}$. Thus (157) is more general than need be, and a more informative and useful form of it is

$$\frac{\psi}{v}\left(\frac{Mx}{\rho v^2}\right)^{-1/3} = f(\eta), \qquad \eta = \frac{My}{\rho v^2}\left(\frac{Mx}{\rho v^2}\right)^{-2/3},$$

or
$$\psi = \left(\frac{Mvx}{\rho}\right)^{1/3} f(\eta), \qquad \eta = \left(\frac{M}{\rho v^2 x^2}\right)^{1/3} y, \qquad (160)$$

from which

$$u = \psi_y = \left(\frac{M^2}{\rho^2 vx}\right)^{1/3} f'(\eta), \qquad v = -\psi_x = \frac{1}{3}\left(\frac{Mv}{\rho x^2}\right)^{1/3}(2\eta f' - f). \quad (161)$$

Substituting (161) into (142), we have

$$3f''' + f'^2 + ff'' = 0. \qquad (162)$$

The boundary conditions

$$u = \begin{cases} 0 & \text{for } y = \pm\infty, \\ 0 = v & \text{at } x = \infty, \end{cases}$$

and
$$v = 0 = \frac{\partial u}{\partial y} \qquad \text{at } y = 0$$

can all be written in terms of f as

$$f'(\pm\infty) = 0, \qquad f(0) = 0 = f''(0). \qquad (163)$$

Integration of (162) yields
$$3f'' + ff' = 0,$$

the constant of integration being zero because of (163). Another integration yields

$$3f' + \tfrac{1}{2}f^2 = c^2. \qquad (164)$$

The constant of integration c^2 has to be determined by (153) later. For convenience, we use the substitutions

$$f = \sqrt{2}\, F \qquad \text{and} \qquad \eta = 3\sqrt{2}\, \zeta,$$

which put (164) into the form

$$\frac{dF}{d\zeta} + F^2 = c^2.$$

The solution, by treating F is the independent variable, is

$$c\zeta = \tanh^{-1}\frac{F}{c},$$

the constant of integration being zero because $f(0) = 0 = F(0)$. Thus

$$f = \sqrt{2}\,F = \sqrt{2}\,c\,\tanh c\zeta = \sqrt{2}\,c\,\tanh \frac{c\eta}{3\sqrt{2}}\,. \tag{165}$$

Substituting (165) into the expression for u in (161) and the result into (153), we have

$$M = M\,\frac{c^3\sqrt{2}}{3}\int_{-\infty}^{\infty} \operatorname{sech}^4 c\zeta\,d(c\zeta) = \frac{4\sqrt{2}\,c^3}{9}\,M,$$

which gives

$$c^3 = \frac{9}{4\sqrt{2}}\,.$$

Thus

$$f = (\tfrac{9}{2})^{\frac13}\tanh\,[(\tfrac{1}{48})^{\frac13}\eta].$$

We can finally put

$$\xi = \left(\frac{M}{48\rho\nu^2}\right)^{\frac13}\frac{y}{x^{\frac23}} \tag{166a}$$

and write

$$\psi = \left(\frac{9M x\nu}{2\rho}\right)^{\frac13}\tanh\,\xi, \qquad u = \left(\frac{3M^2}{32\rho^2\nu x}\right)^{\frac13}\operatorname{sech}^2\,\xi,$$

$$v = \left(\frac{M\nu}{6\rho x^2}\right)^{\frac13}(2\xi\,\operatorname{sech}^2\,\xi - \tanh\,\xi). \tag{166b}$$

This solution was given by Bickley (1939), and was previously attained numerically by Schlichting (1933).

7. AXISYMMETRIC BOUNDARY LAYERS: MANGLER'S TRANSFORMATION

By arguments like those advanced in Sec. 5, and using x again as the coordinate along the boundary in the direction of flow and y as the coordinate normal to the boundary, the boundary-layer equation for axisymmetric flow past a body can again be written as

$$u\,\frac{\partial u}{\partial x} + v\,\frac{\partial u}{\partial y} = U\,\frac{dU}{dx} + \nu\,\frac{\partial^2 u}{\partial y^2}, \tag{167}$$

provided ρ and μ are constant, the radius of curvature of the boundary is not so small as to be of the order of the boundary-layer thickness, and the Reynolds number is large.   The u and v are the velocity components in the directions of x and y, respectively, as before, and U is the velocity just outside the boundary layer.   This is identical in form with (141). The equation of continuity, however, is different.   If r denotes the radius

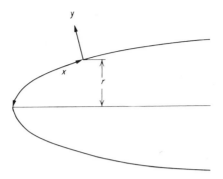

FIGURE 20. COORDINATES FOR AXISYM-
METRIC BOUNDARY LAYERS.

of the cross section of the body normal to its axis (Fig. 20), the equation of continuity is

$$\frac{\partial}{\partial x} ru + \frac{\partial}{\partial y} rv = 0. \tag{168}$$

Mangler (1948) used the substitutions

$$\bar{x} = \frac{1}{L^2} \int_0^x r^2(x) \, dx, \qquad \bar{y} = \frac{r(x)}{L} y, \tag{169}$$

$$\bar{u} = u, \qquad \bar{v} = \frac{L}{r(x)}\left[v + \frac{r'(x)}{r(x)} yu\right],$$

in which L is a reference length, to transform (167) and (168) to

$$\bar{u}\bar{u}_{\bar{x}} + \bar{v}\bar{u}_{\bar{y}} = \bar{U}\bar{U}_{\bar{x}} + v\bar{u}_{\bar{y}\bar{y}} \tag{170}$$

and $$\bar{u}_{\bar{x}} + \bar{v}_{\bar{y}} = 0, \tag{171}$$

in which for simplicity subscripts are used to indicate differentiation. It is evident that (170) is the equation of motion and (171) the equation of continuity for a fictitious two-dimensional boundary-layer flow with velocity components $\bar{u}$ and $\bar{v}$ past a two-dimensional body. The $\bar{x}$ and $\bar{y}$ are the x and y in Fig. 17 (for two-dimensional bodies), and the fictitious two-dimensional body is different in shape from the meridian section of the axisymmetric body. Furthermore, $\bar{U}$ is just the original U and is *not* the potential-flow velocity that truly corresponds to the fictitious two-dimensional body. Nevertheless, Mangler's transformation reduces the equations governing axisymmetric boundary layers formally to those governing two-dimensional boundary layers, and thus all methods of solution of the latter equations also apply to the former.

8. AXISYMMETRIC LAMINAR JET

Even if no solid body is present, (167) can still be used, if the Reynolds number (based on some length and some velocity) is high, so that longitudinal viscous diffusion is still negligible in comparison with transverse viscous diffusion of momentum.   For a round jet, the same arguments advanced in Sec. 5 lead to the equation

$$u \frac{\partial w}{\partial r} + w \frac{\partial w}{\partial z} = \frac{\nu}{r} \frac{\partial}{\partial r}\left(r \frac{\partial w}{\partial r}\right),$$ (172)

in which cylindrical coordinates are used, with the z axis coinciding with the axis of symmetry.   The density and viscosity are again assumed constant.   The equation of continuity is

$$\frac{\partial(ru)}{\partial r} + \frac{\partial(rw)}{\partial z} = 0.$$ (173)

The axisymmetric counterpart of the jet treated in Sec. 6.2 is a jet due to a point source of momentum at the origin.   The z axis has the same direction as the momentum.   The flow of the jet induced by the momentum source is governed by (172) and (173), and the boundary conditions are

$$w = 0 \quad \text{for} \quad r = \infty, \qquad w = 0 = u \quad \text{for} \quad z = \infty,$$

and $$u = 0 \quad \text{and} \quad \frac{\partial w}{\partial r} = 0 \qquad \text{for } r = 0.$$

The method of solution is similar to that employed in Sec. 6.2 and need not be described again step by step.   The momentum flux

$$M = 2\pi\rho \int_0^\infty w^2 r \, dr$$ (174)

can again be shown to be independent of z by integration [by parts and using (173) whenever necessary] of (172).   The appropriate transformation is (ψ being now the Stokes stream function)

$$\psi = 4\nu z f(\eta), \qquad \eta = \frac{1}{8}\left(\frac{3M}{\pi\rho\nu^2}\right)^{1/2} \frac{r}{z},$$ (175)

which (since $ru = -\psi_z$ and $rw = \psi_r$) transforms (172) into

$$\left(\frac{4ff'}{\eta}\right)' = \left(\frac{f' - \eta f''}{\eta}\right)'$$ (176)

and the boundary conditions to

$$f'(\infty) = 0, \qquad f(0) = 0 = f'(0).$$ (177)

Equation (176) can be integrated once to give

$$4ff' = f' - \eta f'',$$

the solution of which is $f(\eta) = \dfrac{\alpha\eta^2}{1 + \alpha\eta^2}$,

in which α is determined by (174) to be unity.   Hence

$$f(\eta) = \frac{\eta^2}{1 + \eta^2} , \tag{178}$$

which gives $u = -\dfrac{1}{r}\dfrac{\partial\psi}{\partial z} = \dfrac{1}{2}\left(\dfrac{3M}{\pi\rho z^2}\right)^{1/2}\dfrac{\eta(1 - \eta^2)}{(1 + \eta^2)^2}$,

$$\tag{179}$$

$$w = \frac{1}{r}\frac{\partial\psi}{\partial r} = \frac{3}{8\pi}\frac{M}{\rho\nu z}\frac{1}{(1 + \eta^2)^2} .$$

This solution is due to Schlichting (1933).

It should be noted that the term $\nu\partial^2 w/\partial z^2$, neglected in (172), can be included in the present problem without destroying the possibility of a similarity solution.   It is therefore desirable to see whether it is worthwhile to include it and to find the condition under which it can be neglected without introducing much error.   In view of (179) the right-hand side of (172) is

$$\frac{\nu}{r}\frac{\partial}{\partial r}\left(r\frac{\partial w}{\partial r}\right) = \frac{9M^2}{512\pi^2\rho^2\nu^2 z^3}\frac{1}{\eta}(\eta W')', \tag{180}$$

whereas $\nu\dfrac{\partial^2 w}{\partial z^2} = \dfrac{3}{8\pi}\dfrac{M}{\rho z^3}\left[2(\eta W)' + \dfrac{1}{\eta}(\eta^2 W')'\right]$,   (181)

in which $W(\eta) = \dfrac{1}{(1 + \eta^2)^2}$.

It is then evident that

$$\frac{\nu}{r}\frac{\partial}{\partial r}\left(r\frac{\partial w}{\partial r}\right) \gg \nu\frac{\partial^2 w}{\partial z^2} \tag{182}$$

if $\dfrac{3M}{64\pi\rho\nu^2} \gg 1.$ (183)

Otherwise the term (181) is not negligible.   It must not be inferred, however, that the mere inclusion of the term (181), giving rise to an equation slightly different from (176), which we shall call (A), would provide an *exact* solution of the Navier-Stokes equations, because the term involving the pressure p has been neglected and the inclusion of p would necessitate consideration of a second equation of motion.   Furthermore the plane $z = 0$, outside the region of validity of the boundary-layer equation, must be included if the solution is to be an exact solution of the Navier-Stokes equations, and at this plane both u and w found from Eq. (A) are infinite.

The reader may have noticed that the exact solution of the problem treated here is Squire's solution presented in Sec. 4.1. Squire showed that when his a is small, Schlichting's solution is not far different from his exact solution. [Note the similarity of (178) to (120).] That condition is equivalent to (183). If (183) is not satisfied, or if the function of η in (181) is much larger than the function of η in (180), which can happen outside the main part of the jet stream, (182) is not satisfied.

9. BOUNDARY-LAYER FLOW ALONG AN ARBITRARILY SHAPED TWO-DIMENSIONAL BODY

We now return to the discussion of steady two-dimensional boundary layers. If the body is symmetric and the flow is symmetric about the stagnation point,

$$U = u_1 x + u_3 x^3 + u_5 x^5 + u_7 x^7 + \cdots , \tag{184}$$

in which x is again measured along the boundary and u_1, u_2, etc., are constants for each particular flow, with different dimensions. These constants are either obtained by solving for the potential flow outside the boundary layer or, since the separation of the boundary layer from the body forms a wake that cannot be accounted for by the solution for potential flow about the body, by experiment. If y is again the normal distance from the boundary and

$$\eta = y \sqrt{\frac{u_1}{\nu}} , \tag{185}$$

then expansion of the stream function into the series

$$\psi = \sqrt{\frac{\nu}{u_1}} \, [u_1 x f_1(\eta) + 4 u_3 x^3 f_3(\eta) + 6 u_5 x^5 f_5(\eta) + 8 u_7 x^7 f_7(\eta) + \cdots] \tag{186}$$

allows the functions f_1, f_3, etc., to be determined for a given set of values u_1, u_3, etc., that is, for a given symmetric two-dimensional body, as shown by Blasius (1908) and Hiemenz (1911). Howarth (1935) showed that it is possible to resolve the functions f_5, f_7, etc. in terms of universal functions quite independent of the shape of the body. Thus,

$$f_5 = g_5 + \frac{u_3^2}{u_1 u_5} h_5, \qquad f_7 = g_7 + \frac{u_3 u_5}{u_1 u_7} h_7 + \frac{u_3^3}{u_1^2 u_7} k_7, \qquad \text{etc.} \tag{187}$$

From (186) it follows that

$$u = u_1 x f_1' + 4 u_3 x^3 f_3' + 6 u_5 x^5 f_5' + \cdots ,$$

$$v = -\sqrt{\frac{\nu}{u_1}} \, (u_1 f_1 + 12 u_3 x^2 f_3 + 30 u_5 x^4 f_5 + \cdots). \tag{188}$$

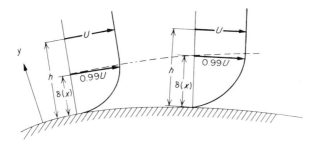

FIGURE 21. DEFINITION OF BOUNDARY-LAYER THICKNESS δ AND
THE QUANTITY h. AT $y = h$ THE VELOCITY IS VERY NEAR U
AND IS ASSUMED TO BE U FOR ALL PRACTICAL PURPOSES.

Substituting these equations into (141), in which $\partial u/\partial t$ is dropped because
the flow is steady, and equating the coefficients of equal powers of x on
both sides of (141) produce

$$f_1^2 - f_1 f_1'' = 1 + f_1''' , \tag{189}$$

$$4f_1' f_3' - 3f_1'' f_3 - f_1 f_3'' = 1 + f_3''' , \tag{190}$$

$$6f_1' g_5' - 5f_1'' g_5 - f_1 g_5'' = 1 + g_5''' , \tag{191}$$

and so forth. The boundary conditions at the solid boundary are

$$u = 0 = v \qquad \text{at } y = 0.$$

As to the condition at the edge of the boundary layer, the situation is a
little ambiguous. It is true that u approaches U as y increases. But to
impose the boundary condition at $y = \infty$ would be an absurd procedure,
if infinity is taken literally, for as we penetrate into the ambient flow outside
the boundary layer, certainly U will significantly vary with y as y increases,
except for the flow along a flat plate, and $U(x)$ will lose its meaning. In
fact, for a concave body lines normal to the boundary will intersect in the
region of ambient flow. To make things definite, we shall agree to define
a boundary-layer thickness $\delta(x)$ by (Fig. 21)

$$u = 0.99U \qquad \text{at } y = \delta, \tag{192}$$

and we shall agree that at $y = h$, where h is, say, 3 or 4 times δ, u is
indistinguishable from U. Thus the boundary condition at the place
where the boundary layer merges into the free stream can be written as
(Fig. 21)

$$u = U \qquad \text{at } y = h,$$

or $$u = U \qquad \text{at } \eta = h\sqrt{\frac{u_1}{\nu}} = \eta_h.$$

The last condition is replaced by

$$u = U \qquad \text{at } \eta = \infty$$

purely for convenience. This is allowable only because at a finite η, u is indistinguishable from U, and we really mean to claim validity of the calculation up to η_h. With this understanding the three boundary conditions demand that

$$f_1(0) = f_1'(0) = f_3(0) = f_3'(0) = g_5(0) = g_5'(0) = \cdots = 0,$$

$$f_1'(\infty) = 1, \qquad f_3'(\infty) = \tfrac{1}{4}, \qquad g_5'(\infty) = \tfrac{1}{6}, \qquad h_5'(\infty) = 0, \qquad \text{etc.}$$

(193)

It should be noted that (189) and its boundary conditions are precisely those for two-dimensional flow against an infinite plate, namely, (125) and (127). This is not surprising, since the Hiemenz-Howarth expansion is an expansion about the stagnation line, and the first term of the expansion should correspond precisely to stagnation-line flow. Another point to be noted is that although (189) is nonlinear, the subsequent equations are all linear. The calculation of Howarth (1935) was further extended by Frössling (1940), Ulrich (1949), and most recently by Tifford (1954). Table 2 shows the values obtained by Tifford (1954).

TABLE 2 Values of universal functions for two-dimensional symmetric flows

η	f_1'	f_3'	g_5'	h_5'	g_7'	h_7'	k_7'
0	0	0	0	0	0	0	0
0.2	0.2266	0.1251	0.1072	0.0141	0.0962	0.0173	0.0016
0.4	0.4145	0.2129	0.1778	0.0117	0.1563	0.0030	0.0044
0.6	0.5663	0.2688	0.2184	−0.0011	0.1879	−0.0286	0.0096
0.8	0.6859	0.2997	0.2366	−0.0177	0.1994	−0.0637	0.0174
1.0	0.7779	0.3125	0.2399	−0.0331	0.1980	−0.0925	0.0271
1.2	0.8467	0.3133	0.2341	−0.0442	0.1896	−0.1102	0.0369
1.4	0.8968	0.3070	0.2239	−0.0499	0.1782	−0.1159	0.0452
1.6	0.9323	0.2975	0.2123	−0.0504	0.1665	−0.1114	0.0506
1.8	0.9568	0.2871	0.2012	−0.0468	0.1558	−0.0997	0.0525
2.0	0.9732	0.2775	0.1916	−0.0406	0.1469	−0.0839	0.0510
2.2	0.9839	0.2695	0.1839	−0.0332	0.1400	−0.0669	0.0466
2.4	0.9905	0.2632	0.1781	−0.0257	0.1349	−0.0507	0.0402
2.6	0.9946	0.2586	0.1740	−0.0189	0.1313	−0.0367	0.0330
2.8	0.9970	0.2554	0.1712	−0.0133	0.1288	−0.0254	0.0257
3.0	0.9984	0.2532	0.1694	−0.0089	0.1273	−0.0168	0.0191
3.2	0.9992	0.2519	0.1682	−0.0057	0.1263	−0.0107	0.0135
3.4	0.9996	0.2510	0.1675	−0.0035	0.1257	−0.0065	0.0091
3.6	0.9998	0.2506	0.1671	−0.0021	0.1254	−0.0038	0.0059
3.8	0.9999	0.2503	0.1669	−0.0012	0.1252	−0.0021	0.0036
4.0	1.0000	0.2501	0.1668	−0.0006	0.1251	−0.0011	0.0021
η	f_1''	f_3''	g_5''	h_5''	g_7''	h_7''	k_7''
0	1.2326	0.7244	0.6347	0.1192	0.5792	0.1829	0.0076

For unsymmetric bodies

$$U = u_1 x + u_2 x^2 + u_3 x^3 + \cdots, \qquad (194)$$

and the expression for u corresponding to that given in (188) for symmetric flow is

$$u = u_1 f_1' x + 3 u_2 f_2' x^2 + 4\left(u_3 g_3' + \frac{u_2^2}{u_1} h_3'\right) x^3$$

$$+ 5\left(u_4 g_4' + \frac{u_2 u_3}{u_1} h_4' + \frac{u_2^3}{u_1^2} k_4'\right) x^4 + \cdots, \qquad (195)$$

the prime indicating differentiation with respect to η. The values of the universal functions given in Table 3 are those of Howarth (1935).

For bluff bodies a few terms in the series expansion have been found to produce satisfactory results, whereas for elongated bodies the series converges too slowly for the method to be useful.

TABLE 3   Values of universal functions for two-dimensional unsymmetric flows, according to Howarth (1935)†

η	f_2	f_2'	f_2''	g_3	g_3'	g_3''	h_3	h_3'	h_3''	k_4	k_4'	k_4''
0.0	0.000	0.000	0.7982	0.000	0.000	0.725	0.00	0.00	0.166	0.00	0.00	−0.019
0.1	0.004	0.075	0.699	0.004	0.068	0.625	0.00	0.01	0.12	0.00	0.00	−0.02
0.2	0.015	0.140	0.602	0.013	0.125	0.529	0.00	0.02	0.07	0.00	0.00	−0.02
0.3	0.032	0.195	0.509	0.028	0.174	0.438	0.01	0.03	+0.03	0.00	−0.01	−0.01
0.4	0.054	0.242	0.423	0.048	0.213	0.354	0.01	0.03	−0.01	0.00	−0.01	−0.01
0.5	0.080	0.280	0.344	0.071	0.245	0.278	0.01	0.03	−0.04	0.00	−0.01	0.00
0.6	0.109	0.311	0.273	0.096	0.269	0.211	0.01	0.02	−0.05	0.00	−0.01	0.00
0.7	0.142	0.336	0.210	0.124	0.287	0.153	0.02	0.02	−0.07	−0.01	0.00	+0.01
0.8	0.175	0.353	0.156	0.154	0.300	0.104	0.02	+0.01	−0.07	−0.01	0.00	0.02
0.9	0.212	0.367	0.109	0.184	0.308	0.063	0.02	0.00	−0.07	−0.01	0.00	0.03
1.0	0.249	0.375	0.070	0.215	0.313	0.029	0.02	−0.01	−0.07	−0.01	0.00	0.03
1.1	0.287	0.381	0.037	0.246	0.314	+0.003	0.02	−0.01	−0.06	−0.01	0.00	0.04
1.2	0.325	0.384	+0.012	0.277	0.313	−0.017	0.01	−0.02	−0.05	−0.01	+0.01	0.04
1.3	0.363	0.384	−0.007	0.309	0.311	−0.031	0.01	−0.02	−0.04	0.00	0.01	0.04
1.4	0.402	0.382	−0.021	0.340	0.307	−0.041	0.01	−0.03	−0.03	0.00	0.01	0.04
1.5	0.440	0.380	−0.032	0.370	0.302	−0.048	+0.01	−0.03	−0.02	0.00	0.02	0.03
1.6	0.477	0.377	−0.039	0.400	0.298	−0.051	0.00	−0.03	0.00	0.00	0.02	0.03
1.7	0.515	0.373	−0.042	0.429	0.293	−0.052	0.00	−0.03	0.00	0.00	0.02	0.02
1.8	0.552	0.368	−0.043	0.458	0.288	−0.051	0.00	−0.03	+0.01	+0.01	0.03	0.01
1.9	0.588	0.364	−0.043	0.487	0.283	−0.048	−0.01	−0.03	0.02	0.01	0.03	+0.01
2.0	0.625	0.361	−0.041	0.515	0.278	−0.045	−0.01	−0.03	0.03	0.01	0.03	0.00
2.1	0.661	0.357	−0.038	0.543	0.273	−0.040	−0.01	−0.02	0.03	0.02	0.03	−0.01
2.2	0.696	0.353	−0.034	0.570	0.269	−0.036	−0.01	−0.02	0.03	0.02	0.03	−0.01
2.3	0.731	0.350	−0.030	0.597	0.266	−0.032	−0.01	−0.02	0.03	0.02	0.02	−0.02
2.4	0.766	0.346	−0.026	0.623	0.263	−0.028	−0.01	−0.02	0.04	0.02	0.02	−0.02
2.5	0.801	0.344	−0.022	0.649	0.259	−0.023	−0.02	−0.01	0.03	0.03	0.02	−0.03
2.6	0.835	0.342	−0.019	0.676	0.258	−0.020	−0.02	−0.01	0.03	0.03	0.02	−0.03
2.7	0.870	0.340	−0.016	0.701	0.257	−0.017	−0.02	−0.01	0.02	0.03	0.02	−0.03
2.8	0.904	0.338	−0.014	0.726	0.256	−0.013	−0.02	−0.01	0.02	0.03	0.01	−0.02
2.9	0.938	0.337	−0.011	0.751	0.254	−0.011	−0.02	0.00	0.02	0.03	0.01	−0.02
3.0	0.972	0.336	−0.009	0.777	0.253	−0.010	−0.02	0.00	0.02	0.03	0.00	−0.01
3.1	1.006	0.335	−0.007	0.802	0.252	−0.008	−0.02	0.00	0.02	0.03	0.00	0.00
3.2	1.040	0.335	−0.006	0.828	0.251	−0.007	−0.02	0.00	0.01	0.03	0.00	0.00
3.3	1.073	0.335	−0.005	0.853	0.251	−0.005	−0.02	0.00	0.01	0.03	0.00	0.00
3.4	1.106	0.334	−0.003	0.879	0.251	−0.004	−0.02	0.00	0.00	0.03	0.00	0.00
3.5	1.139	0.334	−0.003	0.904	0.250	−0.003	−0.02	0.00	0.00	0.03	0.00	0.00
3.6	1.172	0.334	−0.002	0.929	0.250	−0.002	−0.02	0.00	0.00	0.03	0.00	0.00
3.7	1.205	0.334	−0.001	0.954	0.250	−0.002	−0.02	0.00	0.00	0.03	0.00	0.00
3.8	1.238	0.333	−0.001	0.979	0.250	−0.001	−0.02	0.00	0.00	0.03	0.00	0.00
3.9	1.271	0.333	−0.001	1.004	0.250	−0.001	−0.02	0.00	0.00	0.03	0.00	0.00
4.0	1.304	0.333	−0.001	1.029	0.250	−0.001	−0.02	0.00	0.00	0.03	0.00	0.00
4.1	1.337	0.333	0.000	1.054	0.250	0.000	−0.02	0.00	0.00	0.03	0.00	0.00
4.2	1.370	0.333	0.000									
4.3	1.403	0.333	0.000									

† By permission from S. Goldstein (ed.), "Modern Developments in Fluid Dynamics," vol. 1, p. 153, Oxford University Press, 1938.

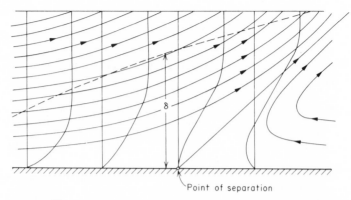

Point of separation

FIGURE 22. SEPARATION AND REVERSE FLOW VERTICAL SCALE EXAGGERATED. [*By permission from Hunter Rouse (ed.), "Advanced Fluid Mechanics," p.* 308, *John Wiley & Sons, Inc., New York,* 1959.]

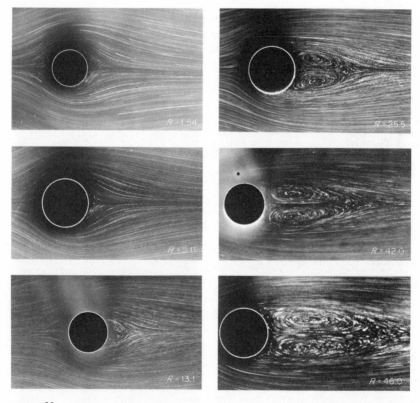

FIGURE 23. FLOW PATTERNS SHOWING WAKES BEHIND CIRCULAR CYLINDERS AT VARIOUS REYNOLDS NUMBERS. THE REYNOLDS NUMBER IS BASED ON THE DIAMETER AND THE SPEED AT INFINITY. THESE PICTURES ARE DIFFERENT FROM THOSE IN TANEDA (1956a) IN THE REYNOLDS NUMBERS OF THE FLOW. (*Courtesy of Prof. S. Taneda.*)

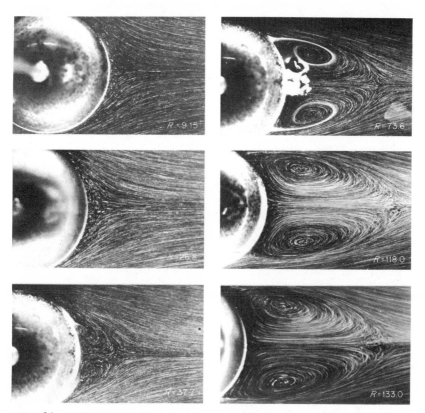

FIGURE 24. FLOW PATTERNS SHOWING WAKES BEHIND SPHERES IN A MERIDIANAL PLANE. THE REYNOLDS NUMBER IS BASED ON THE DIAMETER AND THE SPEED AT INFINITY. [*From Kyūshū Daigaku Seisangaku Kenkyūsho Hōkoku*, 4:99 (1956), *courtesy of Prof. S. Taneda.*]

Where the normal gradient of the tangential velocity is zero on the boundary, reverse flow near the boundary (Fig. 22), and hence separation of the boundary-layer flow from the boundary, are incipient. Hiemenz (1911) measured the velocity distribution in the potential flow around a circular cylinder and found

$$u_1 = 7.151, \quad u_3 = -0.04497, \quad u_5 = -0.00033$$

for the series (184), with the x therein measured in centimeters. That these do not agree entirely with the result obtained from the classical solution of potential flow around a circular cylinder is perhaps due to the existence of the wake. At any rate Hiemenz used these figures and series (186) to carry the calculation up to the value of x where

$$\frac{\partial u}{\partial y} = 0 \quad \text{at } y = 0,$$

FIGURE 25. KÁRMÁN VORTEX STREET BEHIND AN ELLIPTIC CYLINDER. [*From G. J. Richards, Phil. Trans. Roy. Soc. London Ser. A.*, **233** (1935), *courtesy of the Royal Society of London.*]

namely, where separation begins.   His calculation showed that the separation point is 82° from the forward stagnation point.   This is also the figure he obtained by experiment.

Wakes behind a circular cylinder and a sphere at various Reynolds numbers are shown in Figs. 23 and 24, respectively.   At large Reynolds numbers the eddies do not remain stationary with the cylinder but shed alternately to form the Kármán street [von Kármán (1911), von Kármán and Rubach (1912)] in the wake, which consists of two staggered rows of counterrotating vortices (Fig. 25).

10. BOUNDARY-LAYER FLOW ALONG AN ARBITRARILY SHAPED AXISYMMETRIC BODY

The method of series expansion used by Blasius, Hiemenz, and Howarth for two-dimensional boundary layers has been extended by Frössling (1940) for application to axisymmetric boundary layers, for which the governing equations are (167) and (168).   Equation (168) allows the use of a stream function ψ (which is neither the stream function for two-dimensional flow nor the Stokes stream function) in terms of which

$$u = \frac{1}{r}\frac{\partial \psi}{\partial y}, \qquad v = -\frac{1}{r}\frac{\partial \psi}{\partial x}. \tag{196}$$

Frössling described the shape of the body by

$$r = r_1 x + r_3 x^3 + r_5 x^5 + \cdots \tag{197}$$

and used the expansion

$$U = u_1 x + u_3 x^3 + u_5 x^5 + \cdots \tag{198}$$

for the exterior flow. For the boundary-layer flow, Frössling used two alternative expansions, one for the ψ in (196) and the other for ψ/r, which will be denoted by Ψ here. Since Frössling gave numerical results for the latter expansion only, it alone will be presented. In terms of Ψ,

$$u = \Psi_y, \qquad v = -\Psi_x - \frac{1}{r}\frac{dr}{dx}\Psi, \tag{199}$$

in which subscripts on Ψ indicate partial differentiation. Substitution of these into (167) yields

$$\Psi_y \Psi_{xy} - \left(\Psi_x + \frac{1}{r}\frac{dr}{dx}\right)\Psi_{yy} = U\frac{dU}{dx} + \nu\Psi_{yyy}. \tag{200}$$

Frössling took

$$\Psi = \sqrt{\frac{\nu}{2u_1}}\,[u_1 x f_1(\eta) + 2u_3 x^3 f_3(\eta) + 3u_5 x^5 f_5(\eta) + \cdots], \tag{201}$$

in which

$$\eta = y\sqrt{\frac{2u_1}{\nu}} \tag{202}$$

and

$$f_3 = g_3 + \frac{r_3 u_1}{r_1 u_3}h_3,$$

$$f_5 = g_5 + \frac{r_5 u_1}{r_1 u_5}h_5 + \frac{u_3{}^2}{u_1 u_5}k_5 + \frac{r_3 u_3}{r_1 u_5}j_5 + \frac{r_3{}^2 u_1}{r_1{}^2 u_5}q_5, \tag{203}$$

and so on. The functions g_3, h_3, g_5, etc., are universal functions quite independent of the shape of the body. The u in the boundary layer is obtained from (199) and (201) as

$$u = u_1 x f_1' + 2u_3 x^3 f_3' + 3u_5 x^5 f_5' + \cdots,$$

the primes indicating differentiation with respect to η.

Substitution of (203) into (201) and then (201) into (200) yields the equations

$$f_1''' = -f_1 f_1'' + \tfrac{1}{2}(f_1'^2 - 1),$$

$$g_3''' = -f_1 g_3'' + 2f_1' g_3' - 2f_1'' g_3 - 1,$$

$$h_3''' = -f_1 h_3'' + 2f_1' h_3' - 2f_1'' h_3 - \tfrac{1}{2} f_1 f_1'',$$

$$g_5''' = -f_1 g_5'' + 3f_1' g_5' - 3f_1'' g_5 - 1,$$

$$h_5''' = -f_1 h_5'' + 3f_1' h_5' - 3f_1'' h_5 - \tfrac{2}{3} f_1 f_1'',$$

and so on. The boundary conditions are

$$\Psi = 0 = \Psi_y \text{ at } y = 0 \quad \text{and} \quad \Psi_y = U \text{ at } y = \infty.$$

TABLE 4   Values of universal functions for axisymmetric flows, according to Frössling (1940)

η	f_1'	g_3'	h_3'	g_5'	h_5'	k_5'	j_5'	q_5'	$\eta = 0$
0.0	0.0	0.0	0.0	0.0	0.0	0.0	0.0	0.0	$f_1'' = 0.9277$
0.2	0.1755	0.1896	0.0090	0.1612	0 0101	0.0255	0.0058	−0.0049	$g_3'' = 1.0475$
0.4	0.3311	0.3400	0.0176	0.2838	0.0198	0.0324	0.0107	−0.0096	$h_3'' = 0.0448$
0.6	0.4669	0.4535	0.0254	0.3709	0.0285	0.0241	0.0137	−0.0140	$g_5'' = 0.9054$
0.8	0.5833	0.5334	0.0316	0.4270	0.0354	0.0051	0.0134	−0.0176	$h_5'' = 0.0506$
1.0	0.6811	0.5842	0.0358	0.4576	0.0399	−0.0195	0.0096	−0.0204	$k_5'' = 0.1768$
1.2	0.7614	0.6110	0.0377	0.4683	0.0417	−0.0447	0.0025	−0.0222	$j_5'' = 0.0291$
1.4	0.8258	0.6193	0.0372	0.4645	0.0409	−0.0665	−0.0064	−0.0229	$q_5'' = -0.0244$
1.6	0.8761	0.6144	0.0346	0.4515	0.0379	−0.0819	−0.0156	−0.0226	
1.8	0.9142	0.6015	0.0306	0.4335	0.0334	−0.0897	−0.0234	−0.0212	
2.0	0.9422	0.5845	0.0258	0.4139	0.0279	−0.0899	−0.0287	−0.0192	
2.2	0.9622	0.5666	0.0207	0.3952	0.0223	−0.0838	−0.0310	−0.0167	
2.4	0.9760	0.5500	0.0158	0.3787	0.0170	−0.0733	−0.0304	−0.0139	
2.6	0.9853	0.5358	0.0116	0.3652	0.0124	−0.0605	−0.0275	−0.0111	
2.8	0.9912	0.5245	0.0081	0.3549	0.0086	−0.0474	−0.0232	−0.0085	
3.0	0.9949	0.5161	0.0054	0.3473	0.0058	−0.0352	−0.0184	−0.0062	
3.2	0.9972	0.5102	0.0035	0.3420	0.0037	−0.0249	−0.0137	−0.0044	
3.4	0.9985	0.5061	0.0022	0.3385	0.0023	−0.0168	−0.0097	−0.0029	
3.6	0.9992	0.5036	0.0013	0.3363	0.0013	−0.0109	−0.0065	−0.0019	
3.8	0.9996	0.5020	0.0007	0.3350	0.0008	−0.0067	−0.0042	−0.0012	
4.0	0.9998	0.5011	0.0004	0.3342	0.0004	−0.0040	−0.0025	−0.0007	
4.2	0.9999	0.5006	0.0002	0.3338	0.0002	−0.0022	−0.0015	−0.0004	
4.4	0.9999	0.5003	0.0002	0.3336	0.0001	−0.0012	−0.0008	−0.0002	
4.6	1.0000	0.5001	0.0000	0.3334	0.0000	−0.0006	−0.0004	−0.0001	
4.8	1.0000	0.5000		0.3334	0.0000	−0.0003	−0.0002	0.0000	
5.0				0.3334		−0.0001	−0.0001	0.0000	
5.2				0.3334		0.0000	0.0000		
5.4				0.3333					

These, in terms of the universal function, become

All universal functions and their first derivatives $= 0$ at $\eta = 0$;

$$f_1' = 1, \qquad g_3' = \tfrac{1}{2}, \qquad g_5' = \tfrac{1}{3},$$

$$h_3' = h_5' = k_5' = j_5' = q_5' = 0, \qquad \ldots \qquad \text{at } \eta = \infty.$$

At $\eta = \infty$, the first derivatives of only the leading functions in the expressions for f_3, f_5, etc., are different from zero. This system of equations and boundary conditions was solved numerically by Frössling. Note that apart from the coefficient of f_1''', which can be accounted for by the definition of η, the first of Eqs. (203) is the same as that for the flow studied by Homann (1936) and described in Table 1. This is to be expected, since the series (199) is an expansion about the stagnation point, and its first term corresponds to the case of zero curvature of the boundary, which is the case with the Homman flow. Table 4 shows the values of the universal functions obtained by Frössling.

11. THE KÁRMÁN-POHLHAUSEN METHOD FOR APPROXIMATE SOLUTION

For any arbitrarily shaped body, provided the flow outside the boundary layer is known, an approximate solution by step-by-step integration is possible. While the method is not elegant, it manages to satisfy the requirements of continuity and momentum, albeit in integrated forms only. The power of the method lies in its simplicity. We shall discuss only steady two-dimensional boundary layers here. The method can be extended for application to steady axisymmetric boundary layers.

To start with, two thicknesses of the boundary layer other than δ will be defined. First, the displacement thickness δ^* defined by

$$U\delta^* = \int_0^h (U - u)\,dy \tag{204}$$

is the distance through which the outside (and, in the vast majority of cases, irrotational) flow has been displaced. The thickness θ defined by

$$U^2\theta = \int_0^h u(U - u)\,dy \tag{205}$$

is a measure of the momentum deficiency due to the presence of the boundary layer.

The equations governing steady two-dimensional boundary layers are

$$u\,\frac{\partial u}{\partial x} + v\,\frac{\partial u}{\partial y} = U\,\frac{dU}{dx} + v\,\frac{\partial^2 u}{\partial y^2}, \tag{206}$$

$$\frac{\partial u}{\partial x} + \frac{\partial v}{\partial y} = 0. \tag{207}$$

By virtue of (207), (206) can be written as

$$\frac{\partial u^2}{\partial x} + \frac{\partial(uv)}{\partial y} = U\frac{dU}{dx} + \nu\frac{\partial^2 u}{\partial y^2}. \tag{208}$$

Since
$$v_h = \int_0^h \frac{\partial v}{\partial y}\, dy = -\int_0^h \frac{\partial u}{\partial x}\, dy \tag{209}$$

and the shear τ_0 on the boundary is equal to $\mu\partial u/\partial y$ evaluated at $y = 0$, integration of (208) from $y = 0$ to $y = h$ gives

$$\frac{d}{dx}\int_0^h u^2\, dy - U\int_0^h \frac{\partial u}{\partial x}\, dy = hU\frac{dU}{dx} - \frac{\tau_0}{\rho}, \tag{210}$$

which is the Kármán integral condition [von Kármán (1921)].

It is convenient to express (210) in terms of δ^* and θ. Since

$$U\int_0^h \frac{\partial u}{\partial x}\, dy = \frac{d}{dx}\int_0^h Uu\, dy - \frac{dU}{dx}\int_0^h u\, dy,$$

(210) can be written as

$$\frac{\tau_0}{\rho} = \frac{d}{dx}\int_0^h (U - u)u\, dy + \frac{dU}{dx}\int_0^h (U - u)\, dy$$

$$= \frac{d}{dx}U^2\theta + \frac{dU}{dx}\delta^* U,$$

or
$$\frac{\tau_0}{\rho} = U^2\frac{d\theta}{dx} + (2 + \delta^*)U\frac{dU}{dx}, \tag{211}$$

which is really just the equation of motion in integral form.

Seeking a simple method for an approximate solution of (211), Pohlhausen (1921) decided to make the boundary conditions

$$u = U \quad \text{and} \quad \frac{\partial u}{\partial y} = 0 = \frac{\partial^2 u}{\partial y^2} \quad \text{at } y = \delta_p(x), \tag{212}$$

in addition to the rigorous conditions

$$u = 0 \quad \text{and} \quad \nu\frac{\partial^2 u}{\partial y^2} = -UU' \quad \text{at } y = 0. \tag{213}$$

The second condition in (213) follows from (206) since $u = 0 = v$ at $y = 0$. The prime on U denotes differentiation with respect to x, since U is a function of x only. The conditions in (212) are imposed at a finite y ($= \delta_p$), and δ_P is not the δ defined in (192), the subscript being the first letter of "Pohlhausen." The simplest form for u capable of satisfying (212) and (213) is

$$\frac{u}{U} = a\eta + b\eta^2 + c\eta^3 + d\eta^4, \qquad \eta = \frac{y}{\delta_P}, \tag{214}$$

where η varies from zero to 1, since y varies from zero to δ_P. The conditions (212) and (213) then demand

$$a + b + c + d = 1, \qquad a + 2b + 3c + 4d = 0,$$

$$2b + 6c + 12d = 0, \qquad b = -\frac{\lambda}{2}, \tag{215}$$

in which

$$\lambda = \frac{\delta_P{}^2}{\nu} U'. \tag{216}$$

The solution of (215) is

$$a = 2 + \frac{\lambda}{6}, \qquad b = -\frac{\lambda}{2}, \qquad c = -2 + \frac{\lambda}{2}, \qquad d = 1 - \frac{\lambda}{2},$$

so that

$$\frac{u}{U} = F(\eta) + \lambda G(\eta), \tag{217}$$

in which $\quad F(\eta) = \eta(2 - 2\eta^2 + \eta^3), \qquad G(\eta) = \frac{1}{6}\eta(1 - \eta)^3. \tag{218}$

The task is to determine λ, hence δ_P, step by step. For this purpose (211) must be called to service. Since $\lambda = -12$ corresponds to $\partial u/\partial y = 0$ at $y = 0$, that is, to incipient separation, and since for $\lambda > 12$ u would be greater than U in the boundary layer, the range of variation of λ is

$$-12 \leq \lambda \leq 12.$$

Since (211) involves δ^* and θ, these two quantities should first be evaluated in terms of λ. From (204), (205), and (217), if follows that

$$\frac{\delta^*}{\delta_P} = \frac{3}{10} - \frac{\lambda}{120}, \qquad \frac{\theta}{\delta_P} = \frac{37}{315} - \frac{\lambda}{945} - \frac{\lambda^2}{9{,}072}. \tag{219}$$

The right-hand side of (211) is

$$\frac{\tau_0}{\rho} = \nu \left.\frac{\partial u}{\partial y}\right|_{y=0} = \nu \left(2 + \frac{\lambda}{6}\right)\frac{U}{\delta_P}. \tag{220}$$

The rest of the development becomes rather tedious. Equation (211) can be written as

$$\frac{\tau_0}{\mu}\frac{\theta}{U} = \frac{U\theta\theta'}{\nu} + \left(2 + \frac{\delta^*}{\theta}\right)\frac{U'\theta^2}{\nu}. \tag{221}$$

For convenience, two parametric dependent variables

$$\kappa = \frac{\theta^2}{\nu} U' \quad \text{and} \quad Z = \frac{\theta^2}{\nu} \tag{222}$$

will be used instead of the single dependent variable δ_P. From (222) it follows that

$$\kappa = ZU', \tag{223}$$

which provides one of the equations needed. The other equation is to be provided by (221), in some convenient form. From (216) and (219)

$$\kappa = \lambda \left(\frac{37}{315} - \frac{\lambda}{945} - \frac{\lambda^2}{9,072} \right)^2, \tag{224}$$

Then (219) permits us to write

$$\frac{\delta^*}{\theta} = \frac{\delta^*/\delta_P}{\theta/\delta_P} = g(\lambda) = f_1(\kappa), \tag{225}$$

where $g(\lambda)$ is given by (219) and has not been written out in full. Furthermore, upon the use of (220) and (219) we have

$$\frac{\tau_0}{\mu} \frac{\theta}{U} = \frac{\tau_0}{\mu} \frac{\delta_P}{U} \frac{\theta}{\delta_P} = \frac{12 + \lambda}{6} \left(\frac{37}{315} - \frac{\lambda}{945} - \frac{\lambda^2}{9,072} \right) = f_2(\kappa).$$

Thus (221) becomes

$$\tfrac{1}{2} U \frac{dZ}{dx} + [2 + f_1(\kappa)]\kappa - f_2(\kappa) = 0,$$

or

$$\frac{dZ}{dx} = \frac{F(\kappa)}{U}, \tag{226}$$

with

$$F(\kappa) = 2f_2(\kappa) - 4\kappa - 2f_1(\kappa)\kappa.$$

Equations (223) and (226) are the two equations to be used for a step-by-step solution, U and U' being given functions of x.

The beginning value for κ is obtained by solving

$$F(\kappa) = 0, \tag{227}$$

since at the stagnation point, where the calculation starts, $U = 0$. The solution of (227) is

$$\kappa = \kappa_0 = 0.0770 \qquad \text{corresponding to} \qquad \lambda = \lambda_0 = 7.052. \tag{228}$$

Thus

$$Z_0 = \frac{0.0770}{U_0'}, \tag{229}$$

and by de l'Hospital's rule

$$\left(\frac{dZ}{dx} \right)_0 = -0.0652 \frac{U_0''}{U_0'^2}, \tag{230}$$

the zero subscript indicating the stagnation point. Equations (229) and (230) provide the starting values for a numerical integration of (223) and (226), by standard methods, up to the point $\lambda = -12$.

12. SEPARATION AND DRAG

The drag on a body submerged in a stream arises from two sources, the shear stress and the pressure distribution along the body. The part of the drag due to the former is called the *skin friction*, and the part due to the pressure distribution is called the *form drag*.   For a flat plate the drag is entirely due to the shear stress.   For an elongated body the shear stress may account for the major part of the drag.   For a blunt body most of the drag is due to the unsymmetric pressure distribution.   Note that $\partial v/\partial y$, being equal to $-\partial u/\partial x$ for incompressible fluids, is equal to zero on the boundary, x and y being tangential and normal coordinates.   Thus the normal stress $-p + 2\mu\, \partial v/\partial y$ is exactly $-p$.   Hence the form drag is really due to the distribution of p.

If the boundary layer remains thin and attached to the boundary throughout, the pressure distribution on the body will be the same as that calculated from the solution for potential flow around the body, since the pressure variation throughout the thickness of the boundary layer is negligible.   If this flow has constant velocity at infinity and the influence of any other boundaries present is negligible, i.e., if the body moves with constant velocity in a large body of otherwise quiescent fluid, the form drag must then be zero.   The separation of the boundary layer (Fig. 22) creates a wake (Figs. 23 and 24) and changes all this.   A wake region with prevailing low pressures gives rise to the form drag.

It is possible to prevent separation and thus to reduce the form drag by sucking away the retarded fluid in the boundary layer.   A detailed discussion of the subject is outside of the scope of this book.

At a sufficiently high Reynolds number the boundary layer becomes turbulent at some distance from the stagnation point, and the increased mixing due to turbulence is able to carry the slowly moving fluid near the boundary further downstream before separation occurs, thus reducing the extent of the low-pressure region in the wake. The drag is therefore abruptly reduced at this critical Reynolds number, which is shown in Fig. 26 to be about 2×10^5 for a sphere.   The coefficient of drag C_D in this figure is defined by

$$C_D = \frac{F}{\frac{1}{2}\rho U^2 A}\,,$$

F being the drag and A the projectional area of the body on a plane normal to the direction of its velocity (or, equivalently, of the velocity of the fluid at infinity if the body is at rest), which has the magnitude U.   For the Reynolds number Ua/ν ($a =$ radius) less than 0.1,

$$C_D = \frac{24}{R}\,,$$

which is (248) in another form, and which will be derived in the next section.

Separation can also be delayed at Reynolds numbers lower than the critical R by turbulence artificially created by roughness at the boundary. Experiments have shown that sandpaper attached to the body near the front stagnation point can reduce drag.

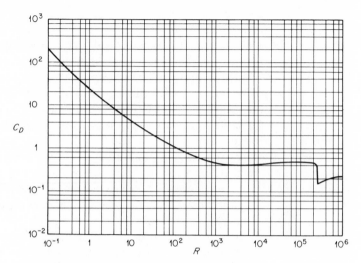

FIGURE 26.   COEFFICIENT OF DRAG C_D ON A SPHERE AS A FUNCTION OF
THE REYNOLDS NUMBER R.

So far boundary layers of a compressible fluid have been left un-
touched. They will be treated briefly in the next chapter (on heat
transfer), since the energy equation must be considered. We conclude
our discussion of the boundary-layer theory for incompressible fluids
here and turn now to a study of very slow motion of a viscous fluid.

13.   VERY SLOW MOTION OF A VISCOUS FLUID: STOKES' SOLUTION FOR THE FALLING SPHERE

If a sphere of radius a is falling slowly in a fluid of viscosity μ, and if $\Delta\gamma$
is the difference in specific weight of the sphere and the fluid and U is the
terminal speed, a dimensional analysis shows that the only dimensionless
parameter is the Stokes number $S = U\mu/(a^2\Delta\gamma)$, which must therefore be
a constant if inertial forces are neglected. This constant must be found
by mathematical analysis.

It is convenient to consider the sphere to be stationary and the
velocity of the flow relative to the sphere to be U at infinite distance from
the sphere. The Navier-Stokes equations are then

$$\mu\nabla^2(u,v,w) = \left(\frac{\partial}{\partial x}, \frac{\partial}{\partial y}, \frac{\partial}{\partial z}\right)p_d, \qquad (231)$$

from which, on elimination of p_d by cross differentiation, one obtains

$$\nabla^2\xi = 0, \qquad \nabla^2\eta = 0, \qquad \nabla^2\zeta = 0, \qquad (232)$$

in which ξ, η, and ζ are Cartesian components of the vorticity. If cylindrical coordinates are used, with the z axis vertical, then

$$u = c \cos \varphi, \qquad v = c \sin \varphi, \qquad w = w, \tag{233}$$

in which c (the velocity component in the r direction) and w are functions of r and z only. It follows from (233) that

$$\xi = -2\omega \sin \varphi, \qquad \eta = 2\omega \cos \varphi, \qquad \zeta = 0, \tag{234}$$

in which 2ω is the resultant vorticity given by

$$2\omega = \frac{\partial c}{\partial z} - \frac{\partial w}{\partial r}. \tag{235}$$

From (234) one obtains

$$\frac{1}{2} \nabla^2 \xi = -\left(\frac{\partial^2}{\partial z^2} + \frac{\partial^2}{\partial r^2} + \frac{1}{r} \frac{\partial}{\partial r} + \frac{1}{r^2} \frac{\partial^2}{\partial \varphi^2} \right) \omega \sin \varphi = -\sin \varphi \left(\frac{\partial^2}{\partial z^2} + \frac{\partial^2}{\partial r^2} \right)$$

$$+ \frac{1}{r} \frac{\partial}{\partial r} - \frac{1}{r^2} \right) \omega = - \frac{\sin \varphi}{r} \left(\frac{\partial^2}{\partial z^2} + \frac{\partial^2}{\partial r^2} - \frac{1}{r} \frac{\partial}{\partial r} \right) r\omega = - \frac{\sin \varphi}{r} L(r\omega)$$

$$\tag{236}$$

if
$$L = \frac{\partial^2}{\partial z^2} + \frac{\partial^2}{\partial r^2} - \frac{1}{r} \frac{\partial}{\partial r}.$$

Similarly,
$$\frac{1}{2} \nabla^2 \eta = \frac{\cos \varphi}{r} L(r\omega). \tag{237}$$

Thus (232) can be represented by the single equation (since $\zeta = 0$):

$$L(r\omega) = 0, \tag{238}$$

but
$$w = \frac{1}{r} \frac{\partial \psi}{\partial r}, \qquad c = -\frac{1}{r} \frac{\partial \psi}{\partial z},$$

and
$$-\omega = \frac{1}{2} \frac{1}{r} \left(\frac{\partial^2}{\partial z^2} + \frac{\partial^2}{\partial r^2} - \frac{1}{r} \frac{\partial}{\partial r} \right) \psi = \frac{1}{2} \frac{1}{r} L\psi. \tag{239}$$

Thus (238) becomes
$$L^2 \psi = 0. \tag{240}$$

If spherical coordinates (R, θ, φ) are now introduced, with the polar axis pointing to the direction opposite to the direction of motion of the sphere,

$$z = R \cos \theta, \qquad r = R \sin \theta,$$

and (240) becomes

$$\left[\frac{\partial^2}{\partial R^2} + \frac{\sin \theta}{R^2} \frac{\partial}{\partial \theta} \left(\frac{1}{\sin \theta} \frac{\partial}{\partial \theta} \right) \right]^2 \psi = 0. \tag{241}$$

This is satisfied by $\qquad \psi = \sin^2 \theta\, f(R)$ $\hfill (242)$

if $\qquad\qquad \left(\dfrac{d^2}{dR^2} - \dfrac{2}{R^2}\right)^2 f(R) = 0.$ $\hfill (243)$

The solution of (243) is

$$f(R) = \frac{A}{R} + BR + CR^2 + DR^4,$$

in which $\qquad\qquad C = \tfrac{1}{2}U \quad$ and $\quad D = 0$

because of the condition at $R = \infty$ $(\psi = \tfrac{1}{2}UR^2 \sin^2 \theta)$, and

$$A = \tfrac{1}{4}Ua^3, \qquad B = -\tfrac{3}{4}Ua$$

because of the nonslip condition at the surface of the sphere. Thus, with v_R and v_θ denoting the velocity components in the directions of increasing R and θ to avoid confusion,

$$v_R = \frac{1}{R^2 \sin \theta} \frac{\partial \psi}{\partial \theta} = U \cos \theta + 2\left(\frac{A}{R^3} + \frac{B}{R}\right) \cos \theta,$$

$$v_\theta = -\frac{1}{R \sin \theta} \frac{\partial \psi}{\partial R} = -U \sin \theta + \left(\frac{A}{R^3} - \frac{B}{R}\right) \sin \theta. \qquad (244)$$

The resultant force acting on the surface of the sphere is by symmetry in the z direction and is given by

$$F = 2\pi a^2 \int_0^\pi \tau_{Rz} \sin \theta\, d\theta, \qquad (245)$$

in which τ_{Rz} is the force (not just the shear force) per unit area of the spherical surface in the z direction and can be obtained from τ_{zz}, τ_{yz}, and τ_{xz} by resolution:

$$\tau_{Rz} = \frac{z}{R}\tau_{zz} + \frac{y}{R}\tau_{yz} + \frac{x}{R}\tau_{xz}. \qquad (246)$$

From the expressions for τ_{zz}, τ_{yz}, and τ_{xz} it follows that

$$\tau_{Rz} = \cos \theta\left(-p_d + 2\mu \frac{\partial w}{\partial z}\right) + \mu \frac{y}{R}\left(\frac{\partial w}{\partial y} + \frac{\partial v}{\partial z}\right) + \mu \frac{x}{R}\left(\frac{\partial u}{\partial x} + \frac{\partial u}{\partial z}\right)$$

$$= -p_d \cos \theta + \mu \frac{\partial w}{\partial R} - \frac{\mu w}{R} + \frac{\mu}{R}\frac{\partial}{\partial z}(xu + yv + zw)$$

$$= -p_d \cos \theta + \mu \frac{\partial w}{\partial R} - \frac{\mu w}{R} + \frac{\mu}{R}\frac{\partial}{\partial z} Rv_R.$$

But

$$w = v_R \cos \theta - v_\theta \sin \theta, \qquad \frac{\partial v_R}{\partial z} = \frac{\partial v_R}{\partial R}\cos \theta - \frac{\partial v_R}{\partial \theta}\frac{\sin \theta}{R},$$

so that[1]

$$\tau_{Rz} = -p_d \cos\theta + 2\mu\cos\theta\,\frac{\partial v_R}{\partial R} - \mu\sin\theta\left(\frac{\partial v_\theta}{\partial R} + \frac{1}{R}\frac{\partial v_R}{\partial\theta} - \frac{v_\theta}{R}\right). \quad (247)$$

If (247) is substituted into (245), the first term in τ_{Rz} contributes a term $2\pi\mu aU$, the second term in τ_{Rz} contributes nothing, and the third term contributes a term equal to $4\pi\mu aU$. Thus

$$F = 6\pi\mu aU, \quad (248)$$

which is only the *hydrodynamic* part of the force acting on the sphere, since p_d instead of p has been used to obtain F. If the sphere is not accelerating, this force is balanced by the weight of the sphere minus the buoyancy if the mean density of the solid ρ' is greater than the fluid density:

$$\frac{4}{3}\pi a^3\,\Delta\gamma = \frac{4}{3}\pi a^3 g(\rho' - \rho) = 6\pi\mu aU.$$

From this we deduce

$$U = \frac{2a^2\,\Delta\gamma}{9\mu}, \qquad \Delta\gamma = g(\rho' - \rho),$$

or

$$S = \frac{U\mu}{a^2\,\Delta\gamma} = \frac{2}{9}. \quad (249)$$

This solution, due to Stokes (1851), has been well verified experimentally, as shown in Fig. 27, where $C_D = R/24$, or (248) holds on solid line.

The flow pattern, to a stationary observer, is shown in Fig. 28 and is quite different from the pattern of irrotational flow of an inviscid fluid caused by a moving sphere.

For two-dimensional flows Eqs. (231) have no solution. This is not surprising when one recalls that many two-dimensional problems in heat conduction governed by the Laplace equation also have no solution, e.g., the problem of temperature distribution in an infinite solid if the temperature at infinity is finite and the surface of a circular cylinder is maintained at a different but constant temperature. The lack of a solution of (231) for two-dimensional flow is called the Whitehead paradox.

---

[1] We have computed τ_{Rz} from the stress components in Cartesian coordinates for the benefit of those readers who do not wish to go through the general tensor analysis in Appendix 2. Those who prefer the systematic approach and have mastered the material in Appendix 2 should note that (247) is just

$$\tau_{Rz} = \tau_{RR}\cos\theta - \tau_{R\theta}\sin\theta = (-p_d + 2\mu e_{RR})\cos\theta - \mu e_{R\theta}, \quad (247a)$$

with τ_{RR} and $\tau_{R\theta}$ given in terms of e_{RR} and $e_{R\theta}$ by (A2.76) and e_{RR} and $e_{R\theta}$ given in (A2.78). Note that p_d can be found once the velocity field is known.

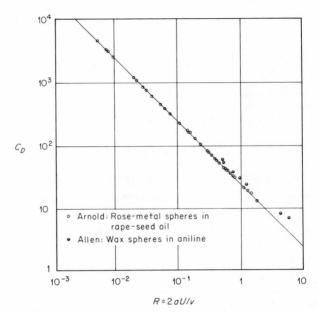

FIGURE 27. EXPERIMENTAL VERIFICATION OF THE STOKES' LAW
$C_D = R/24$, WHICH IS REPRESENTED BY THE STRAIGHT LINE. C_D is
defined on p. 361. [*By permission from Hunter Rouse (ed.),
"Advanced Fluid Mechanics," p.* 240, *John Wiley & Sons, Inc.,
New York*, 1959.]

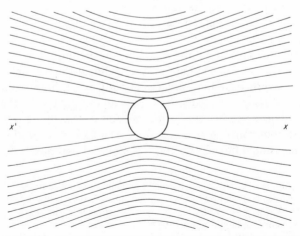

FIGURE 28. STREAMLINES IN THE FLOW CAUSED BY A SPHERE MOVING
WITH A SMALL CONSTANT VELOCITY: STOKES' SOLUTION. (*From
H. Lamb, "Hydrodynamics," p.* 599. *By permission of Cam-
bridge University Press.*)

14. VERY SLOW MOTION OF A VISCOUS FLUID: OSEEN'S APPROXIMATION

In Stokes' solution for the moving sphere, which has just been presented, the inertial terms are entirely neglected. In that sense Stokes' solution is an approximation to the solution of the Navier-Stokes equation at small Reynolds numbers. In a frame of reference moving with the sphere, the fluid near the sphere is almost at rest. Hence the neglect of the inertial terms there is well justified. But far away from the sphere the acceleration of the fluid is not as small. In order to account for the inertial terms in a simple way, Oseen (1910) made the substitutions

$$u = U + u_1, \qquad v = v_1, \qquad w = w_1 \tag{250}$$

in the Navier-Stokes equations and neglected quadratic terms in the quantities with subscripts. The fluid is assumed to move with velocity U in the x direction, in the frame of reference moving with the sphere. The resulting equations are, after the subscripts have been dropped,

$$U \frac{\partial}{\partial x}(u,v,w) = -\frac{1}{\rho}\left(\frac{\partial}{\partial x}, \frac{\partial}{\partial y}, \frac{\partial}{\partial z}\right)p_d + \nu\nabla^2(u,v,w). \tag{251}$$

The three equations in (251) are called *Oseen's equations*, and the approximation they stand for is called *Oseen's approximation*. Away from the sphere (251) is a better approximation to the Navier-Stokes equations than (231), but near the sphere (231) is better. The equation of continuity

$$\frac{\partial u}{\partial x} + \frac{\partial v}{\partial y} + \frac{\partial w}{\partial z} = 0 \tag{252}$$

furnishes the fourth equation needed. The boundary conditions are, in the frame of reference moving with the sphere,

$$u = v = w = 0 \qquad \text{at infinity,}$$

$$u = -U, \qquad v = 0 = w \qquad \text{on the surface of the sphere.} \tag{253}$$

Note that u, v, and w are really u_1, v_1, and w_1, and no longer the u, v, and w in (250) or (231).

Oseen gave a solution for the system consisting of (251) to (253), but we shall present the simpler solution of Lamb (1911). First, (251) and (252) give

$$\nabla^2 p_d = 0. \tag{254}$$

We shall write

$$p_d = -\rho U \frac{\partial \phi}{\partial x}, \tag{255}$$

in which

$$\nabla^2 \phi = 0. \tag{256}$$

Then a particular solution of (251) is

$$u = \frac{\partial \phi}{\partial x}, \qquad v = \frac{\partial \phi}{\partial y}, \qquad w = \frac{\partial \phi}{\partial z}.$$

The complete solution is

$$u = \frac{\partial \phi}{\partial x} + u', \qquad v = \frac{\partial \phi}{\partial y} + v', \qquad w = \frac{\partial \phi}{\partial z} + w', \tag{257}$$

in which u', v', and w' satisfy the equations

$$\left(\nabla^2 - 2k\frac{\partial}{\partial x}\right)(u', v', w') = 0, \tag{258}$$

with $k = U/2\nu$, as substitution of (257) into (251) will show.

Since the motion is symmetric with respect to the x axis, the x component of the vorticity, ξ, is zero, and therefore

$$\frac{\partial \eta}{\partial y} + \frac{\partial \zeta}{\partial z} = 0,$$

since the divergence of the vorticity vector is always zero. This admits the use of a vorticity function χ, in terms of which

$$\eta = \frac{\partial \chi}{\partial z}, \qquad \zeta = -\frac{\partial \chi}{\partial y}. \tag{259}$$

From (258) we deduce that

$$\left(\nabla^2 - 2k\frac{\partial}{\partial x}\right)(\eta, \zeta) = 0. \tag{260}$$

In view of (259) and (260), we have

$$\left(\nabla^2 - 2k\frac{\partial}{\partial x}\right)\chi = F(x). \tag{261}$$

The function $F(x)$ can be eliminated by adding to χ a suitable function in x. Such an addition affects neither η nor ζ, and since the components u', v', and w' correspond strictly to rotational flow, the irrotational part of the flow being accounted for by ϕ, they are unaffected if η and ζ are unaffected. We can therefore assign the value zero to $F(x)$ and write

$$\left(\nabla^2 - 2k\frac{\partial}{\partial x}\right)\chi = 0. \tag{262}$$

Then it is obvious that

$$2k \frac{\partial u'}{\partial x} = \nabla^2 u' = \frac{\partial \eta}{\partial z} - \frac{\partial \zeta}{\partial y} = \frac{\partial^2 \chi}{\partial y^2} + \frac{\partial^2 \chi}{\partial z^2} = -\frac{\partial^2 \chi}{\partial x^2} + 2k \frac{\partial \chi}{\partial x}, \quad (263)$$

$$2k \frac{\partial v'}{\partial x} = \nabla^2 v' = \frac{\partial \zeta}{\partial x} - \frac{\partial \xi}{\partial z} = -\frac{\partial^2 \chi}{\partial x \partial y}, \quad (264)$$

$$2k \frac{\partial w'}{\partial x} = \nabla^2 w' = \frac{\partial \xi}{\partial y} - \frac{\partial \eta}{\partial x} = -\frac{\partial^2 \chi}{\partial x \partial z}, \quad (265)$$

and $\quad u' = -\dfrac{1}{2k} \dfrac{\partial \chi}{\partial x} + \chi, \qquad v' = -\dfrac{1}{2k} \dfrac{\partial \chi}{\partial y}, \qquad w' = -\dfrac{1}{2k} \dfrac{\partial \chi}{\partial z}. \quad (266)$

Note that integrations of (263) to (265) produce additive functions $F_1(y,z)$, $F_2(y,z)$, and $F_3(y,z)$ for the three equations in (266), respectively. But $\xi = 0$ demands that

$$\frac{\partial F_2}{\partial z} = \frac{\partial F_3}{\partial y}, \qquad \text{or} \qquad F_2 = \frac{\partial G}{\partial y}, F_3 = \frac{\partial G}{\partial z}.$$

Thus, a harmonic function G of y and z should be added to χ. Let that function be absorbed in ϕ [see (257)]. Since χ is so far undetermined, there is no loss of generality if we then absorb $F_1(y, z)$ in χ and simply write (266). Thus

$$u = \frac{\partial \phi}{\partial x} - \frac{1}{2k} \frac{\partial \chi}{\partial x} + \chi,$$

$$v = \frac{\partial \phi}{\partial y} - \frac{1}{2k} \frac{\partial \chi}{\partial y}, \quad (267)$$

$$w = \frac{\partial \phi}{\partial z} - \frac{1}{2k} \frac{\partial \chi}{\partial z}.$$

Now (262) can be written as

$$(\nabla^2 - k^2)e^{-kx}\chi = 0, \quad (268)$$

of which $\chi = $ const is one obvious solution and

$$e^{-kx}\chi = \frac{Ce^{-kR}}{R}$$

another. For χ, we choose as the solution the combination

$$\chi = C(e^{-k(R-x)} R^{-1} + k), \quad (269)$$

which for small kR can be expanded in the form

$$\chi = C\left(\frac{1}{R} + \frac{kx}{R} + \cdots\right). \quad (270)$$

As to ϕ, it is harmonic, and since the x direction is the "privileged direction" and the condition at infinity in (253) must be satisfied, the proper form for it is

$$\phi = \frac{A_0}{R} + A_1 \frac{\partial}{\partial x}\frac{1}{R} + A_2 \frac{\partial^2}{\partial x^2}\frac{1}{R} + \cdots, \tag{271}$$

in which each term except the first shows the privileged role of the x axis. Substitution of (270) into (267) produces

$$\frac{1}{2k}\frac{\partial\chi}{\partial x} - \chi = -\frac{C}{2k}\left(\frac{4}{3}\frac{k}{R} - \frac{\partial}{\partial x}\frac{1}{R} + \frac{1}{3}kR^2\frac{\partial^2}{\partial x^2}\frac{1}{R} + \cdots\right),$$

$$\frac{1}{2k}\frac{\partial\chi}{\partial y} = -\frac{C}{2k}\left(-\frac{\partial}{\partial y}\frac{1}{R} + \frac{1}{3}kR^2\frac{\partial^2}{\partial x\,\partial y}\frac{1}{R} + \cdots\right), \tag{272}$$

$$\frac{1}{2k}\frac{\partial\chi}{\partial z} = -\frac{C}{2k}\left(-\frac{\partial}{\partial z}\frac{1}{R} + \frac{1}{3}kR^2\frac{\partial^2}{\partial x\,\partial z}\frac{1}{R} + \cdots\right).$$

It is evident from (267), (271), and (272) that the conditions at infinity are satisfied. Application of the three boundary conditions at the surface of the sphere, where $R = a$, shows that

$$C = -\tfrac{3}{2}Ua, \qquad A_0 = -\tfrac{3}{2}va, \qquad A_1 = \tfrac{1}{4}Ua^3, \qquad \text{etc.,} \tag{273}$$

approximately, provided ka, which is just the Reynolds number Ua/v, is small.

The velocity components are obtained by substituting (271) and (272), with the constants given by (273), into (267). Recalling that these are really u_1, v_1, and w_1 and that the actual velocity components are the u, v, and w in (250), one can show that the velocity distribution near the sphere is in agreement with (244) if the x axis is identified with the polar axis. Hence the hydrodynamic drag remains equal to $6\pi\mu aU$.

The actual radial velocity is

$$v_R = \frac{\partial\phi}{\partial R} - \frac{1}{2k}\frac{\partial\chi}{\partial R} + \chi\cos\theta + U\cos\theta, \tag{274}$$

where θ is defined as in Sec. 13. Hence

$$\psi = R^2\int_0^\theta v_R \sin\theta\,d\theta,$$

and, from (274), (270), and (271),

$$-\psi = \tfrac{3}{2}va(1 + \cos\theta)(1 - e^{-kR(1-\cos\theta)}) - \frac{1}{4}\frac{Ua^3}{R}\sin^2\theta - \tfrac{1}{2}UR^2\sin^2\theta, \tag{275}$$

which for small kR becomes

$$-\psi = \tfrac{3}{4}Ua\left(R - \frac{1}{3}\frac{a^2}{R}\right)\sin^2\theta - \tfrac{1}{2}UR^2\sin^2\theta, \tag{276}$$

which agrees with (242).

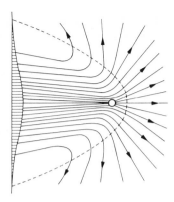

FIGURE 29. STREAMLINES IN THE
FLOW CAUSED BY A SPHERE MOVING
WITH A SMALL CONSTANT VELOC-
ITY: OSEEN'S SOLUTION. [*From
H. Schlichting, "Boundary Layer
Theory," 6th ed. Copyright 1968,
McGraw-Hill Book Company.
Used by permission.*]

If kR is not small, (276) and (242) do not agree, and the flow pattern corresponding to (275) is not symmetric with respect to the plane $x = 0$, as shown in Fig. 29. In other words, there is now a wake. Lamb (1911) showed that for small ka the discrepancy of Oseen's solution from the true solution near the sphere and far from it must be small.

In the Oseen approximation the inertia of the fluid is partially taken into account. If one accepts Oseen's equations as a starting point, one can proceed to find the dependence of the drag F on the Reynolds number Ua/v. Doing that, Oseen (1910) obtained

$$F = 6\pi\mu aU(1 + \tfrac{3}{8}R), \qquad (277)$$

where R now denotes the Reynolds number Ua/v and not the radial distance. To the order of approximation implied by (277), this turns out to be correct. But in the neighborhood of the sphere the Oseen approximation does not represent the inertia effects correctly, and the correctness of (277) is fortuitous. Proudman and Pearson (1957), in an excellent paper, gave the solution of the Navier-Stokes equation in the neighborhood of the sphere and the solution of the Navier-Stokes equation at infinity, i.e., the improved Stokes solution and the improved Oseen solution for the two regions, respectively, and matched them in a supposed common region of their validity. In this way they obtained

$$F = 6\pi\mu aU[1 + \tfrac{3}{8}R + \tfrac{9}{40}R^2 \ln R + O(R^2)]. \qquad (278)$$

To the order R, (277) and (278) agree because the misrepresentation of inertia effects near the sphere by Oseen's approximation, while affecting the stream function, does not affect the drag, the discrepancy between Oseen's result for the stream function and the Proudman-Pearson result having no contribution to the drag up to the order of R of the approximation. In fact, the rather astounding thing about the term $3R/8$ in (278) is that it can be obtained once the *Stokes solution* is known, because the improvement on (248) depends only on a further application of Stokes' solution if only the term of order $O(R)$ is wanted in (278). Chester (1962) showed that this remarkable fact is true of other symmetric (with respect to a plane normal to the direction of motion of the body) three-dimensional bodies. For further details, see Proudman and Pearson (1957) and Chester (1962), in which further references can be found.

Although the Stokes equations (231) do not possess solutions for two-dimensional flows, the Oseen equations (251), in which acceleration is partially represented, do have solutions. In Proudman and Pearson (1957) the case of a circular cylinder is also treated.

15. FORCE ON AN OSCILLATING SPHERE

It is worthwhile to consider the flow due to a sphere oscillating in a viscous fluid, not only because it is interesting in its own right, but also because the solution for this flow leads to the solution of the problem of establishment of the Stokes flow.

The sphere is assumed to move rectilinearly with the velocity $U(t) = U_0 e^{-i\omega t}$ (real part understood), and the amplitude of its motion is supposed to be small, so that the convective terms in the Navier-Stokes equations are negligible. If the polar axis points in the general direction of motion of the sphere, the equations of motion are the first two in (A2.83), which under the assumption made can be written in the simpler forms

$$\frac{\partial u}{\partial t} = -\frac{1}{\rho}\frac{\partial}{\partial R}p_d + \nu\left(\nabla^2 u - \frac{2u}{R^2} - \frac{2}{R^2}\frac{\partial v}{\partial \theta} - \frac{2v\cos\theta}{R^2}\right), \qquad (279)$$

$$\frac{\partial v}{\partial t} = -\frac{1}{\rho R}\frac{\partial}{\partial \theta}p_d + \nu\left(\nabla^2 v - \frac{v}{R^2\sin^2\theta} + \frac{2}{R^2}\frac{\partial u}{\partial \theta}\right), \qquad (280)$$

in which the symbols v_R and v_θ have been replaced by u and v, because there is now no danger of confusion with Cartesian components, and ∇^2 is given by (A2.71).

Eliminating p_d from (279) and (280), and using

$$u = \frac{1}{R^2\sin\theta}\frac{\partial \psi}{\partial \theta}, \quad v = -\frac{1}{R\sin\theta}\frac{\partial \psi}{\partial R}, \qquad (281)$$

we obtain, after multiplication by $\sin \theta$,

$$\frac{\partial}{\partial t} L\psi = \nu L^2 \psi, \tag{282}$$

in which L is an operator now defined by

$$L = \frac{\partial^2}{\partial R^2} + \frac{\sin \theta}{R^2} \frac{\partial}{\partial \theta} \left(\frac{1}{\sin \theta} \frac{\partial}{\partial \theta} \right). \tag{283}$$

The derivation of the right-hand side of (282) requires a great deal of patience—perhaps more than most readers are willing to exercise. For this reason they will be reassured of the correctness of (282) in the following way. Since convective terms have been neglected, the vorticity equations in Cartesian coordinates are in the form

$$\frac{\partial \xi_i}{\partial t} = \nu \, \nabla^2 \xi_i. \tag{284}$$

Since ξ_1, ξ_2, and ξ_3 constitute a vector, the form of these equations in spherical coordinates must be identical with the form of the Navier-Stokes equations if we ignore the convective terms and the terms involving p and the body force. Thus the vorticity component in the φ direction, which is the only nonzero component of the vorticity and is given by

$$\zeta = \frac{1}{R} \left[\frac{\partial (Rv)}{\partial R} - \frac{\partial u}{\partial \theta} \right]$$

according to (A2.74), must satisfy the equation

$$\frac{\partial \zeta}{\partial t} = \nu \left(\nabla^2 \zeta - \frac{\zeta}{R^2 \sin^2 \theta} \right), \tag{285}$$

upon inspection of the third equation in (A2.83). It is a simple matter to show that

$$\zeta = - \frac{1}{R \sin \theta} L\psi$$

and that the right-hand side of (285) indeed can be reduced to

$$\frac{\nu}{R \sin \theta} L(\zeta \, R \sin \theta).$$

Thus (285) is (282).

We shall now assume that all dependent variables, and in particular ψ, contain the factor $e^{-i\omega t}$. Eventually, of course, only the real part of the solution for ψ will be used. Thus (282) becomes [ω not being that in (235)]

$$\left(L + \frac{i\omega}{\nu} \right) L\psi = 0. \tag{286}$$

The boundary conditions at $\theta = 0$ and $\theta = \pi$ require that ψ and $\partial\psi/\partial\theta$ vanish on these half lines. Hence it is natural to assume ψ to have the factor $\sin^2 \theta$:

$$\psi = \sin^2 \theta f(R)e^{-i\omega t}. \tag{287}$$

This reduces (286) to the form

$$\left(L_R + \frac{i\omega}{\nu}\right)L_R f = 0, \tag{288}$$

in which

$$L_R = \left(\frac{\partial^2}{\partial R^2} - \frac{R^2}{2}\right).$$

Two of the solutions of f are to be sought from

$$L_R f = 0,$$

and are AR^{-1} and CR^2. But the latter must be discarded because it violates the condition of zero velocity at infinity. The other two are to be sought from

$$\left(L_R + \frac{i\omega}{\nu}\right)f = 0. \tag{289}$$

It is evident that if f is assumed to have the factor e^{ikR}, (289) possesses the solution

$$f_p = B\left(\frac{1}{R} - ik\right)e^{ikR}, \tag{290}$$

in which

$$k = \frac{1+i}{\delta}, \qquad \delta = \sqrt{\frac{2\nu}{\omega}}. \tag{291}$$

We can obtain the other solution by the method of variation of parameters, i.e., that is, by assuming a solution of the form $f_p g(R)$. It turns out that the second solution, which can be found easily in this way, does not satisfy the condition at infinity, because it behaves like e^{-ikR} at infinity. Hence,

$$f = \frac{A}{R} + B\left(\frac{1}{R} - ik\right)e^{ikR}. \tag{292}$$

The constants A and B are determined by the conditions that at $R = a$, a being the radius of the sphere,

$$u = U_0 \cos \theta\, e^{-i\omega t} \qquad \text{and} \qquad v = -U_0 \sin \theta\, e^{-i\omega t}. \tag{293}$$

From (281), (287), (292), and (293), we find

$$A = U_0 a^3 - \frac{2U_0 a}{k^2}(1 - ika), \qquad B = \frac{2U_0 a}{k^2}e^{-ika}. \tag{294}$$

The hydrodynamic force F on the sphere can now be found by using (247a) and integrating over the sphere, as was done in Sec. 13. The result is

$$F = 6\pi\mu a\left(1 + \frac{a}{\delta}\right)U + 3\pi a^2\rho\delta\left(1 + \frac{2a}{9\delta}\right)\frac{dU}{dt}, \tag{295}$$

the force being directed in the direction opposite to the direction of the polar axis if F is positive. If U is constant ($\omega = 0$), (295) reduces to (248). If ω is large ($\delta \ll a$), (295) can be simplified to

$$F = \frac{2}{3}\pi\rho a^3\frac{dU}{dt} + 3\pi\rho a^2\sqrt{2\nu\omega}\,U\,. \tag{296}$$

The first term on the right-hand side is the force needed to accelerate the fluid, and the second term gives the viscous resistance. Note that the added mass is $2\pi\rho a^3/3$, exactly the same as for irrotational motion [see (4.87)]. This indicates that at large frequencies the acceleration of the fluid is that for irrotational flow, except in a boundary layer whose thickness is of the order of δ, vanishing at very large values of ω.

16. FORCE ON A SPHERE MOVING WITH ARBITRARY SPEED IN A STRAIGHT LINE

If $U(t)$ is not periodic but arbitrary, we can build the solution from the solution obtained in Sec. 15, provided $U(t)$ can be represented as a Fourier integral:

$$U(t) = \int_{-\infty}^{\infty} U_\omega e^{-i\omega t}\,d\omega, \qquad U_\omega = \frac{1}{2\pi}\int_{-\infty}^{\infty} U(\tau)e^{i\omega\tau}\,d\tau, \tag{297}$$

in which U_ω is the Fourier transform of $U(t)$, in the sense indicated above. The subscript ω does not indicate differentiation. The subsequent development is that of Landau and Lifshitz (1959, p. 96).

For each Fourier component $U_\omega e^{-i\omega t}$ the hydrodynamic drag is given by (295) and is

$$\pi\rho a^3 U_\omega e^{-i\omega t}\left[\frac{6\nu}{a^2} - \frac{2i\omega}{3} + \frac{3\sqrt{2\nu\omega}}{a}\,(1 - i)\right]. \tag{298}$$

Since $(dU/dt)_\omega = -i\omega U_\omega$, this is equal to

$$\pi\rho a^3 e^{-i\omega t}\left(\frac{6\nu}{a^2}\,U_\omega + \tfrac{2}{3}(\dot{U})_\omega + \frac{3\sqrt{2\nu}}{a}\,(\dot{U})_\omega\,\frac{1 + i}{\sqrt{\omega}}\right), \tag{299}$$

the dot over U meaning d/dt. The forces corresponding to various values of ω, given by (299), will now be integrated over ω from $-\infty$ to ∞. The first two terms in (299) immediately give terms proportional to $U(t)$ and

dU/dt and present no difficulty. As to the third term, in the range of integration where ω is negative, $(1 + i)/\sqrt{\omega}$ must be replaced by $(1 - i)/\sqrt{|\omega|}$, because (295) is for positive ω and for negative ω the velocity U is of the form $U_0 e^{i|\omega|t}$, which demands that the complex conjugate of (295) be taken, after $|\omega|$ has been substituted for ω. Thus, since we want the real part of the integral, we can integrate from 0 to ∞ and multiply the result by 2. In this way the integral[1] for the third term in (299) is found to be, apart from a constant factor,

$$2 \operatorname{Re} (1 + i) \int_0^\infty \left[\frac{(\dot{U})_\omega e^{-i\omega t}}{\sqrt{\omega}} \, d\omega \right] = \frac{1}{\pi} \operatorname{Re} \left[(1 + i) \int_{-\infty}^\infty \int_0^\infty \frac{\dot{U}(\tau) e^{i\omega(\tau - t)}}{\sqrt{\omega}} \, d\omega d\tau \right]$$

$$= \frac{1}{\pi} \operatorname{Re} \left[(1 + i) \int_{-\infty}^t \int_0^\infty \frac{\dot{U}(\tau) e^{-i\omega(t - \tau)}}{\sqrt{\omega}} \, d\omega \, d\tau \right.$$

$$\left. + (1 + i) \int_t^\infty \int_0^\infty \frac{\dot{U}(\tau) e^{i\omega(\tau - t)}}{\sqrt{\omega}} \, d\omega \, d\tau \right]$$

$$= \sqrt{\frac{2}{\pi}} \operatorname{Re} \left[\int_{-\infty}^t \frac{\dot{U}(\tau)}{(t - \tau)^{1/2}} \, d\tau + i \int_t^\infty \frac{\dot{U}(\tau)}{(\tau - t)^{1/2}} \, d\tau \right]$$

$$= \sqrt{\frac{2}{\pi}} \int_{-\infty}^t \frac{\dot{U}(\tau)}{(t - \tau)^{1/2}} \, d\tau.$$

Thus the hydrodynamic drag is

$$F = 2\pi \rho a^3 \left[\frac{1}{3} \frac{dU}{dt} + \frac{3\nu U}{a^2} + \frac{3}{a} \left(\frac{\nu}{\pi} \right)^{1/2} \int_{-\infty}^t \frac{dU}{d\tau} \frac{d\tau}{(t - \tau)^{1/2}} \right]. \tag{300}$$

17. MOTION OF A SPHERE RELEASED FROM REST

If the motion is started from rest, (300) becomes

$$F = 2\pi \rho a^3 \left[\frac{1}{3} \frac{dU}{dt} + \frac{3\nu U}{a^2} + \frac{3}{a} \left(\frac{\nu}{\pi} \right)^{1/2} \int_0^t \frac{dU}{d\tau} \frac{d\tau}{(t - \tau)^{1/2}} \right]. \tag{301}$$

If the mean density of the sphere ρ' is greater than ρ,

$$\frac{4\pi}{3} a^3 (\rho' - \rho) g - F = \frac{4\pi}{3} a^3 \rho' \frac{dU}{dt}. \tag{302}$$

Equation (302), an integrodifferential equation, is the equation to be solved. We shall first write it in the simple form

$$U'(t) + PU(t) + Q \int_0^t \frac{U'(\tau)}{(t - \tau)^{1/2}} \, d\tau = M, \tag{303}$$

[1] Remember that $\int_\infty^0 e^{-x^2} \, dx = \sqrt{\pi}/2$.

in which

$$P = \frac{9\mu}{a^2(2\rho' + \rho)}, \quad Q = \frac{9\rho}{a(2\rho' + \rho)} \left(\frac{\nu}{\pi}\right)^{\frac{1}{2}}, \quad M = \frac{2(\rho' - \rho)g}{2\rho' + \rho}. \quad (304)$$

The next step is to reduce (303) to an ordinary differential equation, i.e., we shall do something about the last term on the left-hand side of (303). To start with we shall prove the following result due to Abel. If

$$F(t) = \int_0^t \frac{U'(t)}{(t - \tau)^{\frac{1}{2}}} d\tau, \quad (305)$$

then

$$\int_0^t \frac{F(\tau)}{(t - \tau)^{\frac{1}{2}}} d\tau = \pi[U(t) - U(0)]. \quad (306)$$

Since

$$\int_0^t \frac{F(\tau)}{(t - \tau)^{\frac{1}{2}}} d\tau = 2\int_0^{\sqrt{t}} F(t - x^2) \, dx = 2\int_0^{\sqrt{t}} \left[dx \int_0^{t - x^2} \frac{U'(\tau)}{(t - x^2 - \tau)^{\frac{1}{2}}} d\tau \right]$$

with

$$y^2 = t - x^2 - \tau,$$

we can write the preceding equation as

$$\int_0^t \frac{F(t)}{(t - \tau)^{\frac{1}{2}}} d\tau = 4\int_0^{\sqrt{t}} \left[dx \int_0^{\sqrt{t - x^2}} U'(t - x^2 - y^2) \, dy \right]$$

$$= 4\int_0^{\pi/2} \left[d\varphi \int_0^{\sqrt{t}} U'(t - r^2)r \, dr \right] = -\pi U(t - r^2) \Big|_0^{\sqrt{t}}$$

$$= \pi[U(t) - U(0)],$$

as is to be proved. It is evident that a change of variable from x and y to r and φ has been made and that the region of integration is

$$0 \leqslant r^2 = x^2 + y^2 \leqslant t, \quad 0 \leqslant \varphi = \tan^{-1}\frac{y}{x} \leqslant \frac{\pi}{2}.$$

If now (303) is multiplied by $(t - \tau)^{-\frac{1}{2}}$ and integrated with respect to τ, (305) and (306) enable us to write

$$\int_0^t \frac{U'(\tau)}{(t - r)^{\frac{1}{2}}} d\tau + \int_0^t \frac{PU(\tau) - M}{(t - \tau)^{\frac{1}{2}}} d\tau + \pi Q[U(t) - U(0)] = 0. \quad (307)$$

The first term can be eliminated by means of (303) to give the equation

$$U'(t) + PU(t) - M - PQ\int_0^t \frac{U(\tau)}{(t - \tau)^{\frac{1}{2}}} d\tau + MQ\int_0^t \frac{d\tau}{(t - \tau)^{\frac{1}{2}}}$$

$$- \pi Q^2[U(t) - U(0)] = 0. \quad (308)$$

The fourth term in this equation is the only troublesome term. But the integral I in that term can be written as

$$I(t) = 2 \int_0^{\sqrt{t}} U(t - x^2)\, dx$$

by the transformation $\tau = t - x^2$. Therefore, by the Leibniz rule of differentiation,

$$\frac{d}{dt} I(t) = \frac{U(0)}{\sqrt{t}} + 2 \int_0^{\sqrt{t}} U'(t - x^2)\, dx = \frac{U(0)}{\sqrt{t}} + \int_0^t \frac{U'(\tau)}{(t - \tau)^{1/2}}\, d\tau. \quad (309)$$

Since $U(0) = 0$ (release from rest) and the last term in (309) can be evaluated from (303), we have

$$\frac{d}{dt} I(t) = \frac{M}{Q} - \frac{U'(t) + PU(t)}{Q}. \quad (310)$$

With this substituted into (308), we have, finally,

$$U''(t) + (2P - \pi Q^2)U'(t) + P^2 U(t) = PM - \frac{QM}{\sqrt{t}}. \quad (311)$$

The complementary solutions of this equation are $e^{\alpha t}$ and $e^{\beta t}$, α and β being the roots of the indicial equation

$$m^2 + (2P - \pi Q^2)m + P^2 = 0,$$

and by the method of variation of parameters we obtain the complete solution to be

$$U(t) = A(t)e^{\alpha t} + B(t)e^{\beta t} + \frac{M}{P}, \quad (312)$$

where

$$A(t) = A_1 - \frac{QM}{\alpha - \beta} \int_0^t \frac{e^{-\alpha \tau}}{\sqrt{\tau}}\, d\tau, \qquad B(t) = B_1 + \frac{QM}{\alpha - \beta} \int_0^t \frac{e^{-\beta \tau}}{\sqrt{\tau}}\, d\tau. \quad (313)$$

After the change of variables

$$\tau = \frac{x^2}{\alpha} \qquad \text{or} \qquad \tau = \frac{x^2}{\beta},$$

as the case may be, we can write (312) as

$$U(t) = \frac{M}{P} + \frac{e^{\alpha t}}{\alpha - \beta} \left(A_1 - \frac{2QM}{(\alpha - \beta)\sqrt{\alpha}} \int_0^{\sqrt{\alpha t}} e^{-x^2}\, dx \right) + \frac{e^{\beta t}}{\alpha - \beta}$$
$$\times \left(B_1 + \frac{2QM}{(\alpha - \beta)\sqrt{\beta}} \int_0^{\sqrt{\beta t}} e^{-x^2}\, dx \right). \quad (314)$$

The initial condition $U(0) = 0$ demands that

$$(\alpha - \beta)\frac{M}{P} + (A_1 + B_1) = 0, \quad (315)$$

and from (303) we deduce
$$U'(0) = M,$$
which demands that
$$(\alpha - \beta)M = A_1\alpha + B_1\beta,$$
as can be verified by a direct calculation. Thus

$$A_1 = M\left(1 + \frac{\beta}{P}\right), \qquad B_1 = -M\left(1 + \frac{\alpha}{P}\right). \tag{316}$$

The α and β may be real or complex, depending on the ratio ρ'/ρ. Whatever they are, (314) and (316) furnish the complete solution. We shall not stop to discuss the various cases regarding the values of α and β. [For more details regarding these cases, the reader is referred to Villat (1943, pp. 218–224), who gave the solution of (303) presented in this section.]

18. FLOW IN POROUS MEDIA: DARCY'S LAW AND ITS CONSEQUENCES

In his two memoires (1856 and 1857), Darcy described his experiments on the seepage of water through sand. These experiments were on unidirectional flows only, and the main result is that the mean velocity is directly proportional to the permeability k of the medium, which has the dimension of area, to the gradient of $-(p + \rho\Omega)$, where Ω is the potential of gravitational attraction on earth and equal to g times elevation, and inversely proportional to the viscosity μ of the fluid. Generalized to three-dimensional flow, Darcy's law has the Cartesian form

$$\frac{\mu}{k}u_i = -\frac{\partial}{\partial x_i}(p + \rho\Omega), \tag{317}$$

provided ρ is constant. It should be emphasized that in (317) u_i is the ith component of the *mean* velocity taken over a volume containing many grains of the porous material. So defined, the velocity components still satisfy the continuity equation

$$\frac{\partial u_\alpha}{\partial x_\alpha} = 0. \tag{318}$$

It is evident that (317) takes no inertial effects of the fluid into account and is therefore valid only for very slow flows dominated by viscous forces. Indeed, in Darcy's experiments water percolated very slowly through sand. If μ and k are also constant, (317) can be written as

$$u_i = \frac{\partial \phi}{\partial x_i}, \tag{319}$$

in which $$\phi = -\frac{k}{\mu}(p + \rho\Omega). \tag{320}$$

Thus the flow of a fluid of constant ρ and μ in a porous medium of constant k is irrotational. Equations (318) and (319) give

$$\nabla^2\phi = 0. \tag{321}$$

Thus, in the absence of free surfaces, problems of seepage flows can be dealt with by the methods described in Chap. 4, which we need not discuss again. The reader is referred to Muskat (1937) for an extensive discussion of the subject.

When there is a free surface in seepage flow, however, the boundary condition on that surface is different from the free-surface condition for irrotational flows of an inviscid fluid described in Chap. 4. This condition and its treatment therefore need some discussion here. On a free surface the normal stress is zero. The viscous part of the stress tensor, which depends on the velocity, can be neglected since the velocity is low. Hence on the free surface $p = 0$, and

$$\phi = -\frac{k}{\mu}\rho\Omega = -\frac{k}{\mu}\rho g Z, \tag{322}$$

in which Z is the elevation. Since the free surface is unknown a priori, (322) is not very helpful. For steady two-dimensional flows, however, it can be transformed to a useful form.

For steady two-dimensional flows, the direction of increasing y can be conveniently chosen to coincide with the direction of the vertical, so that

$$\phi = -\xi_0^{-1}y, \qquad \xi_0 = \frac{\mu}{k\rho g}. \tag{323}$$

Denoting by q the speed of the fluid on the free surface and by s the distance along it, we have

$$\frac{\partial\phi}{\partial s} = q. \tag{324}$$

Differentiation of (323) with respect to s and multiplication by q produce

$$q^2 = -\frac{q}{\xi_0}\frac{\partial y}{\partial s} = -\frac{1}{\xi_0}v, \tag{325}$$

since $q\,\partial y/\partial s$ is v, the vertical component of the velocity. Equation (325) can be written in terms of v and the horizontal velocity component u as

$$u^2 + v^2 + \xi_0^{-1}v = 0, \tag{326}$$

which is a circle in the hodograph plane [see Muskat (1937)]. Thus we know in the plane of $u - iv$ the free surface is a known boundary, and if the other boundaries do not give trouble, the problem can be solved.

The transformation

$$\zeta = \frac{i}{dw/dz},$$ (327)

in which ζ is the complex variable $\xi + i\eta$, w is the complex velocity potential, and z the complex variable $x + iy$, can be written as

$$\xi + i\eta = \frac{i}{u - iv} = \frac{-v + iu}{u^2 + v^2},$$ (328)

since

$$\frac{dw}{dz} = u - iv.$$ (329)

It is immediately evident that on the free surface, on which (326) holds, (328) gives

$$\xi = \xi_0.$$ (330)

Thus in the ζ plane the free surface is the straight line (330), and reflection across this line can be easily constructed to ensure that it is a streamline, as required.

19. STEADY FLOW IN POROUS MEDIA: EFFECT OF VISCOSITY VARIATION

In steady seepage flows the effect of viscosity variation can be ascertained once and for all, provided the diffusivity of μ is negligible [Yih (1961)]. Under that provision, for steady flows

$$\frac{D\mu}{Dt} = u_\alpha \frac{\partial \mu}{\partial x_\alpha} = 0.$$ (331)

If ρ and μ both vary, the obvious generalization of (317) is

$$\frac{\mu}{k} u_i = -\frac{\partial p}{\partial x_i} - \rho \frac{\partial \Omega}{\partial x_i}.$$ (332)

If the velocity components u_i' are defined by

$$u_i' = \frac{\mu}{\mu_0} u_i,$$ (333)

in which μ_0 is a reference viscosity, (332) can be written

$$\frac{\mu_0}{k} u_i' = -\frac{\partial p}{\partial x_i} - \rho \frac{\partial \Omega}{\partial x_i}$$ (334)

by virtue of (331). Thus given a seepage flow u_i' with constant μ, there correspond infinitely many seepage flows u_i for a variable μ with the same flow pattern but different velocity distributions, the latter being given by (333) once the associated flow (u'_i) for constant viscosity is determined. The density distributions for the associated flow u'_i and the actual flow u_i must of course be the same.

20. STEADY TWO-DIMENSIONAL SEEPAGE FLOW OF A FLUID WITH VARIABLE VISCOSITY AND DENSITY

For two-dimensional flows, the direction of increasing y will again be chosen to be that of the vertical. Then, with u and v written for u_1 and u_2, (332) becomes

$$\frac{\mu}{k} u = -\frac{\partial p}{\partial y}, \qquad \frac{\mu}{k} v = -\frac{\partial p}{\partial y} - g\rho, \tag{335}$$

or

$$\frac{\mu_0}{k} u' = -\frac{\partial p}{\partial x}, \qquad \frac{\mu_0}{k} v' = -\frac{\partial p}{\partial y} - g\rho. \tag{336}$$

Since u' and v' satisfy the equation of continuity

$$\frac{\partial u'}{\partial x} + \frac{\partial y'}{\partial y} = 0,$$

we can write

$$u' = \frac{\partial \psi'}{\partial y}, \qquad v' = -\frac{\partial \psi'}{\partial x}.$$

Since the flow is steady,

$$u' \frac{\partial \rho}{\partial x} + v' \frac{\partial \rho}{\partial y} = 0 \tag{337}$$

if mass diffusion is neglected—and it can be neglected if the number UL/α (U being a reference velocity, L a reference length, and α the mass diffusity) is large. Then ρ does not change on a streamline and is a function of ψ' alone. Bearing this in mind and eliminating p in (335), we have

$$\nabla^2 \psi' = \frac{kg}{\mu_0} \frac{d\rho}{d\psi'} \frac{\partial \psi'}{\partial x}, \tag{338}$$

which governs the flows under discussion. [For the solutions of (338), see Yih (1961).]

21. HELE-SHAW CELLS

Very slow flows of a viscous fluid between two closely spaced vertical plates are governed by the equations

$$\mu \frac{\partial^2 u}{\partial z^2} = \frac{\partial p}{\partial x}, \qquad \mu \frac{\partial^2 v}{\partial z^2} = \frac{\partial p}{\partial y} + g\rho, \tag{339}$$

in which y is measured vertically, x horizontally in any plane parallel to the plates, and z in a direction normal to the plates, which are assumed parallel. Except very near the obstacles placed in the space between the plates, which is called a *Hele-Shaw cell*, the fluid can be assumed to flow as in plane Poiseuille flow, and we can assume

$$(u,v) = 6\frac{z}{b}\left(1 - \frac{z}{b}\right)(U,V), \tag{340}$$

in which b is the spacing of the plates. With (340), (339) becomes

$$\frac{12\mu}{b^2}U = -\frac{\partial p}{\partial x}, \qquad \frac{12\mu}{b^2}V = -\frac{\partial p}{\partial z} - g\rho. \tag{341}$$

These are identical in form with (335) if we identify the $b^2/12$ with the permeability k and U and V with u and v in (335). Thus a Hele-Shaw cell can be considered to be a porous medium of permeability $b^2/12$, and all the solutions of two-dimensional seepage flow in porous media apply to Hele-Shaw cells.

PROBLEMS

1 Show that the velocity distribution in rectilinear laminar flow in a pipe with an equilateral-triangular cross section is given by

$$u = -\frac{\sqrt{3}\gamma}{6\mu b}\frac{dh}{dx}\left(z + \frac{b}{2\sqrt{3}}\right)\left(z + \sqrt{3}y - \frac{b}{\sqrt{3}}\right)$$
$$\times\left(z - \sqrt{3}y - \frac{b}{\sqrt{3}}\right),$$

in which b is the length of any one side of the triangle, the origin is at the mass center of the cross section, the y axis is parallel to one side of the triangle, and

$$\gamma h = p + \rho g \sin\beta,$$

β being the angle of inclination of the axis of the pipe to the horizontal.

2 Find the velocity distribution in longitudinal laminar flow in a pipe with an elliptic cross section described by

$$\frac{y^2}{b^2} + \frac{z^2}{c^2} = 1,$$

using the notation of Prob. 1. *Hint:* Do not look for complicated solutions. The solution is extremely simple.

3 Two superposed liquid layers of depth d_2 (lower layer) and d_1 (upper layer), viscosities μ_2 and μ_1, and densities ρ_2 and ρ_1 flow down an inclined plane. The depths are measured along the normal to the plane, which is inclined at an angle β to the horizontal. The top surface is free. Find the velocity distribution and the shear-stress distribution.

4 Find the velocity distribution of parallel flow in an inclined rectangular pipe if the density and viscosity are functions of z, z being defined as in Sec. 2.6.

5 A fluid of constant ρ and μ in a horizontal circular pipe of inner radius a is set in motion by a longitudinal pressure gradient equal to $K \cos \omega t$. Find the velocity distribution as a function of t and r. One length scale is $(\nu/\omega)^{1/2}$. In what way is this length scale manifested?

6 In the preceding problem, if

$$\frac{\partial p}{\partial z} = \begin{cases} 0 & \text{for } t < 0, \\ -K & \text{for } t \geq 0, \end{cases}$$

find the velocity distribution as a function of t and r.

7 Find the velocity and rate of decay of waves on the surface of a semi-infinite viscous liquid, taking into consideration both surface tension and gravity.

8 Study the establishment of Ekman flow by an impulsively started surface traction.

9 Verify the differential system governing the flow for two of the cases listed in Table 1.

10 Formulate the ordinary differential equations and boundary conditions governing laminar flow against a plate, in which the constant-speed lines are ellipses. The flow is in a sense "between" the Hiemenz flow and the Homann flow shown in Table 1.

11 Find the velocity distribution in the boundary layer on a circular cylinder $x^2 + y^2 = a^2$ placed in a stream with velocity U_0 in the x direction at infinity, up to the separation point. Use both the series solution and the Kármán-Pohlhausen method and compare the results.

12 Interpret (301). Does the first term depend on history? Does the third? The first term evidently is the added-mass resistance discussed in Chap. 4. Why is this term unaffected by the nonslip condition which now must be imposed? Does the third term alone reflect the effect of the nonslip condition?

13 Determine the two-dimensional steady flow of a homogeneous liquid into a sink in a porous medium. The upper surface of the liquid is free [Yih (1964)].

14 An ellipsoidal fluid mass of density ρ_2 and viscosity μ_2 is imbedded in an infinite fluid of density ρ_1 and viscosity μ_1, both flowing in a porous medium of permeability k. The semi-axes of the ellipsoid are a, b, and c. Referred to these axes as coordinate axes, the components of the velocity of the ambient fluid are U_1, V_1, and W_1 at infinity, and the direction cosines of gravitational acceleration are α, β, and γ. Show that the ellipsoid moves as a solid body and find its velocity. Note: Use ellipsoidal coordinates [Yih (1963)].

15 Show that in a porous medium, a viscous-liquid sphere of viscosity μ' and density ρ' moves as a solid body with the velocity $\mathbf{U}'$ in an ambient liquid of viscosity μ and density ρ moving with velocity $\mathbf{U}$ at infinity and

$$\mathbf{U}' = \frac{3\mu}{2\mu' + \mu}\,\mathbf{U} + \frac{2k(\rho - \rho')}{2\mu' + \mu}\,\mathbf{g},$$

in which k is the permeability of the porous medium and $\mathbf{g}$ the gravitational acceleration.

16 If the surface tension T on a free surface or interface S is not uniform, show that the direction of the shear force and the magnitudes of the shear-stress components in the fluid just underneath S are given by **grad** T on S.

17 A thin layer of viscous liquid on a horizontal plane flows under the action of a nonuniform surface tension T on its free surface. If h_0 is a representative vertical scale, ρ the density of the liquid, g the gravitational acceleration, and ΔT a representative difference in T, and if

$$\Delta T \gg \rho g h_0^2$$

show the following:

a The equations governing steady motion of the liquid are

$$0 = \mu\,\nabla^2 u, \qquad 0 = \mu\,\nabla^2 v,$$

in which u and v are the horizontal Cartesian components of the velocity, the vertical component w being much less than u and v.

b The solutions of these equations are

$$u = \frac{Uz}{h}, \qquad v = \frac{Vz}{h},$$

in which $h(x,y)$ is the local depth and U and V are the values of u and v on the free surface.

c The equation of continuity is

$$\frac{\partial}{\partial x}\,(Uh) + \frac{\partial}{\partial y}\,(Vh) = 0.$$

d The free-surface conditions are

$$\frac{\mu U}{h} = \frac{\partial T}{\partial x}, \qquad \frac{\mu V}{h} = \frac{\partial T}{\partial y}.$$

18 A solid metal cylinder of specific gravity 7.8, radius 2 in., and length 6 in. is falling concentrically in a long cylinder (with a bottom) of radius $2\frac{1}{16}$ in. filled with water. Find the terminal velocity of its fall. For water, $\rho = 1.98$ slugs/ft.^{3}, $\mu = 2.34 \times 10^{-5}$ lb-sec/ft^2. *Hint:* Do not forget the difference in pressure at the ends of the cylinder.

19 An impervious dam rests on a porous medium, as shown in Fig. P. 19. Show that if $w = \phi + i\psi$, with ϕ given by (320), the conformal mapping giving the solution is

$$z = c \cosh \frac{i\pi\mu w}{k\gamma H},$$

in which $2c$ is the base of the dam, γ is $g\rho$, and H is the height of water behind the dam.

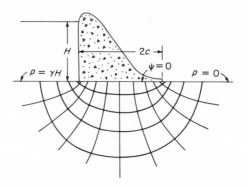

FIGURE P. 19

20 Find the pressure distribution at the base of the dam described in the preceding problem.

REFERENCES

BICKLEY, W., 1939: The Plane Jet, *Phil. Mag.*, (7)**23**: 727.

BLASIUS, H., 1908: Grenzschichten in Flüssigkeiten mit kleiner Reibung, *Z. Math. Phys.*, **56**: 1.

CARSLAW, H. S., and J. C. JAEGER, 1959: "Conduction of Heat in Solids," 2d ed., Oxford University Press, Fair Lawn, N.J.

CHESTER, W., 1962: On Oseen's Approximation, *J. Fluid Mech.*, **13**: 557–569.

DARCY, H. P. G., 1856: Les Fontaines publique de la ville de Dijon, Paris.

DARCY, H. P. G., 1857: Recherches expérimentales relatives au mouvement de l'eau dans les tuyaux, Paris.

EKMAN, V. W., 1905: On the Influence of the Earth's Rotation on Ocean-currents, *Arkiv Mat. Astr. Fys.*, **2**(11): 1–52.

FRÖSSLING, N., 1940: Verdunstung, Wärmeübergang, und Geschwindigkeitsverteilung bei zweidimensionaler und rotationssymmetrischer laminarer Grenzschichtströmung, *Lunds Univ. Arsskr. Avd. 2*, **36**(4): 1–32.

HIEMENZ, K., 1911: Die Grenzschicht an einem in den gleichförmigen Flüssigkeitsstrom eingetauchten geraden Kreiszylinder, thesis, Göttingen.

HOMANN, F., 1936: Der Einfluss grösser Zähigkeit bei der Strömung um den Zylinder und um die Kugel, *Z. Angew. Math. Mech.*, **16**: 153–164.

HOWARTH, L., 1935: On the Calculation of Steady Flow in the Boundary Layer near the Surface of a Cylinder in a Stream, *Aeron. Res. Council (Britain) Rept. Mem.* 1962.

KÁRMÁN, T. VON, 1911: Über den Mechanismus des Widerstandes etc., 1. Teil, *Nachr. Wiss. Ges. Göttingen, Math. Phys. Kl.*, **1911**: 509–517.

KÁRMÁN, T. VON, 1912: Über den Mechanismus des Widerstandes, etc., 2. Teil, *Nachr. Wiss. Ges. Göttingen, Math. Phys. Kl.*, **1912**: 547–556.

KÁRMÁN, T. VON, 1921: Über laminare und turbulente Reibung. *Z. Angew. Math. Mech.*, **1**: 233–252.

KÁRMÁN, T. VON, and H. RUBACH, 1912: Über der Mechanismus des Flüssigkeits- und Luftwiderstandes, *Phys. Z.*, **13**: 49–59.

LAMB, H., 1911: On the Uniform Motion of a Sphere through a Viscous Fluid, *Phil. Mag.*, (6)**21**: 112.

LANDAU, L. E., and E. M. LIFSHITZ, 1959: "Fluid Mechanics," Pergamon Press, New York.

MANGLER, W., 1948: Zusammenhang zwischen ebenen und rotationssymmetrischen Grenzschichten in kompressiblen Flüssigkeiten, *Z. Angew. Math. Mech.*, **28**: 97–103.

MUSKAT, M., 1937: "The Flow of Homogeneous Fluids through Porous Media," McGraw-Hill Book Company, New York. (See especially chaps. 6 and 8.)

OSEEN, C. W., 1910: Über die Stokessche Formel und über die verwandte Aufgabe in der Hydrodynamik, *Arkiv Mat., Astron., Fysik*, **6**(29).

POHLHAUSEN, K., 1921: Zur näherungsweisen Integration der Differentialgleichung der laminaren Reibungsschicht, *Z. Angew. Math. Mech.*, **1**: 252–268.

PRANDTL, L., 1904: Über Flüssigkeitsbewegung bei sehr kleiner Reibung, *Proc. 3d Intern. Math. Congr., Heidelberg.*

PROUDMAN, I., and J. R. A. PEARSON, 1957: Expansions at Small Reynolds Number for the Flow Past a Sphere and a Circular Cylinder, *J. Fluid Mech.*, **2**: 237–262.

RICHARDS, G. J., 1934: On the Motion of an Elliptic Cylinder Through a Viscous Fluid, *Phil. Trans Roy. Soc. London Ser. A*, **233**: 279–302.

SCHLICHTING, H., 1933: Laminare Strahlausbreitung, *Z. Angew. Math. Mech.*, **13**: 260.

SCHLICHTING, H., 1968: "Boundary Layer Theory," 6th ed., McGraw-Hill Book Company, New York.

SQUIRE, H. B., 1951: The Round Laminar Jet, *Quart. J. Mech. Appl. Math.*, **4**: 321–329.

STOKES, G. G., 1851: On the Effect of the Internal Friction of Fluids on the Motion of Pendulums, *Cambridge Trans.*, **9**.

SYNGE, J. L., 1938: Hydrodynamic Stability, *Semi-centennial Publ. Am. Math. Soc.*, **2**: 227–269.

TANEDA, S., 1956a: *J. Phys. Soc. Japan*, **11**: 302.

TANEDA, S., 1956b: *Kyūshū Daigaku Seisankagaku Kenkyūsho Hōkoku*, **4**: 99.

TIFFORD, A. N., 1954: Heat Transfer and Frictional Effects in Laminar Boundary Layers, pt. 4, Universal Series Solutions, *Wright Air Development Center Tech. Rept.*, pp. 53–288.

TOLLMIEN, W., 1931: "Handbuch der experimental Physik," Springer, Leipzig, vol. 4, pt. 1, p. 255.

ULRICH, A., 1949: Die ebene laminare Reibungsschicht an einem Zylinder, *Arch. Math.*, **2**: 33.

VILLAT, H., 1943: "Leçons sur les fluides visqueux," Gauthier-Villars, Paris.

YIH, C.-S., 1961: Flow of a Non-homogeneous Fluid in a Porous Medium, *J. Fluid Mech.*, **10**: 133–140.

YIH, C.-S., 1963: The Velocity of a Fluid Mass Imbedded in Another Fluid Flowing in a Porous Medium, *Phys. Fluids*, **6**: 1403–1407.

YIH, C.-S., 1964: A Transformation for Free-surface Flow in Porous Media, *Phys. Fluids*, **7**: 20–24.

ADDITIONAL READING

BATCHELOR, G. K., 1967: "An Introduction to Fluid Mechanics," Cambridge University Press, New York.

GOLDSTEIN, S. (ed.), 1938: "Modern Developments in Fluid Dynamics," vols. 1 and 2, Oxford University Press, Fair Lawn, N.J.

KAPLUN, S., and P. A. LAGERSTROM, 1957: Asymptotic Expansions of Navier-Stokes Solutions for Small Reynolds Number, *J. Math. Mech.*, **6**: 585–593.

LAMB, H., 1932: "Hydrodynamics," 6th ed., The Macmillan Company, New York; reprinted by Dover Publications, Inc., New York, 1945 (with the same pagination).

POLUBARINOVA-KOCHINA, P. YA., and S. B. FALKOVICH, 1951: *Advan. Appl. Mech.*, **2**: 153. (This concerns flow in porous media.)

STEWARTSON, K., 1951: On the Impulsive Motion of a Flat Plate in a Viscous Fluid, *Quart. J. Mech. Appl. Math.*, **4**: 182–198.

CHAPTER EIGHT

HEAT TRANSFER
AND
BOUNDARY LAYERS
OF A GAS

1. SOME GENERAL CONSIDERATIONS
OF CONVECTION

Heat transfer is a universal phenomenon. Although this alone is sufficient reason for studying it, often there are other compelling reasons. The efficiency in machine operations often depends on cooling, the comfort of a dwelling on the adequacy of heating, and, on a much larger scale, the workings of the atmosphere are intimately related to its heat budget. As is well known, heat can be transferred by conduction, convection, or radiation. In this book we are understandably concerned primarily with the effects of fluid flow on heat transfer. Conduction of heat in solids in contact with flowing fluid and radiation from fluid surfaces affect only the boundary conditions for the differential equations governing heat transport by fluids.

The reader may have noticed that in the preceding chapter boundary-layer flows of a compressible fluid have not been touched upon. The reason for the delay is that, because of the compressibility of the fluid and the effects of the temperature variation that arise therefrom, such flows cannot be discussed apart from heat transfer. Near the end of this chapter, we shall endeavor to discuss just enough of boundary-layer flows of a compressible fluid to appreciate the difficulties and complexities involved. For a fuller discussion the reader must be referred to more comprehensive treatises on the subject.

Heat transfer from solid boundaries depends crucially on the temperature distribution in the fluid in contact with them, and this is further dependent on the velocity distribution in the fluid. Often the velocity distribution is only slightly affected by the temperature distribution. Then the convection of heat, and more generally the phenomenon of heat transfer in which convection plays a part, is called *forced convection*. However, when there is no imposed flow (such as one caused by the translation of a body or by wind blowing past a body), the nonuniformity of the temperature in the fluid will induce a nonuniformity of its density, and this, in a gravitational field, can be the main cause of fluid motion. In that case the velocity and the temperature of the fluid are interdependent, and the ensuing convection is called *free* or *natural convection*.

We shall consider forced convection first. The energy equation, or the equation of heat convection, is (2.80) or (2.84) if thermal units are used for the specific heats c_v and c_p and for the thermal conductivity k. The factor J should be dropped if energy units are used for these quantities. To avoid having too many things to consider all at once and to focus attention on the relative importance of convection and conduction, we shall consider an incompressible fluid and ignore the energy dissipation. This implies that the flow is at a very low Mach number and that the temperature field is predominantly determined by imposed temperature differences between the boundary and the fluid or between parts of the boundary, the temperature rise due to energy dissipation playing a minor role. The latter situation is assured if $U^2/(c_v \Delta T)$ is small, in which U is a representative velocity and ΔT is a representative imposed temperature difference. This can be seen from (2.79) or (2.80) by considering the relative magnitudes of the last two terms in either equation. In fact, this comparison produces the more precise criterion $\sigma U^2/(c_v \Delta T) \ll 1$, in which σ is the Prandtl number defined by

$$\sigma = \frac{\nu}{\kappa}, \tag{1}$$

κ being the thermal diffusivity defined by

$$\kappa = \frac{k}{\rho c_v}. \tag{2}$$

If, further, we assume k and ρ to be constant, (2.79) or (2.80) can be written as

$$\frac{DT}{Dt} = \kappa \nabla^2 T, \tag{3}$$

valid for incompressible fluid with constant ρ, c_v, and k provided energy dissipation is negligible.

If L is a reference length, T_0 a reference temperature, and ΔT a reference temperature difference and the substitutions

$$t' = \frac{tU}{L}, \qquad x_i' = \frac{x_i}{L}, \qquad u_i' = \frac{u}{U}, \qquad \theta = \frac{T - T_0}{\Delta T} \qquad (4)$$

are made and then the primes dropped, (3) becomes

$$\frac{\partial \theta}{\partial t} + u_\alpha \frac{\partial \theta}{\partial x_\alpha} = \frac{1}{P\acute{e}} \nabla^2 \theta, \qquad (5)$$

where ∇^2 is now also dimensionless. The number

$$P\acute{e} = \frac{UL}{\kappa} \qquad (6)$$

is the *Péclét number* and is obviously the product of the Reynolds number and the Prandtl number. Comparison of (5) with (7.140) reveals at least a partial structural similarity, and the arguments advanced in establishing the boundary-layer theory can be used in the same way to reach the conclusion that for large values of the Péclét number the effect of thermal diffusivity is confined to a layer of thickness proportional to $(P\acute{e})^{-1/2}$ and diffusion along boundaries or longitudinal diffusion (in the general direction of the flow) is negligible compared with transverse diffusion. There is then a thermal boundary layer in which the temperature distribution is governed by the boundary-layer form of (5). For very small values of the Péclét number the convective terms on the left-hand side of (5) are negligible, and the temperature field is governed by

$$P\acute{e} \frac{\partial \theta}{\partial t} = \nabla^2 \theta \qquad (7)$$

if the flow is unsteady or by

$$\nabla^2 \theta = 0 \qquad (8)$$

if it is steady. The temperature field is then, in either case, analogous to the velocity field of Stokes' flow discussed in Chap. 7. Note that the left-hand side of (7) is not necessarily negligible even if $P\acute{e}$ is small, for $\partial \theta / \partial t$ may be large.

In the case of free convection the flow is induced by nonuniformity of density in a gravitational field. The intensity of the flow depends on the motive force, which is the term ρX_i in (2.6). The density ρ can be expressed as

$$\rho = \rho_0 + \rho_1, \qquad (9)$$

where ρ_0 is the density of the fluid far away from any heat source and ρ_1 in most practical situations is much less than ρ_0. Using (9), the first two terms on the right-hand side of (2.6) can be written as

$$\frac{\partial}{\partial x_i}\left(-\rho_0\Omega - p + \lambda\frac{\partial u_\alpha}{\partial x_\alpha}\right) + \rho_1 X_i, \tag{10}$$

in which $\partial u_\alpha/\partial x_\alpha$ is written for dilatation to avoid the conflicting usage of θ. If the direction of increasing x_3 is the direction of the vertical, this becomes, for $i = 3$,

$$\frac{\partial}{\partial x_3}\left(-\rho_0\Omega - p + \lambda\frac{\partial u_\alpha}{\partial x_\alpha}\right) - \rho_1 g. \tag{10a}$$

We shall now endeavor to make (2.6) dimensionless. To do so, we need a representative velocity. Since there is no imposed flow, there is no representative velocity. We can, for our purpose, use v/L for U in (4), and use (4) to make (2.6) dimensionless. The expression for θ in (4) can at present be replaced by

$$\rho' = \frac{\rho_1}{\Delta\rho}, \tag{11}$$

where $\Delta\rho$ is a representative density difference. We need not write out the resulting dimensionless form of (2.6), except to note that the term $\rho_1 g$ in (10a) then becomes $(Gr)\rho'$, with Gr defined by

$$Gr = \frac{g\,\Delta\rho\,L^3}{\rho_0 v^2}. \tag{12}$$

The parameter Gr, for insufficient reasons called the *Grashof number* in the literature, is a measure of the relative importance of buoyancy versus viscous forces. The larger it is, the stronger the convective current. Provided that $\rho_1 \ll \rho_0$ and the effect of pressure variation is negligible,

$$\rho = \rho_0 + \rho_1 = \rho_0[1 - \alpha(T - T_0)],$$

where α is the coefficient of expansivity. Thus

$$\Delta\rho = -\alpha\rho_0\,\Delta T, \tag{13}$$

in which ΔT is a representative temperature difference corresponding to $\Delta\rho$. Thus Gr can be written as (with the minus sign dropped).

$$Gr = \frac{g\alpha\,\Delta T\,L^3}{v^2}. \tag{14}$$

For ideal gases,

$$\frac{\Delta T}{T_0} = -\frac{\Delta\rho}{\rho_0}, \tag{15}$$

as can be derived from the equation of state (A1.5). Hence

$$\alpha = \frac{1}{T_0} \tag{16}$$

for ideal gases, T_0 being the ambient temperature of the gas far from any heat source.

The local rate of heat transferred to or from a surface per unit area per unit time is $k \, \partial T/\partial n$, in which n is the distance normal to the surface. If a reference temperature difference is ΔT, the area of heat transfer is A, and the heat transferred by unit time is q, the coefficient of heat transfer h is defined by

$$q = hA \, \Delta T.$$

If L is a reference length, the Nusselt number Nu is defined by

$$Nu = \frac{hL}{k}.$$

In problems of forced convection, the Nusselt number is sought as a function of the Reynolds number and the Prandtl number or of the Reynolds number and Péclét number. In problems of free convection, we seek to determine the Nusselt number as a function of the Grashof number and the Prandtl number. Of course, the heat transferred at the boundary depends directly on the temperature gradient of the fluid at the solid surface. For this reason solution of a problem of heat transfer always involves a determination of the temperature field. Sometimes the temperature field is interesting in its own right.

2. MASS DIFFUSION

Because of the analogy between mass diffusion and heat diffusion, we shall briefly discuss the former before going on to present some detailed results on forced and free convections.

In Secs. 1 and 5.1 of Chap. 1, it was stated that whenever more than one fluid is involved, the continuity equation in the form (1.12) is valid for the mixture if the velocity vector is defined by (1.1); or, what amounts to the same thing, if only two fluids are in the mixture, (1.12a) should be used. The disadvantage of using (1.12) or (1.12a) is that we do not know the density of each gas in the mixture at any particular time and place. We are then left with two alternatives. The first is to write an equation of continuity, a set of equations of motion, an equation of state, and an energy equation for each gas, taking care of any interaction there may be between one gas and another in the mixture, and endeavor to solve the resulting equations. This seems a hopeless task. The other alternative is for a dilute solution of one or more fluids in a solvent fluid and is much more useful. It consists of the following procedure:

1 Retain the equation of continuity in the usual form, whether the velocity is defined by (1.1) or (1.2). Since the solution is dilute, the error is small even if the velocity is defined by (1.2).

2 In dilute solutions the main interest, as far as concentrations are concerned, is in the concentration of each of the solutes. It is therefore not only natural but advantageous to focus our attention on it. When we do so, we find that in the equation of mass conservation for each solute the velocity of the flow can be either defined by (1.1) or (1.2), there being no appreciable difference in the consequence. In this mass-conservation equation, however, there is a diffusive term that accounts for the possibility that the molecules of any solute may move with a mean velocity different from that of the solvent. In other words, the second alternative neglects whatever slight error there is in using the ordinary form of the equation of continuity for the mixture and focuses attention on the mass conservation of any of the solutes. This is a valid approach for dilute solutions and is advantageous because attention is focused on the less abundant material to gain greater accuracy.

The actual derivation of the diffusion equation according to the second approach is quite simple. First the concentration c of a solute can simply be defined as the density of that solute. Second, with κ' denoting mass diffusivity the mass flux of the solute is, on *experimental evidence* for dilute solutions, $-\kappa'\,\mathbf{grad}\,c$. Thus the net influx of solute mass per unit volume is

$$\frac{\partial}{\partial x_\alpha}\left(\kappa'\,\frac{\partial c}{\partial x_\alpha}\right),$$

and this must be equal to Dc/Dt. Hence the mass diffusion equation is

$$\frac{Dc}{Dt} = \frac{\partial}{\partial x_\alpha}\left(\kappa'\,\frac{\partial c}{\partial x_\alpha}\right), \tag{17}$$

or, if κ' is constant,
$$\frac{Dc}{Dt} = \kappa'\,\nabla^2 c. \tag{18}$$

It is immediately evident that apart from the terms involving pressure and dissipation in the heat-diffusion equations (2.80) and (2.81), (17) is analogous to (2.80) and (18) is analogous to (2.81).

If there is an imposed flow, and if the Péclét number UL/κ' for mass diffusion is large, there will be diffusion boundary layers of a thickness proportional to $(P\acute{e})^{-\frac{1}{2}}$, and longitudinal diffusion is negligible as compared with transverse diffusion. The class of problems is analogous to the problems involving forced convection of heat. If there is no imposed flow, and if the mean density of the fluid varies with the concentration (or concentrations) of the solute (or solutes), the nonuniformity of the solute (or solutes) in a gravitational field will provide a motive force for convection, and the diffusion equation and the equations of motion are then interdependent. The resulting flow is also called free convection, since the cause is quite the same as for the free convection of heat. The Grashof number is given by (12), with ρ_0 indicating the ambient density (often just the density of the solvent, since the solution is dilute) and $\Delta\rho$ replaced by a representative concentration increment Δc, or the sum of such increments, if two or more solutes are present. Since there is such close analogy between mass diffusion and heat diffusion, we need not treat (17) or (18) if we intend to treat the heat-diffusion equation in some detail.

3. FORCED CONVECTION FROM BOUNDARIES AT LARGE PÉCLÉT NUMBERS

If the flow is steady and laminar and the Péclét number is very large compared with unity, we can use the boundary-layer form of (2.79) or (2.80). If the fluid is a liquid and the effect of energy dissipation negligible, and if the thermal conductivity can be considered constant, either of these equations can be written as

$$\frac{DT}{Dt} = \kappa \, \nabla^2 T, \qquad \kappa = \frac{k}{\rho c_v} \, . \tag{19}$$

The slight thermal expansion of the liquid is so small that the term $p\theta$ in (2.79) or (2.80) can be neglected. If the fluid is a gas, thermal expansion is not negligible even if the fluid can often (at low Mach numbers) be considered incompressible as far as continuity is concerned. In such cases, if viscous dissipation is again negligible and the thermal conductivity can be considered constant, and if the variation of p over the field of flow is negligibly small, (2.84) can be written as

$$\frac{DT}{Dt} = \kappa \, \nabla^2 T, \qquad \kappa = \frac{k}{\rho c_p} \, . \tag{19a}$$

For two-dimensional problems and with x and y indicating boundary-layer coordinates as in Chap. 7, (19) or (19a) can be written as

$$u \frac{\partial T}{\partial x} + v \frac{\partial T}{\partial y} = \kappa \frac{\partial^2 T}{\partial y^2} \, , \tag{20}$$

provided $Pé \gg 1$. The κ in (20) is defined in (19) or (19a), as the case may be. The u and v denote the tangential and normal components in the boundary layer.

It should be explained why the fluid can be considered incompressible even when the temperature is not uniform. As the fluid is heated or cooled, its density will change accordingly. But if the change in temperature relative to the absolute temperature of any part of the fluid is small, then whether the fluid is a liquid or gas, the change in density is small and can be neglected as far as continuity is concerned. This is implied also if we assume the κ in (19) or (20) to be constant. In free convection the motive force is the nonuniformity in specific weight. Hence the density variation must be accounted for in the term presenting the body force. As far as continuity and inertia are concerned, however, the density can still be assumed constant for free convection. This is called the *Boussinesq approximation*. The equation of continuity for two-dimensional flows, under the stated restriction, can then still be written as

$$\frac{\partial u}{\partial x} + \frac{\partial y}{\partial y} = 0. \tag{21}$$

3.1. Forced Convection from a Heated Plate in Blasius Flow

As an example of forced convection we shall consider the transfer of heat from a plate heated to the temperature T_w to the fluid flowing parallel to it, with velocity U and temperature T_0 far from the plate.

If

$$\theta = \frac{T_w - T}{T_w - T_0}, \tag{22}$$

(20) can be written as

$$u\frac{\partial \theta}{\partial x} + v\frac{\partial \theta}{\partial y} = \kappa \frac{\partial^2 \theta}{\partial y^2}, \tag{23}$$

the x and y being now Cartesian coordinates. Equation (23) is to be solved with the boundary conditions

$$\theta = \begin{cases} 0 & \text{at } y = 0, \\ 1 & \text{at } y = \infty. \end{cases} \tag{24}$$

Here, since the pressure is very nearly uniform, it is natural to define κ as $k/\rho c_p$. Although (21) is valid approximately as far as continuity is concerned, the fluid can expand with temperature; and while the use of c_p or c_v makes little difference for liquids, for gases it is more accurate to use c_p in the definition of κ.

The flow has been discussed in Sec. 6.1 of Chap. 7. Using the u and v given there and the η defined there to be the independent variable and assuming θ to be a function of η alone, we obtain from (23) the simple equation

$$2\theta'' + \sigma f \theta' = 0. \tag{25}$$

Comparing this with (7.148), we see that

$$\theta' = C(f'')^\sigma \tag{26}$$

is a solution.   Integration gives

$$\theta = C \int_0^\eta (f'')^\sigma \, d\eta. \tag{27}$$

The first condition in (24) is satisfied. The second condition is satisfied if

$$C^{-1} = \int_0^\infty (f'')^\sigma \, d\eta. \tag{28}$$

Since f'' is given by the Blasius solution, (27) and (28) give the solution for (23), due to Pohlhausen (1921).

Pohlhausen also considered the temperature rise due to energy dissipation, which involves the solution of a nonhomogeneous differential equation; it is (23), with the term

$$- \frac{\mu}{\rho c_p (T_w - T_0)} \left(\frac{\partial u}{\partial y} \right)^2 \tag{29}$$

added to the right-hand side, the minus sign being in accordance with the definition of θ in (22). (Here c_p is in energy units; if in thermal units, it should be multiplied by the joule's value J.) In terms of η, that nonhomogeneous equation becomes

$$2\theta'' + \sigma f \theta' = A\sigma(f'')^2, \tag{30}$$

in which

$$A = \frac{2U^2}{c_p(T_w - T_0)}.$$

Since $f = -2f'''/f''$, (30) can be written as

$$2(\theta'' f'' - \sigma\theta' f''') = A\sigma(f'')^3,$$

or

$$2[\theta''(f'')^{-\sigma} - \sigma\theta'(f'')^{-1-\sigma} f'''] = A\sigma(f'')^{2-\sigma},$$

integration of which gives, after multiplication by $(f'')^\sigma$,

$$2\theta' = A\sigma(f'')^\sigma \int_\infty^\eta (f'')^{2-\sigma} \, d\eta.$$

Another integration gives

$$\theta = \frac{A\sigma}{2} F(\eta) + C, \tag{31}$$

where
$$F(\eta) = \int_0^\eta \left[(f'')^\sigma \int_\infty^\eta (f'')^{2-\sigma} \, d\eta \right] d\eta,$$

$$C = 1 - \frac{A\sigma}{2} F(\infty).$$

It can be seen that (31) satisfies (24). If A is large compared with unity, the temperature in the boundary layer may be higher than the temperature on the plate.

3.2. Frössling's Series for Steady Two-dimensional Thermal Boundary Layers

In Chap. 7 we have seen how the velocity distribution in boundary layers on curved bodies can be found by a series expansion in terms of universal functions. Frössling (1940) has shown how a similar expansion can be used to determine the temperature distribution in the thermal boundary layer. The initial stage of the development to be given here is only slightly different in notation from Frössling's treatment (which makes all quantities dimensionless ab initio) in order to achieve consistency with the series treated by Blasius, Hiemenz, and Howarth. Only the case of steady boundary layers in symmetric flows over symmetric two-dimensional bodies will be discussed. The case for unsymmetric two-dimensional bodies or for axisymmetric bodies can be treated similarly, as Frössling has shown.

Using x and y as curvilinear boundary-layer coordinates, treating the fluid as incompressible, and ignoring both the effects of energy dissipation and of work done by the fluid in thermal expansion, we find the governing equation to be (20), which can be written as

$$u \frac{\partial \theta}{\partial x} + v \frac{\partial \theta}{\partial y} = \kappa \frac{\partial^2 \theta}{\partial y^2}, \tag{32}$$

with
$$\theta = \frac{T - T_0'}{T_w - T_0}, \qquad \kappa = \frac{k}{\rho c_v}, \tag{33}$$

in which T_0 is the temperature far from the boundary and T_w the wall temperature assumed to be constant.

Frössling's series has the form

$$\theta = \sum_{n=0} \theta_{2n}(\eta) x^n, \tag{34}$$

in which

$$\theta_0 = F_0(\eta), \qquad \theta_2 = \frac{4u_3}{u_1} F_2(\eta), \qquad \theta_4 = \frac{6u_5}{u_1} \left[G_4(\eta) + \frac{u_3{}^2}{u_1 u_5} H_4(\eta) \right],$$

$$\theta_6 = \frac{8u_7}{u_1} \left[G_6(\eta) + \frac{u_3 u_5}{u_1 u_7} H_6(\eta) + \frac{u_3{}^3}{u_1{}^2 u_7} K_6(\eta) \right], \tag{35}$$

and so forth. The η in (34) and (35) are given by (7.185), and u_1, u_3, u_5, etc., are the same quantities as those in (7.186). Using the u and v as given by (7.188) and substituting (34) and (35) into (32), we have, upon collecting the coefficients of equal powers in x, the series of equations

$$\sigma^{-1}F_0'' = -f_1 F_0'$$
$$\sigma^{-1}F_2'' = -f_1 F_2' + 2f_1' F_2 - 3f_3 F_0',$$
$$\sigma^{-1}G_4'' = -f_1 G_4' + 4f_1' G_4 - 5g_5 F_0',$$
$$\sigma^{-1}H_4'' = -f_1 H_4' + 4f_1' H_4 - 5h_5 F_0' + \tfrac{8}{3}(2f_3' F_2 - 3f_3 F_2'),$$
(36)

and so forth. These are to be solved with the boundary conditions consistent with

$$\theta_0(0) = 1 \quad \text{and} \quad \theta_{2n}(0) = 0, \quad \text{for } n = 1, 2, \text{ etc.},$$
$$\theta_{2n}(\infty) = 0, \quad \text{for} \quad n = 0, 1, 2, \text{ etc.}$$

The solution of the first equation in (36) is straightforward. The others were integrated numerically by Frössling, who tabulated his results. We shall not reproduce his table but content ourselves with the following results directly useful for calculation of heat transfer from the boundary:

$$F_0'(0) = -0.4959, \qquad F_2'(0) = -0.1119,$$
$$G_4'(0) = -0.0977, \qquad H_4'(0) = 0.0318.$$

3.3. Forced Convection in Conduits

In conducits, where the flow is parallel, so that $v = 0 = w$, (19) becomes, for steady flows,

$$u\frac{\partial T}{\partial x} = \kappa\left(\frac{\partial^2 T}{\partial y^2} + \frac{\partial^2 T}{\partial z^2}\right) \tag{37}$$

if the term $\partial^2 T/\partial x^2$ in the Laplacian of T is neglected. Cartesian coordinates are used in (37). If the temperature distribution is given at $x = 0$ and the thermal boundary conditions are specified downstream, the method of separation of variables can be used to solve (37) and to satisfy the boundary conditions. When these conditions are such that a solution dependent only on y and z exists as x approaches infinity, the solution of (37) approaches that solution exponentially with respect to x. All this is quite clear after some thought, and we shall not endeavor to present an example.

4. TEMPERATURE DISTRIBUTION IN SQUIRE'S JET

If the momentum source in Squire's flow is also a heat source, and if the effects of gravity are neglected, the equation governing the temperature distribution is, in the notation of Sec. 4.1 of Chap. 7,

$$u\frac{\partial T}{\partial R} + \frac{v}{R}\frac{\partial T}{\partial \theta} = \kappa \nabla^2 T, \tag{38}$$

in which T is the temperature, R and θ are spherical coordinates, and $\kappa = k/\rho c_p$. Thus the pressure variation in the jet is neglected. If we assume, for convenience, that $T = 0$ at $R = \infty$ and

$$T = \frac{1}{R} G(\theta), \tag{39}$$

then

$$\nabla^2 T = \frac{1}{R^2 \sin \theta} \frac{\partial}{\partial \theta} \left(\sin \theta \frac{\partial T}{\partial \theta} \right),$$

and (38) becomes, if we write $g(\mu)$ for $G(\theta)$,

$$f'(\mu)g(\mu) + f(\mu)g'(\mu) = \frac{1}{\sigma} \frac{d}{d\mu} [(1 - \mu^2)g'(\mu)], \tag{40}$$

where μ is used for $\cos \theta$ and $f(\mu)$ is associated with Stokes' stream function as indicated by (7.104). The boundary conditions on G are

$$\frac{d}{d\theta} G(\theta) = 0 \qquad \text{at } \theta = 0 \text{ and } \pi. \tag{41}$$

These are satisfied if $G(\theta)$ is a function only of μ, which turns out to be the case. One integration of (39) produces

$$\sigma f g = (1 - \mu^2) g', \tag{42}$$

the constant of integration being zero since $f(0) = 0$. With f given by (7.119), integration of (42) yields

$$g(\mu) = A \left(\frac{a}{a + 1 - \mu} \right)^{2\sigma} = G(\theta). \tag{43}$$

This is the result of Squire (1951).

The constant A is related to the heat flux H from the source, which is defined by

$$H = 2\pi \rho c_p \int_0^\pi R^2 u T \sin \theta \, d\theta. \tag{44}$$

From (7.110), (39), (43) and, (44), we obtain

$$\frac{H}{2\pi \rho c_p \nu} = A \int_{-1}^{1} f'(\mu)g(\mu) \, d\mu. \tag{45}$$

With $f(\mu)$ given by (7.119) and $g(\mu)$ by (43), this can be integrated to give

$$\frac{H}{4\pi \rho c_p \nu} = A \left\{ \frac{a + 2}{2\sigma + 1} \left[1 - \left(\frac{a}{a + 2} \right)^{2\sigma + 1} \right] - \frac{a}{2\sigma - 1} \left[1 - \left(\frac{a}{a + 2} \right)^{2\sigma - 1} \right] \right\}. \tag{46}$$

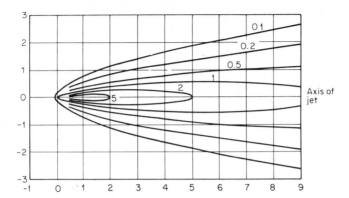

FIGURE 1. TEMPERATURE CONTOURS FOR HEATED ROUND LAMINAR JET. $a = 10^{-2}$, $\sigma = 0.72$. [*From H. B. Squire, Quart. J. Mech. Appl. Math.*, **4** (1951), *by permission of the Quarterly Journal of Mathematics.*]

For $a \ll 1$,
$$\frac{H}{4\pi \rho c_p \nu} = \frac{2A}{2\sigma + 1}.$$ (47)

The isotherms for $Pr = 0.72$ and $a = 10^{-2}$ are shown in Fig. 1.

5. TEMPERATURE DISTRIBUTION IN A STEADY LAMINAR PREHEATED JET

As another example of forced convection we may consider [Yih (1950)] the temperature distribution in the flows discussed in Secs. 6.2 and 8 of Chap. 7, when the fluid issuing from the opening is preheated, i.e., when the origin is a heat source as well as a momentum source. The air issuing from the opening is assumed to have a temperature difference from the air surrounding the jet, and the resulting temperature distribution is sought. If the additional velocity induced by the nonuniformity of temperature distribution is small compared with the velocity obtained for isothermal jets, it can be neglected and the velocity distribution can be assumed to be practically the same as that obtained by Schlichting and Bickley. Under this assumption, the equation for temperature distribution can be solved easily.

It should be noted that the solution is directly applicable not only to precooled air jets but also to problems where some property other than heat is undergoing diffusion. For instance, if oxygen is being discharged through a small slit or a small hole into an open space filled with nitrogen, the concentration of oxygen at different points in the jet can be obtained directly from the corresponding solution by making the necessary changes of physical constants.

Although the true line or point source cannot be realized, the solutions are still valid at a distance sufficiently large compared with the dimension of the opening.

In Secs. 6.2 and 8 of Chap. 7, the pressure in the fluid is assumed constant.[1] Therefore in the energy equation c_p should be used instead of c_v. If, furthermore, viscous dissipation of energy is neglected, Eq. (2.84) can be written

$$\frac{DT}{Dt} = \kappa \nabla^2 T, \tag{48}$$

in which $\kappa = k/\rho c_p$ is the thermal diffusivity based on the specific heat c_p at constant pressure, k being the thermal conductivity.

If T_0 denotes the original temperature of the free space into which the heated air jet is discharged, T the temperature at any point of the free space, and θ the quantity $(T - T_0)/T_0$, the equation for temperature distribution for the two-dimensional jet can be written

$$u \frac{\partial \theta}{\partial x} + v \frac{\partial \theta}{\partial y} = \kappa \left(\frac{\partial^2 \theta}{\partial x^2} + \frac{\partial^2 \theta}{\partial y^2} \right), \tag{49}$$

the coordinates being defined in Sec. 6.2 of Chap. 7. Assuming, as in the case of velocity, that $\partial^2 \theta/\partial x^2$ is uniformly small compared with $\partial^2 \theta/\partial y^2$ (the assumption being justifiable a posteriori), we can write (49) in the form

$$u \frac{\partial \theta}{\partial x} + v \frac{\partial \theta}{\partial y} = \kappa \frac{\partial^2 \theta}{\partial y^2}. \tag{50}$$

The flux of temperature difference per unit length of the slit, with T_0 as the datum temperature, is defined to be

$$H = \int_{-\infty}^{\infty} u(T - T_0) \, dy = T_0 \int_{-\infty}^{\infty} u\theta \, dy. \tag{51}$$

Integration of (50) with respect to y, by parts for the second term, produces

$$\int_{-\infty}^{\infty} u \frac{\partial \theta}{\partial x} dy + v\theta \Big|_{-\infty}^{\infty} - \int_{-\infty}^{\infty} \theta \frac{\partial v}{\partial y} dy = \frac{\partial \theta}{\partial y} \Big|_{-\infty}^{\infty}. \tag{52}$$

Since $\quad\quad \theta = 0 \quad$ at $y = \pm\infty$

and the equation of continuity is

$$\frac{\partial u}{\partial x} + \frac{\partial v}{\partial y} = 0,$$

[1] This actually implies the neglect of gravity effects, justifiable when M is sufficiently large.

(52) can be written as

$$\frac{d}{dx} \int_{-\infty}^{\infty} u\theta \, dy = 0.$$

This means that H is a constant.

A dimensional analysis shows that θ must be a function of the three dimensionless quantities $H/T_0\nu$, $\rho\nu^2/Mx$, and the ξ defined in (7.166a). In order that the right side of (51) can be reduced to H, it is readily seen that θ must be of the functional form

$$\theta = \frac{H}{T_0 \nu} \left(\frac{\rho\nu^2}{Mx} \right)^{1/3} t(\xi), \tag{53}$$

where the function $t(\xi)$ is to be determined from (50).

Substitution of (7.166b) and (53) in (50) gives, after simplification,

$$-2\sigma[t(\xi) \operatorname{sech}^2 \xi + t'(\xi) \tanh \xi] = t''(\xi), \tag{54}$$

where $\sigma = \nu/\kappa$ is the Prandtl number. As $t'(0) = 0$, integration of (54) gives

$$t'(\xi) = -2\sigma t(\xi) \tanh \xi.$$

A second integration then yields

$$t(\xi) = C \operatorname{sech}^{2\sigma} \xi, \tag{55}$$

where C is to be determined from (51). Substitution of (7.166) and (55) into (51) gives

$$C = \frac{1}{36^{1/3} \displaystyle\int_0^{\infty} \operatorname{sech}^{2+2\sigma} \xi \, d\xi}. \tag{56}$$

Equation (53) then becomes

$$\theta = \frac{CH}{T_0} \left(\frac{\rho}{Mvx} \right)^{1/3} \operatorname{sech}^{2\sigma} \xi,$$

or

$$T - T_0 = CH \left(\frac{\rho}{Mvx} \right)^{1/3} \operatorname{sech}^{2\sigma} \xi, \tag{57}$$

where C is given by (56). For $\sigma = 1$, $C = 0.455$. For air under normal conditions $\sigma = 0.733$, and by numerical integration $C = 0.421$.

When $\sigma = 1$, the solution for θ is of exactly the same functional form with respect to ξ as that for u, as it should be, since (50) and (7.142), on which (7.166) is based, are exactly the same for $\nu = \kappa$. From (57) the use of the boundary-layer equation (50) can be justified a posteriori.

We now turn to the temperature distribution in the round laminar jet, discussed in Sec. 8 of Chap. 7, when the fluid is preheated. The coordinates are defined in that section. With T_0 and θ denoting the same things as in the two-dimensional case, the boundary-layer form of the energy equation is

$$u \frac{\partial \theta}{\partial r} + w \frac{\partial \theta}{\partial z} = \frac{\kappa}{r} \frac{\partial}{\partial r} \left(r \frac{\partial \theta}{\partial r} \right). \tag{58}$$

The total flux of temperature difference, with T_0 as the datum temperature, is defined to be

$$H = \int_0^\infty 2\pi r w (T - T_0) \, dr = 2\pi T_0 \int_0^\infty r w \theta \, dr. \tag{59}$$

Integration of (58) between $r = 0$ and $r = \infty$ and utilization of the boundary conditions

$$\frac{\partial \theta}{\partial r} = 0 \quad \text{and} \quad u = 0 \quad \text{at} \quad r = 0$$

$$\theta = 0 \quad \text{at} \quad r = \infty$$

again indicate that H is constant. A dimensional analysis shows that θ must be a function of the three dimensionless quantities

$$\frac{H}{T_0 v z}, \quad \frac{M}{\rho v^2}, \quad \text{and} \quad \eta,$$

with η defined by (7.175). In order that the right side of (59) can be reduced to H, it is readily seen that θ must be of the functional form

$$\theta = \frac{H}{T_0 v z} t(\eta), \tag{60}$$

where the function $t(\eta)$ is to be determined from (58).

Substitution of (7.179) and (60) into (58) produces, after simplification,

$$-\frac{8\eta t}{(1 + \eta^2)^2} - \frac{4\eta^2 t'}{1 + \eta^2} = \frac{1}{\sigma} (t' + \eta t''),$$

which can be immediately integrated to

$$-\frac{4\eta^2 t}{1 + \eta^2} = \frac{1}{\sigma} \eta t',$$

the constant of integration being zero, since $t(\eta)$ and $t'(\eta)$ are both finite for $\eta = 0$. A second integration yields

$$t(\eta) = \frac{C}{(1 + \eta^2)^{2\sigma}}, \tag{61}$$

in which C is determined from (59) to be $(1 + 2\sigma)/8\pi$. Thus

$$\frac{T - T_0}{T_0} = \theta = \frac{1 + 2\sigma}{8\pi} \frac{H}{T_0 v z} \frac{1}{(1 + \eta^2)^{2\sigma}}. \tag{62}$$

This was given by Yih (1950). For $\sigma = 1$, θ is of exactly the same functional form as w given by (7.179), as it should be, since (58) is of the same form as (7.172) for $\nu = \kappa$. From (62) the use of the boundary-layer equation (58) can again be justified. But this again requires the satisfaction of (7.183).

Comparison of (43) and (61) reveals that for small a in (43) or, equivalently, for large values of

$$\frac{1}{8}\left(\frac{3M}{\pi\rho\nu^2}\right)^{\!\!1/2}$$

in the definition of η in (7.175), (43) and (61) are quite close, at least for those values of r/z corresponding to the main region of the jet. Thus, if (7.183) is satisfied, (62) is very near the exact solution of Squire (1951).

6. FREE CONVECTION FROM A HEATED VERTICAL PLATE

Consider a semi-infinite vertical plate heated to a temperature T_1 and surrounded by a fluid at a temperature $T_0 < T_1$ far from the plate and the resulting temperature distribution and the steady flow induced by it. The lower edge of the plate, or rather its trace in the plane of flow, will be taken as the origin, the direction of increasing x will be the direction of the vertical, and y will be measured along the direction normal to the plate.

The parameter θ defined by

$$\theta = \frac{T - T_0}{T_1 - T_0} \tag{63}$$

satisfies the conditions

$$\theta = 1 \text{ at } y = 0 \quad \text{and} \quad \theta = 0 \text{ at } y = \infty. \tag{64}$$

We shall accept the Boussinesq approximation and assume the ρ, whether associated with the acceleration or the specific heat or as far as the continuity equation is concerned, to be constant, and shall consider it variable only so far as the body-force term in the equation of motion is concerned. The heat conductivity and the specific heat will also be considered constant. Since in this problem the pressure does not vary much, we shall use (2.84) for the energy equation, in which Dp/Dt will be equated to zero. The energy-dissipation term can be neglected. Thus the boundary-layer form of (2.84) can be written as

$$u\frac{\partial T}{\partial x} + v\frac{\partial T}{\partial y} = \kappa\frac{\partial^2 T}{\partial y^2}, \tag{65}$$

in which $\kappa = k/\rho c_p$ and u and v are Cartesian velocity components. This can be written in terms of θ as

$$u\frac{\partial\theta}{\partial x} + v\frac{\partial\theta}{\partial y} = \kappa\frac{\partial^2\theta}{\partial y^2}. \tag{66}$$

The pressure distribution outside (and therefore also inside) the boundary layer is given by

$$\frac{\partial p}{\partial x} = -g\rho_0, \tag{67}$$

where ρ_0 corresponds to T_0 and is constant, because the fluid outside the boundary layer can be assumed to be of constant temperature and density, the vertical extent considered being too small to necessitate a change in the density ρ_0. Thus the buoyant force per unit volume is simply

$$-g\,\Delta\rho = \alpha g\rho_0(T - T_0), \tag{68}$$

according to (13). (For ideal gases α is just $1/T_0$.) Thus the equation of motion is

$$u\frac{\partial u}{\partial x} + v\frac{\partial u}{\partial y} = \nu\frac{\partial^2 u}{\partial y^2} + g\alpha(T_1 - T_0)\theta. \tag{69}$$

The equation of continuity is still

$$\frac{\partial u}{\partial x} + \frac{\partial v}{\partial y} = 0.$$

Equations (66) and (69) admit a similarity solution. Since we have seen several times in Chap. 7 how one can search for a similarity solution, we shall not make a systematic search again and shall only give the results. Assuming

$$\eta = C\frac{y}{x^{1/4}}, \qquad C = \left[\frac{\alpha g(T_1 - T_0)}{4\nu^2}\right]^{1/4}, \tag{70}$$

and $\qquad \psi(x,y) = 4\nu Cx^{3/4}f(\eta), \qquad \theta(x,y) = g(\eta),$

Pohlhausen (1921) obtained

$$u = 4\nu C^2 x^{1/2}f'(\eta), \qquad v = \nu Cx^{-1/4}(\eta f' - 3f) \tag{71}$$

and transformed (69) and (66) to

$$f''' + 3ff'' - 2f'^2 + g = 0, \tag{72}$$

$$g'' + 3\sigma fg' = 0, \tag{73}$$

with g denoting $g(\eta)$ for brevity. The boundary conditions are

$$f(0) = 0, \qquad f'(0) = 0, \qquad g(0) = 1,$$
$$f'(\infty) = 0, \qquad g(\infty) = 0.$$

The differential system was solved by Pohlhausen for an ideal gas with $\sigma = 0.733$. He found that the local Nusselt number at $x = d$ is given by

$$Nu = \frac{hd}{k} = 0.508 \left[\frac{gd^3(T_1 - T_0)}{4v^2 T_0} \right]^{\frac{1}{4}} = 0.359(Gr)^{\frac{1}{4}}, \qquad (74)$$

in which h is the heat-transfer coefficient, k is the thermal conductivity, and

$$Gr = \frac{gd^3(T_1 - T_0)}{v^2 T_0}. \qquad (75)$$

7. FREE CONVECTION DUE TO A POINT SOURCE OF HEAT

The problem of free convection due to a point source of heat is interesting because it is an idealized model of many convection phenomena found in nature. Specifically, the point source is considered to be situated in an infinite plane above which the atmosphere was originally isothermal and at rest, and the resulting mean temperature and velocity distributions are sought. The vertical variation of density in the ambient atmosphere is negligible in phenomena involving a small vertical scale. As can be observed from the behavior of smoke from a burning cigarette, the flow caused by the heat source is laminar at first, then at some height above it becomes unstable and the subsequent flow is turbulent. Only the solution for laminar convection will be presented here.

. In the analysis of the present problem it will be assumed that the temperature variation should be small enough for the physical properties of the fluid to be considered constant. Under this assumption, if T denotes the absolute temperature and γ the specific weight in this section and the next and not the ratio of specific heats, it can be shown from the well-known equation of state for perfect gases

$$\frac{p}{\gamma} = \frac{RT}{g} \qquad (76)$$

that

$$\frac{\Delta T}{T_0} = -\frac{\Delta \gamma}{\gamma_0}, \qquad (77)$$

where T_0 is the ambient temperature, γ_0 the corresponding specific weight, $\Delta T = T - T_0$, and $\Delta \gamma = \gamma - \gamma_0$. In fact the proportionality between ΔT and $\Delta \gamma$ for small variations in T is quite independent of the equation of state, and can be obtained for any fluid. Since ΔT and $\Delta \gamma$ are connected by (77), $\Delta \gamma$ can be used as a dependent variable instead of ΔT, for convenience.

Taking the heat source as origin and x and r as vertical and radial coordinates,[1] one denotes the vertical velocity by u and the radial velocity

---

[1] If cylindrical coordinates r and z are used instead of r and x, the u used here would be written as w and the v here would be u. The present usage should not cause confusion if this is remembered.

by v. Since the convectional heat flux must be constant for all values of x because of continuity, and since ΔT and $\Delta \gamma$ are related by (77), the quantity

$$G = -\int_0^\infty 2\pi r u \, \Delta \gamma \, dr \tag{78}$$

must be independent of x and is in fact a measure of the strength of the heat source. The constancy of G can be demonstrated by integrating the energy equation with the aid of the equation of continuity and the boundary condition at infinity for ΔT or $\Delta \gamma$. We have already performed this integration several times in Sec. 5, and since it is intuitively quite clear, a demonstration is not necessary. With G properly defined, the independent variables can be taken to be G, ρ, μ, x, r, and κ, where ρ is the density, μ the dynamic viscosity, and κ the thermal diffusivity defined as $k/\rho c_p$. The dependent variables are $\Delta \gamma$, u, and v, but since u and v are connected by the equation of continuity, it is sufficient to study $\Delta \gamma$ and u or the stream function ψ of Stokes.

Assuming the hydrostatic pressure of the atmosphere to be essentially undisturbed by the flow caused by the heat source and $\partial^2 u/\partial x^2$ to be negligible compared with

$$\frac{1}{r}\frac{\partial}{\partial r}\left(r\frac{\partial u}{\partial r}\right),$$

we can write the equation of motion as

$$u\frac{\partial u}{\partial x} + v\frac{\partial u}{\partial r} = \frac{v}{r}\frac{\partial}{\partial r}\left(r\frac{\partial u}{\partial r}\right) - g\frac{\Delta \gamma}{\gamma_0}, \tag{79}$$

where g is the gravitational acceleration and v is the kinematic viscosity. By virtue of (77), the equation for heat diffusion (the energy equation) can be written

$$u\frac{\partial \Delta \gamma}{\partial x} + v\frac{\partial \Delta \gamma}{\partial r} = \frac{\kappa}{r}\frac{\partial}{\partial r}\left(r\frac{\partial \Delta \gamma}{\partial r}\right) \tag{80}$$

if $\partial^2 \Delta \gamma/\partial x^2$ is neglected in comparison with

$$\frac{1}{r}\frac{\partial}{\partial r}\left(r\frac{\partial \Delta \gamma}{\partial r}\right).$$

The equation of continuity is

$$\frac{\partial}{\partial x}ru + \frac{\partial}{\partial r}rv = 0. \tag{81}$$

Equations (78) to (81) are to be solved with the boundary conditions that u, v, and $\Delta \gamma$ vanish at $r = \infty$,

$$v, \frac{\partial \Delta \gamma}{\partial r}, \text{ and } \frac{\partial u}{\partial r} \text{ vanish at } r = 0.$$

Equation (81) permits the use of the Stokes stream function ψ such that

$$u = \frac{1}{r} \frac{\partial \psi}{\partial r}, \qquad v = -\frac{1}{r} \frac{\partial \psi}{\partial x}. \tag{82}$$

Then, with the substitutions

$$\Delta\gamma = -\frac{G\rho}{x\mu} \theta(\eta) \tag{83}$$

and

$$\psi = 4\nu x f(\eta), \tag{84}$$

in which

$$\eta = \left(\frac{\rho^2 G}{4\mu^3}\right)^{\!1/4} \frac{r}{x^{1/2}}, \tag{85}$$

(79) and (80) become, after one integration in the case of (80),

$$(1 - 4f) \frac{d}{d\eta} \frac{f'}{\eta} = f''' + \eta\theta \tag{86}$$

and

$$f = -\frac{1}{4\sigma} \frac{\theta'}{\theta} \eta. \tag{87}$$

The boundary conditions are substituted by the following:

$$\theta(\infty) = 0,$$
$$f(0) = f'(0) = \theta'(0) = 0, \tag{88}$$
$$f(\infty) = \text{a finite number.}$$

The integral condition for the heat flux as expressed by (78) now assumes the form

$$\int_0^\infty f'\theta \, d\eta = \frac{1}{8\pi}. \tag{89}$$

The differential system consisting of (86) to (89) possesses solutions in closed forms for $\sigma = 1$ and $\sigma = 2$. For $\sigma = 1$,

$$\frac{\psi}{4\nu x} = f(\eta) = \frac{3}{2} \frac{\eta^2}{6\sqrt{2\pi} + \eta^2}, \tag{90}$$

and

$$-\frac{\mu x \, \Delta\gamma}{G\rho} = \theta(\eta) = \frac{1}{3\pi(1 + \eta^2/6\sqrt{2\pi})^3}. \tag{91}$$

From (82), (85), and (90), one obtains

$$\left(\frac{\mu}{G}\right)^{\!1/2} u = \frac{2f'}{\eta} = \frac{1}{\sqrt{2\pi}(1 + \eta^2/6\sqrt{2\pi})^2}, \tag{92}$$

$$\left(\frac{\rho^2 x^2}{\mu G}\right)^{\!1/4} v = \sqrt{2}\left(f' - \frac{2f}{\eta}\right) = -\frac{3\sqrt{2}\,\eta^3}{(6\sqrt{2\pi} + \eta^2)^2}. \tag{93}$$

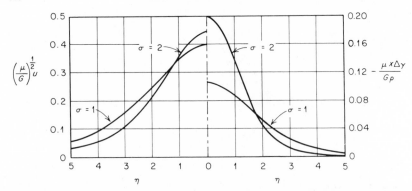

FIGURE 2. VELOCITY AND TEMPERATURE (REPRESENTED BY $\Delta\gamma$) DISTRIBUTIONS FOR LAMINAR CONVECTION OVER A POINT SOURCE. (*From C.-S. Yih, in "Symposium on the Use of Models in Geophysical Fluid Mechanics," Government Printing Office, Washington, 1956.*)

It is interesting to note that for $\sigma = 1$ the temperature distribution is more concentrated than the vertical velocity distribution, as can be seen from (91) and (92). This marks a sharp distinction between laminar free convection due to a point source of heat and laminar forced convection of a heated jet. In the latter case, the ratio of the spread of the longitudinal velocity to that of the temperature rise is, roughly speaking, equal to σ, as can be seen from (62) and (7.179).

For $\sigma = 2$,

$$\frac{\psi}{4\nu x} = f(\eta) = \frac{\sqrt{5}\eta^2}{8\sqrt{2\pi} + \sqrt{5}\eta^2}, \tag{94}$$

$$\frac{-\mu x\,\Delta\gamma}{G\rho} = \theta(\eta) = \frac{5}{8\pi[1 + (\sqrt{5}/8\sqrt{2\pi})\eta^2]^4}, \tag{95}$$

$$\left(\frac{\mu}{G}\right)^{1/2} u = \frac{2f'}{\eta} = \frac{\sqrt{5}}{2\sqrt{2\pi}[1 + (\sqrt{5}/8\sqrt{2\pi})\eta^2]^2}, \tag{96}$$

and $$\left(\frac{\rho^2 x^2}{\mu G}\right)^{1/4} v = \sqrt{2}\left(f' - \frac{2f}{\eta}\right) = -\frac{10\sqrt{2}\eta^3}{(8\sqrt{2\pi} + \sqrt{5}\eta^2)^2}. \tag{97}$$

In this case the exponent in (95) is 4, and the corresponding one in (96) is 2, so that the ratio of the exponents is 2, which happens to be equal to σ, as in the case of forced convection.

From a comparison of the two solutions with previous ones for forced convection, it is seen that for σ less than 2 the effect of buoyancy is to sharpen the curve for temperature distribution in relation to that for longitudinal-velocity distribution. It should be borne in mind, of course, that the solutions given are valid only in the main part of the jet far enough from the heat source, in the neighborhood of which the boundary-layer equations are not valid. Thus we do not require u and v to vanish at $x = 0$, although u happens to vanish there.

Equations (91), (92), (95), and (96) are plotted in Fig. 2. It can

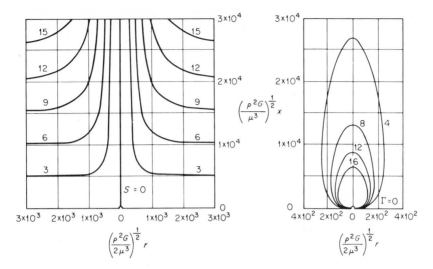

FIGURE 3. CONVECTION PATTERN OVER A POINT SOURCE. (a) $S = 10^{-4}(\rho^4 G/\mu^5)^{1/4}\psi$; (b) $\Gamma = -10^6(\mu^5/\rho^4 G^3)^{1/2} \Delta\gamma$. (From C.-S. Yih, in "Symposium on the Use of Models in Geophysical Fluid Mechanics," Government Printing Office, Washington, 1956.)

be seen that the temperature and velocity distributions are more concentrated for the larger Prandtl number and that the effect of Prandtl number on the temperature is more pronounced. The patterns of streamlines and isotherms for laminar free convection with Prandtl number 1 are shown in Fig. 3. The parameters used are dimensionless.

The solutions given by (90) to (97) seem so artificial that some description of their discovery may be helpful. One may note that (88) demands that f start with the term η^2 in a series expansion in η and that (86) and (87) admit an even f and an even θ for a solution. Thus expansions of f and θ in powers of η^2 are in order, with the expansion of f starting with η^2. Furthermore, integration of (87) yields

$$\ln \theta = -\int \frac{4\sigma f}{\eta} d\eta, \qquad (98)$$

which shows that an exact solution is hopeless unless the integral in (98) is the logarithm of something simple. Since f is a function of η^2, this narrows its admissible form to

$$f(\eta) = \frac{A\eta^2}{B + \eta^2}, \qquad (99)$$

which incidentally behaves acceptably at $\eta = \infty$, since for infinite η^2 both u and θ should vanish, although f does not. Trying (99), one finds that closed solutions are possible only for $\sigma = 1$ and $\sigma = 2$.

The solutions in closed form given by Yih (1951) contain some numerical errors. The corrected forms of the solutions, as presented here, were given by Yih (1953), together with some exact solutions for the two-dimensional case. The solutions were reproduced exactly by Brand and Lahey (1967). The author of this book learned much later than 1953 that the possibility of a similarity solution was known to Zel'dovich (1937), who gave no solutions, and that Gutman (1949) gave a numerical solution for the case of $\sigma = 1$.

8. FREE CONVECTION DUE TO A LINE SOURCE OF HEAT

Under the same assumptions as for the axisymmetric case, two mathematical solutions in closed form for $\sigma = \frac{5}{9}$ and $\sigma = 2$ have been found for the laminar case.

In a vertical plane perpendicular to the line source, the trace of the line source is taken as the origin, and the vertical and horizontal lines issuing therefrom are taken as the x and y axes, respectively. If u and v are used to denote the velocity components in the x and y directions, the equation of motion and the diffusion equation can be written in their boundary-layer forms:

$$u \frac{\partial u}{\partial x} + v \frac{\partial u}{\partial y} = \nu \frac{\partial^2 u}{\partial y^2} - g \frac{\Delta \gamma}{\gamma_0}, \tag{100}$$

$$u \frac{\partial \Delta \gamma}{\partial x} + v \frac{\partial \Delta \gamma}{\partial y} = \kappa \frac{\partial^2 \Delta \gamma}{\partial y^2}. \tag{101}$$

The equation of continuity is

$$\frac{\partial u}{\partial x} + \frac{\partial v}{\partial y} = 0,$$

which permits the use of the stream function ψ such that

$$u = \frac{\partial \psi}{\partial y} \quad \text{and} \quad v = -\frac{\partial \psi}{\partial x}. \tag{102}$$

The boundary conditions are

$$v = 0, \quad \frac{\partial u}{\partial y} = 0, \quad \text{and} \quad \frac{\partial \Delta \gamma}{\partial y} = 0, \quad \text{for } y = 0,$$

$$u = 0 \quad \text{and} \quad \Delta \gamma = 0, \quad \text{for } y = \pm \infty,$$

the conditions $u = v = 0$ at $x = 0$ being relaxed since one is chiefly concerned with the core of the jet, where the boundary-layer equations are applicable. One integral condition is furnished by

$$G = -\int_{-\infty}^{\infty} u \, \Delta \gamma \, dy, \tag{103}$$

which should be independent of x, and is a measure of the strength of the line source per unit length. Taking

$$\left(\frac{\rho^3}{Gx^3\mu^2}\right)^{1/5}\psi = f(\xi), \tag{104}$$

$$\left(\frac{\mu^2x^3}{\rho^3G^4}\right)^{1/5}\Delta\gamma = -\tfrac{1}{125}\theta(\xi), \tag{105}$$

in which
$$\xi = \frac{1}{5}\left(\frac{\rho^2G}{\mu^3x^2}\right)^{1/5}y, \tag{106}$$

and using (102), we can transform (100), (101), and (103) into the following forms:

$$f'f' - 3ff'' = f''' + \theta, \tag{107}$$

$$\theta f' + f\theta' = -\frac{\theta''}{3\sigma}, \tag{108}$$

$$\int_{-\infty}^{\infty} f'\theta \, d\xi = 125. \tag{109}$$

The boundary conditions can now be written

$$f(0) = f''(0) = \theta'(0) = 0, \quad f'(\pm\infty) = \theta(\pm\infty) = 0. \tag{110}$$

The differential system consisting of (107) to (110) is satisfied by

$$f(\xi) = A \tanh B\xi, \quad \theta(\xi) = C \operatorname{sech}^n B\xi, \tag{111}$$

in which

$$n = 2, \; A = (405\!/\!8)^{1/5}, \quad B = \frac{5A}{6}, \quad C = \frac{50A^4}{27} \quad \text{for } \sigma = 5\!/\!9, \tag{112}$$

$$n = 4, \; A = 5(240)^{-1/5}, \quad B = \frac{3A}{2}, \quad C = 9A^4, \quad \text{for } \sigma = 2. \tag{113}$$

With the functions $f(\xi)$ and $\theta(\xi)$ given, (102) and (104) to (106) constitute the solution. In particular

$$5\left(\frac{\rho\mu}{G^2x}\right)^{1/5}u = f'(\xi) = AB \operatorname{sech}^2 B\xi, \tag{114}$$

in which A and B are given by (112) for $\sigma = 5\!/\!9$ and by (113) for $\sigma = 2$.

For $\sigma = 0.733$, which is the case for air under normal conditions, Bendor[1] has found an approximate solution by numerical integration.

[1] Personal communication; see Yih (1953).

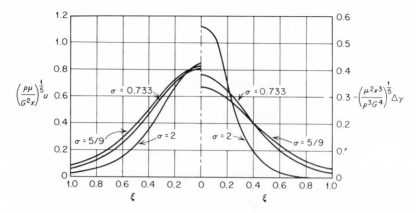

FIGURE 4a. VELOCITY AND TEMPERATURE (REPRESENTED BY $\Delta\gamma$) DISTRIBUTIONS FOR LAMINAR CONVECTION OVER A LINE SOURCE. (*From C.-S. Yih, in "Symposium on the Use of Models in Geophysical Fluid Mechanics," Government Printing Office, Washington,* 1956.)

The results are plotted in Fig. 4a together with those for Prandtl numbers $5\!/\!9$ and 2. It is seen that although Bendor's value for the velocity parameter at $\xi = 0$ is not between the corresponding values for Prandtl numbers $5\!/\!9$ and 2, as it should be, his solution shows the correct general trend. Here again, as in the case of the point source, the larger the Prandtl number, the more concentrated the temperature and velocity distributions, the effect of Prandtl number on the temperature distribution being the more pronounced.

From a comparison of the three known solutions obtained with the corresponding solutions for forced convection, (57) and (7.166b), it can be concluded that, up to $\sigma = 2$ at least, the effect of buoyancy is again to sharpen the temperature distribution curve in comparison with the velocity distribution curve.

The patterns of isotherms and streamlines can be easily obtained, but since they do not differ greatly in general character from the corresponding ones for the case of the point source, they are omitted. Instead, we show the temperature distribution of a convection plume in air, obtained by using a laser interferometer.

Like the solutions for the axisymmetric case, the solutions (111) and (112) also seem artificial at first sight. Therefore some explanation and description of their discovery seem appropriate. Equations (107) and (108), together with the boundary conditions (110) and the obvious condition that u should be even in ξ, reveal that f is odd and θ even in ξ. Equation (108) can be integrated twice to produce, in succession,

$$\theta f = -\frac{\theta'}{3\sigma}, \qquad \ln\theta = -3\sigma\int f\,d\xi. \tag{115}$$

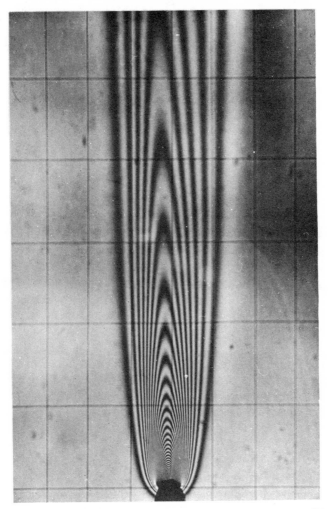

FIGURE 4b. PLUME ABOVE AN ELECTRICALLY HEATED WIRE IN AIR AT
1 ATM. WIRE LENGTH IS 6 IN.; WIRE DIAMETER IS 0.005 IN.; UPPER
WIRE SPACING IS $\frac{1}{2} \times \frac{1}{4}$ IN.; ONE FRINGE IS 6.7°F. (*Courtesy of
Mr. L. Pera and Professor B. Gebhart.*)

A solution in closed form seems impossible unless the integral in (115) is
the logarithm of something. Since u, or $f'(\xi)$, must be zero at $\xi = \pm\infty$,
an odd function in ξ that approaches constant values at $\pm\infty$ would be
appropriate for $f(\xi)$. One such function that makes the integral in (115)
a logarithm of something simple is

$$f = A \tanh B\xi. \qquad (116)$$

Thanks to the amenable forms this function and its derivatives have, solutions were found possible for $\sigma = \frac{5}{9}$ and $\sigma = 2$. The solution for $\sigma = \frac{5}{9}$ is especially welcome, since the Prandtl number for air (0.733) is not far from $\frac{5}{9}$. The solutions found by Yih (1952) contained some numerical errors. The solutions given here are the corrected solutions, given by Yih (1953) a year later. They do not seem to have been well known, for as late as 1967 exact reproductions were published by Brand and Lahey.

9. COMPRESSIBLE BOUNDARY LAYERS: BASIC EQUATIONS

The rest of the chapter will be devoted to the discussion of boundary-layer flows of a compressible fluid. Since the velocity and temperature fields of a compressible fluid are interdependent, it seems reasonable for such a discussion to follow the discussion of heat transfer and thermal diffusion just presented in the first half of the chapter.

Following the same order-of-magnitude analysis as applied to boundary-layer flows of an incompressible fluid and neglecting gravity effects, we obtain as the equation of motion for two-dimensional flows

$$\rho\left(\frac{\partial u}{\partial t} + u\frac{\partial u}{\partial x} + v\frac{\partial u}{\partial y}\right) = -\frac{\partial p}{\partial x} + \frac{\partial}{\partial y}\left(\mu\frac{\partial u}{\partial y}\right), \tag{117}$$

in which x and y are the same curvilinear coordinates as for incompressible boundary layers; that is, x is measured along the boundary and y along the normal to it. Again the $\partial p/\partial x$ is the same as in the flow outside the boundary layer, and is given by

$$-\frac{1}{\rho_1}\frac{\partial p}{\partial x} = \frac{\partial u_1}{\partial t} + u_1\frac{\partial u_1}{\partial x}, \tag{118}$$

in which the subscript 1 indicates the main stream and the u_1 corresponds to U for incompressible boundary layers, in order to save the letter U for other uses. The form of (117) is quite independent of the volume viscosity, because the term associated with it is not the dominant term. The equation of continuity is

$$\frac{\partial \rho}{\partial t} + \frac{\partial(\rho u)}{\partial x} + \frac{\partial(\rho v)}{\partial y} = 0, \tag{119}$$

and the energy equation (2.84) takes the boundary-layer form

$$\rho\left(\frac{\partial}{\partial t} + u\frac{\partial}{\partial x} + v\frac{\partial}{\partial y}\right)c_p T - \left(\frac{\partial p}{\partial t} + u\frac{\partial p}{\partial x}\right) = \frac{\partial}{\partial y}\left(k\frac{\partial T}{\partial y}\right) + \mu\left(\frac{\partial u}{\partial y}\right)^2, \tag{120}$$

$\partial p/\partial y$ being still of the order of v, and therefore $v\partial p/\partial y$ negligible.

For steady two-dimensional flows, (117) and (120) become

$$\rho\left(u\frac{\partial u}{\partial x} + v\frac{\partial u}{\partial y}\right) = -\frac{\partial p}{\partial x} + \frac{\partial}{\partial y}\left(\mu\frac{\partial u}{\partial y}\right) \tag{121}$$

and $\quad \rho\left[u\frac{\partial}{\partial x}(c_p T) + v\frac{\partial}{\partial y}(c_p T)\right] = u\frac{\partial p}{\partial x} + \frac{\partial}{\partial y}\left(k\frac{\partial T}{\partial y}\right) + \mu\left(\frac{\partial u}{\partial y}\right)^2,$ (122)

in which $\qquad\qquad \dfrac{\partial p}{\partial x} = -\rho_1 u_1 \dfrac{\partial u_1}{\partial x}.$ (123)

Equation (119) becomes

$$\frac{\partial(\rho u)}{\partial x} + \frac{\partial(\rho v)}{\partial y} = 0, \tag{124}$$

which allows us to write

$$\rho u = \rho_0 \frac{\partial \psi}{\partial y}, \qquad \rho v = -\rho_0 \frac{\partial \psi}{\partial x}, \tag{125}$$

ψ being the stream function and ρ_0 a reference density.

For steady axisymmetric flows, with x and y still denoting the same things in a meridional plane containing the axis of symmetry and with r denoting the radius of the cross section of the body cut by a plane normal to that axis (Fig. 7.20), the equations of motion and the energy equations are, respectively,

$$\rho\left(u\frac{\partial u}{\partial x} + v\frac{\partial u}{\partial y}\right) = -\frac{\partial p}{\partial x} + \frac{\partial}{\partial y}\left(\mu\frac{\partial u}{\partial y}\right), \tag{126}$$

$$\rho\left[u\frac{\partial}{\partial x}(c_p T) + v\frac{\partial}{\partial y}(c_p T)\right] = u\frac{\partial p}{\partial x} + \frac{\partial}{\partial y}\left(k\frac{\partial T}{\partial y}\right) + \mu\left(\frac{\partial u}{\partial y}\right)^2, \tag{127}$$

and the equation of continuity is

$$\frac{\partial}{\partial x}\rho r u + \frac{\partial}{\partial y}\rho r v = 0, \tag{128}$$

permitting us to write

$$\rho r u = \rho_0 \frac{\partial \psi}{\partial y}, \qquad \rho r v = -\rho_0 \frac{\partial \psi}{\partial x}, \tag{129}$$

where ψ is a stream function.

10. MANGLER'S TRANSFORMATION

Mangler (1948) showed that with the transformations

$$\bar{x} = c^2 \int_0^x \frac{r^2}{d^2}\,dx, \qquad \bar{y} = c\frac{r}{d}y, \qquad \bar{\psi}(\bar{x},\bar{y}) = \frac{c}{d}\psi(x,y),$$

$$\bar{p}(\bar{x}) = p(x), \qquad \bar{T}(\bar{x},\bar{y}) = T(x,y), \qquad \bar{\rho}(\bar{x},\bar{y}) = \rho(x,y), \tag{130}$$

$$\bar{\mu}(\bar{x},\bar{y}) = \mu(x,y), \qquad \bar{k}(\bar{x},\bar{y}) = k(x,y),$$

Eqs. (126) to (128) can be transformed to Eqs. (121), (122), and (124), with a bar over all variables in these latter equations. This can easily be verified. The coordinates are as shown in Fig. 7.20, d is a fixed reference length, and c is an arbitrary scaling factor. Hence methods for solving two-dimensional boundary-layer flows of a compressible fluid can be applied for determining flows in axisymmetric boundary layers. Therefore, we shall henceforth consider only two-dimensional boundary layers.

11. THE LAW OF VARIATION OF VISCOSITY OF GASES WITH TEMPERATURE

To continue the discussion of compressible boundary layers we must consider the variation of viscosity of the fluid with temperature. For ideal gases Sutherland's formula

$$\mu = \frac{\text{const} \cdot T^{3/2}}{T + C} \tag{131}$$

gives a good approximation, in which T is measured in degrees Kelvin, and for air the best value for C is $114°K$. For convenience in theoretical considerations (131) is often further approximated by

$$\mu = \text{const} \cdot T^{\omega}. \tag{132}$$

The formula

$$\mu = 1.15 T^{8/9},$$

in which μ is in micropoises (a poise is a unit of viscosity in the cgs system), has been suggested by Cope and Hartree as applicable to air in the range from 90 to $300°K$.

12. THE CASE OF $\sigma = 1$

It was first noticed by Busemann (1935) that if the Prandtl number is equal to unity, then, whatever the variation of viscosity with temperature, $c_p T$ is of the form $Au^2 + C$, u being the same u in (121) and A and C being constants, if the boundary is insulated and if c_p is constant. For the special case of the layer on a flat plate at zero angle of incidence,

$$c_p T = Au^2 + Bu + C,$$

provided only $\sigma = 1$. If the plate is not insulated, B is not zero, because $u = 0$ at $y = 0$, and

$$\left.\frac{\partial T}{\partial y}\right|_0 = Au \left.\frac{\partial u}{\partial y}\right|_0 + B \left.\frac{\partial u}{\partial y}\right|_0 = B \left.\frac{\partial u}{\partial y}\right|_0.$$

That the statements made in the last paragraph concerning the relationship between T and u are true can be seen by a direct inspection of (121) and (122).   Treating c_p as constant and remembering that

$$\sigma = \frac{\mu c_p}{k},$$

we see immediately that if $\sigma = 1$, then

$$c_p T = -\frac{u^2}{2} + C \tag{133}$$

makes (122) and (121) equivalent.   If $\sigma = 1$ and $\partial p/\partial x = 0$, which holds for the flow along a flat plate at zero angle of incidence, then

$$c_p T = -\frac{u^2}{2} + Bu + C \tag{134}$$

reduces (122) to an equation equivalent to (121)—in fact, to (121) multiplied by $B - u$.   Throughout the rest of this chapter, σ is assumed constant, whether or not equal to 1.

13.   THE TRANSFORMATION OF STEWARTSON

Great simplifications of the boundary-layer equations (121), (122), and (124) are possible if the Prandtl number σ and the ω in (132) are equal to unity.   Since for air they are not far from unity, it is not useless to consider these simplifications.   For $\sigma = 1 = \omega$ and an insulated boundary, Howarth (1948) showed that (121) can be reduced to the equation which is almost the boundary-layer equation of motion for incompressible fluids, and Stewartson (1949) showed that (121) can be reduced exactly to that boundary-layer equation of motion.   At first sight (121), with the term involving p given by (123), seems already identical in form to the equation of motion governing boundary-layer flow of an incompressible fluid. But it must be realized that in (121) both ρ and μ are functions of T. In the boundary layer the pressure p is the same as in the local main stream, i.e., just outside the boundary layer.   Hence according to the equation of state for ideal gases ρ varies inversely as T.   Thus it is necessary to consider (122) and (124) also.

Since Stewartson's transformation gives stronger results than Howarth's, only the former will be given here.   The assumptions that $c_p = \text{const}$, $\sigma = 1$, and the boundary is insulated enable us to use (133), which we shall write in the form

$$\frac{T}{T_1} = 1 + \frac{\gamma - 1}{2c_1^2}\,(u_1^2 - u^2), \tag{135}$$

in which the subscript 1 indicates that the quantity is taken in the main stream just outside the boundary layer. Note that

$$\frac{\gamma - 1}{c_1{}^2} T_1 = \frac{\gamma - 1}{\gamma} \frac{T_1 \rho_1}{p_1} = \frac{\gamma - 1}{\gamma R} = \frac{1}{c_p}, \tag{136}$$

as it should be if (135) is to agree with (133).

Instead of y, we shall use the new coordinate

$$Y = \frac{c_1}{c_0} \int_0^y \frac{\rho \, dy}{\rho_0}, \tag{137}$$

in which the subscript zero indicates some standard state which will be taken here to be the state at the stagnation point, where $x = 0$. (For the case of the flat plate there is no stagnation point, and we can take the state in the free stream as the standard state, and the modification is slight. At any rate for that special case a more general solution is possible, as we shall show.) The definition of Y given by (137) differs from Stewartson's by a constant factor $\sqrt{\nu_0}$ only. The difference is inconsequential, and (137) is dimensionally more appealing.

Then (125) gives

$$u = \frac{c_1}{c_0} \frac{\partial \psi}{\partial Y}, \qquad v = -\frac{\rho_0}{\rho} \left[\left(\frac{\partial \psi}{\partial x} \right)_Y + \left(\frac{\partial \psi}{\partial Y} \right)_x \left(\frac{\partial Y}{\partial x} \right)_y \right], \tag{138}$$

$$\left(\frac{\partial u}{\partial x} \right)_y = \frac{\partial \psi}{\partial Y} \frac{1}{c_0} \frac{dc_1}{dx} + \frac{c_1}{c_0} \frac{\partial^2 \psi}{\partial Y \, \partial x} + \frac{c_1}{c_0} \frac{\partial^2 \psi}{\partial Y^2} \left(\frac{\partial Y}{\partial x} \right)_y,$$

and

$$\left(\frac{\partial u}{\partial y} \right)_x = \frac{c_1{}^2 \rho}{c_0{}^2 \rho_0} \frac{\partial^2 \psi}{\partial Y^2}.$$

Hence

$$u \frac{\partial u}{\partial x} + v \frac{\partial u}{\partial y} = \left(\frac{c_1}{c_0} \right)^2 \left(\frac{\partial^2 \psi}{\partial Y \, \partial x} \frac{\partial \psi}{\partial Y} - \frac{\partial \psi}{\partial x} \frac{\partial^2 \psi}{\partial Y^2} \right) + \frac{c_1}{c_0{}^2} \frac{dc_1}{dx} \left(\frac{\partial \psi}{\partial Y} \right)^2. \tag{139}$$

As to the term involving viscosity, since $\mu_0 T = \mu T_0$,

$$\mu \left(\frac{\partial u}{\partial y} \right)_x = \frac{\mu_0 T}{T_0} \frac{c_1{}^2 \rho}{c_0{}^2 \rho_0} \frac{\partial^2 \psi}{\partial Y^2} = \frac{p c_1{}^2 \mu_0}{p_0 c_0{}^2} \frac{\partial^2 \psi}{\partial Y^2}, \tag{140}$$

and therefore

$$\frac{\partial}{\partial y} \left(\mu \frac{\partial u}{\partial y} \right) = \mu_0 \frac{p \rho c_1{}^3}{p_0 \rho_0 c_0{}^3} \frac{\partial^3 \psi}{\partial Y^3} = \nu_0 \frac{p \rho c_1{}^3}{p_0 c_0{}^3} \frac{\partial^3 \psi}{\partial Y^3}. \tag{141}$$

Note that the pressure p is p_1, according to the boundary-layer theory, and hence is independent of y. Another preparatory result needed is

$$\frac{1}{\rho} \frac{\partial p}{\partial x} = \frac{p_1 T}{p \rho_1 T_1} \frac{\partial p}{\partial x} = \frac{1}{\rho_1} \frac{T}{T_1} \frac{dp_1}{dx}, \tag{142}$$

since $p = p_1$.

For the main stream the Euler equation is

$$\frac{1}{\rho_1} \frac{dp_1}{dx} = -u_1 \frac{\partial u_1}{\partial x}, \tag{143}$$

and the Bernoulli equation or energy equation is

$$c_1{}^2 + \frac{\gamma - 1}{2} u_1{}^2 = \text{const} = c_0{}^2 + \frac{\gamma - 1}{2} u_0{}^2, \tag{144}$$

which, in fact, is obtained by integrating (143), using the isentropic relation

$$p_1 = \text{const} \cdot \rho_1{}^\gamma \tag{145}$$

for the main stream. Thus

$$c_1 \frac{dc_1}{dx} = -\frac{\gamma - 1}{2} u_1 \frac{\partial u_1}{\partial x} = \frac{\gamma - 1}{2} \frac{1}{\rho_1} \frac{dp_1}{dx}. \tag{146}$$

If now (135) is substituted into (142) and (146) into (139) and the resulting equations are substituted into the boundary-layer equation of motion (121) together with (141), we obtain

$$\left(\frac{c_1}{c_0}\right)^2 \left(\frac{\partial \psi}{\partial Y} \frac{\partial^2 \psi}{\partial x\, \partial Y} - \frac{\partial \psi}{\partial x} \frac{\partial^2 \psi}{\partial Y^2}\right) + \frac{\gamma - 1}{2} \left(\frac{\partial \psi}{\partial Y}\right)^2 \frac{1}{\rho_1 c_0{}^2} \frac{dp}{dx}$$

$$= \frac{\nu_0 \rho_1}{p_0} \left(\frac{c_1}{c_0}\right)^3 \frac{\partial^3 \psi}{\partial Y^3} - \frac{1}{\rho_1} \frac{dp}{dx}\left\{1 + \frac{\gamma - 1}{2 c_1{}^2}\left[u_1{}^2 - \left(\frac{c_1}{c_0} \frac{\partial \psi}{\partial Y}\right)^2\right]\right\}. \tag{147}$$

The terms involving $(\partial \psi / \partial Y)^2$ on both sides of this equation cancel each other. Furthermore, because of (145) and

$$c_1{}^2 = \frac{\gamma p_1}{\rho_1},$$

we have

$$\frac{p_1}{p_0} = \left(\frac{c_1}{c_0}\right)^{2\gamma/(\gamma - 1)}. \tag{148}$$

If we agree to take the stagnation point as the reference point (or the virtual stagnation point if the flow has no actual stagnation point), then $u_0 = 0$ and (144) becomes

$$1 + \frac{\gamma - 1}{2 c_1{}^2} u_1{}^2 = \left(\frac{c_0}{c_1}\right)^2. \tag{149}$$

Substituting (148) and (149) into (147), we have, after division by a factor,

$$\nu_0 \frac{\partial^3 \psi}{\partial Y^3} - \left(\frac{c_0}{c_1}\right)^{(3\gamma - 1)/(\gamma - 1)} \left(\frac{\partial^2 \psi}{\partial x\, \partial Y} \frac{\partial \psi}{\partial Y} - \frac{\partial \psi}{\partial x} \frac{\partial^2 \psi}{\partial Y^2}\right) = \frac{1}{\rho_1}\left(\frac{c_0}{c_1}\right)^{(7\gamma - 5)/(\gamma - 1)} \frac{dp}{\partial x}. \tag{150}$$

If we define a new variable X by

$$X = \int_{x_0}^{x} \left(\frac{c_1}{c_0}\right)^{(3\gamma-1)/(\gamma-1)} dx,$$

the right-hand side, by virtue of (146), is

$$\frac{2}{\gamma-1}\left(\frac{c_0}{c_1}\right)^{(7\gamma-5)/(\gamma-1)} c_1 \frac{dc_1}{dx} = \frac{2c_0^2}{\gamma-1}\left(\frac{c_1}{c_0}\right)^{-(6\gamma-4)/(\gamma-1)} \frac{d}{dx}\left(\frac{c_1}{c_0}\right)$$

$$= \frac{2c_0^2}{\gamma-1}\left(\frac{c_1}{c_0}\right)^{-3} \frac{d}{dX}\left(\frac{c_1}{c_0}\right) = -\frac{c_0^2}{\gamma-1}\frac{d}{dX}\left(\frac{c_0}{c_1}\right)^2 = -U_1 \frac{d}{dX} U_1,$$

in which

$$U_1 = \frac{c_0^2}{c_1}\left(\frac{2}{\gamma-1}\right)^{\frac{1}{2}}. \tag{151}$$

Hence (150) becomes

$$\nu_0 \frac{\partial^3 \psi}{\partial Y^3} - \left(\frac{\partial \psi}{\partial Y}\frac{\partial^2 \psi}{\partial X\,\partial Y} - \frac{\partial \psi}{\partial X}\frac{\partial^2 \psi}{\partial Y^2}\right) = -U_1 \frac{d}{dX} U_1, \tag{152}$$

which is the equation governing the flow of an incompressible fluid of kinematic viscosity ν_0 in a boundary layer with main-stream velocity U_1. Thus Stewartson achieved his goal.

In the same paper Stewartson also applied the Kármán-Pohlhausen method to the boundary layers of a compressible fluid. Because of limitation of space the details of the method will not be presented here, since the method has been so explicitly described in Chap. 7, in connection with boundary layers of an incompressible fluid.

It should be remembered that Stewartson's transformation is possible only if $\sigma = 1 = \omega$, because (135) depends on $\sigma = 1$ and (140) on $\omega = 1$. Furthermore, (135), which plays a crucial rule, implies that $\partial T/\partial y = 0$ at $y = 0$, where $u = 0$, as already mentioned in Sec. 12. Thus the boundary must be insulated.

14. VON MISES' TRANSFORMATION

A transformation by von Mises (1927), applicable to boundary-layer flows in general, is especially useful to the study of boundary layers of a compressible fluid. This transformation consists simply in the substitution of the variables x and stream function ψ (for two-dimensional flows) for x and y as independent variables. Thus

$$\left(\frac{\partial}{\partial x}\right)_y = \left(\frac{\partial}{\partial x}\right)_\psi + \frac{\partial \psi}{\partial x}\left(\frac{\partial}{\partial \psi}\right)_x = \left(\frac{\partial}{\partial x}\right)_\psi - \frac{\rho}{\rho_0}v\left(\frac{\partial}{\partial \psi}\right)_x, \tag{153}$$

$$\left(\frac{\partial}{\partial y}\right)_x = \frac{\partial \psi}{\partial y}\left(\frac{\partial}{\partial \psi}\right)_x = \frac{\rho u}{\rho_0}\left(\frac{\partial}{\partial \psi}\right)_x,$$

in which (125) has been used. The equations of motion (121) and the energy equation (122), after substitution of (123) into them, become, respectively,

$$u \frac{\partial u}{\partial x} - \frac{\rho_1 u_1}{\rho} \frac{du_1}{dx} = \frac{u}{\rho_0{}^2} \frac{\partial}{\partial \psi}\left(\mu \rho u \frac{\partial u}{\partial \psi}\right), \tag{154}$$

$$c_p \frac{\partial T}{\partial x} + \frac{\rho_1 u_1}{\rho} \frac{du_1}{dx} = \frac{1}{\rho_0{}^2} \frac{\partial}{\partial \psi}\left(k \rho u \frac{\partial T}{\partial \psi}\right) + \frac{\mu \rho u}{\rho_0{}^2}\left(\frac{\partial u}{\partial \psi}\right)^2, \tag{155}$$

provided that c_p can be treated as constant. The usefulness of (154) and (155) will be demonstrated in the next section.

15. THE BOUNDARY LAYER ON A FLAT PLATE AT ZERO ANGLE OF INCIDENCE

For flow past a flat plate at zero angle of incidence $\partial p/\partial x = 0$, and (121) and (122) become

$$\rho\left(u \frac{\partial u}{\partial x} + v \frac{\partial u}{\partial y}\right) = \frac{\partial}{\partial y}\left(\mu \frac{\partial u}{\partial y}\right), \tag{156}$$

$$\rho\left(u \frac{\partial}{\partial x} c_p T + v \frac{\partial}{\partial y} c_p T\right) = \frac{\partial}{\partial y}\left(k \frac{\partial T}{\partial y}\right) + \mu\left(\frac{\partial u}{\partial y}\right)^2. \tag{157}$$

Since $p = \text{const}$, ρ varies as $1/T$. The physical constants μ and k are functions of T only. If c_p is assumed to be also only a function of T, then (156) and (157) possess a similarity solution of the form

$$u = Uf(\eta), \qquad \frac{T - T_1}{T_1} = \theta(\eta), \qquad \eta = \left(\frac{U}{\nu_1 x}\right)^{1/2} y, \tag{158}$$

in which U is the free-stream velocity (which has been denoted by u_1 but **is here denoted by U to emphasize its independence of x**) and ν_1 and T_1 the kinematic viscosity and temperature of the free stream, respectively. The truth of this statement becomes evident if we recall that (158) implies

$$\rho = r(\eta), \qquad \psi = (U\nu_1 x)^{1/2} F(\eta), \qquad v = -\frac{\rho_0}{\rho}\frac{\partial \psi}{\partial x} = x^{-1/2} g(\eta), \tag{159}$$

$F(\eta)$ and $g(\eta)$ being related to $f(\eta)$ and $r(\eta)$, and $r(\eta)$ being further related to $\theta(\eta)$. Thus (156) and (157) can be reduced to two equations involving $F(\eta)$ and $\theta(\eta)$, and a similarity solution is possible.

If we wish, we can proceed as just described, but it is much more convenient to employ von Mises' transformation. For the flow under consideration, $\partial u_1/\partial x = 0$, and (154) and (155) respectively become, upon choosing the reference density ρ_0 to be the free-stream density ρ_1 for convenience,

$$\rho_1{}^2 \frac{\partial u}{\partial x} = \frac{\partial}{\partial \psi}\left(\mu \rho u \frac{\partial u}{\partial \psi}\right), \tag{160}$$

$$\rho_1{}^2 c_p \frac{\partial T}{\partial x} = \frac{\partial}{\partial \psi}\left(k \rho u \frac{\partial T}{\partial \psi}\right) + \mu \rho u\left(\frac{\partial u}{\partial \psi}\right)^2. \tag{161}$$

We now consider the special cases $\sigma = 1$ and $\omega = 1$ separately. If $\sigma = 1$, whether or not there is heat transfer from or to the plate, (134) holds, provided c_p is constant. Whatever the variation of μ with T, μ can then be expressed as a function of u. The same is true of ρ, since ρ is proportional to $1/T$. Thus (160) is an equation in u alone, and it is nonlinear. We have already seen the possibility of a similarity solution in the form of (158). It is not difficult to see from the boundary conditions on u and T and the form of (160) that it admits a similarity solution of the form

$$u = Uf(\eta), \qquad \rho = \rho_1 r(\eta), \qquad T = T_1\theta(\eta), \qquad \mu = \mu_1 m(\eta), \quad (162)$$

in which $\qquad\qquad\qquad \eta = \psi(\nu_1 Ux)^{-\frac{1}{2}}.$ $\qquad\qquad\qquad\qquad$ (163)

The resulting equation is

$$-\frac{\eta}{2}\frac{df}{d\eta} = \frac{d}{d\eta}\left(rmf\frac{df}{d\eta}\right), \qquad\qquad (164)$$

which was obtained by von Kármán and Tsien (1938). The functions in (162) are related. For instance, (134) relates f and θ, and B and C in (134) being determined from T_1 and the wall temperature T_w, as well as U. Furthermore

$$\frac{\rho}{\rho_1} = \frac{T_1}{T}.$$

Hence $r = \theta^{-1}$. Also, m is related to θ through the dependence of μ on T. With these connections in mind, (164) can be solved with the boundary conditions

$$f(0) = 0, \qquad f(\infty) = 1. \qquad\qquad (165)$$

Once $f(\eta)$ is determined, T is determined from (134). The case of no heat transfer, for which $B = 0$ in (134), and the case $T_w = T_1/4$ were treated by von Kármán and Tsien, who solved (164) iteratively for the viscosity-temperature relation

$$\mu = \mu_1\left(\frac{T}{T_1}\right)^{0.76}.$$

Their results are shown in Figs. 5 and 6, respectively, for the two cases mentioned.

So much for the case $\sigma = 1$. If we assume, for simplicity, that $\omega = 1$, which is not far from the recommended value $\frac{8}{9}$, great simplification is again possible. For in that case $\mu\rho$ is a constant, and (160) and (161) are not explicitly coupled. This does not mean that the velocity field is independent of the temperature field, for to determine u as a function of x and y we need to know ψ as a function of x and y, and we can know that only if we know ρ (or T). However, in seeking a solution for u in terms of x and ψ we need not know anything about T. Herein lies the simplification. We can, as it were, overcome our difficulties one by one, which

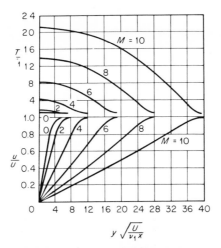

FIGURE 5. VELOCITY AND TEMPERATURE DISTRIBUTIONS WITH NO HEAT TRANSFER AT WALL. [*T. von Kármán and H. S. Tsien, J. Aeron. Sci.*, **5** (1938), *by permission of the American Institute of Aeronautics and Astronautics.*]

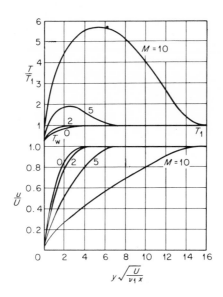

FIGURE 6. VELOCITY AND TEMPERATURE DISTRIBUTIONS WHEN THE WALL TEMPERATURE IS ONE-FOURTH THE FREE-STREAM TEMPERATURE. [*T. von Kármán and H. S. Tsien, J. Aeron. Sci.*, **5** (1938), *by permission of the American Institute of Aeronautics and Astronautics.*]

is always a much easier task. For $\omega = 1$, (160) becomes

$$\rho_1{}^2 \frac{\partial u}{\partial x} = \frac{\partial}{\partial \psi} \left(\mu_1 \rho_1 u \frac{\partial u}{\partial \psi} \right),$$

or

$$\frac{\partial u}{\partial x} = \frac{\partial}{\partial \psi} \left(\nu_1 u \frac{\partial u}{\partial \psi} \right). \tag{166}$$

Again using (162) and (163), we can transform (166) into

$$-\frac{\eta}{2} \frac{df}{d\eta} = \frac{d}{d\eta} \left(f \frac{df}{d\eta} \right). \tag{167}$$

Similarly, (161) is transformed into

$$-\frac{\eta}{2} \frac{d\theta}{d\eta} = \frac{1}{\sigma} \frac{d}{d\eta} \left(f \frac{d\theta}{d\eta} \right) + \frac{U^2}{c_p T_1} f \left(\frac{df}{d\eta} \right)^2. \tag{168}$$

Equation (167) is to be solved with the boundary conditions (165). After f is determined as a function of η, (168), being linear, can easily be solved with the boundary conditions

$$\theta(0) = \frac{T_w}{T_1}, \qquad \theta(\infty) = 1. \tag{169}$$

Equation (167) looks unfamiliar at first sight, but if we define a new independent variable ξ and a new dependent variable $F(\xi)$ by

$$\frac{d\eta}{d\xi} = f(\eta) \qquad \text{and} \qquad \eta = F(\xi), \tag{170}$$

a simple calculation shows that (167) assumes the form

$$2 \frac{d^3 F}{d\xi^3} + F \frac{d^2 F}{d\xi^2} = 0, \tag{171}$$

which is just Blasius' equation, to be solved with the boundary conditions

$$F(0) = 0 = F'(0), \qquad F'(\infty) = 1. \tag{172}$$

The solution is well known, and was given in Chap. 7 in graphical form. Schlichting (1968, p. 129) tabulates the numerical values. Our ξ here is there written η, and our F is there written f. Then (170) gives f as a function of η, and (168) can be solved, as mentioned before.

Alternatively we can use ξ instead of η and write (168) as

$$-\frac{F}{2} \frac{d\theta}{d\xi} = \frac{1}{\sigma} \frac{d^2\theta}{d\xi^2} + \frac{U^2}{c_p T_1} (F'')^2, \tag{173}$$

which can easily be solved. In fact it has exactly the same form as (30), the solution of which is known. Thus, the solution of the boundary-layer equations for the case of a gas with $\omega = 1$ can be reduced to the solution of the boundary-layer equations (for u and T) for an incompressible fluid. This certainly is not obvious from (156) and (157). Therefore the von Mises transformation has been necessary.

Whether we use ξ or η, the return to the xy plane still takes a little more calculation, albeit quite straightforward. The reader may have noticed that η is not explicitly defined in terms of y, and ξ, defined in terms of f and η, is therefore also not explicitly expressed in y. To see, at any chosen value of x, what value of y corresponds to a value of η, we have to use the relation

$$\rho u = \rho_1 \frac{\partial \psi}{\partial y}. \tag{174}$$

With u and T (and therefore ρ) given as functions of η, which is proportional to ψ at any fixed value of x, the y-η relationship is given by (174). If ξ has been used as the independent variable, the determination of the y-ξ relationship is also through the use of (174).

As has been said, (156) and (157) admit a similarity solution even if neither σ nor ω is equal to 1. We can derive the simultaneous ordinary differential equations governing the flow and temperature fields either from (156) and (157) or from (160) and (161). We leave that for the problems. Instead, we shall present Crocco's transformation and his equations for the boundary-layer flow over a flat plate at zero angle of incidence. The reason is historical: it was Crocco who actually gave the results for the general case (in which neither σ nor ω is equal to 1), and it seems reasonable to relate his equations to the general case, although we shall only refer to his numerical results for the general case without presenting them.

16. CROCCO'S TRANSFORMATION

Crocco's transformation (1939) consists in changing the independent variables from (x,y) to (x,u) and the dependent variables from (u,T) to (τ,I), where

$$\tau = \mu\left(\frac{\partial u}{\partial y}\right)_x = \mu\left(\frac{\partial y}{\partial u}\right)^{-1}, \qquad I = c_p T. \tag{175}$$

The replacement of T by I is, of course, the trivial part of the transformation. If c_p is still considered constant, by Crocco's transformation we can write (121), (124), and (122) as

$$-\rho u \frac{\partial y}{\partial x} + \rho v = -\frac{\partial p}{\partial x}\frac{\partial y}{\partial u} + \frac{\partial \tau}{\partial u}, \tag{176}$$

$$\frac{\partial y}{\partial u}\frac{\partial}{\partial x}\rho u - \frac{\partial y}{\partial x}\frac{\partial}{\partial u}\rho u + \frac{\partial}{\partial u}\rho v = 0 \tag{177}$$

$$(1 - \sigma)\frac{\partial \tau}{\partial u}\frac{\partial I}{\partial u} + \left(\frac{\partial^2 I}{\partial u^2}\right)\tau + \sigma\tau - \rho u\sigma\frac{\partial I}{\partial x}\frac{\partial y}{\partial u} + \sigma\left(\frac{\partial I}{\partial u} + u\right)\frac{\partial p}{\partial x}\frac{\partial y}{\partial u} = 0, \tag{178}$$

in which $\partial/\partial x$ means $(\partial/\partial x)_u$. If ρv is eliminated between (176) and (177), we have

$$u \frac{\partial}{\partial x}\rho y_u + \tau_{uu} - p_x y_{uu} = 0, \tag{179}$$

in which subscripts are now used to indicate partial differentiation. By virtue of the first equation in (175), (179) and (178) can be written as

$$u \frac{\partial}{\partial x} \frac{\rho\mu}{\tau} + \tau_{uu} - p_x \frac{\partial}{\partial u} \frac{\mu}{\tau} = 0, \tag{180}$$

$$(1 - \sigma)I_u\tau\tau_u + (I_{uu} + \sigma)\tau^2 - \sigma u\rho\mu I_x + \mu\sigma(I_u + u)p_x = 0. \tag{181}$$

Equation (181) is linear in τ^2, and can be expressed in terms of I, u, and x, since μ depends only on I and ρ depends on I and p, the latter being a function of x only. Thus in principle the solution of (181) for τ^2 can be substituted into (180) to obtain a single equation for I. The resulting equation, however, is in general too complicated to solve.

17. CROCCO'S TRANSFORMATION APPLIED TO THE BOUNDARY LAYER ALONG A FLAT PLATE

For flow along a flat plate at zero incidence, (180) and (181) become

$$u \frac{\partial}{\partial x} \frac{\rho\mu}{\tau} + \tau_{uu} = 0, \tag{182}$$

$$(1 - \sigma)I_u\tau\tau_u + (I_{uu} + \sigma)\tau^2 - \sigma\rho\mu u I_x = 0. \tag{183}$$

If we seek a similarity solution and write

$$\tau = \phi(x)g(u), \tag{184}$$

we find from (182) and (183) that $\phi(x)$ is proportional to $x^{-1/2}$. It is convenient to write

$$\tau = \frac{g(u)}{\sqrt{2x}}. \tag{185}$$

Then (182) and (183) become

$$gg'' + u\rho\mu = 0, \tag{186}$$

$$(I'' + \sigma)g + (1 - \sigma)I'g' = 0. \tag{187}$$

If we write

$$\frac{u}{U} = \eta, \quad \frac{I}{I_1} = \theta(\eta), \quad \frac{\rho}{\rho_1} = r(\eta), \quad \frac{\mu}{\mu_1} = m(\eta), \tag{188}$$

and

$$F(\eta) = \left(\frac{2}{\rho_1\mu_1 U^3}\right)^{1/2} g(u), \tag{189}$$

(186) and (187) become

$$FF'' + 2\eta rm = 0, \tag{190}$$

$$\left(\theta'' + \frac{\sigma U^2}{I_1}\right) F + (1 - \sigma)\theta'F' = 0, \tag{191}$$

with the boundary conditions

(i) $F'(0) = 0$, $\theta(0) = \dfrac{I_w}{I_1}$,

(ii) $F(1) = 0$, $\theta(1) = 1$.

The boundary condition $F'(0) = 0$ corresponds to

$$\left(\frac{\partial \tau}{\partial y}\right)_0 = \mu\left(\frac{\partial^2 u}{\partial y^2}\right)_x = 0 \qquad \text{at } y = 0, \text{ where } u = 0.$$

The other boundary conditions are obvious. Note that U^2/I_1 is just $(\gamma - 1)M_1^2$, M_1 being the Mach number of the free stream.

We now discuss the cases $\sigma = 1$ and $\omega = 1$ separately. If $\sigma = 1$, (191) gives

$$I_1\theta'' = -U^2, \tag{192}$$

integration of which yields (135). Since $r(\eta)$ and $m(\eta)$ are explicit functions of $\theta(\eta)$ given by the equation of state in one case and the equation specifying the variation of μ with T in the other, (190) is an ordinary equation involving F and η only:

$$FF'' + h(\eta) = 0, \tag{193}$$

This can be solved iteratively and was so solved by Crocco (1946). With F a known function of η, $g(u)$ is a known function of u, and (185) gives τ. Then (175) gives y as a function of u for any fixed x, and (135) gives the temperature distribution since the distribution of u in the xy plane is known. The method parallels that of von Kármán and Tsien based on von Mises' transformation, presented in Sec. 15.

For $\omega = 1$, Eq. (190) becomes

$$FF'' + 2\eta = 0, \tag{194}$$

since $rm = 1$. If we define a new independent variable ξ and dependent variable f by the relations

$$F = 2\frac{d^2f}{d\xi^2}, \qquad \eta = \frac{df}{d\xi}, \tag{195}$$

(194) reduces (primes indicating differentiation with respect to ξ) to

$$2f''' + ff'' = 0, \tag{196}$$

which is Blasius' equation. The boundary conditions on f are

$$f(0) = 0 = f'(0), \qquad f'(\infty) = 1. \tag{197}$$

The first two conditions are obtained by noting that

$$f = \int_0^\xi \eta \, d\xi$$

and that we can take $\eta = 0$ at $\xi = 0$, since (195) does not define the origin of ξ. The last condition corresponds to $\eta = 1$ at $\xi = \infty$. Note that since, for a given x,

$$F \sim \tau \quad \text{and} \quad \tau \sim \left(\frac{\partial \eta}{\partial y}\right)_x,$$

(195) shows that ξ varies like y and therefore ranges from zero to infinity.

But (196) and (197) constitute the Blasius system, the solution of which is known. We can therefore translate the solution in terms of the variables F and η, and with $F(\eta)$ known, two integrations of (191) determine θ. Translation back to the variables x and y is achieved by the first equation in (175). Only y need be calculated, since x is unaffected by Crocco's transformation. The solution for $\omega = 1$, so similar to the method based on von Mises' transformation, was obtained by Crocco (1946) by using his transformation.

As to the general case in which neither σ nor ω is 1, Crocco (1946) integrated (190) and (191) by laborious iteration. [For detailed results the reader is referred to Crocco's work or to Howarth (1953, vol. II, pp. 419–424).]

In concluding the brief discussion of compressible boundary layers in this chapter, we may mention the case of the impulsively started infinite plate bounding a semi-infinite space above filled with air. Howarth (1951), neglecting gravity effects and forced to linearization of the governing equations, was able to obtain a solution for the special case $\sigma = \frac{3}{4}$. This is not far from the actual Prandtl number of air and is not an unrealistic case. The mathematics for the case of impulsive start of the plate is rather formidable, but for an oscillating plate when the air above has "settled down" to a sort of dynamic equilibrium a long time after the plate started oscillations, the solution is much simpler. The most interesting qualitative feature of Howarth's solution for an infinite plate is that although the plate has no motion in a direction normal to its plane, the air does move in that direction, as a result of the expansion induced by thermal inhomogeneity, which arises either from the higher (or lower) temperature of the plate as compared with air temperature at a great distance from it or from heating by viscous dissipation. We omit the mathematical details of Howarth's solution and the solution of the simpler problem of periodic oscillation.

PROBLEMS

1 A line source of heat (or mass) is placed at the leading edge of a plate in Blasius flow, and the plate is assumed to be insulated. Find the temperature (or concentration) in the thermal (or diffusion) boundary layer.

2 Show that free convection due to a point source of heat and mass is governed by (87), (86) with the term $\eta\theta$ replaced by $\eta\theta + \eta\lambda$, and

$$f = -\frac{1}{4\sigma'}\frac{\eta}{\lambda}\frac{d\lambda}{d\eta},$$

in which $\sigma' = \nu/\kappa'$ is the Prandtl number, κ' being the mass diffusivity, and

$$c - c_0 = -\frac{G\rho}{x\mu}\lambda(\eta),$$

c being the concentration of the diffusing mass related to ρ by

$$\rho = \rho_0[1 - \alpha(T - T_0) - \alpha'(c - c_0)],$$

in which c_0 is the ambient value for c. The boundary conditions are (88) and $\lambda(\infty) = 0 = \lambda'(0)$. The variable η, the functions f and θ, and the flux G are as defined in Sec. 7, and G' is the total flux of the diffusing mass defined by

$$G' = \alpha'\rho_0 g \int_0^\infty 2\pi r u(c - c_0)\, dr.$$

3 Show that the differential system formulated in Prob. 2 possesses a simple solution in closed form if

a $\sigma = \sigma' = 1$,
b $\sigma = \sigma' = 2$,
c $3\sigma' = 4\sigma$,
d $3\sigma = 4\sigma'$,

and that in all cases the solution is of the form

$$f = \frac{A\eta^2}{B + \eta^2}, \qquad \theta = \frac{c}{(B + \eta^2)^{2\sigma A}}, \qquad \lambda = \frac{C'}{(B + \eta^2)^{2\sigma' A}},$$

in which A, B, C, and C' are functions of G'/G. In cases **c** and **d**, σ and σ' are also functions of G'/G.

4 Show that free convection due to a line source of heat *and* mass is governed by (108), (107) with θ replaced by $\theta + \lambda$, and

$$\lambda f' + f\lambda' = -\frac{\lambda''}{3\sigma'},$$

in which $\sigma' = \nu/\kappa' = $ Prandtl number for mass diffusion, κ' being the mass diffusivity, the primes on f and λ indicate differentiation with respect to ξ, and

$$\left(\frac{\mu^2 x^3}{\rho^3 G^4}\right)^{\!1/5}(c - c_0) = -\tfrac{1}{125}\lambda(\xi),$$

c being the concentration of the diffusing mass and c_0 its ambient value, as in Prob. 3. The boundary conditions are (110) and $\lambda'(0) = 0 = \lambda(\pm\infty)$. The variable ξ, the functions f and θ, and the quantity G are as defined in Sec. 8, and

$$G' = \alpha' \rho_0 g \int_{-\infty}^{\infty} u(c - c_0) \, dy.$$

5 Show that the differential system formulated in Prob. 4 possesses a simple solution in closed form if

a $\sigma = \sigma' = 2$,
b $\sigma = \sigma' = \frac{5}{9}$,
c $\sigma = 2\sigma'$,
d $\sigma' = 2\sigma$,

and that the solution is of the form

$$f = A \tanh B\xi, \qquad \theta = C(\text{sech } B\xi)^{3\sigma A/B},$$
$$\lambda = C'(\text{sech } B\xi)^{3\sigma' A/B},$$

in which A, B, C, and C' are functions of G'/G. In cases **c** and **d**, σ and σ' are also functions of G'/G.

6 Consider a thin layer of viscous liquid on a horizontal plate. If the surface tension T is a linear function $(T_0 - k\gamma, k$ being a constant) of the concentration γ of some surface-active material, for steady flows the diffusion equation for T is

$$\frac{\partial}{\partial x} UT + \frac{\partial}{\partial y} VT = \kappa \left(\frac{\partial^2}{\partial x^2} + \frac{\partial^2}{\partial y^2} \right) T,$$

in which U and V are the components of velocity on the surface, which is nearly horizontal. Show from this equation and the answers to Prob. 7.16 that

$$\left(U \frac{\partial}{\partial x} + V \frac{\partial}{\partial y} \right) \left(\frac{T - T_0}{h} - \frac{\mu\kappa}{h^2} \right) = 0,$$

and hence that

$$\frac{T - T_0}{h} - \frac{\mu\kappa}{h^2} = f(\psi),$$

in which ψ is a stream function related to U and V by

$$U = \frac{1}{h} \frac{\partial \psi}{\partial y}, \qquad V = -\frac{1}{h} \frac{\partial \psi}{\partial x}.$$

If T is constant on a constant-h line, show that

$$\frac{T - T_0}{h} - \frac{\mu\kappa}{h^2} = C,$$

and hence that, using parts **c** and **d** of Prob. 7.17,

$$\left(\frac{\partial^2}{\partial x^2} + \frac{\partial^2}{\partial y^2} \right) \left(-\kappa h + \frac{C}{3\mu} h^3 \right) = 0.$$

applicable to the core of the flow, i.e., away from vertical boundaries.

7 The flow obtained for the core in the preceding problem does not satisfy the nonslip condition at vertical boundaries. Develop a theory for the boundary layer in which the velocity distribution satisfies the nonslip condition at vertical boundaries and merges into the velocity in the core. Ignore the curvature of the free surface. *Hint:* In the boundary layer $\partial^2 q/\partial n^2$ is as important as $\partial^2 q/\partial z^2$, q being the speed and n the distance normal to the vertical wall.

8 For the general case, in which neither σ nor ω is equal to unity, show that (156) and (157) admit a similarity solution by deriving from them a pair of ordinary differential equations governing the flow in the boundary layer of a gas along a plate at zero angle of incidence.

9 For the general case, in which neither σ nor ω is equal to unity, show that (160) and (161) admit a similarity solution by deriving from them a pair of ordinary differential equations governing the flow in the boundary layer of a gas along a plate at zero angle of incidence.

REFERENCES

BRAND, R. S., and F. J. LAHEY, 1967: The Heated Laminar Vertical Jet, *J. Fluid Mech.*, **29**: 305.

BUSEMANN, A., 1935: Gasströmung mit laminarer Grenzschicht entlang eine Platte, *Z. Angew. Math. Mech.*, **15**: 23–25.

CROCCO, L., 1939: *Atti di Guidonia XVII*, **7**: 118; translated in *Aeron. Res. Council (Britain) Rept.* 4,582, 1939.

CROCCO, L., 1946: *Monograph. Sci. Aeron.*, **3**.

FRÖSSLING, N., 1940: Verdunstung, Wärmeübergang, und Geschwindigkeits verteilung bei zweidimensionaler und rotationssymmetrischer laminarer Grenzschichtströmung, *Lunds Univ. Arsskr. Avd.* 2, **36**(4): 1–32.

GUTMAN, L. N., 1949: On Laminar Thermal Convection above a Stationary Heat Source, *Prikl. Mat. Mekhan.*, **13**: 435 (in Russian).

HOWARTH, L., 1948: Concerning the Effect of Compressibility on Laminar Boundary Layers and Their Separation, *Proc. Roy. Soc. London Ser. A*, **194**: 16.

HOWARTH, L. (ed.), 1953: "Modern Developments in Fluid Mechanics: High Speed Flow," vols. I and II, Oxford University Press, Fair Lawn, N.J.

KÁRMÁN, T. VON, and H. S. TSIEN, 1938: Boundary Layer in Compressible Fluid, *J. Aeron. Sci.*, **5**: 227–232.

MANGLER, W., 1948: Zusammenhang zwischen ebenen und rotationssymmetris-chen Grenzschichten in kompressiblen Flüssigkeiten, *Z. Angew. Math. Mech.*, **28**: 97–103.

MISES, R. VON, 1927: Bemerkungen zur Hydrodynamik, *Z. Angew. Math. Mech.*, **7**: 425–431.

POHLHAUSEN, E., 1921: Der Wärmeautausch zwischen festen Körpern und Flüssigkeiten mit kleiner Reibung und kleiner Wärmeleitung, *Z. Angew. Math. Mech.*, **1**: 115.

SCHLICHTING, H., 1968: "Boundary Layer Theory," McGraw-Hill Book Company, New York.

SQUIRE, H. B., 1951: The Round Laminar Jet, *Quart. J. Mech. Appl. Math.*, **4**: 321–329.

STEWARTSON, K., 1949: Correlated Compressible and Incompressible Boundary Layers, *Proc. Roy. Soc. London Ser. A*, **200**: 84.

YIH, C.-S., 1950: Temperature Distribution in a Steady, Laminar, Preheated Air Jet, *J. Appl. Mech.*, **17**: 381–382.

YIH, C.-S., 1951: Free Convection Due to a Point Source of Heat, *Proc. 1st U.S. Natl. Congr. Appl. Mech.*, *Chicago*, June, 1950, pp. 941–947.

YIH, C.-S., 1952: Laminar Free Convection Due to a Line Source of Heat, *Trans. Am. Geophys. Union*, **33**: 669–672.

YIH, C.-S., 1953: Free Convection Due to Boundary Sources, in "Symposium on the Use of Models in Geophysical Fluid Mechanics," held at Johns Hopkins University, Baltimore, 1953.   Proceedings published by Government Printing Office, Washington, 1956, pp. 117–133.

ZEL'DOVICH, YA. B., 1937: Limiting Laws of Freely Rising Convective Currents, *Zh. Eksperim. i Teor. Fiz.*, **7**: 1963 (in Russian).

ADDITIONAL READING

GOLDSTEIN, S. (ed.), 1938: "Modern Developments in Fluid Dynamics," Vol. II, chap. 14, Oxford University Press, Fair Lawn, N.J.

OSTRACH, S., 1955: "On the flow, heat transfer, and stability of viscous fluids subject to body forces and heated from below in vertical channels, '50 Jahre Grenzschichtforschung'," Friedr. Vieweg & Sohn, Brunswick, Germany.

CHAPTER NINE

HYDRODYNAMIC STABILITY

1. INTRODUCTION

A fluid at rest may be in equilibrium, and a laminar flow may satisfy the requirements of the governing equations and boundary conditions. But the equilibrium of the flow may be precarious in the sense that a small disturbance may precipitate a change of configuration or state of flow. If so, the fluid or the flow is unstable. Stability therefore depends on whether a small disturbance imposed on the fluid in its primary state will grow or be damped.

If a fluid is unstable, a disturbance may grow to a finite amplitude and cease growing. If the resulting state is itself stable, a new state of laminar flow will persist. If this new flow is not stable against other disturbances, they will grow and bring about still newer states, and if the rates of growth are large enough, the final flow will be turbulent. In this chapter only incipient instability of small disturbances will be discussed, with the emphasis on the various physical causes of instability.

2. GRAVITATIONAL INSTABILITY

It is a common experience that because of the presence of the earth's gravitational field, an incompressible fluid of variable density is stable only if its density decreases with height, provided the effect of surface tension is negligible in the case of superposed immiscible fluids. The physical explanation is simple. If a fluid particle of density ρ_1 at height H_1 is displaced to a lower height H_2, where the density of the fluid in its undisturbed state is ρ_2, the particle will be surrounded by fluid with density ρ_2. If $\rho_2 > \rho_1$, the buoyant force acting on the fluid particle will be greater than its weight and the fluid will return toward its former position. The fluid is then stable. If $\rho_2 < \rho_1$, the reverse is true, and the fluid is unstable. The argument is quite the same if the particle is displaced to a greater height H_2.

For a compressible fluid the situation is more involved. As the particle is brought from elevation H_1 to H_2, its pressure changes from p_1 to p_2 and as a result its density does not remain ρ_1. In fact, if we imagine the particle to be held for a long time at elevation H_2, its temperature will eventually be T_2 and hence, with its pressure equal to p_2, its density must be equal to ρ_2 and neutral stability would seem to be the only possible result. However, if a properly defined Péclét number is large, the effect of thermal conductivity is small. If, further, the displacement of the particle is gradual, as it must be if it is caused by a small disturbance in a linear hydrodynamic theory, then the change of state of the particle can be considered isentropic. Thus, according to (A1.45),

$$\frac{p_1}{\rho_1{}^\gamma} = \frac{p_2}{\tilde{\rho}_1{}^\gamma}, \tag{1}$$

in which $\tilde{\rho}_1$ is the density of the particle at the elevation H_2. If $H_2 < H_1$, the gas is stable if $\tilde{\rho}_1 < \rho_2$ and unstable if $\tilde{\rho}_1 > \rho_2$. That is to say, it is stable if

$$\frac{p_1}{\rho_1{}^\gamma} > \frac{p_2}{\rho_2{}^\gamma}, \tag{2}$$

and unstable if

$$\frac{p_1}{\rho_1{}^\gamma} < \frac{p_2}{\rho_2{}^\gamma}. \tag{3}$$

According to (A1.45), one then concludes that the gas is stable if the entropy increases upward and unstable if it decreases upward. A homentropic gas is neutrally stable.

2.1. Instability of Two Superposed Fluids

To demonstrate gravitational instability we shall consider two superposed fluids of infinite depths with their interface coinciding with the xy plane. The density of the upper fluid is denoted by ρ_1 and that of the lower fluid by ρ_2, and the interfacial tension is denoted by T. The third coordinate z is measured in the direction of the vertical. The analysis to be presented can be easily modified to deal with finite depths. Infinite depths are assumed for the sake of simplicity.

Since the fluids are viscous, the starting point of the analysis should be the Navier-Stokes equations. However, if the parameter $\sqrt{g\lambda}\,\lambda/\nu$ is large, in which g is the gravitational acceleration, λ the wavelength of the distance, and ν the kinematic viscosity, gravity effects will dominate the viscous effects and an analysis based on inviscidness will be a good approximation. If one supposes the fluid to be at rest and a disturbance applied to it, the subsequent motion will then be irrotational, and one can use the velocity potentials ϕ_1 and ϕ_2 for the two fluids. Remembering that the velocities must vanish at $z = \pm\infty$, we demand, without loss of generality, that

$$\phi_1 \to 0 \text{ as } z \to \infty \quad \text{and} \quad \phi_2 \to 0 \text{ as } z \to -\infty.$$

Since the ϕ's must satisfy the Laplace equation, we can assume

$$\phi_1 = C_1 \exp(\sigma t - kz)S(x,y), \tag{4}$$

$$\phi_2 = C_2 \exp(\sigma t + kz)S(x,y), \tag{5}$$

in which
$$\left(\frac{\partial^2}{\partial x^2} + \frac{\partial^2}{\partial y^2} + k^2\right)S = 0. \tag{6}$$

The wave number k is quite unrestricted if the fluids are unbounded, but if they are contained in a vertical tube, the k's can assume only certain definite values. The stability or instability is determined by the sign of σ or, if it is complex, of its real part.

If ζ denotes the displacement of the interface, the interfacial kinematic condition is

$$\frac{\partial\phi_1}{\partial z} = \frac{\partial\phi_2}{\partial z} = \frac{\partial\zeta}{\partial t} \quad \text{at } z = \zeta, \tag{7}$$

which simply states that the vertical velocity of the upper layer must equal that of the lower layer. Actually, it is the velocity components normal to the free surface that must be equal, but the amplitude of the disturbance being small, the normal component is equal to the vertical component if higher-order terms are neglected, as they are in a linear theory.

At the interface, in the absence of disturbances, the pressures must be the same for the two fluids. If p_1 and p_2 denote the pressure perturbations, then

$$p_1 - p_2 = T\left(\frac{1}{R_1} + \frac{1}{R_2}\right) = T\,\Delta_1\zeta, \tag{8}$$

in which T is the interfacial tension, R_1 and R_2 are the principal radii of curvature of the interface (the subscripts on R not denoting the upper or lower fluid), and

$$\Delta_1 = \frac{\partial^2}{\partial x^2} + \frac{\partial^2}{\partial y^2}.\tag{9}$$

In view of the form of (4) and (5), the displacement of the interface can be assumed to have the form

$$\zeta = ae^{\sigma t}S(x,y).\tag{10}$$

Thus when it is applied at $z = 0$, (7) becomes

$$-kC_1 = kC_2 = \sigma a,\tag{11}$$

and (8) becomes

$$p_1 - p_2 = -Tk^2\,\zeta.\tag{12}$$

The application of (7) at $z = 0$ instead of at $z = \zeta$ introduces an error only of a higher order in a (strictly speaking, in ak).

Now the pressures p_1 and p_2 can be found from the Bernoulli equations

$$\frac{p_1}{\rho_1} = -\frac{\partial\phi_1}{\partial t} - g\zeta, \qquad \frac{p_2}{\rho_2} = -\frac{\partial\phi_2}{\partial t} - g\zeta.\tag{13}$$

From (4), (5), (10), (12), and (13), it follows that

$$\rho_1(-\sigma C_1 - ga) - \rho_2(-\sigma C_2 - ga) = -Tk^2a.\tag{14}$$

Elimination of C_1, C_2, and a from (11) and (14) yields

$$\sigma^2 = \frac{g(\rho_1 - \rho_2)k}{\rho_1 + \rho_2} - \frac{Tk^3}{\rho_1 + \rho_2}.\tag{15}$$

Thus the fluid is unstable if

$$k^2 < \frac{g(\rho_1 - \rho_2)}{T}.\tag{16}$$

If the fluids are unlimited laterally, k, being inversely proportional to the wavelength of the disturbance, can be as small as desired. Therefore there are always unstable modes in such a fluid if $\rho_1 > \rho_2$. However, if the fluids are laterally confined, there is a minimum nonzero value of k.

For instance, if the fluids are contained in a vertical circular cylinder and polar coordinates r, φ, and z are used,

$$S(x,y) = \cos n\varphi\, J_n(kr).$$

Since $\partial\phi/\partial r$ must be zero at $r = r_0$ (inner radius of the cylinder),

$$\frac{dJ_n(kr)}{dr} = 0 \qquad \text{at } r = r_0.$$

This determines the values of kr_0 to be

$$kr_0 = \begin{cases} 3.83, 7.02, 10.17, \ldots, & \text{for } n = 0, \\ 1.84, 5.33, 8.53, \ldots, & \text{for } n = 1, \\ 3.05, 6.70, 9.97, \ldots, & \text{for } n = 2. \end{cases}$$

Thus, without consideration of the curvature of the meniscus, the criterion for incipient instability is

$$\left[\frac{g(\rho_1 - \rho_2)}{T} \right]^{\frac{1}{2}} r_0 = 1.84. \tag{17}$$

Maxwell (1890, p. 587) assumed axisymmetry of the disturbance and using 3.83 instead of 1.84 in (17), found agreement with the experimental results of Duprez (1851, 1854). Possibly Duprez carefully introduced only axisymmetric disturbances. Possibly the axial symmetry of the meniscus favors such disturbances. At any rate, existing theory points to (17) as the correct criterion for incipient instability.

The effects of viscosity on the rate of growth of disturbances in an unstable superposition of two fluids have been considered by Chandrasekhar (1955). When $\sigma/\nu_1 k^2$ and $\sigma/\nu_2 k^2$, which are Reynolds numbers for the disturbance, are both very large, viscous effects are small, and the growth rate σ is nearly the same as that given by the potential theory presented here. Thus one can use the σ obtained from the potential theory and see whether it is very large compared with $\nu_1 k^2$ and $\nu_2 k^2$. If so, that σ is not far from the actual growth rate.

2.2. Instability of the Free Surface or Interface Due to Acceleration

It was d'Alembert's idea that the dynamics of a solid particle can be reduced to statics by adopting, in effect, a noninertial system moving with the particle, in which the particle appears to have a body force per unit mass equal and opposite to its acceleration. This idea cannot be generalized to apply to continuum mechanics, since different particles move with different accelerations. Nevertheless it is partially recovered when one seeks the form of the equations of motion relative to an accelerating frame, such as an accelerating vessel containing fluid.

If the Cartesian coordinates in the accelerating frame are denoted by by x_i' and those in a fixed frame by x_i, and if x_i' and x_i coincide momentarily at $t = 0$, t being the time, then

$$x_i' = x_i - \int_0^t \int_{C_i}^{t_1} a_i(t_2) \, dt_2 \, dt_1, \tag{18}$$

in which a_i is the acceleration in the direction of increasing x_i and is a function of t. For clarity we shall denote the time in the moving frame by t'. Although we shall set t' equal to t, the meaning of partial differentiation demands that

$$\frac{\partial}{\partial t} = \frac{\partial}{\partial t'} - \left(\int_{C_i}^{t} a_i\, dt\right)\frac{\partial}{\partial x_i'}. \tag{19}$$

The velocity components in the two frames are related by

$$u_i = u_i' + \int_{C_i}^{t} a_i\, dt. \tag{20}$$

When (19) and (20) are substituted in the Navier-Stokes equations and the primes on u_i' and x_i' are dropped, one obtains

$$\rho\left(\frac{\partial u_i}{\partial t} + u_\alpha \frac{\partial u_i}{\partial x_\alpha}\right) = -\frac{\partial p}{\partial x_i} + \rho(X_i - a_i) + \mu\,\Delta u_i. \tag{21}$$

Equation (21) is for an incompressible fluid with constant viscosity. For the cases of variable viscosity or compressibility only the last term need be modified. The essential point in (21) is that with respect to the accelerating frame the body force per unit mass X_i is reduced by the amount a_i. To this extent d'Alembert's idea is recovered.

Now consider a fluid of density ρ_1 superposed on a fluid of *greater* density ρ_2 and let there be an acceleration of magnitude a in the direction of decreasing z. If $a > g$, the stability or instability of the fluid is exactly the same as that for a fluid of density ρ_2 $(>\rho_1)$ over the fluid of density ρ_1, with the gravity changed to the magnitude $a - g$. Thus when the interface of an originally stable stratified fluid is accelerated downward with an acceleration greater than g, the fluid is unstable unless it is laterally confined and surface tension is taken into account. The degree of instability is even greater if an originally unstable configuration is accelerated *upward*. In general when there is a component of the acceleration in the direction of the density gradient, the acceleration, whether due to the motion of the fluid container or to the motion of the interface, is destabilizing, as indicated by Taylor (1950).

The somewhat analogous case of instability of a rotating liquid film with a free surface was treated by Yih (1960).

2.3. The Bénard Problem

Although in the absence of viscosity and diffusivity a top-heavy fluid or superposition of fluids is in general unstable, viscosity and diffusivity may stabilize the fluid even if its density increases upward. That viscosity can damp out convection is not surprising. The stabilizing effect of diffusivity can be understood if we consider a fluid particle displaced from one location to another. Diffusivity, whether thermal or mass diffusivity, has the effect or the tendency to equalize the temperature or concentration of the particle with that of its new surroundings. This reduces the intensity of any instability or even inhibits it.

Bénard (1900, 1901) experimented with a thin layer of liquid with a free surface and heated from below and observed a hexagonal convection pattern. Problems involving the convection of a fluid layer heated from below, with the upper boundary free or fixed, are now called the *Bénard problem*. We shall consider the Bénard problem with a fixed upper boundary.

Let the boundaries be horizontal and let the temperature of the lower boundary be maintained at T_0 and that at the upper boundary be maintained at T_1, assumed lower than T_0. The primary temperature distribution is then

$$T_m = T_0 + \beta x_3, \tag{22}$$

in which

$$\beta = \frac{T_1 - T_0}{d}, \tag{23}$$

d being the spacing of the plates. Note that the T_m given by (22) satisfies the equation of heat conduction, as it should, since in the absence of disturbances the fluid is quiescent. We have assumed the thermal conductivity to be constant. The Cartesian coordinate x_3 is measured in the direction of the vertical. The other two Cartesian coordinates x_1 and x_2 are measured horizontally.

The density ρ of the fluid is related to the temperature T by

$$\rho = \rho_0[1 - \alpha(T - T_0)], \tag{24}$$

in which ρ_0 is the density of the liquid (or a gas, so long as the pressure does not vary excessively) and α the thermal expansivity, assumed positive. (It is positive except for water below 4°C.) The mean pressure distribution is given by

$$\frac{\partial \bar{p}}{\partial x_3} = -\rho_0 g[1 - \alpha(T_m - T_0)]. \tag{25}$$

If the fluid is slightly perturbed,

$$T = T_m + T', \qquad p = \bar{p} + p', \tag{26}$$

and the velocity perturbations will be simply denoted by u_i, since there is no primary flow. If the quadratic terms in the perturbation quantities are neglected and (22) is taken into account, the equations of motion become

$$\frac{\partial u_i}{\partial t} = (0,0,g\alpha T') - \frac{1}{\rho_0}\frac{\partial p'}{\partial x_i} + \nu \nabla^2 u_i, \tag{27}$$

and the heat equation, or energy equation, becomes

$$\left(\frac{\partial}{\partial t} - \kappa\nabla^2\right)T' = -\beta u_3. \tag{28}$$

In (27) and (28), ∇^2 is the Laplacian operator, the viscosity and the thermal diffusivity are assumed constant, and the density is assumed constant and equal to ρ_0 except in the body-force term, where the density variation is of crucial importance.

The variables u_1, u_2, and p' can be eliminated from (27). Thus, taking the Laplacian of the third equation in (27), we have

$$\frac{\partial}{\partial t}\,\nabla^2 u_3 = -\,\frac{1}{\rho_0}\,\frac{\partial}{\partial x_3}\,\nabla^2 p' + g\alpha\nabla^2 T' + \nu\,\nabla^2\nabla^2 u_3. \tag{29}$$

On the other hand, taking the "divergence" of the three equations (27), differentiating the result with respect to x_3, and using the equation of continuity

$$\frac{\partial u_i}{\partial x_i} = 0,$$

we obtain
$$0 = -\,\frac{1}{\rho_0}\,\frac{\partial}{\partial x_3}\,\nabla^2 p' + g\alpha\,\frac{\partial^2 T'}{\partial x_3{}^2}. \tag{30}$$

The difference of (29) and (30) is

$$\frac{\partial}{\partial t}\,\nabla^2 u_3 = g\alpha\,\nabla_1{}^2 T' + \nu\,\nabla^2\nabla^2 u_3, \tag{31}$$

or
$$\left(\frac{\partial}{\partial t} - \nu\,\nabla^2\right)\nabla^2 u_3 = g\alpha\,\nabla_1{}^2 T', \tag{32}$$

in which
$$\nabla_1{}^2 = \frac{\partial^2}{\partial x_1{}^2} + \frac{\partial^2}{\partial x_2{}^2}.$$

With the origin midway between the plates (or the boundaries), the nonslip condition at the plates is

$$u_i = 0 \qquad \text{at } x_3 = \pm\frac{d}{2}, \qquad i = 1, 2, 3,$$

which, by virtue of the equation of continuity, can be written, as far as u_3 is concerned, as

$$u_3 = 0 = \frac{\partial u_3}{\partial x_3} \qquad \text{at } x_3 = \frac{d}{2}. \tag{33}$$

If the plates are assumed to be much more conductive than the fluid,

$$T' = 0 \qquad \text{at } x_3 = \pm\frac{d}{2}. \tag{34}$$

Equations (28) and (32) to (34) constitute the governing differential system.

We now endeavor to separate the variables x_1 and x_2 from x_3. In order to obtain dimensionless equations, which will be more convenient to use, the dimensionless variables

$$\tau = \frac{t\kappa}{d^2}, \qquad (x,y,z) = \left(\frac{x_1}{d}, \frac{x_2}{d}, \frac{x_3}{d}\right) \tag{35}$$

are introduced. The separation of x and y from z is achieved by assuming both u_3 and T' to contain the factor $f(x,y)$ satisfying

$$f_{xx} + f_{yy} + a^2 f = 0, \tag{36}$$

in which subscripts indicates partial differentiation and a is a constant inversely proportional to a linear dimension of the cells in a horizontal plane and is thus a wave number.[1] Furthermore, examination of the governing differential system reveals that u_3 and T' can be assumed to have an exponential time factor $e^{\sigma\tau}$. We shall then assume

$$u_3 = \frac{\kappa}{d} e^{\sigma\tau} f(x,y)w(z), \qquad T' = \beta d e^{\sigma\tau} f(x,y)\theta(z). \tag{37}$$

Equations (28) and (32) then become

$$[\sigma - (D^2 - a^2)]\theta = -w, \tag{38}$$

$$\left[\frac{\sigma}{Pr} - (D^2 - a^2)\right](D^2 - a^2)w = Ra^2\theta, \tag{39}$$

in which $\qquad D = \dfrac{d}{dz}, \qquad R = -\dfrac{g\alpha\beta d^4}{\kappa\nu}, \qquad Pr = \dfrac{\nu}{\kappa}, \tag{40}$

the R being called the *Rayleigh number*. The boundary conditions are

$$w = 0 = Dw \qquad \text{and} \qquad \theta = 0 \qquad \text{at } z = \pm\tfrac{1}{2}. \tag{41}$$

The convection pattern in a horizontal plane is determined by the solution of (36). For instance, the solution

$$f(x,y) = \cos mx \cos ny, \qquad m^2 + n^2 = a^2, \tag{42}$$

corresponds to rectangular cells. The solution [Christopherson (1940)]

$$f(x,y) = f_0\left[\cos\frac{a}{2}(\sqrt{3}\,x + y) + \cos\frac{a}{2}(\sqrt{3}\,x - \dot{y}) + \cos ay\right] \tag{43}$$

corresponds to the hexagonal pattern of Bénard's experiments (1901), with the length L of a side given by

$$L = \frac{4\pi d}{3a}. \tag{44}$$

---

[1] For any assigned value of the wave number, the horizontal pattern of the convection cells is not determined by the linear theory. For the tendency toward forming hexagonal cells, see a survey article by Segel (1966).

The sign of the real part of σ ($= \sigma_r + i\sigma_i$) determines whether the fluid is stable or unstable. It will now be shown, after Pellew and Southwell (1940), that σ is real for a fluid with the mean temperature T_m given by (22), if β is negative, so that the fluid is heated from below. If (38) is multiplied by θ^* (the complex conjugate of θ) and integrated (by parts if necessary) from $-\frac{1}{2}$ to $\frac{1}{2}$, there results the equation

$$\sigma I_0 + I_1 = -\int w\theta^* \, dz, \tag{45}$$

in which $\quad I_0 = \int |\theta|^2 \, dz, \qquad I_1 = \int (|D\theta|^2 + a^2 |\theta|^2) \, dz,$

the limits of the integrals being understood to be $\pm\frac{1}{2}$. Similarly, multiplication of (39) by w^* and integration produce

$$\frac{\sigma}{Pr} J_1 + J_2 = -Ra^2 \int w^*\theta \, dz, \tag{46}$$

with $\quad J_1 = \int (|Dw|^2 + a^2 |w|^2) \, dz,$

$$J_2 = \int (|D^2w|^2 + 2a^2 |Dw|^2 + a^4 |w|^2) \, dz.$$

From (45) and (46) we have

$$-Ra^2(\sigma^* I_0 + I_1) + [\sigma(Pr)^{-1}J_1 + J_2] = 0, \tag{47}$$

or $\quad \sigma_r[-Ra^2 I_0 + (Pr)^{-1}J_1] + (-Ra^2 I_1 + J_2) = 0, \tag{48}$

$$\sigma_i[Ra^2 I_0 + (Pr)^{-1}J_1] = 0. \tag{49}$$

If $\sigma_i \neq 0$, (49) demands $R < 0$, and from (48) it follows that $\sigma_r < 0$. Physically, this is to say that if $\sigma_i \neq 0$, the fluid is heated from above and is therefore stable. Thus we know that if the density increases upward, σ is real. But from (47) one cannot conclude that if $R < 0$ then σ_i must be different from zero. This is, however, likely to be the case.

The result given in the preceding paragraph is often called the *principle of exchange of stabilities*. Such an appellation is not without ambiguity, in spite of efforts to justify its use. This welcome result simplifies the search for the criterion for neutral stability, because to obtain it we can merely set σ equal to zero in (38) and (39). Doing so and eliminating θ between (38) and (39), we obtain

$$[(D^2 - a^2)^3 + Ra^2]w = 0. \tag{50}$$

The boundary conditions are then

$$w = 0 = Dw \quad \text{and} \quad (D^2 - a^2)^2w = 0 \quad \text{at } z = \pm \frac{1}{2}. \tag{51}$$

Examination of the differential system (38) to (41), or the one for neutral stability, (50) and (51), reveals that the system possesses solutions which are either even or odd in z. To have a clearer understanding of the behavior of a disturbance which is neither even nor odd, let the six independent solutions $w_1, w_2, \ldots, w_6$ be mixed. Since each of these can be expressed as the sum of an even and an odd function in z, we shall write, without loss of generality,

$$w_1 = E_1(z) + O_1(z), \qquad w_2 = E_2(z) + O_2(z), \qquad w_3 = E_3(z) + O_3(z),$$
$$(52)$$
$$w_4 = E_1(z) - O_1(z), \qquad w_5 = E_2(z) - O_2(z), \qquad w_6 = E_3(z) - O_3(z),$$

in which E stands for an even function and O an odd one. For simplicity we shall write Dw as w'. Then $\theta_1, \theta_2, \ldots, \theta_6$ are given by

$$e_i(z) \pm o_i(z), \qquad i = 1, 2, 3, \tag{53}$$

with $\qquad e_i(z) = (D^2 - a^2)^2 E_i(z), \qquad o_i(z) = (D^2 - a^2)^2 O_i(z). \tag{54}$

Since (summation convention observed)

$$w = a_i w_i, \qquad \theta = a_i \theta_i, \qquad i = 1, 2, \ldots, 6,$$

satisfy (41), and since the a_i must not vanish for all i, it is necessary that

$$\begin{vmatrix} E_1(\tfrac{1}{2}) + O_1(\tfrac{1}{2}) & E_2(\tfrac{1}{2}) + O_2(\tfrac{1}{2}) & \cdots & E_3(\tfrac{1}{2}) - O_3(\tfrac{1}{2}) \\ E_1'(\tfrac{1}{2}) + O_1'(\tfrac{1}{2}) & E_2'(\tfrac{1}{2}) + O_2'(\tfrac{1}{2}) & \cdots & E_3'(\tfrac{1}{2}) - O_3'(\tfrac{1}{2}) \\ e_1(\tfrac{1}{2}) + o_1(\tfrac{1}{2}) & e_2(\tfrac{1}{2}) + o_2(\tfrac{1}{2}) & \cdots & e_3(\tfrac{1}{2}) - o_3(\tfrac{1}{2}) \\ E_1(-\tfrac{1}{2}) + O_1(-\tfrac{1}{2}) & E_2(-\tfrac{1}{2}) + O_2(-\tfrac{1}{2}) & \cdots & E_3(-\tfrac{1}{2}) - O_3(-\tfrac{1}{2}) \\ E_1'(-\tfrac{1}{2}) + O_1'(-\tfrac{1}{2}) & E_2'(-\tfrac{1}{2}) + O_2'(-\tfrac{1}{2}) & \cdots & E_3'(-\tfrac{1}{2}) - O_3'(-\tfrac{1}{2}) \\ e_1(-\tfrac{1}{2}) + o_1(-\tfrac{1}{2}) & e_2(-\tfrac{1}{2}) + o_2(-\tfrac{1}{2}) & \cdots & e_3(-\tfrac{1}{2}) - o_3(-\tfrac{1}{2}) \end{vmatrix} = 0.$$
$$(55)$$

Using
$$E_i(\tfrac{1}{2}) = E_i(-\tfrac{1}{2}), \qquad O_i(\tfrac{1}{2}) = -O_i(\tfrac{1}{2}),$$
$$E_i'(\tfrac{1}{2}) = -E_i'(-\tfrac{1}{2}), \qquad O_i'(\tfrac{1}{2}) = O_i'(\tfrac{1}{2}),$$

and so forth, we can add or subtract columns and rows to obtain the final result

$$\begin{vmatrix} E_1(\tfrac{1}{2}) & E_2(\tfrac{1}{2}) & E_3(\tfrac{1}{2}) & 0 & 0 & 0 \\ E_1'(\tfrac{1}{2}) & E_2'(\tfrac{1}{2}) & E_3'(\tfrac{1}{2}) & 0 & 0 & 0 \\ e_1(\tfrac{1}{2}) & e_2(\tfrac{1}{2}) & e_3(\tfrac{1}{2}) & 0 & 0 & 0 \\ 0 & 0 & 0 & O_1(\tfrac{1}{2}) & O_2(\tfrac{1}{2}) & O_3(\tfrac{1}{2}) \\ 0 & 0 & 0 & O_1'(\tfrac{1}{2}) & O_2'(\tfrac{1}{2}) & O_3'(\tfrac{1}{2}) \\ 0 & 0 & 0 & o_1(\tfrac{1}{2}) & o_2(\tfrac{1}{2}) & o_3(\tfrac{1}{2}) \end{vmatrix} = 0. \tag{55a}$$

This shows that the even and odd modes are separable. At any given Rayleigh number and any wave number, the even and odd modes will have different rates of growth or decay. A mixed disturbance will sort itself out into odd and even modes, which will grow or be damped at their proper rates.

It can be shown that the most unstable mode has a convection pattern, in the vertical plane, of one row of cells. This corresponds to the gravest even mode. (The gravest odd mode has two rows.) The neutral modes, even or odd, are determined by (50) and (51). For a given wave number a, there are infinitely many Rayleigh numbers for either the even or the odd modes. The smallest Rayleigh number of the even modes is smaller than the smallest R for the odd modes. As the Rayleigh number increases from one of its values for neutral stability to the next, the even modes have $1, 3, 5, \ldots$ rows of cells and the odd modes have $2, 4, 6, \ldots$ rows of cells.

Now the even solution of (50) is

$$w = A_0 \cos q_0 z + A_1 \cosh q z + A_2 \cosh q^* z, \qquad (56)$$

in which $-q_0^2$, q^2, and q^{*2} (the complex conjugate of q) are roots of the equation

$$(X - a^2)^3 + Ra^2 = 0, \qquad (57)$$

and are given by

$$q_0 = a\left[\left(\frac{R}{a^4}\right)^{1/3} - 1\right]^{1/2}, \qquad q^2 = a^2\left[1 + \frac{1}{2}\left(\frac{R}{a^4}\right)^{1/3}(1 + i\sqrt{3})\right]. \qquad (58)$$

Since the A's must not all be zero, we obtain the secular equation

$$\begin{vmatrix} \cos\dfrac{q_0}{2} & \cosh\dfrac{q}{2} & \cosh\dfrac{q^*}{2} \\[2ex] -q_0 \sin\dfrac{q_0}{2} & q \sinh\dfrac{q}{2} & q^* \sinh\dfrac{q^*}{2} \\[2ex] (q_0^2 + a^2)^2 \cos\dfrac{q_0}{2} & (q^2 - a^2)^2 \cosh\dfrac{q}{2} & (q^{*2} - a^2)^2 \cosh\dfrac{q^*}{2} \end{vmatrix} = 0. \qquad (59)$$

This is a relation between R and a. The smallest R is 1,707.76, corresponding to $a = 3.117$. For these values, the three cofactors of any one row of (59) are proportional to A_0, A_1, and A_2. If $A_0 = 1$, then

$$A_1 = -\frac{1 - i\sqrt{3}}{2} \qquad \text{and} \qquad A_2 = A_1^*. \qquad (60)$$

2.3.1. *Chandrasekhar's Method*

Although an exact solution of Bénard's problem has been presented, a method used fruitfully by Chandrasekhar (1954*a* and *b*) to solve Bénard's problem and many mathematically similar problems of hydrodynamic stability will be presented here because it is powerful and can be applied where no exact solution can be found.

With $\sigma = 0$, the equations governing neutral stability are

$$(D^2 - a^2)\theta = w, \tag{38a}$$

$$(D^2 - a^2)^2 w = -Ra^2\theta. \tag{39a}$$

A convenient substitution is

$$\theta_1 = -Ra^2\theta.$$

When this is made in (38*a*) and (39*a*) and the subscript on θ_1 is subsequently dropped, one has

$$(D^2 - a^2)\theta = -Ra^2 w, \tag{38b}$$

$$(D^2 - a^2)^2 w = \theta. \tag{39b}$$

It is also convenient to shift the origin to the lower boundary, so that the boundary conditions become

$$w = 0 = Dw \qquad \text{at } z = 0 \text{ and } z = 1, \tag{61}$$

$$\theta = 0 \qquad \text{at } z = 0 \text{ and } z = 1, \tag{62}$$

in which the θ is θ_1 after the subscript is dropped. To satisfy (62), one assumes

$$\theta = \sum_{n=1}^{\infty} A_n \sin n\pi z. \tag{63}$$

Substituting this into (39*b*), one can solve for w and put the solution in the form

$$w = \sum_{n=1}^{\infty} A_n \Big\{ B_{n1} \cosh az + B_{n2} \sinh az + B_{n3} z \cosh az + B_{n4} z \sinh az$$

$$+ \frac{1}{[(n\pi)^2 + a^2]^2} \sin n\pi z \Big\}. \tag{64}$$

The constants B in (64) can be solved for each n to satisfy the conditions (61). Then (64) is substituted into (38*b*), which is then multiplied by

$\sin m\pi z$ on both sides and integrated from zero to 1, with m taking the values 1, 2, ... in turn. We then obtain infinitely many equations of the type

$$\sum_{n=1}^{\infty} A_n a_{nm} = 0, \qquad m = 1, 2, \ldots, \tag{65}$$

which demands that

$$\begin{vmatrix} a_{11} & a_{12} & a_{13} & \cdots \\ a_{21} & a_{22} & a_{23} & \cdots \\ a_{31} & a_{32} & a_{33} & \cdots \\ \cdots\cdots\cdots\cdots\cdots\cdots \end{vmatrix} = 0. \tag{66}$$

Chandrasekhar showed that taking only one element a_{11} of the determinant one obtains for $a = 3.13$ the value 1,715.1 for R. A 3×3 determinant gives $R = 1,708.0$, which is very near 1,707.76 indeed.

2.4. The Salt-finger Phenomenon of Stommel

If the fluid is a liquid, and if in addition to a temperature gradient there is a gradient of the concentration of a solute in the liquid, then, instead of (24) we have

$$\rho = \rho_0[1 - \alpha(T - T_0) + \alpha'(c - c_0)], \tag{67}$$

in which ρ_0 is the density of the liquid at temperature T_0 and concentration c_0, c is the concentration, and α' is just $1/\rho_0$ if c is measured in mass per unit volume and if the liquid does not change its volume appreciably when some solute is added to it to form a dilute solution. According to (67), it is quite possible for a liquid heated from above to be lighter on top even if the concentration *increases* upward. Stommel et al. (1956) describe how such a fluid may be unstable. The explanation must be sought in the difference in thermal and mass diffusivity. For the case of common salt, the mass diffusivity is so much smaller than the thermal diffusivity that a displaced fluid particle will harmonize in temperature with its surroundings much more readily than in concentration. As a result the fluid particle, if it has come from above, will become colder but as salty as before the displacement. Or, if it has come from below, will become warmer but as fresh as before. In either case instability can result.

Let the mean concentration distribution be

$$\bar{c} = c_0 + \beta' x_3, \tag{68}$$

so that

$$\bar{\rho} = \rho_0[1 - (\alpha\beta - \alpha'\beta')x_3] \tag{69}$$

and

$$\frac{d\bar{\rho}}{dx_3} = -g\bar{\rho}. \tag{70}$$

Equation (68), in which β' is the concentration gradient, satisfies the concentration equation

$$\frac{Dc}{Dt} = \kappa' \, \nabla^2 c, \tag{71}$$

in which κ' is the mass diffusivity, assumed constant. The linearized form of (71), corresponding to (28), is

$$\left(\frac{\partial}{\partial t} - \kappa'\nabla^2\right)c' = -\beta' u_3, \tag{72}$$

in which c' is the perturbation in c, equal to $c - \bar{c}$. The body-force term $g\alpha T'$ in (27) must now be replaced by $g\alpha T' - gc'/\rho_0$. After this is done and the substitutions (35),

$$c' = \beta' d e^{\sigma\tau} f(x,y)\gamma(z), \tag{73}$$

and (37) are made, a development similar to that in Sec. 2.3 produces the governing differential equations

$$[\sigma - (D^2 - a^2)]\theta = -w, \tag{74}$$

$$[\sigma - k(D^2 - a^2)]\gamma = -w, \tag{75}$$

$$[\sigma(Pr)^{-1} - (D^2 - a^2)](D^2 - a^2)w = -Ra^2\theta + R'a^2\gamma, \tag{76}$$

in which

$$k = \frac{\kappa'}{\kappa}, \qquad R' = \frac{g\beta' d^4}{\rho_0 \nu \kappa}, \tag{77}$$

and now (since β is positive)

$$R = \frac{g\alpha\beta d^4}{\nu\kappa}. \tag{78}$$

(*Caution:* k is no longer the thermal conductivity.)

If the boundary conditions on T' and c' are

$$T' = 0 = c' \qquad \text{at } x_3 = \pm\frac{d}{2},$$

with the origin of the coordinates midway between the boundaries, then

$$\theta = 0 = \gamma \qquad \text{at } z = \pm\tfrac{1}{2}. \tag{79}$$

Then it is clear from (74), (75), and (79) that $\theta = \gamma$ if $k = 1$, and from (74) and (76) it can again be shown that $\sigma_r < 0$ if $\sigma_i \neq 0$. But $k = 1$ is not necessary for this conclusion on σ to hold. Yih (unpublished) *see errata* has shown that this conclusion does hold, and for neutral stability we can deal with

$$(D^2 - a^2)\theta = w,$$
$$(D^2 - a^2)\gamma_1 = w, \tag{80}$$
$$(D^2 - a^2)^2 w = a^2\left(R\theta - \frac{R'\gamma_1}{k}\right),$$

in which $\gamma_1 = k\gamma$. Since the first two of these equations are similar and

$$\theta = 0 = \gamma_1 \qquad \text{at } z = \pm\tfrac{1}{2},$$

we have $\theta = \gamma_1$, and (80) can be replaced by

$$(D^2 - a^2)\theta = w$$
$$(D^2 - a^2)^2 w = a^2(R - R_1)\theta, \tag{81}$$

in which
$$R_1 = \frac{R'}{k} = \frac{g\beta'd^4}{\rho_0\nu\kappa'}. \tag{82}$$

Equation (81) is similar to (38a) and (39a), with $R - R_1$ replacing $-R$. Thus the critical condition is given by [Stern (1960)]

$$a = 3.117, \qquad R_1 - R = 1{,}707.76.$$

Even if $\bar{\rho}$ decreases upward, R_1 may exceed R by more than 1,707.76 if κ' is much smaller than κ. If so, the fluid is unstable—a most interesting situation indeed.

Stommel's instability may explain the mixing of water near the surface of the sea, where evaporation renders the water saltier than below but the sun may keep it so much warmer that it is lighter than the water below. In spite of this situation mixing can still occur as a result of the instability just described.

The instability of revolving fluids of variable density is analogous to Stommel's instability, and has been treated by Yih (1961).

3. INERTIAL INSTABILITY

The free-surface instability due to the acceleration of the container discussed in Sec. 2.2 is a kind of inertial instability. We have discussed it in connection with gravitational instability only because it is analogous to it. There are, however, other kinds of inertial instability the analogy of which to gravitational instability is much less complete. In this section we shall consider mainly the stability of a viscous fluid between concentric cylinders the study of which was initiated by Taylor (1923).

As shown in Chap. 8, the mean velocity in the annular space bounded by

$$r = R_1 \qquad \text{and} \qquad r = R_2$$

is
$$V = A_1 r + \frac{B_1}{r}, \tag{83}$$

in which $\quad A_1 = \dfrac{\Omega_2 R_2{}^2 - \Omega_1 R_1{}^2}{R_2{}^2 - R_1{}^2},\qquad B_1 = \dfrac{(\Omega_1 - \Omega_2)R_1{}^2 R_2{}^2}{R_2{}^2 - R_1{}^2},$

Ω_1 and Ω_2 being the angular speeds of the inner and outer cylinders, respectively, and cylindrical coordinates (r, φ, z) being used.

As is easily verified, if axial symmetry is assumed, the linearized equations of motion are

$$\frac{\partial u'}{\partial t} - 2\left(A_1 + \frac{B_1}{r^2}\right)v' = -\frac{1}{\rho}\frac{\partial p'}{\partial r} + \nu\left(\nabla^2 u' - \frac{u'}{r^2}\right), \qquad (84)$$

$$\frac{\partial v'}{\partial t} + 2A_1 u' = \nu\left(\nabla^2 v' - \frac{v'}{r^2}\right), \qquad (85)$$

$$\frac{\partial w'}{\partial t} = -\frac{1}{\rho}\frac{\partial p'}{\partial z} + \nu\,\nabla^2 w', \qquad (86)$$

in which the primes indicate the perturbation quantities. The equation of continuity is

$$\frac{\partial(ru')}{\partial r} + \frac{\partial(rw')}{\partial z} = 0. \qquad (87)$$

If we let

$$r' = \frac{r}{R_1},\qquad z' = \frac{z}{R_1},\qquad t' = t\Omega_1,$$

drop the primes, and assume

$$u' = u(r)\cos \lambda z\, e^{\sigma t},\qquad v' = v(r)\cos \lambda z\, e^{\sigma t},\qquad w' = w(r)\sin \lambda z\, e^{\sigma t},$$

$$(88)$$

elimination of p' from Eqs. (84) and (86) gives

$$(L - \lambda^2 - \sigma R)(L - \lambda^2)u = 2\lambda^2 R\left(A + \frac{B}{r^2}\right)v \qquad (89)$$

and (85) becomes

$$(L - \lambda^2 - \sigma R)v = 2RAu, \qquad (90)$$

in which $\qquad L = D^2 + \dfrac{D}{r} - \dfrac{1}{r^2},\qquad D = \dfrac{d}{dr}, \qquad (91)$

$$A = \frac{A_1}{\Omega_1},\qquad B = \frac{B_1}{R_1{}^2\Omega_1},\qquad R = \frac{\Omega_1 R_1{}^2}{\nu}. \qquad (92)$$

The equation of continuity is

$$\lambda rw = -D(ru) \qquad (93)$$

The boundary conditions are

$$u = Du = 0 = v \qquad \text{at } r = 1 \text{ and at } r = \alpha = \frac{R_2}{R_1}. \qquad (94)$$

Equations (89), (90), and (94) constitute the differential system governing stability. If the flow is unstable, circular vortices with axial symmetry will occur first before their symmetry is destroyed at larger values of R. These vortices, called *Taylor vortices*, have coil-spring streamlines.

3.1. A Sufficient Condition for Stability

If we let the viscosity approach zero, so that $R \to \infty$, and ignore any boundary layer that may exist, (89) and (90) become

$$(L - \lambda^2)u = -2\lambda^2\sigma^{-1}\left(A + \frac{B}{r^2}\right)v,$$

$$-\sigma v = 2Au,$$

which can be combined into

$$(L - \lambda^2)v - \frac{4\lambda^2 A}{\sigma^2}\left(A + \frac{B}{r^2}\right)v = 0. \tag{95}$$

If we write

$$\omega = A + \frac{B}{r^2},$$

(95) can be written as

$$D\left[\frac{1}{r}D(rv)\right] - \lambda^2\left(\frac{1}{\sigma^2}\frac{F}{r} + \frac{1}{r}\right)rv = 0, \tag{96}$$

in which
$$F = \frac{1}{r^3}\frac{d}{dr}\Gamma^2, \qquad \Gamma = \omega r^2. \tag{97}$$

The boundary conditions are

$$rv = 0 \qquad \text{for } r = 1 \text{ and } r = \alpha. \tag{98}$$

Equations (96) and (98) constitute a Sturm-Liouville system. If F is positive, so that Γ^2 increases outward, σ^2 must be real and negative, so that the flow is stable. If F is negative, σ^2 can only be positive and one of the roots of σ is positive, giving instability. If F is positive in certain regions and negative in others, σ^2 has, according to the Sturm-Liouville theory, both positive and negative eigenvalues, so that the flow is again unstable. Thus under the assumption of inviscidness, the flow is unstable if

$$\frac{d}{dr}\Gamma^2 < 0 \tag{99}$$

in any region and stable if

$$\frac{d}{dr}\Gamma^2 > 0 \tag{100}$$

in the entire region. For a viscous fluid, (99) is no longer sufficient to ensure instability, whether it holds in the entire region or only in a part of it. But, as will be seen, (100) remains a sufficient condition for stability even for viscous fluids, provided it holds in the entire region.

Rayleigh (1916) gave a physical interpretation of the stability criterion (100). If the fluid is considered inviscid, the circulation $2\pi r v$ along any fluid ring is constant, according to Kelvin's circulation theorem. Thus

$$rv = k = \text{const for a given ring.}$$

Suppose now the radius of this ring is varied, so that the ring finds itself at various positions. The centripetal acceleration of any part of the ring at the radius r is

$$\frac{v^2}{r} = \frac{k^2}{r^3},$$

which is to say that the centripetal force k^2/r^3 (per unit mass) is necessary to keep that part from centrifugal motion, in the same way that a buoyant force is necessary to keep a particle of a fluid at rest from falling in the direction of gravity. Hence k^2/r^3 corresponds to a "centrifugal" gravitational force per unit mass, and $\rho k^2/2r^2$ corresponds then to the potential energy of the fluid ring. With k different for different rings, it is evidently necessary that $k(r)$ for the undisturbed flow increase outward in order to have the total potential energy of the fluid a minimum. In fact ρk^2 corresponds to the "specific weight" of the fluid in a field where the potential energy per unit "weight" is $1/2r^2$.

Von Kármán (1934) also gave a physical interpretation of (100). If a ring situated at $r = r_1$, with $k_1 = r_1 v_1$, is displaced to $r = r_2 > r_1$, the centripetal acceleration it experiences at its new position is

$$\frac{\rho k_1^2}{r_2^3}$$

if it is stationary there, whereas the pressure gradient prevailing at $r = r_2$ is still

$$\frac{\rho k_2^2}{r_2^3},$$

where $k_2 = r_2 v_2$ is proportional to the circulation along the ring originally situated at $r = r_2$. If

$$k_1^2 < k_2^2,$$

the pressure gradient is more than sufficient to supply the centripetal acceleration of the displaced ring, considered as stationary at $r = r_2$, and will therefore push it back toward $r = r_1$. Hence if $k^2(r)$ increases with r, the fluid is stable.

We shall now present a much simplified version of Synge's proof (1938a) that (100) is a sufficient condition for stability for a viscous fluid.

Multiplying (89) by ru^* (u^* is the complex conjugate of u) and integrating between 1 and α, by parts and using (94) whenever necessary, we have

$$I_2 + (2\lambda^2 + \sigma R)I_1 + \lambda^2(\lambda^2 + \sigma R)I_0 = 2\lambda^2 R \int r\left(A + \frac{B}{r^2}\right)vu^*, \quad (101)$$

in which $\quad I_0 = \int r\,|u|^2, \qquad I_1 = \int \frac{1}{r}\,|D(ru)|^2, \qquad I_2 = \int r\,|Lu|^2,$

the limits of integration and dr having been omitted for brevity. If (90) is multiplied by rv^* and integrated between 1 and α, the result is

$$J_1 + (\lambda^2 + \sigma R)J_0 = -2RA \int ruv^*, \quad (102)$$

in which $\qquad J_0 = \int r\,|v|^2, \qquad J_1 = \int \frac{1}{r}\,|D(rv)|^2.$

Finally, multiplying (90) by v^*/r and integrating, we have

$$\int 2\frac{v}{r}D\frac{v^*}{r} + J_3 + (\lambda^2 + \sigma R)J_2 = -2RA \int \frac{uv^*}{r}, \quad (103)$$

in which $\qquad J_3 = \int r\left|D\frac{v}{r}\right|^2, \qquad J_2 = \int \frac{1}{r}\,|v|^2.$

In these integrations, it is best to write

$$L = D\frac{1}{r}Dr(\ ), \qquad L^2 = D\frac{1}{r}DrD\frac{1}{r}Dr(\ ),$$

successive operation starting at the far right being always understood. Then, as if by magic, integration by parts moves sections of the "snake" for L^2 from u to ru^*, for instance, when one wishes to obtain the left-hand side of (101). If the reader tries to obtain (101), he will see what is meant by the last sentence.

Combining (101) to (103) in such a way as to make the right-hand sides cancel out, we have

$$I_2 + (2\lambda^2 + \sigma^* R)I_1 + \lambda^2(\lambda^2 + \sigma^* R)I_0 + \lambda^2[J_1 + (\lambda^2 + \sigma R)J_0]$$

$$+ \frac{B\lambda^2}{A}\left[\int 2\frac{v}{r}D\frac{v^*}{r} + J_3 + (\lambda^2 + \sigma R)J_2\right] = 0. \quad (104)$$

Now since, upon integration by parts and using the boundary conditions for v in (94),

$$\int \frac{v}{r}D\frac{v^*}{r} = -\int \frac{v^*}{r}D\frac{v}{r},$$

the real part of

$$\int \frac{v}{r} D \frac{v^*}{r}$$

is zero. Taking the real part of (104), we have then

$$I_2 + (2\lambda^2 + \sigma_r R)I_1 + \lambda^2(\lambda^2 + \sigma_r R)I_0 + \lambda^2[J_1 + (\lambda^2 + \sigma_r R)J_0]$$

$$+ \frac{B\lambda^2}{A} [J_3 + (\lambda^2 + \sigma_r R)J_2] = 0. \quad (105)$$

If $\Omega_1 R_1{}^2 = \Omega_2 R_2{}^2$, then $A = 0$ and (105), after multiplication by A, shows that $\sigma_r < 0$, since R is positive. Understanding Ω_1 to be positive, if

$$\Omega_1 R_1{}^2 < \Omega_2 R_2{}^2,$$

so that the cylinders are rotating in the same direction but the circulation increases outward, then A is positive. In that case either B is positive, or, if negative,

$$|B| < A. \quad (106)$$

In case B is also positive, it is obvious from (105) that $\sigma_r < 0$. If B is negative, the same conclusion follows if one uses (106) and

$$J_2 < J_0 \quad \text{and} \quad J_3 < J_1. \quad (107)$$

It remains to show the truth of (107). The first inequality is obviously true. The second inequality becomes evident if we note

$$J_3 = \int \left(\frac{|Dv|^2}{r} - \frac{|v|^2}{r^3} \right), \quad J_1 = \int \left(r |Dv|^2 + \frac{|v|^2}{r} \right), \quad (108)$$

which follow from the definitions of J_1 and J_3, upon expansion, some integrations by parts, and utilization of the boundary conditions on v in (94). Hence, if

$$\Omega_1 R_1{}^2 \leq \Omega_2 R_2{}^2, \quad (109)$$

the fluid is stable. This is in agreement with (100), since we have assumed the cylinders to rotate in the same direction.

3.2. The Adjoint System of Roberts and Chandrasekhar

With ω defined by

$$\omega = A + \frac{B}{r^2},$$

if we write v for the $2\lambda^2 Rv$ in (89), then without using a new symbol for v we obtain

$$(L - \lambda^2 - \sigma R)(L - \lambda^2)u = \omega v, \quad (110)$$

$$(L - \lambda^2 - \sigma R)v = 4\lambda^2 R^2 Au. \quad (111)$$

The boundary conditions are still given by (94). Now we can assume that

$$v = \sum_{j=1}^{\infty} A_j v_j, \qquad v_j = Z_1(\alpha_j r), \tag{112}$$

in which $Z_1(\alpha_j r)$ is a linear combination of Bessel functions J_1 and Y_1 (or N_1) satisfying

$$(L + \alpha_j^2)Z_1 = 0, \qquad Z(\alpha_j) = Z(\alpha_j \alpha) = 0, \tag{113}$$

and $\alpha_1, \alpha_2, \ldots$ are the eigenvalues determined by (113). With (112), we can in principle apply Chandrasekhar's method described in Sec. 2.3.1 to solve the problem, i.e., to obtain, for given A and B, a relationship between R and λ. Unfortunately it is impossible to find the particular solutions of (110) in a finite number of terms when (112) is substituted into it, because of the term B/r^2 in ω. But a technique due to Roberts (1960) and Chandrasekhar (1961) rescued the method. A slightly more general version of the technique is given below, in which σ is not assumed to be zero to start with.

Consider the adjoint system

$$(L - \lambda^2 - \sigma'R)(L - \lambda^2)u' = v', \tag{114}$$

$$(L - \lambda^2 - \sigma'R)v' = 4\lambda^2 R^2 A \omega u', \tag{115}$$

$$u' = 0 = Du' \quad \text{and} \quad v' = 0 \quad \text{at } r = 1 \text{ and } r = \alpha. \tag{116}$$

It will now be proved that whether σ is considered as the eigenvalue for given λ and R or R is considered to be the eigenvalue for given λ and σ, the eigenvalues of the system (110), (111), and (94) are exactly the same as those of the system (114) to (116), even though the eigenfunctions are different. The proof for the general case follows closely the path blazed by Chandrasekhar.

Substituting (112) in (110) and solving for u (in principle), we have

$$u = \sum_{j=1}^{\infty} A_j u_j, \tag{117}$$

in which $\qquad (L - \lambda^2 - \sigma R)(L - \lambda^2)u_j = \omega v_j \qquad (118)$

and u_j satisfies the four conditions concerning u in (94). Similarly, let

$$v' = \sum_{j=1}^{\infty} B_j v_j', \qquad v_j' = Z_1(\alpha_j r), \tag{119}$$

and write the solution of (114) in the form

$$u' = \sum_{j=1}^{\infty} B_j u_j', \tag{120}$$

where $\qquad (L - \lambda^2 - \sigma R)(L - \lambda^2)u_j' = v_j'. \qquad (121)$

Now (110), (94), (114), and (116) are all satisfied. We need only take care of (111) and (115). To satisfy (111), multiply it by rv_k', replace

v_k' by its equivalent in (121) on the right-hand side of the equation at hand, and integrate it, by parts and using the boundary conditions on u whenever necessary. The result, with the summation over j understood to be from 1 to ∞, is

$$\sum A_j \int r v_k'(L - \lambda^2 - \sigma R)v_j$$

$$= 4\lambda^2 R^2 A \sum A_j \int r[(L - \lambda^2)u_j(L - \lambda^2)u_k' + \sigma R(D_1 u_j D_1 u_k' + \lambda^2 u_j u_k')]$$

$$= 4\lambda^2 R A \sum A_j M_{jk}, \tag{122}$$

in which $D_1 = (1/r)Dr$, M_{jk} is defined by the last equality sign, and the limits of integration, understood to be 1 and α, are omitted together with dr. Equation (122) is of the form (δ_{jk} being the Kronecker delta)

$$\sum A_j[N_j^2 \delta_{jk}(\alpha_j^2 + \lambda^2 + \sigma R) + 4\lambda^2 R^2 A M_{jk}] = 0. \tag{123}$$

As we vary k through $1, 2, 3, \ldots$, we obtain infinitely many equations. The condition for the nonvanishing of all A's is then

$$|N_j^2 \delta_{jk}(\alpha_j^2 + \lambda^2 + \sigma R) + 4\lambda^2 R^2 A M_{jk}| = 0, \tag{124}$$

where $$N_j^2 = \int r Z_1^2(\alpha_j r)$$

and the vertical lines denote the determinant whose element at the jth column and the kth row is the sum written between them. On the other hand, multiplying (115) by $r v_k$, using (118) on the right-hand side, and integrating, we obtain equations for $k = 1, 2, 3, \ldots$ similar to (123), and the equation

$$|N_j^2 \delta_{jk}(\alpha_j^2 + \lambda^2 + \sigma R) + 4\lambda^2 R^2 A M_{jk}'| = 0, \tag{125}$$

in which

$$M_{jk}' = \int r[(L - \lambda^2)u_j'(L - \lambda^2)u_k + \sigma R(D_1 u_j' D_1 u_k + \lambda^2 u_j' u_k)]. \tag{126}$$

From the M_{jk} and M_{jk}' defined in (123) and (126), it can be seen immediately that M_{jk}' would be the same as M_{kj} but for the σ' and σ in them, and (124) and (125) have the forms $F(\sigma, \sigma') = 0$ and $F(\sigma', \sigma) = 0$. Upon elimination of σ' one has $G(\sigma) = 0$ and upon elimination of σ one has $G(\sigma') = 0$. (If λ and σ are specified, the proof for the sameness of R is strictly similar.) Thus system 1 consisting of (110), (111), and (94) is equivalent to system 2 consisting of (114), (115), and (116). But now the particular solution of (114), when (119) is substituted into it, is easy to obtain. It has only one term for each A_j, which is

$$\frac{Z_1(\alpha_j r)}{(\alpha_j^2 + \lambda^2 + \sigma R)(\alpha_j^2 + \lambda^2)}. \tag{127}$$

It should be remembered that although systems 1 and 2 have the same eigenvalues, they do not have the same eigenfunctions; i.e., v is quite different from v' and u from u'. Calculations for Taylor's stability problem were performed by Chandrasekhar and Elbert (1962). The method was also applied by Debler (1966) to convectional instabilities.

3.3. Taylor's Result

The system consisting of (89), (90), and (94) was solved by Taylor (1923), who assumed σ to be zero for neutral stability. Yih (1972) has proved that if the cylinders rotate in the same direction, then $\sigma_i = 0$ if $\sigma_r \geq 0$. Taylor's procedure is as follows.

1 Expand u and v in a series of the normalized functions $\phi_n(\kappa_n r)$, which is a linear combination of Bessel functions of the first order, as demanded by the form of L in (91), and of both kinds. The ϕ_n is the same as the v_j in (112), with $j = n$ and $\kappa_n = \alpha_j$, apart from a numerical factor. Each of ϕ_n vanish at $r = 1$ and $r = \alpha$, so that $u = 0 = v$ at these values. Let

$$u = \sum_{n=1}^{\infty} b_n \phi_n(\kappa_n r), \qquad v = \sum_{n=1}^{\infty} c_n \phi_n(\kappa_n r).$$

2 Multiply (90) by $r\phi_m$ and integrate, by parts and using the boundary conditions on v if necessary, to obtain

$$-(\kappa_m{}^2 + \lambda^2 + \sigma R)c_m = 2RAb_m,$$

since the functions $\phi_1, \phi_2, \ldots$ are orthogonal. Thus c_m is simply related to b_m. If one does the same to (89), one obtains, on the left-hand side,

$$\int_1^\alpha r\phi_m(L - \lambda^2 - \sigma R)(L - \lambda^2)u \, dr = -\left[\frac{d\phi_m}{dr} r \frac{d^2 u}{dr^2}\right]_1^\alpha$$

$$+ \int_1^\alpha ru(L - \lambda^2 - \sigma R)(L - \lambda^2)\phi_m \, dr$$

$$= -\left[\frac{d\phi_m}{dr}\Psi\right]_1^\alpha + (\kappa_m{}^2 + \lambda^2 + \sigma R)(\kappa_m{}^2 + \lambda^2)b_m,$$

where $\Psi = r \, d^2u/dr^2$. On the right-hand side one obtains an expression in terms of the c's, which are simply related to the b's. Thus the result, for $m = 1, 2, \ldots$, is a system of infinitely many equations in terms of the b's and Ψ_1 and Ψ_2, which are the values of Ψ at $r = 1$ and $r = \alpha$, respectively.

3 The boundary conditions $Du = 0$ at $r = 1$ and $r = \alpha$ provide, through term-by-term differentiation of the expansion for u, two equations involving the b's.

4 Treating Ψ_1, Ψ_2, and the b's as unknowns or for convenience treating Ψ_1, Ψ_2, and the coefficients

$$e_n = (\kappa_n{}^2 + \lambda^2 + \sigma R)b_n$$

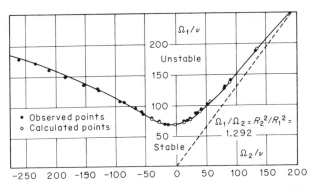

FIGURE 1. TAYLOR'S THEORETICAL AND EXPERIMENTAL RESULTS
FOR $R_1 = 3.55$ CM, $R_2 = 4.035$ CM. THE COORDINATES ARE
IN UNITS OF CM^{-2}. [*G. I. Taylor, Phil. Trans. Roy. Soc. London
Ser. A*, **223** (1923), *by permission of the Royal Society of London.*]

as the unknowns, Taylor obtained the secular equation (in determinant form) the vanishing of which is necessary for the existence of a nontrivial solution. This secular equation determines σ for each λ and R. If one sets σ equal to zero, the secular equation gives a relation between λ and R. The minimum value of R, denoted by R_c, depends only on α and Ω_1/Ω_2, since A, B, and the boundary conditions involve only these and λ. So does the corresponding wave number λ_c.

Taylor's experimental and theoretical results, obtained from a 6×6 determinant, for a set of special values of R_1 and R_2, are given in Fig. 1.

4. INSTABILITY DUE TO SURFACE TENSION

Since the effect of surface tension is in general to minimize the area of the surface, it can be expected to be stabilizing if the mean position of the surface is flat. The situation is not so obvious if the surface is curved. In the following section we shall consider a case in which surface tension is destabilizing to a fluid cylinder with a circular cross section. Sometimes it is not the surface tension but its variation from place to place that is destabilizing. In Sec. 4.2 the instability due to a temperature-induced nonuniformity in surface tension is described.

4.1. Instability of a Circular Liquid Jet

Consider a circular liquid jet of radius a. If viscous effects are ignored for the moment and the motion assumed irrotational, the velocity potential ϕ satisfies the Laplace equation. If cylindrical coordinates r, φ, and

z are used, if the surface corrugation is assumed to be axially symmetric and to have a factor sin mz as well as the exponential time factor $e^{\sigma t}$, and if the method of separation of variables is employed, the velocity potential is readily found to have the form

$$\phi = Ae^{\sigma t} \sin mz J_0(imr), \tag{128}$$

in which J_0 is the Bessel function of zeroth order. Of course a uniform cylindrical surface is impossible if gravity effects are not neglected. In a vertical jet the cross section decreases downward. The analysis is based on the neglect of this nonuniformity of cross section and therefore on the neglect of gravity.

The pressure p in the liquid and the pressure p_0 outside differ by an amount

$$T\left(\frac{1}{R_1} + \frac{1}{R_2}\right), \tag{129}$$

in which T is the surface tension and R_1 and R_2 the principal radii of curvature of the surface. If η is the displacement of the free surface from its undisturbed position $r = a$, and if powers in η higher than the first are neglected,

$$p - p_0 = T\left(\frac{1}{a} - \frac{\eta}{a^2} - \frac{\partial^2 \eta}{\partial z^2}\right). \tag{130}$$

If p' is the pressure perturbation in the liquid, it follows from an examination of (130) that

$$p' = -T\left(\frac{\eta}{a^2} + \frac{\partial^2 \eta}{\partial z^2}\right). \tag{131}$$

Now $\qquad\qquad\qquad \eta = \eta_0 e^{\sigma t} \sin mz. \tag{132}$

Hence $\qquad\qquad p' = -T\eta_0 e^{\sigma t}\left(\frac{1}{a^2} - m^2\right) \sin mz. \tag{133}$

Since the kinematic condition at the free surface is

$$\frac{\partial \eta}{\partial t} = \frac{\partial \phi}{\partial r}, \quad \text{and} \quad \frac{d}{dx}J_0(x) = -J_1(x), \tag{134}$$

we have

$$-imAJ_1(ima) = \sigma\eta_0. \tag{135}$$

On the other hand, with the square of the velocity and the gravity term neglected, the Bernoulli equation is

$$\frac{\partial \phi}{\partial t} + \frac{p'}{\rho} = 0. \tag{136}$$

Equations (128), (133), (135), and (136) together give [Rayleigh (1879)]

$$\sigma^2 = -\frac{T}{\rho}\frac{imJ_1(ima)}{J_0(ima)}\left(\frac{1}{a^2} - m^2\right),$$

or $\qquad\qquad \sigma^2 = \frac{T}{\rho a^3}\frac{kI_1(k)}{I_0(k)}\,(1 - k^2),$ $\qquad\qquad$ (137)

in which $\qquad\qquad\qquad k = ma,$

$$I_1(k) = -iJ_1(ik) = \frac{k}{2} + \frac{k^2}{2^2\cdot 4} + \frac{k^3}{2^2\cdot 4^2\cdot 6} + \cdots,$$ $\qquad$ (138)

$$I_0(k) = J_0(ik) = 1 + \frac{k^2}{2^2} + \frac{k^4}{2^2\cdot 4^2} + \cdots.$$

The maximum value of the square root of

$$\frac{kI_1(k)}{I_0(k)}\,(1 - k^2)$$

is about 0.34, attained at $k = 0.679$. This is the critical unstable mode. Rayleigh also showed that all axially unsymmetric disturbances are stable. It can also be shown that whenever the fluid is unstable ($\sigma^2 > 0$) according to (137), the surface area is decreased by the surface waviness.

The Reynolds number based on the representative surface-tension wave velocity $(T/\rho a)^{\frac{1}{2}}$ is

$$R = \left(\frac{Ta\rho}{\mu^2}\right)^{\frac{1}{2}}.$$

Hence the assumption of inviscidness does not introduce serious errors if

$$\frac{\mu^2}{Ta\rho} \ll 1.$$

The effect of viscosity was considered by Rayleigh (1892), Weber (1931), Tomotika (1935), and Goren (1962). It is surprising that even for small values of R the experimental data of Goren give values for k not far from 0.6. Goren's conclusions indicate that Weber's analysis gives far too small values for k when R is small.

4.2. Convection Cells Induced by Surface Tension

Bénard's experiments (1901) were actually on a liquid layer with the upper surface free. According to an analysis of Jeffreys (1928), the critical value of the Rayleigh number as defined in (40) for the free-surface case is 571. Pearson (1958) estimated the Rayleigh number attained in Bénard's experiments to be below this figure and showed that the actual cause of instability demonstrated by these experiments is probably the nonuniformity of surface tension induced by any nonuniformity of the surface temperature. His analysis will be briefly outlined below.

Taking the extreme case in which surface-tension effects greatly dominate gravitational effects, Pearson dropped the right-hand side in (39). The boundary conditions (to be given presently) do not allow the use of the Pellow-Southwell approach to show that σ is zero at neutral stability. If σ is assumed to be zero at neutral stability, (38) and (39) become

$$(D^2 - a^2)\theta = w, \tag{140}$$

$$(D^2 - a^2)^2 w = 0. \tag{141}$$

The notation of Sec. 2.3 is used here, but for convenience z will be given the values zero and 1 at the lower and upper boundaries, respectively. If the lower boundary is again assumed to be much more thermally conductive than the fluid, the thermal boundary condition there is

$$\theta(0) = 0. \tag{142}$$

The nonslip conditions are still

$$w(0) = 0 = Dw(0). \tag{143}$$

The boundary conditions at the free surface are more complicated. The vertical velocity can certainly be assumed zero at the free surface since σ is assumed zero and the surface displacement is assumed infinitesimal. Thus one of the free-surface conditions is

$$w(1) = 0. \tag{144}$$

The nonuniformity of the surface tension Γ (not indicating circulation throughout the rest of the chapter) is connected to the nonuniformity of the temperature by

$$\Gamma = \Gamma_1 - \gamma_1 T', \tag{145}$$

in which Γ_1 is the value of Γ at $T = T_1$ and

$$\gamma_1 = \frac{\partial \Gamma}{\partial T} \qquad \text{at} \qquad T = T_1. \tag{146}$$

The condition concerning the shear stress in the x_1 direction just below the surface, since $u_3 = 0 = w$ at $z = 1$, is

$$\mu \frac{\partial u_1}{\partial x_3} = \frac{\partial \Gamma}{\partial x_1} = -\gamma_1 \frac{\partial T'}{\partial x_1}. \tag{147}$$

Similarly, $$\mu \frac{\partial u_2}{\partial x_3} = -\gamma_1 \frac{\partial T'}{\partial x_2}. \tag{148}$$

If (147) and (148) are differentiated with respect to x_1 and x_2, respectively, the results added, and the equation of continuity invoked,

$$\mu \frac{\partial^2 u_3}{\partial x_3{}^2} = \gamma_1 \left(\frac{\partial^2}{\partial x_1{}^2} + \frac{\partial^2}{\partial x_2{}^2} \right) T'. \tag{149}$$

In dimensionless terms, this is, after (37) has been used.

$$D^2 w = a^2 B \theta, \quad \text{at } z = 1 \text{ with } B = -\frac{\beta \gamma_1 d^2}{\mu \kappa}. \tag{150}$$

The final boundary condition concerns the heat transfer at the free surface. With Q denoting the rate of heat transfer per unit area of the free surface,

$$Q = k\beta + qT',$$

where

$$q = \left(\frac{\partial Q}{\partial T} \right)_{T=T_1}.$$

Thus the rate of heat transfer per unit area due to the temperature T' is

$$-k \frac{\partial T'}{\partial x_3} = qT'.$$

In dimensionless terms, this is

$$D\theta = -L\theta \quad \text{at } z = 1 \text{ with } L = \frac{qd}{k}. \tag{151}$$

The governing differential system consists of (140) to (144), (150), and (151).

The solution of (141) that satisfies the boundary conditions (143) and (144) is

$$w = 4 \left(\sinh az + \frac{aA - C}{C} z \sinh az - az \cosh az \right), \tag{152}$$

in which $A = \cosh a, \quad C = \sinh a.$

With (152), the solution of (140) satisfying (142) and (151) is

$$\theta = \frac{3}{a} z \cosh az + \frac{aA - C}{aC} z^2 \cosh az - z^2 \sinh az - \frac{aA - C}{a^2 C} z \sinh az$$

$$- \frac{a^2 A^2 + aAC + C^2 + L(a^2 + aAC + C^2)}{a^2 C(aA + LC)} \sinh az. \tag{153}$$

Then (150) demands

$$B = \frac{8a(aA + LC)(a - AC)}{a^3 A - C^2}. \tag{154}$$

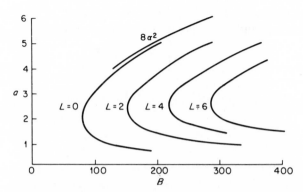

FIGURE 2.   NEUTRAL STABILITY CURVES, CONDUCTING CASE.
[*After J. R. A. Pearson, J. Fluid Mech,* **4** (1958), *by permission of the Cambridge University Press.*]

Equation (154) is plotted in Fig. 2, for various values of L.   The critical value of B is about 80 for $L = 0$, and this was shown by Pearson to be attained in Bénard's experiments.

Pearson also considered the "insulating" case, for which

$$\frac{\partial T'}{\partial x_3} = 0 \qquad \text{at } z = 0. \tag{155}$$

This seems strange at first sight, for the lower boundary cannot really be insulated if a mean temperature gradient β is to be obtained.   However, if the lower boundary is much less conductive than the liquid, (155) is a good approximation.   Note that (155) is a condition on the temperature perturbation only, not on the total temperature.

5. PRELIMINARIES ON THE STABILITY OF PARALLEL FLOWS OF HOMOGENEOUS FLUIDS

Aside from the gravitational instability discussed in Sec. 2 and the inertial instability discussed in Sec. 3, the bulk of the literature on the theory of hydrodynamic stability is concerned with the possibility of instability of parallel flows.   The physical causes for the several kinds of instability discussed so far in this chapter are quite clear.   In contrast, the cause of instability of parallel flows is much subtler.

Except in Sec. 4.2, the primary flows considered are all unidirectional, with the velocity in the x direction and varying with y only. The linearized equations governing stability of such flows will be presented first.   They will be followed by a discussion of the relationship between the behavior of three-dimensional disturbances and that of two-dimensional disturbances.   This relationship, which will be generalized in the last subsection to apply to flows with velocity varying with y not only in magnitude but also in direction, permits us to consider only two-dimensional disturbances in the rest of the chapter.

5.1. The Equations Governing Stability

For simplicity we shall assume for the moment that the density ρ and the viscosity μ are constant. Then, with u, v, and w representing the velocity components, V a reference velocity, and L a reference length, the substitutions

$$(u_1, v_1, w_1) = \left(\frac{u}{V}, \frac{v}{V}, \frac{w}{V} \right), \qquad p_1 = \frac{p}{\rho V^2},$$

$$(x, y, z) = \left(\frac{x_1}{L}, \frac{x_2}{L}, \frac{x_3}{L} \right), \qquad \text{and} \qquad \tau = \frac{tV}{L} \tag{156}$$

will be used to make all quantities discussed dimensionless. Note that the use of u_1 involves an inconsistency in notation with previous usage in Chaps. 1 to 3. Once this is noted and tolerated, no confusion will arise. The use of x, y, and z to denote only dimensionless coordinates may also contradict previous usage. This, too, should be noted in order to avoid confusion. We shall assume that the gravitational acceleration makes angles α_1, α_2, and α_3 with the coordinate axis. The Navier-Stokes equations are then

$$\frac{Du_1}{D\tau} = -\frac{\partial p_1}{\partial x} + F^{-2} \cos \alpha_1 + \frac{1}{R} \nabla^2 u_1,$$

$$\frac{Dv_1}{D\tau} = -\frac{\partial p_1}{\partial y} + F^{-2} \cos \alpha_2 + \frac{1}{R} \nabla^2 v_1. \tag{157}$$

$$\frac{Dw_1}{D\tau} = -\frac{\partial p_1}{\partial z} + F^{-2} \cos \alpha_3 + \frac{1}{R} \nabla^2 w_1,$$

in which ∇^2 is now the dimensionless Laplacian operator and

$$\frac{D}{D\tau} = \frac{\partial}{\partial \tau} + u_1 \frac{\partial}{\partial x} + v_1 \frac{\partial}{\partial y} + w_1 \frac{\partial}{\partial z}, \qquad F^2 = \frac{V^2}{gL}, \qquad R = \frac{VL}{\nu}. \tag{158}$$

The equation of continuity is

$$\frac{\partial u_1}{\partial x} + \frac{\partial v_1}{\partial y} + \frac{\partial w_1}{\partial z} = 0. \tag{159}$$

The criterion of stability of a flow or a configuration is whether, when it is slightly disturbed, the disturbance will or will not grow. If u', v', and w' are the deviations of u_1, v_1, and w_1 from their mean values, and hence represent the velocity perturbation, then

$$u_1 = U(y) + u', \qquad v_1 = v', \qquad w_1 = w', \tag{160a}$$

in which $U(y)$ is the velocity of the undisturbed flow, assumed to be a function of y only. If the corresponding perturbation in p_1 is p', then

$$p_1 = P + p', \tag{160b}$$

in which P is the pressure of the primary flow. If (160a) and (160b) are substituted in (157) and (159) and quadratic terms in the perturbation quantities are neglected, we have, with subscripts indicating the variables of partial differentiation,

$$u'_\tau + U u'_x + U_y v' = -p'_x + \frac{1}{R} \nabla^2 u', \tag{161}$$

$$v'_\tau + U v'_x = -p'_y + \frac{1}{R} \nabla^2 v', \tag{162}$$

$$w'_\tau + U w'_x = -p'_z + \frac{1}{R} \nabla^2 w', \tag{163}$$

and $\qquad\qquad u'_x + v'_y + w'_z = 0. \tag{164}$

These four equations are the equations governing stability. The boundary conditions are the nonslip conditions at a solid boundary and the stress conditions at a free surface or an interface. These, too, are linear, and will be presented when specific problems are considered. For the moment we merely wish to note that Eqs. (161) to (164) as well as the boundary conditions permit a solution containing a factor of the type

$$\exp\left[i(mx + nz - mc\tau)\right] \quad \text{or} \quad \exp\left[i(mx - nz - mc\tau)\right] \tag{165}$$

and representing a sinusoidal disturbance. A general disturbance can be obtained by superposition from sinusoidal disturbances, in the form of a Fourier integral. In (165), m is the wave number in the x direction, n is the wave number in the z direction, and c is the dimensionless complex wave velocity, equal to $c_r + ic_i$. It is implicit that the wave velocity is expressed in units of the reference velocity V. Inspection of (165) reveals that the disturbance will grow if c_i is positive and be damped if c_i is negative. The stability of the flow then depends on the sign of c_i.

5.2. Relationship between the Behavior of Two-dimensional Disturbances and That of Three-dimensional Disturbances

Except for non-Newtonian fluids, there is a relationship between the behavior of two-dimensional disturbances and that of three-dimensional ones for incompressible fluids which will obviate calculation for the stability of three-dimensional disturbances. This relationship was first found by Squire (1933) for the case of a homogeneous fluid flowing between parallel plates, but it can be generalized to apply to a much larger class of flows. We shall adopt an approach due to Lin (1954a) to show how Squire's result can be obtained in a simple way.

As has been shown in the preceding subsection, the differential system governing stability admits of a solution of the form

$$\psi' = \phi(y) \exp \left[i(mx + nz - mc\tau) \right]. \tag{166}$$

If a rotation of coordinates about the y axis is performed so that the x' axis has the direction numbers $(m,0,n)$ with respect to the original coordinates, then

$$mx + nz = m'x', \tag{167}$$

in which
$$m' = (m^2 + n^2)^{\frac{1}{2}}. \tag{168}$$

If c' is defined by
$$m'c' = mc, \tag{169}$$

(166) can be written as

$$\psi' = \phi(y) \exp \left[im'(x - c'\tau) \right]. \tag{170}$$

{Note that c' is the wave velocity in the x' direction, still in units of the original reference velocity V. If the reference velocity is changed from V to mV/m', which is the reference velocity corresponding to the x' direction, then c' would be equal to c. This point should be kept in mind when comparing the present treatment with, say, Lin [1955, Eq. (3.1.1), p. 27].} But (170) represents a two-dimensional disturbance of wave number m' progressing in the x' direction with wave velocity c'. Similarly,

$$\psi'' = \phi(y) \exp \left[i(mx - nz - mc\tau) \right] = \phi(y) \exp \left[im'(x'' - c't) \right].$$

represents a two-dimensional disturbance with the same wave number progressing with the same wave velocity c' in the x'' direction with direction numbers $(m,0,-n)$.

Now
$$\psi = \psi' + \psi'' = 2\phi(y) \cos nz \exp \left[im(x - c\tau) \right] \tag{171}$$

represents a three-dimensional disturbance progressing in the x direction with wave velocity c and having wave numbers m and n in the x and z directions, respectively. If ψ' is a solution of the differential system, so is ψ'', by symmetry, and so is ψ, since that system is linear and homogeneous. Since ψ' and ψ'' are symmetric and physically equivalent, it follows that the flow is stable or unstable for ψ according as it is stable or unstable for ψ'. This is often stated in the following equivalent form:

The behavior of three-dimensional disturbances can be deduced from the behavior of two-dimensional disturbances in a parallel flow of an incompressible fluid.

It is important to remember that ψ' corresponds to a two-dimensional disturbance because the *cross flow*, or flow in the z' direction, contributes nothing to the differential system. The flow with disturbance ψ' present is independent of z', and the boundary conditions, whether kinematical or dynamical, are in the forms they would take if the cross flow did not exist. The approach is so general that the conclusion is generally valid for incompressible fluids except for non-Newtonian fluids whose nonlinear constitutive equation, though contributing only linear[1] terms in the final linearized equations governing stability, nevertheless produces additional linear terms in these equations for three-dimensional disturbances which are absent for two-dimensional disturbances. For Newtonian incompressible fluids the primary temperature or density distribution in the unidirectional flow, being a function of y only, is quite independent of the flow, and the aforementioned correspondence between two- and three-dimensional disturbances is fully valid. In a brief note [Yih (1955)] it was shown that a fluid stratified in density and viscosity and flowing down a slope s with Reynolds number R under a pressure gradient $\partial p/\partial x$ is stable or unstable with respect to three-dimensional disturbances according as the same fluid flowing down a slope s' with Reynolds number R' under a pressure gradient $\partial p'/\partial x'$ is stable or unstable, with the primed quantities related to the unprimed ones by

$$m'R' = mR, \qquad m's' = ms, \qquad \frac{\partial p'}{\partial x'}\csc\beta' = \frac{\partial p}{\partial x}\csc\beta, \qquad (172)$$

in which β and β' are the angles of inclination of the x and x' axes to the horizontal, respectively, so that

$$s = \sin\beta \text{ and } s' = \sin\beta'. \qquad (173)$$

Note that the second equation in (172) cannot be obtained by the geometric consideration of direction resolution alone, but results from the fact that the component of gravitational acceleration normal to the inclined surface (or surfaces) is $g\cos\beta$. Equations (172) are otherwise direct consequences of the theorem of correspondence just presented. They state that the primary flow is stable or unstable for a three-dimensional disturbance according as it is stable or unstable for a two-dimensional disturbance at a lower Reynolds number, a milder slope, and a reduced pressure gradient.

The theorem of correspondence also applies fully to unsteady unidirectional flows, but here one must guard against the tempting intuitive conclusion that three-dimensional disturbances are generally[2] more stable, i.e., that their instability occurs generally at larger R than the instability of two-dimensional disturbances. This would be true if with increasing R the flow became generally more unstable, but if we may take a hint from the theories of Floquet, Mathieu, Hill, and Liapounoff on the stability of a system under the action of a periodic force, which

---

[1] In the determination of the primary flow the nonlinear constitutive equation of course contributes nonlinearly.

[2] This does not mean that for a given R a three-dimensional disturbance is necessarily more stable than a two-dimensional one.

indicate that the frequency and amplitude of that force influence the stability in a complicated and often not readily or intuitively predictable way, we can understand how a three-dimensional disturbance in a periodically unsteady flow might well be less stable than a two-dimensional disturbance.

One more word of caution should be added. For compressible fluids the formalism of the theorem of correspondence is still there, as indicated by Lin (1955, pp. 76–78) and as the reader can with some patience demonstrate to his own satisfaction. However, the interpretation of this correspondence is fraught with danger. The chief obstacle is that the temperature distribution, which is preserved in the rotation of the x and z axes, depends on the actual free-stream velocity and not merely its x' component. Thus while the cross flow does not affect anything else, it does affect the temperature distribution. For this reason the differential system governing the stability of the disturbance ψ' is not quite the same as that governing the stability of a two-dimensional disturbance at reduced Reynolds and Mach numbers, in the absence of the cross flow. As a result it is not possible to dispense with the calculation for three-dimensional disturbances. For a given flow a new calculation must be performed for each direction of x'—or of the propagation of the disturbance ψ'. Dunn and Lin (1953) have shown that three-dimensional disturbances can in fact be more unstable.

5.3. The Orr-Sommerfeld Equation

We can now concentrate on two-dimensional disturbances, for which (164) becomes

$$u'_x + v'_y = 0, \tag{174}$$

permitting the use of a stream function ψ, in terms of which

$$u' = \psi_y, \qquad v' = -\psi_x. \tag{175a}$$

Equations (161) and (162) can then be written

$$\psi_{y\tau} + U\psi_{xy} - U_y\psi_x = -p'_x + \frac{1}{R}\nabla^2\psi_y,$$

$$\psi_{x\tau} + U\psi_{xx} = p'_y + \frac{1}{R}\nabla^2\psi_x. \tag{175b}$$

For two-dimensional disturbances we shall assume

$$\psi = \phi(y)\exp[i\alpha(x - c\tau)], \qquad p' = f(y)\exp[i\alpha(x - c\tau)]. \tag{176}$$

With these, (175b) becomes

$$i\alpha(U - c)\phi' - i\alpha U'\phi = -i\alpha f + \frac{1}{R}(\phi''' - \alpha^2\phi'), \tag{177}$$

$$\alpha^2(c - U)\phi = f' + \frac{i\alpha}{R}(\phi'' - \alpha^2\phi). \tag{178}$$

The elimination of f from these equations then produces the Orr-Sommerfeld equation [Orr (1907) and Sommerfeld (1908)]

$$\phi^{\text{iv}} - 2\alpha^2\phi'' + \alpha^4\phi = i\alpha R[(U - c)(\phi'' - \alpha^2\phi) - U''\phi]. \qquad (179)$$

The density and viscosity have been assumed constant in the derivation of (179). When these vary with y, equations similar to (179) can be derived in much the same way.

6. INSTABILITY DUE TO VORTICITY DISTRIBUTION IN AN INVISCID FLUID

If we ignore the viscosity of the fluid, (179) becomes

$$(U - c)(\phi'' - \alpha^2\phi) - U''\phi = 0, \qquad (180)$$

which can be called the *Rayleigh equation*. Since the study of this equation illuminates the understanding of the instability due to a vorticity distribution even in a viscous fluid, it will be discussed in some detail. General results will be given first.

6.1. Rayleigh's Theorem

If c is complex, so is ϕ. Let ϕ^* denote the complex conjugate of ϕ and consider a flow between two plane boundaries at $y = y_1$ and $y = y_2$. At these boundaries v' is zero, since the normal velocity must vanish there. But since the fluid is considered inviscid, u' is not zero. From (175) and (176) it follows that

$$\phi(y_1) = 0 \qquad \text{and} \qquad \phi(y_2) = 0. \qquad (181)$$

If now (180) is multiplied by

$$-\frac{\phi^*}{U - c}$$

and integrated, by parts if necessary, between y_1 and y_2, using (181) whenever possible, we have

$$\int (|\phi'|^2 + \alpha^2 |\phi|^2) \, dy + \int \frac{U'' |\phi|^2}{U - c} \, dy = 0, \qquad (182)$$

the limits of integration being understood. These limits can also be at infinity. The imaginary part of (182) is

$$c_i \int \frac{U'' |\phi|^2}{|U - c|^2} \, dy = 0, \qquad (183)$$

from which follows Rayleigh's theorem (1880):

Parallel flows of an inviscid fluid are stable if the velocity profile has no point of inflection.

It is quite obvious that if that profile has no point of inflection, U'' is of one sign throughout and c_i must be zero according to (183). Note that for stability c must be real; for if c is an eigenvalue and ϕ is an eigenfunction, c^* is also an eigenvalue with the eigenfunction ϕ^*, as is evident by taking the complex conjugate of (180). Thus so long as c is complex, the imaginary part of one eigenvalue is positive and the flow unstable. This is true of inviscid fluids only. For viscous fluids the governing equation is (179), which has an i in it, and the argument does not hold.

6.2. Fjørtoft's Theorem

Rayleigh's theorem gives a necessary condition for instability or a sufficient condition for stability for inviscid fluids. This theorem is sharpened by another one, discovered fully 70 years later by Fjørtoft (1950).

The real part of (182) is

$$\int \frac{U''(U - c_r) |\phi|^2}{|U - c|^2} \, dy = -\int (|\phi'|^2 + \alpha^2 |\phi|^2) \, dy. \tag{184}$$

Suppose that $c_i \neq 0$, so that according to the Rayleigh theorem there is a point of inflection. Let the velocity U at this point be denoted by U_s. Then (183) can be written as

$$(c_r - U_s) \int \frac{U'' |\phi|^2}{|U - c|^2} \, dy = 0. \tag{185}$$

Addition of (185) to (184) produces

$$\int \frac{U''(U - U_s) |\phi|^2}{|U - c|^2} \, dy = -\int (|\phi'|^2 + \alpha^2 |\phi|^2) \, dy < 0. \tag{186}$$

Thus for instability not only must U'' change sign according to Rayleigh's theorem but, if U is monotonic,
$$U''(U - U_s),$$

which is either positive throughout or negative throughout, must be negative throughout except at the point of inflection, according to (186). This constitutes Fjørtoft's theorem:

For instability, the absolute value of the vorticity of the primary flow must have a maximum in the domain of flow, if the velocity is monotonic.

For instance, if
$$U = \tanh y,$$

the flow may possibly be unstable.   We say "possibly" because Fjørtoft's theorem also gives only a *necessary* condition for instability.

6.3.   Howard's Semicircle Theorem

If the displacement η of a material line from its mean position $y = $ const is written as

$$\eta = F(y)[\exp i\alpha(x - c\tau)], \tag{187}$$

then the kinematic relationship regarding η,

$$\frac{\partial \eta}{\partial \tau} + U \frac{\partial \eta}{\partial x} = v',$$

can be written as

$$F = \frac{\phi}{c - U} \tag{188}$$

and (180) can be written as

$$[(U - c)^2 F']' - \alpha^2 (U - c)^2 F = 0. \tag{189}$$

If this is multiplied by F^* and integrated between y_1 and y_2, by parts if necessary and using (181) whenever desirable, then

$$\int (U - c)^2 (|F'|^2 + \alpha^2 |F|^2) \, dy = 0. \tag{190}$$

Clearly, if F is nonsingular, so that (190) has an unambiguous meaning, c cannot be real.   This means that c cannot be real and lie beyond the range of U, for then F would be nonsingular and c real, contradicting (190).
     Let

$$Q = |F'|^2 + \alpha^2 |F|^2,$$

so that Q is positive definite.   The real and imaginary parts of (190) are

$$\int [(U - c_r)^2 - c_i^2] Q \, dy = 0, \tag{191a}$$

$$2c_i \int (U - c_r) Q \, dy = 0, \tag{191b}$$

from the second of which, if c_i is not zero,

$$\int UQ \, dy = \int c_r Q \, dy. \tag{192}$$

Using (192), we can write (191a) as

$$\int U^2 Q \, dy = \int (c_r^2 + c_i^2) Q \, dy. \tag{193}$$

But the inequality

$$0 \geq \int (U - U_{min})(U - U_{max})Q \, dy$$

obviously holds. With (192) and (193), this can be written as

$$\int [(c_r{}^2 + c_i{}^2) - (U_{min} + U_{max})c_r + U_{min}U_{max}]Q \, dy \leq 0.$$

Since Q is positive throughout, this leads to

$$c_i{}^2 + c_r{}^2 - (U_{min} + U_{max})c_r + U_{min}U_{max} \leq 0,$$

or $\qquad [c_r - \tfrac{1}{2}(U_{min} + U_{max})]^2 + c_i{}^2 \leq [\tfrac{1}{2}(U_{max} - U_{min})]^2.$ \hfill (194)

Thus in the complex plane the point representing c lies in the circle with diameter along the real axis, extending from the point $(U_{min}, 0)$ to $(U_{max}, 0)$. Since only positive values of c_i are significant, one can say that all eigenvalues c of the unstable modes must lie in the semicircle defined by (194) and $c_i > 0$. This is *Howard's semicircle theorem* [Howard (1961)], as pretty a gem as ever has been polished in the hands of hydrodynamicians. (Howard's theorem of 1961 was derived for a stratified fluid. The present results are obtained upon letting the stratification vanish.)

6.4. Tollmien's Investigation on Neutral Modes

Equation (190) shows that c cannot be real and outside the range of U, for then, according to (180) and (188), both ϕ and F would be regular in the entire domain of flow and (190) would present a contradiction. The question then is: Can c be real and fall within the range of U? The mode with such a c is called a *singular neutral mode* if $\phi(y)$ and U'' do not vanish where U is equal to c.

Tollmien (1935) showed that:

1 A special regular neutral mode, with wave number zero, is possible if U has the same value at the boundaries.

2 Regular neutral modes are possible if the velocity profile is symmetric and if there are points of inflection in that profile.

3 U cannot be equal to c anywhere in the flow field for monotone velocity profiles or symmetric velocity profiles divisible into monotonic halves, if there are no points of inflection in the velocity profile.

The proof of case 1 is simple. Inspection of (180) shows that

$$\phi = U - c, \qquad \alpha = 0 \tag{195}$$

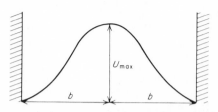

FIGURE 3. THE VELOCITY PROFILE CON-
SIDERED BY TOLLMIEN.

is a solution, and ϕ satisfies the conditions

$$\phi(y_1) = 0 \quad \text{and} \quad \phi(y_2) = 0$$

if
$$c = U(y_1) = U(y_2). \tag{196}$$

In fact, if the velocity profile has no point of inflection, (195) is the only regular solution, provided $U(y_1) = U(y_2)$. For c cannot be complex, according to Rayleigh's theorem, and c cannot fall outside the range of U, by virtue of (190), as already mentioned. Furthermore, c cannot be within the range of U if the solution is to be regular.

The proof of case 2 depends on the well-known fact that the solution of an ordinary differential equation with nonsingular coefficients varies continuously at any point as any one of its coefficients is made to vary continuously. In the following we shall assume $c = U_s$, where $U_s = U(y_s)$, y_s being the value of y at which U'' vanishes. We shall assume U analytic and $U - c$ to have only a simple zero at y_s. Then (180) has regular coefficients, and ϕ for any y will vary continuously as α does so. The velocity profile considered by Tollmien is symmetric, as shown in Fig. 3. [Actually the proof to be given applies equally well to a more complicated profile, so long as the velocity U has the same value $U(y_1)$ at the end points and is equal to the same value U_s at *all* the points of inflection. But for simplicity we shall assume the velocity profile to be as shown in Fig. 3.] There are two symmetrically placed points of inflection, and U vanishes at the boundaries. It is convenient to assign the value zero to y at the midpoint, so that $y_2 = -y_1$ as far as case 2 is concerned.

The solution (195) remains a solution of (180), and since $U(y)$ is even, that solution is also even. But now we are investigating the possibility of a solution of (180) with $c = U_s$, so that $U - c$ does not satisfy the boundary conditions since U now vanishes at the boundaries and $U_s \neq 0$. We shall compare $U - U_s$, which satisfies (180) for $\alpha = 0$, with an even $\phi(y)$, which satisfies (180) for some nonzero α. The solution $U - U_s$ vanishes at $\pm y_s$ and has zero slope at $y = 0$. The solution $\phi(y)$, assumed even, also has zero slope at $y = 0$. Its own value at $y = 0$ is immaterial but will be assumed to be equal to the value of

$U - U_s$ there, that is, to $U(0) - U_s$. Now (180), having regular coefficients, is a differential equation of the Sturm-Liouville type, and as α^2 increases continuously from zero, the zero of ϕ, being at y_s for $\alpha = 0$, will move continuously outward, until for some positive α^2 the function $\phi(y)$ will vanish at y_1 (hence also at y_2). This proves case 2.

The statement just made about the outward movement of the zero of ϕ is the consequence of Sturm's first oscillation theorem, which is presented here for convenience. Consider two systems

$$(K_1 f_1')' - G_1 f_1 = 0,$$
$$f_1(a) = \alpha_1, \qquad f_1'(a) = \beta_1,$$

(197)

and

$$(K_2 f_2') - G_2 f_2 = 0,$$
$$f_2(a) = \alpha_2, \qquad f_2'(a) = \beta_2,$$

(198)

in which K_1, K_2, K_1', K_2', G_1, and G_2 are bounded functions of y and

$$K_1 \geq K_2 \qquad \text{and} \qquad G_1 \geq G_2.$$

(199)

It is assumed that

$$|\alpha_1| + |\beta_1| \neq 0, \qquad |\alpha_2| + |\beta_2| \neq 0.$$

Furthermore, if $\alpha_1 \neq 0$, it is assumed that $\alpha_2 \neq 0$, and

$$\frac{K_1(a)\beta_1}{\alpha_1} \geq \frac{K_2(a)\beta_2}{\alpha_2}.$$

(200)

If $\alpha_1 = 0$, no supplemental assumption need be made. Then if f_1 has a certain number of zeros in the half-open interval $a < y \leq b$, f_2 must have at least as many zeros in the same interval, and if the zeros of f_1 in that interval are denoted by $z_1, z_2, \ldots$ and those of f_2 by $z_1', z_2', \ldots$, then

$$z_i' < z_i.$$

(201)

This theorem has a trivial exception, which can be ruled out by demanding that $G_1 = 0 = G_2$ never hold in any subinterval. In our case $\alpha_1 \neq 0 \neq \alpha_2$, $K_1 = K_2 = 1$, and G increases with α^2, and we identify a with $(y_1 + y_2)/2$ and b with y_2. Since we are considering an even mode, both β_1 and β_2 are equal to zero. We can take α_1 and α_2 to be any nonzero numbers, and the equality sign in (200) holds. Thus all the assumptions of the theorem hold, and we can compare the solutions ϕ_1 and ϕ_2 satisfying respectively the systems

$$\phi_1'' - \frac{U''}{U - U_s} \phi_1 = 0,$$
$$\phi_1(a) = \alpha_1, \qquad \phi_1'(a) = 0,$$

and

$$\phi_2'' - \left(\alpha^2 + \frac{U''}{U - U_s}\right)\phi_2 = 0,$$
$$\phi_2(a) = \alpha_2, \qquad \phi_2'(a) = 0.$$

The desired conclusion then follows.

We shall give a more direct proof of case 2 by a variational method, due to Friedrichs (1942). For the kind of velocity profiles considered by Tollmien

$$K(y) = -\frac{U''}{U - U_s} > 0.$$

If λ denotes $-\alpha^2$, the Rayleigh equation can be written as

$$\phi'' + K(y)\phi + \lambda\phi = 0, \tag{180a}$$

with boundary conditions

$$\phi(y_1) = 0 = \phi(y_2).$$

If (180a) is multiplied by ϕ and integrated from y_1 to y_2 and the boundary conditions are utilized whenever possible, then

$$\lambda = \frac{\int [(\phi')^2 - K\phi^2]\, dy}{\int \phi^2\, dy}, \tag{202}$$

in which the limits of integration are understood and therefore omitted for brevity. Now if ϕ really satisfies (180a) as well as the boundary conditions, then the λ determined from (202) has the well-known remarkable property that it is stationary if ϕ is given a slight variation $\delta\phi$ satisfying only the boundary conditions. This is proved in the following way. If $\delta\lambda$ corresponds to $\delta\phi$, then

$$\delta\lambda = \frac{2I\int (\phi'\, \delta\phi' - K\phi\, \delta\phi)\, dy - 2J\int \phi\, \delta\phi\, dy}{I^2}, \tag{203}$$

in which J and I are respectively the numerator and denominator of the right-hand side of (202). Now, by virtue of the boundary conditions on $\delta\phi$,

$$\int \phi'\, \delta\phi'\, dy = -\int \phi''\, \delta\phi\, dy,$$

since $\delta\phi' = (\delta\phi)'$. Thus (203) can be written as

$$-\frac{I}{2}\delta\lambda = \int (\phi'' + K\phi + \lambda\phi)\, \delta\phi\, dy, \tag{204}$$

with λ replacing J/I on the right-hand side. If (180a) is satisfied, then

$$\delta\lambda = 0. \tag{205}$$

Conversely, if $\delta\lambda$ vanishes for all variations $\delta\phi$ satisfying the boundary conditions and possessing a derivative which, like ϕ', is square-integrable, then (204) shows that (180a) is satisfied.[1] Equation (205) merely shows

[1] The square-integrability of ϕ and ϕ' is implied by (202) and that of $\delta\phi$ and $\delta\phi'$ by (203); the square-integrability of ϕ' and $\delta\phi'$ implies that of ϕ and $\delta\phi$.

that the eigenvalue λ is stationary. In books on variational principles it is also proved that it is a minimum. Thus we can write

$$\lambda = \min \lambda_1, \qquad \lambda_1 = \frac{\displaystyle\int [(f\,')^2 - Kf^2]\,dy}{\displaystyle\int f^2\,dy}, \tag{206}$$

in which f is any function possessing a square-integrable derivative and satisfying the boundary conditions. One way of determining λ is to assume

$$f = \sum_{i=1}^{h} A_n f_n, \qquad f_n = \sin\frac{n\pi(y - y_1)}{y_2 - y_1}. \tag{207}$$

Substituting (207) into (206), varying the A's, and demanding that

$$\frac{\partial \lambda}{\partial A_1} = \frac{\partial \lambda}{\partial A_2} = \cdots = 0,$$

we obtain a series of equations involving the A's and λ. The condition that the A's not all vanish then produces a secular equation which determines λ, and thereby the A's, up to a multiplicative constant.

Returning to the proof of case 2, we note that since $K(y) > 0$, the test function U replacing f in (206) gives, when

$$\int U'^2\,dy = -\int U''U\,dy$$

is utilized,[1]

$$I_1\lambda_1 = \int \frac{U''UU_s}{U - U_s}\,dy = -\int UU_sK\,dy < 0, \qquad I_1 = \int U^2\,dy,$$

provided U is positive throughout, $K(y)$ being assumed positive. Thus λ is negative a fortiori, and α is real, as required. There is therefore a neutrally stable mode. This completes the alternative proof of case 2.

The proof of case 3 is more complicated; but it is also very revealing and therefore rewarding. First, if c is real and $U = c$ in the domain of flow, a power-series solution of (180) around the critical point $y = y_c$, where $U = c$, gives the two independent solutions

$$\phi_1 = \eta + \frac{U''_c}{2U'_c}\eta^2 + \cdots, \tag{208}$$

$$\phi_2 = 1 + \cdots + \frac{U''_c}{U'_c}\phi_1(\eta)\ln\eta, \tag{209}$$

in which $\qquad\qquad\qquad \eta = y - y_c.$

[1] Remember that U vanishes on the boundaries.

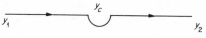

FIGURE 4.   THE CONTOUR FOR EVALUATING ϕ_2.

The form of (209) shows immediately that if $\ln \eta$ is taken to be real for positive η, for negative η it will be complex. The phase (or argument) of η when it is negative depends on the path of reaching negative η from positive η. The correct path will be discussed later. At the moment we merely note that for negative η, if the path is taken below the *lower* one of the critical points where $y = y_c$, as shown in Fig. 4, then

$$\phi_2 = 1 + \cdots + \frac{U_c''}{U_c'} \phi_1(\eta)(\ln |\eta| - i\pi). \tag{210}$$

Note that the imaginary part of ϕ_2 for negative η is proportional to $\phi_1(\eta)$ and is therefore obviously a solution of (180). This is to be expected since (180) has real coefficients.

The functions ϕ_1, ϕ_2, and ϕ_1' are continuous at $\eta = 0$, at which ϕ_2' has a jump

$$-i\frac{U_c''}{U_c'} \pi. \tag{211}$$

From the definition of $\phi(y)$ given by (176), the true meaning of (175a) is

$$2u' = \phi' \exp [i\alpha(x - c\tau)] + \phi'^* \exp [-i\alpha(x - c^*\tau)],$$
$$-2v' = i\alpha\phi \exp [i\alpha(x - c\tau)] - i\alpha\phi^* \exp [-i\alpha(x - c^*\tau)],$$

since the real parts are meant by (175a). Thus, for neutral stability $(c_i = 0)$,

$$-4\overline{u'v'} = i\alpha(\phi\phi'^* - \phi'\phi^*) \tag{212}$$

if the bar indicates the average, at any y, over a full wavelength in the x direction or locally over a period in time. If

$$\phi = A\phi_1(\eta) + B\phi_2(\eta),$$

then

$$\phi(0) = B,$$

and the jump of the right-hand side of (212) when η *decreases* through $\eta = 0$ is, by virtue of (211),

$$-2\alpha \frac{U_c''}{U_c'} \pi |\phi_c|^2,$$

in which

$$\phi_c = \phi(0).$$

The quantity $-\rho\overline{u'v'}$ is the Reynolds stress discussed in the next chapter. Hence, with

$$\tau = -\rho\overline{u'v'},$$

the jump in τ/ρ is

$$\left[\frac{\tau}{\rho}\right] = \frac{\alpha\pi}{2U'_c} U''_c |\phi_c|^2 \tag{213}$$

as η increases through $\eta = 0$.

Now ϕ and ϕ^* are two independent solutions of (180). Hence the quantity within the parentheses on the right-hand side of (212) is the Wronskian, and it is well known that so long as the singularity is not crossed, the Wronskian of a second-order differential equation like (180) must be constant. All one need do to reassure oneself on this point is to differentiate the Wronskian and to use (180). Thus τ is constant if a singular point is not crossed. For a monotone velocity profile, there is only one critical point, and τ is zero at both boundaries. Thus unless ϕ_c or U''_c equals zero, U cannot be equal to c anywhere in the fluid. For the symmetric velocity profile described in Fig. 3 the disturbance can be separated into odd and even modes, and for either mode τ is zero at the midpoint of the channel, which divides the velocity profile into two monotone halves. For either half τ is zero at the extremities, and thus again U cannot be equal to c anywhere in the fluid unless U''_c or ϕ_c is zero. If $U''_c = 0$ at all points where $U = c$, the mode is not singular. We exclude that situation in case 3, and shall assume U''_c different from zero. We then need only prove that ϕ_c is not equal to zero. To do so, suppose for the moment that $\phi_c = 0$. Then $B = 0$, $\phi = A\phi_1$, and ϕ' exists even at y_c. Integration of (180) between y_c and any y gives

$$(U - c)\phi' - U'\phi = -\phi^2\left(\frac{U - c}{\phi}\right)' = \alpha^2 \int_{y_c}^{y} (U - c)\phi \, dy, \tag{214}$$

in which both c and ϕ are real. We shall use the higher y_c if there are two points at which $U = c$. Since a change of sign for ϕ is always allowable, we can assume that

$$(U - c)\phi > 0$$

for y slightly greater than y_c. Then (214) shows

$$-\left(\frac{U - c}{\phi}\right)' > 0$$

for y slightly greater than y_c. The argument can be repeated step by step, and since for the velocity profiles under consideration $U - c$ is monotone for y equal to or greater than y_c at least, the two inequalities show that $U - c$ and ϕ are always of the same sign and that $(U - c)/\phi$ decreases as y increases, so that ϕ cannot vanish at y_2, as it should. Hence $\phi_c \neq 0$, and the proof of case 3 is now complete.

Note that we could have compared the equations

$$\phi'' - \left(\alpha^2 + \frac{U''}{U - c}\right)\phi = 0$$

and

$$f'' - \frac{U''}{U - c} f = 0,$$

the latter of which has the solution $f = U - c$, and used Sturm's first oscillation theorem but for the fact that $U''/(U - c)$ has a singularity at y_c and the functions G_1 and G_2 in Sturm's theorem are supposed to be bounded.

If there is no point of inflection in the velocity profile, there are no unstable modes, according to Rayleigh's theorem, no stable modes with c outside of the range of U, according to (190), and no stable modes with c inside that range, according to Tollmien's result (case 3). If U has the same value U_1 at both boundaries, the only mode is the stable mode $\phi = U - U_1$, $\alpha = 0$, $c = U_1$. Thus for an inviscid fluid with no point of inflection in the velocity profile, the normal-mode analysis gives only one trivial result, and one has to consider the stability problem as an initial-value problem.

6.5. Instability of an Inviscid Fluid with a Point of Inflection in Its Velocity Profile

In Sec. 6.4 it was shown that if U vanishes on the boundaries and its profile has a point of inflection in the domain of flow, then a neutral mode exists. Let the wave number for this mode be denoted by α_s, the corresponding eigenvalue of c by U_s, as before, and the corresponding eigenfunction by ϕ_s. We shall compare

$$\phi_s'' - \left(\alpha_s^2 + \frac{U''}{U - U_s}\right)\phi_s = 0 \tag{215}$$

with

$$\phi'' - \left(\alpha^2 + \frac{U''}{U - c}\right)\phi = 0 \tag{180b}$$

and show that as α decreases slightly, c will have a positive imaginary part provided that throughout the domain of flow

$$K(y) = -\frac{U''}{U - U_s} > 0, \tag{216}$$

as assumed in the alternative proof of case 2 in Sec. 6.4. Note that (216) implies that at the point of inflection

$$-\frac{U'''}{U'} > 0. \tag{217}$$

Instead of presenting Tollmien's heuristic demonstration (1935), we give here Lin's shorter proof[1] (1945, pp. 223–224; 1955, pp. 122–123).

[1] In both Lin's works cited here the proof is based on the existence of ϕ_s with a positive α_s^2. On p. 122 of his book (1955), he seems to offer a demonstration of ϕ_s for positive $K(y)$. There he says that if $K(y)$ is positive, the oscillation theorem of Sturm, applied to (215), guarantees the existence of a positive α_s^2. It certainly guarantees *negative* values for α_s^2 but not a positive one. An independent demonstration, such as Tollmien's, presented in Sec. 6.4, is needed; but since we know ϕ_s exists, Lin's proof of the existence of an amplified mode in the neighborhood of $\alpha = \alpha_s$, as given on p. 123 of his book, is correct. We emphasize also that Tollmien's proof of the existence of ϕ_s is based on the assumption that at *all* the points where $U = c \, (= U_s)$ the quantity U'' is zero.

Multiplying (215) by ϕ and (180b) by ϕ_s, integrating between y_1 and y_2, utilizing the boundary conditions (that ϕ_s and ϕ vanish on the boundaries) whenever possible, and taking the difference of the two resulting equations, we have

$$(\alpha^2 - \alpha_s{}^2)\int \phi\phi_s \, dy = \overset{\downarrow}{(c} - U_s)\int \frac{U''\phi\phi_s}{(U - c)(U - U_s)} \, dy.$$

Now let α, c, and ϕ approach α_s, U_s, and ϕ_s, respectively. Then, as c_i approaches zero through positive values,

$$\frac{d\alpha^2}{dc}\int \phi_s{}^2 \, dy = \lim_{c \to U_s} \left[\int \frac{(U - c_r)K\phi_s{}^2}{(U - c_r)^2 + c_i{}^2} \, dy + i \int \frac{c_i K\phi_s{}^2}{(U - c_r)^2 + c_i{}^2} \, dy \right]$$

$$= -\int \frac{U''\phi_s{}^2}{(U - U_s)^2} \, dy - i\pi\left(\frac{U'''}{U' |U'|}\right)_s \phi_s{}^2. \tag{218}$$

It follows from (217) and (218) that

$$\frac{dc_i}{d\alpha^2} < 0,$$

which means that as α decreases slightly from α_s, the disturbance will be unstable. [Use the principal value of the penultimate term in (218).]

6.6. Stability of Inviscid Fluids with No Point of Inflection in the Velocity Profile

The results of Sec. 6.4 and the Rayleigh theorem together lead to a puzzling situation. If the velocity profile does not have a point of inflection, the flow cannot be unstable, according to Rayleigh's theorem. It also cannot have stable modes except the very special one described in case 1 of Sec. 6.4, since (190) rules out any c outside the range of U and case 3 of Sec. 6.4 rules out any c within the range of U. How, then, is a disturbance having components other than that for which $\alpha = 0$ going to behave as time goes on? The answer is that although normal modes with a discrete spectrum for c no longer exist, the stability problem can be formulated as an initial-value problem. In fact an arbitrary disturbance satisfying certain conditions at $x = \pm \infty$ can still be expressed in terms of normal modes, except that these now have a continuous spectrum. Since the most notorious flow without discrete normal modes for the relevant Rayleigh equation is the plane Couette flow, we shall limit the following discussion to that flow.

Let the plane Couette flow be described by

$$U = y, \qquad -1 \leq y \leq 1,$$

the reference length L in (156) being one-half the depth of the fluid. Then the linearized vorticity equation for ψ (which is used here, as in

Sec. 5.3, to denote the perturbation stream function) is

$$\left(\frac{\partial}{\partial \tau} + y\frac{\partial}{\partial x}\right)\nabla^2\psi = 0, \qquad \nabla^2 = \frac{\partial^2}{\partial x^2} + \frac{\partial^2}{\partial y^2}. \tag{219}$$

Following essentially the treatment of Orr (1907, pp. 26–27), we have from (219)

$$\nabla^2\psi = F(x - y\tau, y). \tag{220}$$

Since the disturbance is assumed free of sources and sinks, the boundaries constitute the same streamline for the disturbance, i.e., ψ is the same constant at $y = \pm 1$. For convenience we shall take ψ to be zero there, since the actual value of that constant is immaterial. Then the initial ψ can be expanded as

$$\psi(x,y,0) = \int_{-\infty}^{\infty} d\alpha \cos \alpha x \sum_{n=1}^{\infty} b_n \sin \left[\tfrac{1}{2}n\pi(y + 1)\right], \tag{221}$$

in which b_n is such a function of α approaching zero as either n or $|\alpha|$ approaches infinity that (221) is convergent. Thus $F(x,y)$ and hence $F(x - y\tau, y)$ are given by (221) and (220). Equation (220) with the boundary conditions $\psi = 0$ at $y = \pm 1$ has the solution[1]

$$\psi = \int_{-\infty}^{\infty} d\alpha \sum_{n=1}^{\infty} \tfrac{1}{2}b_n(\alpha^2 + \tfrac{1}{4}n^2\pi^2)(\text{cosech } 2\alpha)[F(n) - F(-n)], \tag{222}$$

in which

$$F(n) = [\alpha^2 + (\tfrac{1}{2}n\pi - \alpha\tau)^2]^{-1}\{\sinh 2\alpha \sin [\alpha x - \alpha y\tau + \tfrac{1}{2}n\pi(y + 1)]$$
$$- \sinh \alpha(1 - y) \sin \alpha(x + \tau) - \sinh \alpha(y + 1) \sin (\alpha x - \alpha\tau + n\pi)\}.$$

If ψ is evaluated for large t, it can be shown that it is of the order of t^{-1}. Thus plane Couette flow is stable.

   Case (1960) used the Laplace transform with respect to time and reached the same conclusion as Orr. Eliassen, Høiland, and Riis (1953) also showed how the initial-value problem for plane Couette flow can be solved by the use of eigenfunctions with a continuous spectrum. The Rayleigh equation is

$$(y - c)(\phi'' - \alpha^2\phi) = 0, \tag{223}$$

which can also be written as

$$\phi'' - \alpha^2\phi = \delta(y - c), \tag{224}$$

in which δ is the Dirac delta function. This equation has the solution

$$\phi(\alpha,y,c) = \begin{cases} (\alpha \sinh 2\alpha)^{-1} \sinh \alpha(c - 1) \sinh \alpha(y + 1) & \text{for } -1 \leq y \leq c, \\ (\alpha \sinh 2\alpha)^{-1} \sinh \alpha(c + 1) \sinh \alpha(y - 1) & \text{for } c \leq y \leq 1. \end{cases} \tag{225}$$

---

[1] The solution given in Drazin and Howard (1966, p. 28) appears to be in error, since it does not satisfy (220).

This is, in fact, the Green's function associated with the equation

$$\phi'' - \alpha^2\phi = 0$$

and has the property that

$$\phi(\alpha,c_1,c_2) = \phi(\alpha,c_2,c_1).$$

The solution (225) enables one to write

$$\psi(x,y,\tau) = \int_{-\infty}^{\infty} d\alpha \int_{-1}^{1} f(\alpha,c)\phi(\alpha,y,c)e^{i\alpha(x-c\tau)}\,dc,$$

in which $f(\alpha,c)$ is determined from the initial disturbance $\psi(x,y,0)$. The behavior of ψ for large t can be shown to be still $\psi = O(t^{-1})$, in agreement with the results of Orr and Case.

But the choice of plane Couette flow as an example suffers from the peculiarity of this flow: its vorticity is constant throughout. If we write ζ for the total vorticity, $\bar{\zeta}$ for the vorticity of the primary flow, and ζ' for the vorticity of the disturbance, then

$$\zeta = \bar{\zeta} + \zeta', \tag{226}$$

and for an inviscid fluid in two-dimensional unidirectional flow

$$\frac{\partial \zeta}{\partial \tau} + (U + u')\frac{\partial \zeta}{\partial x} + v'\frac{\partial \zeta}{\partial y} = 0. \tag{227}$$

For plane Couette flow $\bar{\zeta}$ is constant, and (227) becomes

$$\frac{\partial \zeta'}{\partial \tau} + (U + u')\frac{\partial \zeta'}{\partial x} + v'\frac{\partial \zeta'}{\partial y} = 0. \tag{228}$$

This means that along a path line ζ' is constant. Let M be the maximum of $|\zeta'(x,y,\tau)|$. Then

$$M = \max |\zeta'(x,y,0)|,$$

by virtue of (228); i.e., the maximum of the magnitude of the vorticity of the disturbance remains the same with time. Irrotational flow is ruled out because it is required to be free of singularities, and if so, the boundaries and the segments bounding the flow region at infinity must have the same value for ψ and ψ must be constant throughout the flow region. If the flow due to the disturbance is rotational, if the vorticity is bounded in magnitude by the initial maximum of $|\zeta'|$, and if ζ' is preserved particle by particle, the flow must be considered as stable, even in a nonlinear treatment.

It might be said that (227) would always lead to the conclusion of stability in the same way for any two-dimensional flow of an inviscid fluid. But if $\bar{\zeta}$ is not constant, there can be transfer of vorticity from the primary flow to the disturbance, because $\bar{\zeta}$ and ζ' are no longer separable. Thus there can be instability in spite of the conservation of ζ along a path line.

7. THE STABILITY OF PARALLEL FLOWS OF A VISCOUS FLUID

Since it has been shown that it is sufficient to consider a two-dimensional disturbance, the linearized equations of motion governing stability of parallel flows of a viscous fluid are (161) and (162). Concerning the disturbance we shall assume that either u', v', and p' are all periodic in x or that u' and v' vanish at infinity as well as on the boundaries. Multiplying (161) by u' and (162) by v' and adding, we have

$$\frac{\partial}{\partial \tau} \frac{1}{2} (u'^2 + v'^2) + U \frac{\partial}{\partial x} \frac{1}{2} (u'^2 + v'^2) + U_y u'v'$$

$$= -\left(u' \frac{\partial p'}{\partial x} + v' \frac{\partial p'}{\partial y} \right) + \frac{1}{R} (u' \, \nabla^2 u' + v' \, \nabla^2 v'). \quad (229)$$

Now the parenthesis containing p' can be written as

$$\frac{\partial(u'p')}{\partial x} + \frac{\partial(v'p')}{\partial y}$$

by the use of the equation of continuity. If now (229) is integrated over the entire domain of flow, we have

$$\frac{d}{d\tau} \int \tfrac{1}{2}(u'^2 + v'^2) \, dA = -\int U_y u'v' \, dA - \frac{1}{R} J, \quad (230)$$

in which $\qquad J = \int [(u'_x)^2 + (u'_y)^2 + (v'_x)^2 + (v'_y)^2] \, dA,$

obtained by integration by parts. By virtue of the assumed nature of the disturbance, the integrals of the terms containing U (but not U_y) and p' in (229) are zero. Now (230) can be interpreted as follows. The first term represents the rate of increase of the kinetic energy of the disturbance, the second the rate of work done by the Reynolds stress on the disturbance flow, and the third, without the minus sign, represents the rate of energy dissipation through viscosity. If the fluid is considered entirely inviscid, we have

$$\frac{d}{d\tau} \int \tfrac{1}{2}(u'^2 + v'^2) = -\int U_y u'v' \, dA.$$

For velocity profiles without a point of inflection we have shown in Sec. 6.4 that the Reynolds stress is everywhere zero. Hence for such profiles there can be no increase in the kinetic energy of the disturbance, and thus no instability. However, for a viscous fluid, however small the viscosity, the Reynolds stress is materially modified at the boundary by the nonslip condition, and the first term on the right-hand side of (230) can be more than sufficient to offset the viscous dissipation and thus to cause instability.

We shall now discuss in some detail the instability of plane Poiseuille flow. By Rayleigh's theorem this flow is stable if the fluid is inviscid.   But we shall see that a little viscosity can make it unstable.

7.1.   Instability of Plane Poiseuille Flow of a Viscous Fluid

For the flow considered

$$U = 1 - y^2$$

if the center of the channel is taken to be the origin for y, the reference length L is the half-width of the channel, and the reference velocity is the maximum velocity at the center of the channel.   The Reynolds number is then based on the half-width of the channel and the maximum velocity of the primary flow.   The governing equation is (179).   Since U is even in y, and since the boundary conditions

$$\phi(\pm 1) = 0 \quad \text{and} \quad \phi'(\pm 1) = 0$$

are symmetric, the symmetric modes (odd ϕ) and antisymmetric modes (even ϕ) are separable.   No symmetric modes have been found unstable. For the antisymmetric modes, the boundary conditions can be written as

$$\phi(-1) = 0, \quad \phi'(-1) = 0, \quad \phi'(0) = 0, \quad \phi'''(0) = 0, \quad (231)$$

and the domain of interest is now $-1 \le y \le 0$.

There are four independent solutions of (179): ϕ_1, ϕ_2, ϕ_3, and ϕ_4. For convenience, we shall follow Lin and use a second subscript 1 on any of these to indicate that it is evaluated at $y = -1$ and a second subscript 2 to indicate evaluation at $y = 0$.   Thus

$$\phi_{12} = \phi_1(0), \quad \phi'_{41} = \phi'_4(-1), \quad \text{etc.}$$

The general solution is

$$\phi = C_1\phi_1 + C_2\phi_2 + C_3\phi_3 + C_4\phi_4.$$

If, as we desire, the C's are not all to vanish, (231) demands that

$$F(\alpha, R, c) = \begin{vmatrix} \phi_{11} & \phi_{21} & \phi_{31} & \phi_{41} \\ \phi'_{11} & \phi'_{21} & \phi'_{31} & \phi'_{41} \\ \phi'_{12} & \phi'_{22} & \phi'_{32} & \phi'_{42} \\ \phi'''_{12} & \phi'''_{22} & \phi'''_{32} & \phi'''_{42} \end{vmatrix} = 0. \quad (232)$$

This is the secular equation from which the relationship between α and R for neutral stability or for any value of c_i can be deduced.   Thus, with $c = c_r + ic_i$, (232) is equivalent to two real equations.   If we put $c_i = 0$ or any other constant and eliminate c_r between these two equations, the curve for neutral stability ($c_i = 0$) or for a constant c_i is obtained in the αR plane.

The determination of the fundamental solutions, of course, is not a simple matter. Two of the four solutions are obtained by setting the right-hand side of (179) equal to zero. These are the solutions of the Rayleigh equation relevant to (179) and are useful only because of one important anticipated fact: the flow is unstable only at rather large Reynolds numbers, so great that the left-hand side of (179) can be neglected for two of the solutions. In fact, if the solution of (179) is expanded formally in the asymptotic form

$$\phi(y) = \phi^{(0)}(y) + \frac{1}{\alpha R} \phi^{(1)}(y) + \cdots,$$

$\phi^{(0)}(y)$ is found to satisfy the Rayleigh equation. The Rayleigh equation can be solved by the Frobenius method of expansion in power series of y, as already indicated in Sec. 6.4, as accurately as desired, but it will be necessary to specify U in each case. Heisenberg (1924) sought solutions of the Rayleigh equation by expansion into a convergent series in α^2 and obtained

$$\phi(y) = (U - c)[q_0(y) + \alpha^2 q_1(y) + \cdots + \alpha^{2n} q_n(y) + \cdots], \quad (233)$$

in which, with lower and upper limits understood to be -1 and y,

$$q_0(y) = 1 \quad \text{or} \quad \int (U - c)^{-2} \, dy,$$

$$q_{n+1}(y) = \int (U - c)^{-2} \, dy \int (U - c)^2 q_n(y) \, dy. \quad (234)$$

Note that
$$\int (U - c)^{-2} \, dy$$

is not convergent if $U = c$ in the range of integration if the path of integration is the real axis. The same is true of all the q's obtained therefrom. But (234) has a meaning if the path is along the axis with an indentation near the point where $U = c$. The indentation is below that point, as will be demonstrated later. The indented path is understood for all the q's. The advantage of Heisenberg's expansion is that the solutions are obtained explicitly without specifying the exact form of U. The two solutions given by (233) and (234) will be denoted by ϕ_1 and ϕ_2.

Two other solutions of (179) must be found, and these should be appropriate for large R. Evidently the effect of viscosity must be reflected in these two solutions, however large the Reynolds number (or however small the kinematic viscosity compared with VL). Since R is large, we may expect the effect of viscosity to be concentrated near the boundaries (region 1) and near the critical point (region 2), where according to the theory for inviscid fluids there would be a discontinuous u'. [Recall the form of ϕ_2 in (210).] Region 1 is the ordinary boundary layer, and region 2 is an internal *viscous layer*, sometimes called the critical layer, or internal *friction layer.* If c is small, regions 1 and 2 may well overlap or merge into one single region where (alone) viscous effects must be taken into account. Outside these regions, whether they overlap or not, the solution of (179) is dominated by the solutions of the relevant Rayleigh equation, i.e., by ϕ_1 and ϕ_2. While the inclusion of viscous effects can always satisfy the nonslip condition at solid boundaries, it really cannot remove the logarithmic singularity in ϕ' as calculated from the solution of the Rayleigh equation indicated in (210). This is an inherent difficulty caused by the method of solution only. Fortunately this unsatisfactory situation actually has very little effect on the calculation for the eigenvalues or even the eigenfunction ϕ (though not ϕ'). This example shows that if a part (the inviscid part) of the solution ϕ is a good representation of the true one aside from the neighborhood of the critical point where $U = c$, it may be good enough for use in the determination of eigenvalues.

By using a method reminiscent of the Wentzel-Brillouin-Kramer (WBK) method, Heisenberg (1924) obtained the two "viscous solutions" in the form

$$(\phi_4, \phi_3) = \exp\left(\pm\sqrt{\alpha R Q}\right)\left[f_0(y) + (\alpha R)^{-\frac{1}{2}}f_1(y) + \cdots\right], \quad (235)$$

where Q and f's are obtained by substitution of (235) into (179) and, comparing coefficients, are

$$Q(y) = \int_{y_c}^{y}\sqrt{i(U - c)}\,dy, \quad f_0(y) = (U - c)^{-\frac{5}{4}}, \ldots \quad (236)$$

Note that as y varies across the point y_c, where $U = c$, the path must be properly determined in order to evaluate Q, f_0, etc. As will be discussed later, the path must be below the point y_c, so that the real part of Q increases monotonically along the path from y_1 to y_2 along the indented path in the complex y plane. It was Lin (1945) who first demonstrated convincingly that the correct path must be *below* the critical point. This is a very difficult question in the whole analysis, and in this book we can only give some indication of the reason why the correct path must be below. Accepting this for the moment, for real c

$$U - c = |U - c|, \ \arg Q = \frac{\pi}{4}, \ \text{for } y > y_c,$$

$$U - c = |U - c|\,e^{-i\pi}, \ \arg Q = -\frac{5\pi}{4}, \ \text{for } y < y_c. \quad (237)$$

Further description of the calculation closely follows the development of Lin (1955). If the solution (235) associated with the positive sign in the exponent is denoted by ϕ_4 and that associated with the negative sign by ϕ_3, then both ϕ_3 and ϕ_4 oscillate rapidly as y increases along the real axis, since R is anticipated to be large. Furthermore the amplitude of ϕ_3 decreases exponentially, whereas that of ϕ_4 increases exponentially as y increases. It is useful to exhibit just how ϕ_3 decreases with y:

$$\frac{\phi_{32}}{\phi_{31}} = \exp \left[-(\alpha R)^{1/2} Y\right]\left[\frac{f_0(y_2)}{f_0(y_1)} + \cdots\right], \tag{238}$$

in which $Y = Q(y_2) - Q(y_1)$. Even more useful is the result

$$\frac{\phi_3'}{\phi_3} = -[i\alpha R(U-c)]^{1/2} - \frac{5U'}{4(U-c)} + \cdots . \tag{239}$$

Now if the third column of the determinant in (232) is divided by ϕ_{31} and the fourth column by ϕ_{42}', the result

$$\begin{vmatrix} \phi_{11} & \phi_{21} & 1 & 0 \\[2mm] \phi_{11}' & \phi_{21}' & \dfrac{\phi_{31}'}{\phi_{31}} & 0 \\[2mm] \phi_{12}' & \phi_{22}' & 0 & 1 \\[2mm] \phi_{12}''' & \phi_{22}''' & 0 & \dfrac{\phi_{42}'''}{\phi_{42}'} \end{vmatrix} = 0 \tag{240}$$

is obtained if we keep in mind (238) and an analogous result for ϕ_4 and remember that, for instance, ϕ_{32}'''/ϕ_{31} and ϕ_{32}'/ϕ_{31} are negligibly small for large R. Since ϕ_{42}'''/ϕ_{42}' is large for large R whereas differentiation of ϕ_1 and ϕ_2 does not change their order of magnitude, (240) can be written as

$$\begin{vmatrix} \phi_{11} & \phi_{12}' \\ \phi_{21} & \phi_{22}' \end{vmatrix} : \begin{vmatrix} \phi_{11}' & \phi_{12}' \\ \phi_{21}' & \phi_{22}' \end{vmatrix} = \frac{\phi_{31}}{\phi_{31}'},$$

which, by virtue of (233), can further be written as

$$-\frac{c\phi_{22}'}{U_1'\phi_{22}' + \phi_{12}'c^{-1}} = \frac{\phi_{31}}{\phi_{31}'}, \qquad \text{with } U_1' = U'(-1),$$

or, since $U'(-1) = 2$,

$$1 + \frac{2c\phi_{22}'}{\phi_{12}'} = \left(1 + \frac{2\phi_{31}}{c\phi_{31}'}\right)^{-1}. \tag{241}$$

The left-hand side of (241) depends only on the solutions of the relevant Rayleigh equation, whereas the right-hand side depends only on ϕ_3. From (239) it follows that

$$\frac{c}{U_1'}\frac{\phi_{31}'}{\phi_{31}} = \frac{5}{4} - Z^{3/2}e^{-\pi i/4}, \tag{242}$$

where

$$Z = c\left(\frac{\alpha R}{U_1'^2}\right)^{1/3}.$$

As to the left-hand side of (241), (233) gives

$$\phi'_{12} = (1 - c)^{-1}[\alpha^2 H_1(c) + \alpha^4 H_3(c) + \cdots],$$

$$\phi'_{22} = (1 - c)^{-1}[1 + \alpha^2 K_2(c) + \alpha^4 K_4(c) + \cdots], \tag{243}$$

in which, with $y_1 = -1$, $y_2 = 0$, and $w_c = U - c$,

$$H_1(c) = \int_{-1}^{0} w_c^2 \, dy,$$

$$H_3(c) = \int_{-1}^{0} w_c^2 \, dy \int_{-1}^{y} w_c^{-2} \, dy \int_{-1}^{y} w_c^2 \, dy,$$

$$\cdots\cdots\cdots\cdots\cdots\cdots\cdots\cdots\cdots\cdots\cdots\cdots, \tag{244}$$

$$K_2(c) = \int_{-1}^{0} w_c^2 \, dy \int_{-1}^{y} w_c^{-2} \, dy,$$

$$K_4(c) = \int_{-1}^{0} w_c^2 \, dy \int_{-1}^{y} w_c^{-2} \, dy \int_{-1}^{y} w_c^2 \, dy \int_{-1}^{y} w_c^{-2} \, dy,$$

$$\cdots\cdots\cdots\cdots\cdots\cdots\cdots\cdots\cdots\cdots\cdots\cdots\cdots$$

If we omit terms involving α^4, the imaginary part of ϕ'_{22} / ϕ'_{12} is simply the imaginary part of $K_2(c)/H_1(c)$. To evaluate $K_2(c)$, Lin (1945) used the easily verifiable transformation

$$K_2(c) = K_1(c)H_1(c) - N_2(c),$$

in which $\quad K_1 = \int_{-1}^{0} w_c^{-2} \, dy, \qquad N_2 = \int_{-1}^{0} w_c^2 \, dy \int_{y}^{0} w_c^{-2} \, dy. \tag{245}$

Since N_2 is real except when $-1 \le y < y_c$, which is a small interval if c is small, the imaginary part of the inner integral for N_2 is small, and this smallness is further strengthened by the factor w_c^2 in the outer integral. Hence the imaginary part of $K_2(c)$ is to be sought mainly in $K_1(c)H_1(c)$. Since $H_1(c)$ is real, this means that the imaginary part of $K_2(c)/H_1(c)$ is that of $K_1(c)$. Now (taking the principal value of the first term)

$$K_1(c) = \int_{-1}^{0} w_c^{-2} \, dy = \frac{1}{U_c'^2} \int_{-1}^{0} \frac{1 - U_c''(U_c')^{-1}(y - y_c) + \cdots}{(y - y_c)^2} \, dy$$

$$= -\frac{1}{U_c' c} + \frac{U_c''}{U_c'^3}(\ln c - i\pi) + O(1),$$

where $O(1)$ is real. Therefore the imaginary part of the left-hand side of (241) is

$$\tilde{v} = -\pi U_1' c \frac{U_c''}{U_c'^3} [1 + O(c^3)], \tag{246}$$

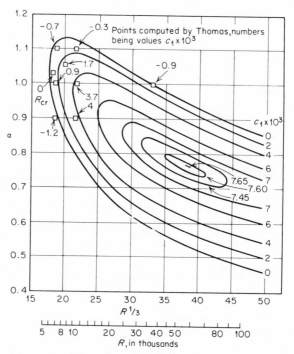

FIGURE 5.   STABILITY DIAGRAM FOR PLANE POISEUILLE FLOW.
[*S. F. Shen, J. Aeron. Sci.*, **21** (1954), *by permission of the American Institute of Aeronautics and Astronautics.*]

the $O(c^3)$ coming from the contribution of $N_2(c)$.   Now (241) can be written as

$$(\tilde{u} - 1) + i\tilde{v}(c) = \mathscr{G}(Z), \quad \text{with } \mathscr{G}(Z) = (Z^{3/2} e^{-\pi i/4} - \tfrac{9}{4})^{-1}, \quad (247)$$

where $\tilde{u} + i\tilde{v}$ is the left-hand side of (241).   The procedure used by Lin to compute the neutral stability curve is as follows.   Assign a real value of c, obtain $\tilde{v}$ from (246), determine Z from (247) by matching the imaginary parts of both sides, and finally obtain $\tilde{u}$ by considering the real parts of that equation.   Now $\tilde{u}$ depends only on α and c.   Since c is assigned, α is known from $\tilde{u}$ and R is computed from Z with the known values of c and α.   We have thus obtained a point on the neutral-stability curve in the αR plane.   In this manner Shen (1954) obtained Fig. 5, in which the curves for constant nonzero c_i were obtained by perturbation from the solution for neutral stability.

All investigations have shown that plane Couette flow is stable. Plane Couette-Poiseuille flow has been investigated by Potter (1966).

7.2. Lin's Improved Theory for the Stability of Plane Poiseuille Flow

From (239) it can be seen that although R is supposed large, the first term on the right-hand side can be less than the second if c is small. This means that for small Z in (242) or small $\tilde{v}(c)$ in (247), an improvement of the calculation is needed. This improvement has little effect on the upper branch of the neutral-stability curve but leads to a better determination of the lower branch, along which c can be small.

Inspection of (179) reveals that for large R and near y_c, where $U = c$, (179) can be approximated by

$$\phi^{iv} = i\alpha R U'_c (y - y_c)\phi'', \qquad (248)$$

for upon making the transformation

$$\zeta = (\alpha R U'_c)^{1/3}(y - y_c)$$

the terms in (179) omitted in (248) can be seen to be much smaller than the retained ones. In terms of the new variable, (248) becomes

$$\frac{d^4}{d\zeta^4}\phi - i\zeta \frac{d^2}{d\zeta^2}\phi = 0, \qquad (248a)$$

and it is a good approximation to (179) near y_c. If c is small, its range of validity includes the boundary $y = -1$, where the nonslip conditions have to be imposed. Now any equation of the form (primes indicating differentiation with respect to ζ)

$$f'' + A\zeta^m f = 0$$

can be transformed to the Bessel equation in the new variable $g(\xi)$ by a transformation of the type

$$f = \zeta^n g(\xi), \qquad \xi = \zeta^k,$$

n and k being determined by m. For $m = 1$,

$$n = \tfrac{1}{2} \quad \text{and} \quad k = \tfrac{3}{2},$$

and (248a) becomes

$$\left(\frac{d^2}{d\xi^2} + \frac{1}{\xi}\frac{d}{d\xi} + \frac{4i^3}{9} - \frac{1}{9\xi^2}\right)\left(\zeta^{-1/2}\frac{d^2}{d\zeta^2}\phi\right) = 0, \qquad (249)$$

the solutions of which are (since $\eta = y - y_c$)

$$\phi_1 = \eta, \qquad \phi_2 = 1,$$

$$(\phi_3, \phi_4) = \int^\zeta d\zeta \int^\zeta \zeta^{1/2} H_{1/3}^{(1,2)}[\tfrac{2}{3}(i\zeta)^{3/2}]\, d\zeta, \qquad (250)$$

in which $H_{1/3}^{(1)}$ and $H_{1/3}^{(2)}$ are Hankel functions of the third order, of the first and the second kind, respectively. Hankel functions instead of Bessel functions are used because the former not only satisfy the Bessel equation but behave exponentially at infinity, as we wish them to. The lower limits of the integrals for ϕ_3 are such that ϕ_3 and ϕ_3' approach zero as $\zeta \to +\infty$. Thus

$$\phi_3 = \int_{+\infty}^{\zeta} d\zeta \int_{+\infty}^{\zeta} \zeta^{\frac{1}{2}} H_{\frac{1}{3}}^{(1)} [\tfrac{2}{3}(i\zeta)^{\frac{3}{2}}] \, d\zeta. \tag{251}$$

Note that $H_{\frac{1}{3}}^{(1)} [\tfrac{2}{3}(i\zeta)^{\frac{3}{2}}]$ approaches zero as $\zeta \to +\infty$. Hence ϕ_3 behaves like Heisenberg's ϕ_3 in (235). In fact, Lin (1945) identified the ϕ_i in (250) with the ϕ_i in Sec. 7.1, for $i = 1, 2, 3, 4$.

Still using (241) for the secular equation, we can now write

$$\frac{U_1'}{c} \frac{\phi_{31}}{\phi_{31}'} = -(1 + \lambda)F(z), \tag{252}$$

in which
$$F(z) = \frac{-\displaystyle\int_{+\infty}^{-z} d\zeta \int_{+\infty}^{\zeta} \zeta^{\frac{1}{2}} H_{\frac{1}{3}}^{(1)} [\tfrac{2}{3}(i\zeta)^{\frac{3}{2}}] \, d\zeta}{z \displaystyle\int_{+\infty}^{-z} \zeta^{\frac{1}{2}} H_{\frac{1}{3}}^{(1)} [\tfrac{2}{3}(i\zeta)^{\frac{3}{2}}] \, d\zeta}, \tag{253}$$

with z and λ defined by

$$z = -\zeta_1 = (\alpha R U_c')^{\frac{1}{3}}(y_c - y_1),$$
$$c = U_1' \frac{(y_c - y_1)}{1 + \lambda}. \tag{254}$$

Thus (241) becomes

$$\tilde{u} + i\tilde{v} = [1 - (1 + \lambda)F(z)]^{-1}. \tag{255}$$

If we define $\mathscr{F}(z)$ as $[1 - F(z)]^{-1}$, (255) can be written as

$$\mathscr{F}(z) = \frac{(1 + \lambda)(\tilde{u} + i\tilde{v})}{1 + \lambda(\tilde{u} + i\tilde{v})}. \tag{256}$$

For small c, λ is small, and (256) becomes

$$\mathscr{F}(z) = \tilde{u} + i\tilde{v}. \tag{257}$$

The function $F(z)$ was first used by Tietjens (1925). Lin (1955) has given values for the real and imaginary parts of $\mathscr{F}(z)$ for various values of z, as shown in Table 1. The procedure is as follows. With these tabulated values, (257) can be used in the same way as (247) for small c. Then λ is found from (254), and (256) can be used to improve the calculation in successive iterations. In this way the neutral-stability curve in Fig. 5 was actually found. As a result of the action of viscosity, the Reynolds stress can be expected to be destabilizing. This was early noted by Taylor (1915). Lin (1954b) actually demonstrated that the Reynolds stress is destabilizing near the boundary.

TABLE 1   The function $\mathscr{F}(z)$ and its first derivative†

z	$\mathscr{F}_r(z)$	$\mathscr{F}_i(z)$	$\mathscr{F}'_r(z)$	$\mathscr{F}'_i(z)$
1.0	0.80630	−2.60557		
1.1				
1.2	1.77012	−2.29854	3.71	2.781
1.3			2.54	
1.4	2.26836	−1.71669	1.505	2.937
1.5			0.850	2.633
1.6	2.44985	−1.18600	0.4245	2.384
1.7			0.1390	2.127
1.8	2.48104	−0.75892	−0.06428	1.907
1.9			−0.2140	1.726
2.0	2.43927	−0.41253	−0.3337	1.573
2.1			−0.4377	1.443
2.2	2.35196	−0.12348	−0.5333	1.325
2.3			−0.6242	1.213
2.4	2.22724	+0.11916	−0.7108	1.1007
2.5			−0.7912	0.9833
2.6	2.06929	0.31558	−0.8625	0.8584
2.7			−0.9209	0.7254
2.8	1.88566	0.46043	−0.9627	0.5853
2.9			−0.9850	0.4414
3.0	1.68938	0.54872	−0.9853	0.2978
3.1			−0.9640	0.1590
3.2	1.49726	0.58082	−0.9223	0.0296
3.3			−0.8629	−0.0867
3.4	1.32516	0.56401	−0.7893	−0.1862
3.5			−0.7054	−0.2695
3.6	1.18429	0.51074	−0.6151	−0.3332
3.7			−0.5222	−0.3786
3.8	1.07982	0.43560	−0.4302	−0.4067
3.9			−0.3422	−0.4193
4.0	1.01118	0.35220	−0.2609	−0.4189
4.1			−0.1865	−0.4060
4.2	0.97361	0.27133	−0.1172	−0.3793
4.3			−0.0625	−0.3538
4.4	0.96056	0.20038	−0.01470	−0.3225
4.5			−0.0061	
4.6	0.95989	0.13601	−0.0034	
4.7			+0.0476	
4.8	0.97659	0.09503	+0.1794	
4.9				
5.0	0.99582	0.07266		

† From C. C. Lin, "The Theory of Hydrodynamic Stability,"
1955, by permission of the Cambridge University Press.

One last point must be discussed before concluding this section. The asymptotic representation of $H_\nu^{(1)}(z)$ is [Watson (1958, p. 198)]

$$H_\nu^{(1)}(z) = \left(\frac{2}{\pi z}\right)^{\frac{1}{2}}\left\{\exp\left[i\left(z - \frac{\nu\pi}{2} - \frac{\pi}{4}\right)\right] \sum_{m=0}^{\infty} \frac{(-)^m(\nu,m)}{(2iz)^m}\right\},$$

with $-\pi < \arg z < 2\pi$ (258)

and $$(\nu,m) = \frac{\Gamma(\nu + m + \frac{1}{2})}{m!\,\Gamma(\nu - m + \frac{1}{2})},$$

Γ being the gamma function. The argument of $H_{\frac{1}{3}}^{(1)}$ being what it is in (250), the range of the argument given in (258) shows that

$$-\pi < \frac{3\pi}{4} + \frac{3}{2}\arg\zeta < 2\pi,$$

or $$-\frac{7\pi}{6} < \arg\zeta < \frac{5\pi}{6}$$ (259)

which in turn shows that the indented path is below the critical point $y = y_c$ as y increases from y_1 to y_2.

7.3. Spatial Growth Rates

The kind of analysis presented in Secs. 7.1 and 7.2 has been applied to the Blasius flow over a plate by Tollmien (1929), Schlichting (1933a and b, 1935), and Lin (1944). The neutral-stability curves obtained by Tollmien and Lin are in good agreement and are in better agreement with the experimental results of Schubauer and Skramstad (1947) than Schlichting's curve. Later, Shen (1954) used Lin's procedure to calculate the curves of constant c_i. His neutral-stability curve agrees closely with that of Lin, and his other results are also in better agreement with Schubauer and Skramstad's experimental data than Schlichting's results. In Fig. 6 Shen's neutral-stability curve is shown in comparison with Schlichting's and with the experimental data. [For a similar comparison of amplification rates, see Lin (1955, p. 70).] The tendency toward three-dimensionality of unstable disturbances has been shown in an interesting paper by Lin and Benney (1962).

In the experiments of Schubauer and Skramstad, a vibrating band with angular frequency ω was placed in the flow, and hot-wire anemometers placed at various distances from it along the direction of wave propagation (also the direction of flow) recorded the signal and indicated whether the disturbance was growing as it traveled in the x direction. Such a procedure does not cause difficulty of interpretation so long as one is interested only in determining the condition of neutral stability. However, if one measures the rate of spatial growth or damping in this way,

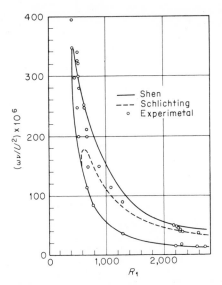

FIGURE 6. NEUTRAL-STABILITY CURVE FOR BLASIUS FLOW. R_1 IS THE REYNOLDS NUMBER BASED ON THE DISPLACEMENT THICKNESS, AND ω IS THE CIRCULAR FREQUENCY OF OSCILLATION. [*S. F. Shen, J. Aeron. Sci.*, **21** (1954), *by permission of the American Institute of Aeronautics and Astronautics.*]

keeping σ equal to ω (real), there arises the question of how to relate this spatial-growth rate with the theoretical temporal-growth rate for the same flow when α is considered real.

When ac in (179) is denoted by σ, the secular equation has the form

$$F(\alpha,\sigma,R) = 0. \tag{260}$$

For temporal growth

$$\alpha = \text{real}, \qquad \sigma = \sigma_r + i\sigma_i,$$

and for spatial growth

$$\sigma = \text{real}, \qquad \alpha = \alpha_r + i\alpha_i.$$

The spatial-growth rate is $-\alpha_i$. Consider now a point on the neutral-stability curve, specified by

$$R = R_0, \qquad \alpha = \alpha_0, \qquad \sigma = \omega,$$

all real. Now let R be increased by dR. For the same real α, there is a $(d\sigma)_\alpha$ corresponding to dR. The imaginary part of $(d\sigma)_\alpha$ is the temporal-growth rate. For the same $\sigma = \omega$, there is a $(d\alpha)_\sigma$ corresponding to dR, the imaginary part of which, with a minus sign added, gives the spatial-growth rate. The task is to relate $(d\alpha)_\sigma$ to $(d\sigma)_\alpha$. If we can do so, we know how to compute $(d\sigma_i)_\alpha$ from $(d\alpha)_\sigma$.

From (260),

$$\frac{\partial F}{\partial \alpha}\, d\alpha + \frac{\partial F}{\partial \sigma}\, d\sigma + \frac{\partial F}{\partial R}\, dR = 0,$$

so that, for $\alpha = \alpha_0$,

$$(d\sigma)_\alpha = -dR\,\frac{\partial F/\partial R}{\partial F/\partial \sigma} \tag{261}$$

and, for $\sigma = \omega$,

$$(d\alpha)_\sigma = -dR\,\frac{\partial F/\partial R}{\partial F/\partial \alpha}. \tag{262}$$

Combination of (261) with (262) gives [Gaster (1962)]

$$(d\sigma)_\alpha = (d\alpha)_\sigma\,\frac{\partial F/\partial \alpha}{\partial F/\partial \sigma} = (d\alpha)_\sigma\left(-\frac{\partial \sigma}{\partial \alpha}\right)_{R=R_0} = -c_g(d\alpha)_\sigma, \tag{263}$$

in which c_g is the group velocity, generally complex. To determine $(d\sigma)_\alpha$ from the experimentally obtained $(d\alpha)_\sigma$, which is complex, one needs to know c_g or $\partial\sigma/\partial\alpha$ at $R = R_0$. This is obtained theoretically in the usual calculation for real α. Equation (263) is the correspondence sought.

We have related $d\sigma$ to $d\alpha$ only in the neighborhood of the neutral-stability curve. This is possible because that curve is the intersection of the two surfaces, both represented by (260), one with α real and the other with σ real. When we depart too far from the neutral-stability curve, confirmation of the temporal growth rate σ_i by experimenting with a vibrating band becomes impossible with only the usual relationship (260) for real α. When experiments are done with a vibrating band for growing or damping disturbances and comparison with a theory is desired, α should be determined with real R and σ according to (260) and the complex values obtained checked against experiments, as done by Mischalke (1965) for jets.

Consideration of spatial growth of a disturbance in parallel flows becomes imperative if one or more of the boundary conditions are x-dependent, such as when the flow is caused by a surface tension varying with x. [For a specific example, see Yih (1967b), with a correction in (Yih 1968a).]

8. INSTABILITY OF AN INVISCID STRATIFIED FLUID IN PARALLEL FLOW

In Secs. 5 to 7, the fluid has been assumed homogeneous. If the density $\bar\rho$ of the primary flow is variable and a function of y, then the stability problem is more complicated to solve. Gravity, assumed to act in the direction of decreasing y, is usually, but not always, stabilizing. In this section Helmholtz's instability for two inviscid homogeneous fluids in relative motion will first be presented. This will be followed by the equations governing the stability of a continuously stratified inviscid fluid and some general results. In Secs. 9 to 11, the fluid is stratified either continuously or discontinuously with respect to the specific weight (ρg), or viscosity, or density, but the viscosity of the fluid will always be taken into account. In this section, there is no need to use dimensionless variables, and we shall use all the usual symbols to denote variables in their appropriate dimensions.

8.1. Helmholtz's Instability

In the latter part of the nineteenth century, Helmholtz (1868) considered the stability of two homogeneous liquids of different densities ρ_1 and ρ_2 ($> \rho_1$) flowing with different x velocities U_1 and U_2, the subscript 1 indicating the upper fluid. Since in each layer the density is constant, the flow with the disturbance, assumed to have started from the flow without the disturbance, must be irrotational since the primary flow is irrotational, U_1 and U_2 being constant. Hence the flow due to the disturbance is irrotational, and the velocity potentials for the layers can be written

$$\phi_1 = U_1 x + \phi_1', \qquad \phi_2 = U_2 x + \phi_2', \tag{264}$$

in which the primes indicate the velocity potentials for the disturbance. The ϕ's, with or without the prime, are harmonic. As has been shown in Sec. 5.2, it is sufficient to consider two-dimensional disturbances. The disturbance is assumed to be periodic in x, since a Fourier integral can then deal with a general disturbance. Since, furthermore, the boundary conditions at infinity are

$$\phi_1' = 0 \text{ at } y = +\infty \qquad \text{and} \qquad \phi_2' = 0 \text{ at } y = -\infty,$$

the appropriate forms of ϕ_1' and ϕ_2' are obtained by the method of separation of variables to be

$$\phi_1' = C_1 \exp\left[-ky + ik(x - ct)\right],$$
$$\phi_2' = C_2 \exp\left[ky + ik(x - ct)\right], \tag{265}$$

in which k is the wave number and c is equal to $c_r + ic_i$. Again the flow is unstable if there is any $c_i > 0$. The form of (265) dictates that the displacement η of the interface from its undisturbed position have the form

$$\eta = a \exp\left[ik(x - ct)\right]. \tag{266}$$

As shown in the chapter on waves, the linearized kinematic conditions on the interface are

$$\frac{\partial \eta}{\partial t} + U_1 \frac{\partial \eta}{\partial x} = \frac{\partial \phi_1'}{\partial y}, \qquad \frac{\partial \eta}{\partial t} + U_2 \frac{\partial \eta}{\partial x} = \frac{\partial \phi_2'}{\partial y}. \tag{267}$$

The dynamical condition at the interface is, in its linearized form,

$$p_1' - p_2' = T \frac{\partial^2 \eta}{\partial x^2}, \tag{268}$$

in which p' is the pressure perturbation and T the surface tension. The pressure p' is to be evaluated from the linearized Bernoulli equation (constant terms or terms containing only t being omitted because only sinusoidal terms are collected)

$$\frac{p'}{\rho} = -\frac{\partial \phi'}{\partial t} - U \frac{\partial \phi'}{\partial x} - gy,$$

in which g is the gravitational acceleration. Substituting this into (268) and setting $y = \eta$, we have

$$\rho_2 \left(\frac{\partial \phi_2'}{\partial t} + U_2 \frac{\partial \phi_2'}{\partial x} + g\eta \right) - \rho_1 \left(\frac{\partial \phi_1'}{\partial t} + U_1 \frac{\partial \phi_1'}{\partial x} + g\eta \right) = T \frac{\partial^2 \eta}{\partial x^2}. \tag{269}$$

Substitution of (265) and (266) into (267) and (269) yields

$$i(U_1 - c)a = -C_1, \qquad i(U_2 - c)a = C_2,$$

$$\rho_2[ik(U_2 - c)C_2 + ga] - \rho_1[ik(U_1 - c)C_1 + ga] = -k^2 Ta.$$

Elimination of C_1, C_2, and a from these equations gives

$$k\rho_2(U_2 - c)^2 + k\rho_1(U_1 - c)^2 = k^2 T + g(\rho_2 - \rho_1),$$

the solutions of which are

$$c = \frac{\rho_2 U_2 + \rho_1 U_1}{\rho_2 + \rho_1} \pm \left[c_0^2 - \rho_1 \rho_2 \left(\frac{U_1 - U_2}{\rho_2 + \rho_1} \right)^2 \right]^{1/2}, \tag{270}$$

in which

$$c_0^2 = \frac{g}{k} \frac{\rho_2 - \rho_1}{\rho_2 + \rho_1} + \frac{Tk}{\rho_2 + \rho_1}, \tag{271}$$

so that c_0 is the speed of waves in the absence of the currents represented by U_1 and U_2. Thus the flow is unstable if

$$\rho_1 \rho_2 \left(\frac{U_1 - U_2}{\rho_2 + \rho_1} \right)^2 > c_0^2,$$

for then one of the two eigenvalues for c given by (270) will have a positive c_i.

The minimum value of c_0^2 is $[2Tg(\rho_2 - \rho_1)/(\rho_2 + \rho_1)^2]^{1/2}$, attained at

$$k^2 = \frac{g(\rho_2 - \rho_1)}{T} = k_{\text{cr}}^2. \tag{272}$$

We have assumed $\rho_2 > \rho_1$. If $\rho_2 < \rho_1$, c_0^2 will be negative for sufficiently small k and the fluid will always be unstable for such k's, no matter what $U_1 - U_2$ is, with the degree of instability increasing with increasing $(U_1 - U_2)^2$.

8.2. Equations Governing the Stability of Inviscid and Continuously Stratified Fluids

The mean density $\bar{\rho}$ and mean velocity U of a continuously stratified fluid will be assumed to be continuous and differentiable functions of y. The linearized equations of motion are

$$\bar{\rho}(u'_t + Uu'_x + v'U_y) = -p'_x, \tag{273}$$

$$\bar{\rho}(v'_t + Uv'_x) = -p'_y - g\rho', \tag{274}$$

in which the primes denote perturbation quantities and subscripts denote the variable with respect to which differentiation is performed. The fluid is assumed incompressible, so that

$$\rho'_t + U\rho'_x + v\bar{\rho}_y = 0 \tag{275}$$

and

$$u'_x + v'_y = 0.$$

The latter of these permits the use of the perturbation stream function ψ, in terms of which

$$u' = \psi_y, \qquad v' = -\psi_x. \tag{276}$$

We shall denote by η the displacement of a line of constant density. Then η is a function not only of x and t but also of y, and if higher-order terms are neglected

$$\eta_t + U\eta_x = v' = -\psi_x. \tag{277}$$

For a disturbance periodic in x,

$$\eta(x,\bar{\rho},t) = F(y)e^{ik(x-ct)}, \tag{278}$$

since $\bar{\rho}$ is a function of y, and (277) and (276) give

$$\psi = (c - U)\eta, \qquad u' = [(c - U)\eta]_y, \qquad v' = ik(U - c)\eta. \tag{279}$$

From (273) and (275) it then follows that

$$p' = \bar{\rho}(U - c)^2\eta_y \quad \text{and} \quad \rho' = -\bar{\rho}_y\eta. \tag{280}$$

With (279) and (280), (274) becomes

$$[\bar{\rho}(U - c)^2F']' + \bar{\rho}[\beta g - k^2(U - c)^2]F = 0, \tag{281}$$

in which

$$\beta = -\frac{\bar{\rho}_y}{\bar{\rho}},$$

and the primes now indicate differentiation with respect to y, unambiguously. Equation (281) is the equation sought. At horizontal rigid boundaries, say $y = 0$ and $y = d$,

$$F(0) = 0 = F(d). \tag{282}$$

At a free surface or interface the boundary condition can easily be formulated.

8.3. Miles' Theorem

Taylor (1931) and Goldstein (1931), considering special velocity profiles, found that if everywhere in the domain of flow

$$g\beta \geq \tfrac{1}{4}U'^2, \tag{283}$$

the flow is stable. Miles (1960) showed that this is generally true. We shall give here the elegant proof by Howard (1961) of Miles' theorem. With

$$W = U - c, \qquad G = W^{1/2}F, \tag{284}$$

(281) becomes

$$(\bar{\rho}WG')' - [\tfrac{1}{2}(\bar{\rho}U')' + k^2\bar{\rho}W + \bar{\rho}W^{-1}(\tfrac{1}{4}U'^2 - g\beta)]G = 0. \tag{285}$$

Both Miles and Howard assumed two rigid boundaries. The boundary conditions (282) can be written in terms of G as

$$G(0) = 0 = G(d). \tag{286}$$

Multiplication of (285) by G^*, integration between 0 and α, and utilization of (286) whenever desirable give

$$\int \left[\bar{\rho}W(|G'|^2 + k^2\,|G|^2) \right.$$

$$\left. + \tfrac{1}{2}(\bar{\rho}U')'\,|G^2| + \bar{\rho}(\tfrac{1}{4}U'^2 - g\beta)W^*\left|\frac{G}{W}\right|^2 \right] dy = 0, \quad (287)$$

the imaginary part of which is, if $c_i \neq 0$,

$$\int \bar{\rho}(|G'|^2 + k^2\,|G|^2)\,dy + \int \bar{\rho}(g\beta - \tfrac{1}{4}U'^2)\left|\frac{G}{W}\right|^2 dy = 0. \tag{288}$$

It follows immediately that if (283) is true everywhere, the flow is stable. If U' is nowhere equal to zero, we can define the Richardson number J as $g\beta/U'^2$ and Miles theorem can be stated thus:

If $J \geq \tfrac{1}{4}$ everywhere, the flow is stable.

8.4. Howard's Semicircle Theorem

If (281) is multiplied by F^* and the result is integrated between 0 and d, and if (282) is used whenever desirable, we have

$$g \int \bar{\rho}\beta\,|F|^2\,dy = \int \bar{\rho}(U - c)^2(|F'|^2 + k^2\,|F|^2)\,dy, \tag{289}$$

provided $c_i \neq 0$. The real part of (289) is

$$\int \bar{\rho}[(U - c_r)^2 - c_i^2](|F'|^2 + k^2\,|F^2|)\,dy - \int g\bar{\rho}\beta\,|F|^2\,dy = 0. \tag{290}$$

Howard (1961) took

$$Q = \bar{\rho}(|F'|^2 + k^2 |F|^2).$$

Then (289) and (290) become

$$\int UQ \, dy = c_r \int Q \, dy,$$

$$\int U^2Q \, dy = (c_r^2 + c_i^2) \int Q \, dy + \int g\bar{\rho}\beta \, |F|^2 \, dy = 0.$$

Subsequent development (very short) is similar to the several steps after (193), and Howard again proved (194), now also for a stratified fluid. Hence Howard's (1961) semicircle theorem:

> **The complex wave velocity c of any unstable mode of a disturbance in parallel flows of a stratified inviscid fluid must lie inside the semicircle in the upper half of the c plane, which has the range of U as its diameter.**

Noting that $|W|^{-2} \le c_i^{-2}$, Howard also obtained from (288) that

$$k^2 \int \bar{\rho} \, |G|^2 \, dy = \int \bar{\rho}(\tfrac{1}{4}U'^2 - g\beta) \left| \frac{G}{W} \right|^2 dy - \int \bar{\rho} \, |G'|^2 \, dy$$

$$\le \frac{1}{c_i^2} \max (\tfrac{1}{4}U'^2 - g\beta) \int \bar{\rho} \, |G|^2 \, dy,$$

from which $\qquad\qquad k^2c_i^2 \le \max (\tfrac{1}{4}U'^2 - g\beta).$ $\qquad\qquad$ (291)

This gives an upper bound[1] for the temporal rate of growth kc_i [Howard (1961)]. In fact, (291) also contains Miles' theorem.

We shall conclude this section by mentioning two results which the reader can easily verify for himself. If the mean flow is horizontal but its velocity varies not only in magnitude but also in direction from layer to layer, so that the velocity of the primary flow has the components $U(y)$, 0, $W(y)$, then (*a*) Squire's transformation can still be applied formally to reduce the problem for three-dimensional disturbances to the problem for two-dimensional disturbances, and (*b*) Howard's semicircle theorem still holds, provided, for any direction (in the xz plane) of propagation of the two-dimensional waves, the maximum and minimum of U in Howard's semicircle theorem presented in Secs. 6.3 and 8.4 are replaced by the maximum and minimum of the velocity of the primary flow in that particular direction. The second result is closely related to the first. Also, one conclusion that can be drawn from the second result is that the eigenvalues c for unstable waves propagating in all directions must lie in the semicircle $c_i > 0$, $|c| \le$ maximum speed of the mean flow.

[1] The result for the case $\beta = 0$ was obtained by Høiland (1953, p. 11) by a different method.

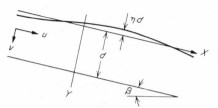

FIGURE 7. DEFINITION SKETCH FOR FLOW
DOWN AN INCLINED PLANE.

9. STABILITY OF A LIQUID LAYER FLOWING DOWN AN INCLINED PLANE

As an example of the instability due to variation of specific weight we shall consider an extreme case of stratification in which the specific weight ρg changes from that of the liquid to zero across the free surface. The flow is that of a layer of viscous liquid of depth d down an inclined plane with the angle of inclination β to the horizontal (Fig. 7). The stability under consideration here differs from that of plane Poiseuille flow between rigid plates in that here the presence of the free surface gives rise to a boundary condition that involves the eigenvalue c, which makes the problem very easy to solve for long waves. The problem was considered by various people in the middle of this century, including Kapitza (1948a and b, 1949), in an approximate way. The exact formulation of the problem was first given by Yih (1954), who demonstrated the existence of instability at low Reynolds numbers, but his eigenvalues were in error. Benjamin (1957) considered the problem anew by using a power-series expansion in y and obtained much more accurate results. Finally Yih (1963) gave a perturbation procedure which rendered the calculations much simpler, especially in the first stages, and Yuan (1966) used it to produce the error-free results, which he also obtained by using Benjamin's expansion in powers of y.

The velocity distribution in the flow under consideration, in the notation of Sec. 5 and in dimensionless form, is

$$U(y) = \tfrac{3}{2}(1 - y^2), \qquad (292)$$

in which U is expressed in terms of the average value $\bar{u}_a$ of the velocity $\bar{u}$ of the primary flow, which is equal to $(gd^2 \sin \beta)/3\nu$. The Reynolds number and Froude number are given by

$$R = \frac{\bar{u}_a d}{\nu} \quad \text{and} \quad F = \frac{\bar{u}_a}{(gd)^{1/2}}, \qquad (293)$$

so that $3F^2 = R \sin \beta.$

The governing differential equation is still the Orr-Sommerfeld equation (179). The boundary conditions at the bottom, where $y = 1$

according to the coordinates chosen, are the nonslip conditions

(i) $\phi(1) = 0$,

(ii) $\phi'(1) = 0$,   (294)

the prime again indicating differentiation with respect to y, with ϕ defined in (176). The condition for the shear stress at the free surface is

(iii) $\dfrac{\partial v_1}{\partial x} + \dfrac{\partial u_1}{\partial y} = 0$ at $y = \eta$,

and the condition that the normal stress at the free surface vanish is

(iv) $\left(-p_1 + \dfrac{2}{R}\dfrac{\partial v_1}{\partial y}\right)\rho \bar{u}_a{}^2 + T\dfrac{\partial^2(\eta d)}{\partial x_1{}^2} = 0$ at $y = \eta$

if ηd is the displacement of the free surface from its mean position and T again denotes surface tension. Condition (iv) can be simplified to

(iv) $-p_1 + \dfrac{2}{R}\dfrac{\partial v_1}{\partial y} + S\dfrac{\partial^2 \eta}{\partial x^2} = 0$,   with $S = \dfrac{T}{\rho\, d\bar{u}_a{}^2}$.

But (iii) and (iv) must be applied at $y = \eta$, not at $y = 0$. The evaluation of the perturbation quantities at $y = 0$ introduces errors only of higher orders, but the primary quantities may change from $y = 0$ to $y = \eta$, and these changes will be of the same order of magnitude as the perturbation quantities themselves. Thus, the (dimensionless) rate of shearing deformation at the free surface contributed by the mean flow and the displacement η is $U''\eta$, and the contribution to the normal stress at the free surface by the mean flow is $-P - P_y\eta$. Since

$$U'' = -3, \qquad P(0) = 0, \qquad \text{and} \qquad P_y(0) = \frac{\cos \beta}{F^2},$$

(iii) and (iv) can be written as

(iii) $-3\eta + \psi_{yy} - \psi_{xx} = 0$,

(iv) $\dfrac{\cos \beta}{F^2}\eta + p' + \dfrac{2}{R}\psi_{xy} - S\eta_{xx} = 0$.

Now at the free surface

$$\eta_\tau + U\eta_x = v = -\psi_x. \qquad (295)$$

With ϕ defined in (176), this becomes

$$\eta = \frac{\phi(0)}{c'}\exp\left[i\alpha(x - c\tau)\right], \qquad c' = c - \tfrac{3}{2}.$$

The boundary conditions (iii) and (iv) in their latest forms can then be applied at $y = 0$, and written in terms of ϕ as follows:

(iii) $\chi(0) \equiv \phi''(0) + \left(\alpha^2 - \dfrac{3}{c'}\right)\phi(0) = 0,$ \hfill (296)

(iv) $\theta(0) \equiv \left(\alpha \dfrac{3\cot\beta + \alpha^2 SR}{c'}\right)\phi(0)$

$$+ \alpha(Rc' + 3\alpha i)\phi'(0) - i\phi'''(0) = 0. \qquad (297)$$

We now proceed to solve the eigenvalue problem defined by (179), (294), (296), and (297). It turns out that for flows with a free surface or interfaces, where discontinuities of one kind or another occur, long waves are often the most unstable. Thus it is appropriate to expand the eigenfunction ϕ and eigenvalue c in powers of α:

$$\phi = \phi_0(y) + \alpha\phi_1(y) + \alpha^2\phi_2(y) + \cdots, \qquad (298)$$

$$c = c_0 + \alpha c_1 + \alpha^2 c_2 + \cdots. \qquad (299)$$

This was in effect done in a paper by Yih (1963). If (298) and (299) are substituted into the governing differential system and the various powers in α are separated out, it is found that ϕ_0 satisfies the system

$$\phi_0{}^{\text{iv}} = 0, \qquad (300)$$

(i) $\phi_0(1) = 0,$

(ii) $\phi_0'(1) = 0,$

(iii) $\phi_0''(0) - \dfrac{3}{c_0'}\phi_0(0) = 0,$

(iv) $\phi_0'''(0) = 0.$

The solution of (300) is

$$\phi_0(y) = A + By + Cy^2 + Dy^3,$$

and the bounday conditions determine c_0' and ϕ_0 to be

$$c_0' = \tfrac{3}{2} \qquad \text{or} \qquad c_0 = 3, \qquad (301)$$

and $$\phi_0 = A(1 - y)^2.$$

Since the solution of a homogeneous linear differential system is determined only up to a multiplicative constant, we shall take A to be unity *once and for all*. That is to say, we shall assume $\phi(0)$, not just $\phi_0(0)$, to be 1. This will have a bearing on the determination of ϕ_1, ϕ_2, etc. Thus

$$\phi_0 = (1 - y)^2. \qquad (302)$$

For the second approximation the governing differential system is

$$\phi_1^{iv} = iR[(U - c_0)\phi_0'' - U''\phi_0], \tag{303}$$

(i) $\phi_1(1) = 0$,

(ii) $\phi_1'(1) = 0$,

(iii) $\phi_1''(0) - 2\phi_1(0) + \dfrac{4}{3}c_1 = 0$,

(iv) $2\cot\beta\,\phi_0(0) + \frac{3}{2}R\phi_0'(0) - i\phi_1'''(0) = 0$.

In (iii) and (iv) $c_0' = \frac{3}{2}$ has been used. With $\phi_0(y)$ given by (302) and $c_0 = 3$, the solution of (303) is easily shown to be of the form

$$\phi_1 = \frac{-iRy^5}{20} + \Delta A + \Delta By + \Delta Cy^2 + \Delta Dy^3. \tag{304}$$

Since we have taken $\phi(0)$ to be 1 once and for all, $\Delta A = 0$, and requirement of (304) to satisfy the boundary conditions produces

$$c_i = i\left(\frac{6R}{5} - \cot\beta\right),$$

$$\phi_1 = \frac{iR}{20}(-y^5 + 10y^3 - 16y^2 + 7y) + \frac{i\cot\beta}{3}(-y^3 + 2y^2 - y). \tag{305}$$

Note that if ΔA is not taken to be zero, ϕ_1 will have the additional part $\Delta A(1 - y)^2$, which does not affect the c_i in (305). Inclusion of that additional part would eventually amount to modifying the value of $\phi(0)$ and nothing else. Thus the flow is unstable for very long waves if

$$\frac{6R}{5} > \cot\beta \tag{306}$$

and stable otherwise—a result found by Benjamin (1957) by expansion of ϕ in powers of y. The present approach is that of Yih (1963). Continuing in this way, Yuan (1966) found that

$$c_2 = -\tfrac{12}{7}R^2 + \tfrac{10}{7}R\cot\beta - 3, \tag{307}$$

$$c_3 = i\left[-\left(\frac{1,293}{224} + \frac{S}{3}\right)R + \frac{9}{5}\cot\beta - \frac{29,134,851}{9,609,600}R^3 \right.$$
$$\left. + \frac{17,363}{5,775}R^2\cot\beta - \tfrac{4}{15}R\cot^2\beta\right]. \tag{308}$$

The results (301), (305), (307), and (308) agree with an independent calculation by Benney (1966), after the difference in the definition of R and of the reference velocity is taken into account. Yuan also used Benjamin's (1957) method and obtained identical results. On comparing Yuan's (and Benney's) results with Benjamin's [Eqs. (4.12) and (4.13)], it is found that Benjamin's results corresponding to c_2 and c_3 contain

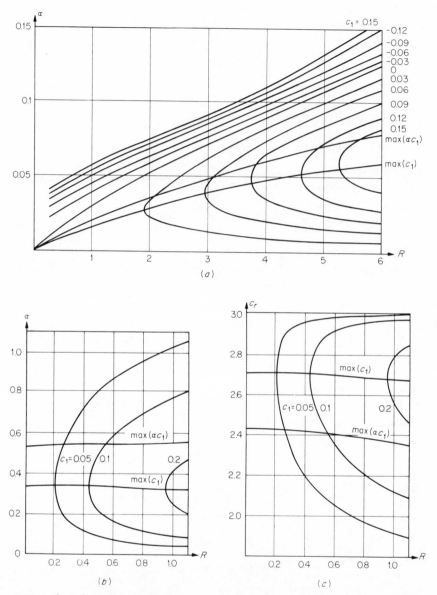

FIGURE 8. (a) CONSTANT-c_i CURVES FOR $\beta = 90°$, $\gamma \equiv \frac{2}{3}S(R^5 \sin \beta)^{1/3} = 3,000$, OBTAINED BY EXPANSION IN POWER SERIES OF αR; (b) CONSTANT-c_i CURVES FOR $\beta = 90°$, $\gamma = 0$; (c) VALUES OF c_r FOR UNSTABLE WAVES, WITH $\beta = 90°, \gamma = 0$. [M. Graef, Mitt. Max-Planck Inst. Strömunsgforschung Aeron. Versuchsaustalt, **26** (1966). *Courtesy of Dr. Graef*.]

several unimportant errors. Fortunately, his results corresponding to the c_0 and c_1 above are correct (if again the difference in the reference velocity is noted). Thus his main conclusion concerning the critical Reynolds number ($R_{cr} = \frac{5}{6} \cot \beta$) remains valid.

The form of (306) is interesting. Since the longitudinal component of the gravitational acceleration is $g \sin \beta$ and the transverse component $g \cos \beta$, (306) indicates that the latter is stabilizing and the former is the chief cause of instability. The kind of instability is different in nature from the instability of confined plane Poiseuille flow treated by Heisenberg (1924) and Lin (1945) in that here the free-surface waves occur at very small Reynolds numbers for β near $\pi/2$. The paradoxical nature of the case $\beta = \pi/2$, for which $R_{cr} = 0$, was explained by Yih [see Benjamin (1957, appendix) or Yih (1963)]. When $R_{cr} = 0$ and $\beta = \pi/2$, the state of neutral stability corresponds to a state of no motion but with surface corrugation. The vestiges of the kind of waves studied by Heisenberg, Tollmien, Schlichting, and Lin, as described in Sec. 7.1 were indicated in Yih (1963).

The foregoing analysis is for long waves. A numerical calculation was made by Graef (1961, 1966) using high-speed computers and without making the assumption that α is small. His results (1966) are given in Figs. 8 and 9.

10. INSTABILITY DUE TO VISCOSITY STRATIFICATION

It has long been known that plane Couette flow is stable for all Reynolds numbers and that confined plane Poiseuille flow is unstable only at high Reynolds numbers provided the fluid is homogeneous. Section 9 has demonstrated that plane Poiseuille flow with a free surface can be unstable at low Reynolds numbers when the slope is steep. The instability is due to a stratification (in an extreme way) in the specific weight ρg and to the longitudinal component of gravity. In this section it will be shown that a stratification in viscosity can also be destabilizing [Yih (1967a)].

Consider, for simplicity, two superposed liquid layers (Fig. 10) with the interface, when undisturbed, at $y = 0$. These layers are each of depth d and confined between rigid horizontal boundaries at $y = \pm 1$, where again y is measured in units of d. The flow is caused by a pressure gradient $-K$ in the x direction, and the rigid boundaries are stationary, so that the flow is a plane Poiseuille flow of a stratified fluid. The densities of the two layers may be different, and if the lower layer has the greater density, the density difference in the presence of gravity is stabilizing, although it can be destabilizing if the boundaries are inclined. In any case the effect of the difference in ρ or in $g\rho$ can be dealt with if desired. But here we wish to demonstrate how a difference in viscosity between the two layers can be the cause of instability. To isolate this cause and to

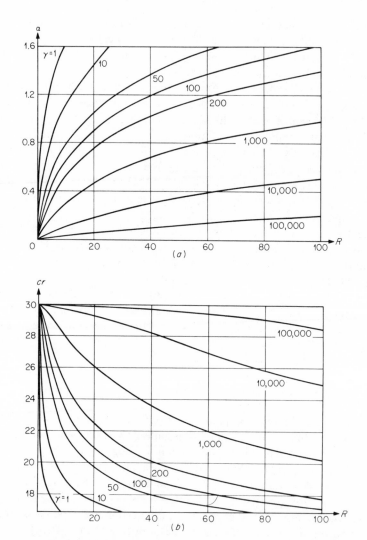

FIGURE 9. (a) NEUTRAL STABILITY CURVES FOR VARIOUS VALUES OF γ AND (b) VALUES OF c_r FOR NEUTRAL WAVES AND VARIOUS VALUES OF R AND γ. $β = 90°$ FOR BOTH (a) AND (b). [M. Graef, Mitt. Max-Planck Inst. Strömungsforschung Aeron. Versuchsaustalt, **26** (1966). Courtesy of Dr. Graef.]

simplify matters, the density of the two layers will be assumed the same. If the subscripts 1 and 2 are associated respectively with the upper and lower layers, we shall use the abbreviations

$$m = \frac{\mu_2}{\mu_1}, \qquad U_1 = \frac{\bar{u}_1}{\bar{u}(0)}, \qquad U_2 = \frac{\bar{u}_2}{\bar{u}(0)}, \qquad (309)$$

in which $\bar{u}(y)$ is the velocity of the mean flow and $\bar{u}(0)$ is used as the

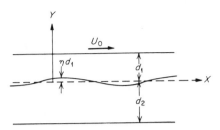

FIGURE 10. DEFINITION SKETCH FOR THE PROBLEM OF STABILITY OF A FLOW WITH VISCOSITY STRATIFICATION.

reference velocity. The Reynolds number is then

$$R = \frac{\rho \bar{u}(0) d}{\mu_1}. \tag{310}$$

It is an easy matter to show that

$$U_1 = 1 + a_1 y + b_1 y^2, \tag{311}$$

$$U_2 = 1 + a_2 y + b_2 y^2, \tag{312}$$

in which $\quad a_1 = \frac{1}{2}(m-1), \qquad b_1 = -\frac{1}{2}(m+1), \tag{313}$

$$a_2 = \frac{1}{2}\left(1 - \frac{1}{m}\right), \qquad b_2 = -\frac{1}{2}\left(1 + \frac{1}{m}\right).$$

The reference velocity $\bar{u}(0)$ is connected with K by

$$\bar{u}(0) = \frac{m}{1+m} \frac{Kd^2}{\mu_2}. \tag{314}$$

The Orr-Sommerfeld equations for the two layers are, respectively,

$$\phi^{\text{iv}} - 2\alpha^2 \phi'' + \alpha^4 \phi = i\alpha R[(U_1 - c)(\phi'' - \alpha^2 \phi) - U_1'' \phi], \tag{315}$$

$$\chi^{\text{iv}} - 2\alpha^2 \chi'' + \alpha^4 \chi = i\alpha R m^{-1}[(U_2 - c)(\chi'' - \alpha^2 \chi) - U_2'' \chi], \tag{316}$$

in which the χ is used for ϕ for the lower layer, to avoid confusion.

The boundary conditions are formulated in much the same way as in Sec. 9. The nonslip conditions at the solid boundaries are

$$\phi(1) = 0 = \phi'(1), \qquad \chi(-1) = 0 = \chi'(-1). \tag{317}$$

The continuity in v' demands that

$$\phi(0) = \chi(0), \tag{318}$$

and the continuity in u' demands that

$$\phi'(0) - \chi'(0) = \frac{\phi(0)}{c'}(a_2 - a_1), \qquad c' = c - 1. \tag{319}$$

Condition (319) is very important, for it turns out to be the only condition that involves c for vanishing α. The continuity in the shear stress, in

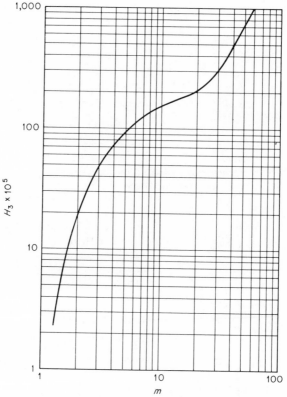

FIGURE 11.   INSTABILITY OF PLANE POISEUILLE FLOW OF TWO
SUPERPOSED LIQUID LAYERS OF EQUAL DEPTH AND DENSITY BUT
OF DIFFERENT VISCOSITY. $m = \mu_2/\mu_1$, $\mu_2 =$ VISCOSITY OF LOWER
LAYER, $\mu_1 =$ VISCOSITY OF UPPER LAYER. H_3 IS PROPORTIONAL
TO THE RATE OF GROWTH, ACCORDING TO (323). [*From C.-S.
Yih, J. Fluid Mech.,* **27** (1967), *by permission of the Cambridge
University Press.*]

contradistinction to that condition in Sec. 9, now does not involve c,
because the mean shear gradient is continuous at the interface. It is

$$\phi''(0) + \alpha^2\phi(0) = m[\chi''(0) + \alpha^2\chi(0)]. \qquad (320)$$

Finally, the continuity of the normal stress at the interface demands that

$$m(\chi''' - 3\alpha^2\chi') - (\phi''' - 3\alpha^2\phi) = \frac{i\alpha^3 RS\phi}{c'} \qquad \text{at } y = 0, \qquad (321)$$

in which $\qquad\qquad S = \dfrac{T}{\rho\,dV^2}, \qquad V = \bar{u}(0). \qquad\qquad (322)$

Using the method presented in Sec. 9, we obtain

$$c_0 = 1 + 2(m - 1)^2(m^2 + 14m + 1)^{-1},$$
$$c_1 = i8RH_3, \qquad\qquad (323)$$

where H_3, which is a function of m only, can be given explicitly in terms of
m but is given only graphically in Fig. 11.   It is seen that for the values of

m covered by the graph—and in fact for the reciprocals of these values—the flow is unstable against long waves for any Reynolds number. The analysis presented here can also be applied to investigate the stability of plane Couette flow of two superposed layers [Yih (1966)]. That flow can also be unstable at low Reynolds numbers. This is rather unexpected because the broken-line profile has been shown by Rayleigh (1894, pp. 384–392) to be stable for inviscid fluids and because plane Couette flow for a single fluid is known to be stable at all Reynolds numbers. The mathematical source of the instability of plane Poiseuille and plane Couette flows of superposed fluids is (319), which enables one to find c_0 at vanishing α and to proceed from there to find c_1, c_2, etc. From a physical point of view, the discontinuity in the vorticity of the primary flow makes it possible for sinusoidal waves to draw energy from the mean flow and thus gives rise to a kind of instability nonexistent in a homogeneous fluid. In a way Taylor's explanation (1915) for the instability of a viscous fluid which is nonexistent if it is considered inviscid still applies here, but it is now at the interface that the nonslip condition really makes things different, not at the solid boundaries, since the Reynolds stresses born at the solid boundaries, so to speak, either cannot cause instability (plane Couette flow) or can do so only at high Reynolds numbers (plane Poiseuille flow). What is being discussed in this section is predominantly an interfacial wave caused by conditions prevailing near the interface. For a discussion of the limiting cases $m = 0$ and $m = \infty$, the reader is referred to the original paper of Yih (1966).

11. SECULAR INSTABILITY DUE TO PERIODIC EXCITATIONS

A fluid contained in an open vessel set in vertical solid-body translation by a periodic force may be unstable if its surface is slightly disturbed. A layer of liquid caused to flow by its lower boundary moving periodically in its own plane may be unstable against surface waves. Laminar flows in a pipe caused by a periodic pressure gradient may be much more unstable than a steady flow in the same pipe. The kind of instability is akin to the instability of the oscillation of a swing when a child excites it by periodic motion of his body coupled in some "advantageous" way to the oscillation, or it is akin to the production of transverse waves in an elastic string fixed at one end and periodically pulled at the other. Since the flow (or the pressure, in the case of solid-body motion) is periodic with respect to time, whether a disturbance is growing or not can be decided only after at least one period of the excitation. The theory involves a differential system with coefficients periodic in time, and is closely related to the classical theory of Hill, Mathieu, Floquet, Liapounov, and Poincaré, who dealt with ordinary differential systems. The system governing many stability problems, however, may involve a partial differential equation.

Since the instability to be discussed is an instability over a long period of time (at least one period of the exciting motion), we shall call it *secular instability* because the Latin root *saeculum* means age, century, or generation. In fact, the term *secular equation*, used before in this chapter, may have its origin in the study of the secular, or long-term, motion of the moon by Hill.

Two representative cases will be discussed. The first deals with an inviscid fluid with a free surface when it is in vertical motion like a solid body. The second deals with the instability of a viscous fluid in periodic motion.

11.1. Instability of the Free Surface of a Liquid in Vertical Periodic Motion

The author can still remember vividly the regular pattern of standing waves on the surface of water in open oil drums left on the decks of steamers plying the waters of the upper Yangtze. These steamers were exceptionally well constructed to cope with the rapid streams of the famous Yangtze Gorges, but the vibrations were quite noticeable, as attested by the wave pattern mentioned above. At the time not only was the existence of these waves incompletely explained, but their frequency remained a matter of some controversy.

Early in the nineteenth century Faraday (1831) experimented with water in a vertically vibrating vessel and observed that the frequency of the waves that appeared on the surface of the water was one-half that of the vibrating vessel. Later, Matthiessen (1868, 1870) observed that the two frequencies were the same. The disagreement led Rayleigh (1883) to perform some experiments. They confirmed Faraday's finding, and, as mentioned by Benjamin and Ursell (1954), Rayleigh already had some idea why the waves might have half the frequency of the vessel. At that time the Mathieu equation, to which many problems of secular instability lead, had not been sufficiently studied. By the time Benjamin and Ursell took up the question again, the Mathieu equation had been well developed, and they were able to give a proper treatment of the matter and thereby resolved a disagreement which should not have existed in the first place.

The equations of motion relative to the vibrating container are (21). If viscosity is neglected, (21) can be written as

$$\rho\left(\frac{\partial u_i}{\partial t} + u_\alpha \frac{\partial u_i}{\partial x_\alpha}\right) = -\frac{\partial p}{\partial x_i} + \rho(X_i - a_i),$$

in which

$$X_1 = 0 = a_1, \qquad X_2 = 0 = a_2, \qquad X_3 = -g, \qquad a_3 = a \cos \omega t$$

if the direction of the gravitational acceleration is the direction of decreasing x_3 (Fig. 12) and if the vessel vibrates vertically with amplitude a and

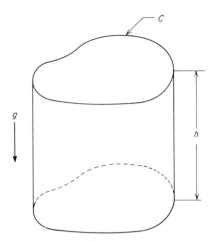

FIGURE 12. DEFINITION SKETCH FOR FREE-
SURFACE INSTABILITY OF A LIQUID IN AN
OSCILLATING CONTAINER.

angular frequency ω. Since the virtual body force $(0, 0, -g - a \cos \omega t)$ is independent of spatial coordinates and therefore conservative, i.e., its curl is zero, and since ρ is constant, irrotationality will persist. If we assume the motion due to any disturbance to have started from rest,

$$u_i = \frac{\partial \phi}{\partial x_i},$$
(324)

in which ϕ is the velocity potential. The Bernoulli equation is

$$\frac{\partial \phi}{\partial t} + \frac{p}{\rho} + \frac{u_\alpha u_\alpha}{2} + (g + a \cos \omega t)x_3 = F(t).$$

Since p on the free surface is given by Eq. (5.6), the free-surface condition in the linearized form is

$$-\frac{T}{\rho}\left(\frac{\partial^2 \zeta}{\partial x_1^2} + \frac{\partial^2 \zeta}{\partial x_2^2}\right) + \frac{\partial \phi}{\partial t} + (g + a \cos \omega t)\zeta = 0$$
(325)

if ζ is the displacement of the free surface from its mean position, so that

$$\frac{\partial \phi}{\partial x_3} = \frac{\partial \zeta}{\partial t}.$$
(326)

If the origin is taken at the bottom of the vessel,

$$\frac{\partial \phi}{\partial x_3} = 0 \qquad \text{at } x_3 = 0.$$
(327)

On the lateral wall, assumed vertical,

$$\frac{\partial \phi}{\partial n} = 0, \qquad \text{for all values of } x_3, \tag{328}$$

in which n is the normal distance from the wall. This condition demands that ϕ have a factor $S_m(x_1,x_2)$ satisfying

$$\left(\frac{\partial^2}{\partial x_1^2} + \frac{\partial^2}{\partial x_2^2} + k_m^2\right) S_m(x_1,x_2) = 0,$$

in which k_m^2 must assume certain (countably infinite) eigenvalues in order that

$$\frac{\partial S_m}{\partial n} = 0. \tag{329}$$

Benjamin and Ursell took

$$\phi(x_1,x_2,x_3,t) = \sum_{m=1}^{\infty} \frac{da_m(t)}{dt} \frac{\cosh k_m x_3}{k_m \sinh k_m h} S_m(x_1,x_2) + G(t) \tag{330}$$

and

$$\zeta = \sum_0^{\infty} a_m(t) S_m(x_1,x_2), \tag{331}$$

which satisfy (326) to (328). Note that $k_0^2 = 0$ is an eigenvalue and the corresponding eigenfunction $S_0(x_1,x_2)$ is simply a constant. Since, with A indicating the cross-sectional area of the container (in a horizontal plane) and C its boundary, along which s is measured, for any $m \neq 0$ we have

$$\int_A S_m(x_1,x_2)\, dA = -\frac{1}{k_m^2} \int_A \nabla^2 S_m\, dA = -\frac{1}{k_m^2} \int_C \frac{\partial S_m}{\partial n}\, ds = 0,$$

by virtue of Green's theorem, and since the definition of ζ requires that its integral over A be zero at all times, $a_0(t)$ in (331) must be zero. A glance at (325) then reveals that the $G(t)$ in (330) can at most be a constant, which we shall assign the value zero without loss of generality. When (330) and (331) are substituted into (325), the result is

$$\sum_{m=1}^{\infty} \frac{S_m(x_1,x_2)}{k_m \tanh k_m h} \left[\frac{d^2 a_m}{dt^2} + k_m \tanh k_m h \left(\frac{k_m^2 T}{\rho} + g + a \cos \omega t\right) a_m \right] = 0. \tag{332}$$

Since the functions are linearly independent, the quantity within the brackets in (332) must vanish, or, with

$$p_m = \frac{4\omega_m^2}{\omega^2}, \qquad q_m = \frac{2ak_m}{\omega^2} \tanh k_m h,$$

$$\omega_m^2 = k_m \tanh k_m h \left(\frac{T}{\rho} k_m^2 + g\right), \qquad \text{and} \qquad \tau = \tfrac{1}{2}\omega t + \frac{\pi}{2},$$

we have

$$\frac{d^2 a_m}{d\tau^2} + (p_m - 2q_m \cos 2\tau) a_m = 0, \tag{333}$$

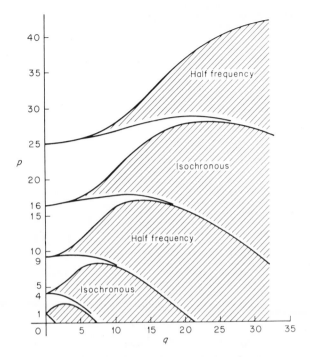

FIGURE 13. THE MATHIEU DIAGRAM.

which is the canonical form of the Mathieu equation. The stability diagram in the pq space for the solution of this equation is shown in Fig. 13.

From Fig. 13 it can be seen that for quite small values of q, hence also of a, the free surface can be unstable. Since $g + a \cos \omega t$ in that case will at no time be negative, the instability is evidently different from the instability due to a downward acceleration greater than g (Taylor instability). The result just obtained also indicates how wrong one would be if one adopted the point of view that the stability or instability of the system could be decided by considering the stability of the system at all moments, at each of which the acceleration of the vessel is treated as if it were constant, and that if the system be stable at all moments, it would be stable. For lack of a truly better term we shall call such an approach the *time-section approach*. The time-section approach is inadequate for dealing with secular instability.

Still, the result just obtained begs for an explanation. We shall offer a qualitative one here based on stepwise variation of the acceleration. So long as the acceleration a_3 is constant, the concept of potential energy with respect to the vessel is still valid. This potential energy is based on the virtual gravity $g + a_3$. Suppose that at time $t = 0$ the acceleration is A

directed upward, and the waves attain their full amplitude. After a while the free surface, according to a linear theory at any rate, will be flat. At that instant let the upward acceleration be decreased to $B < A$ algebraically. The total energy, at this instant consisting entirely of the kinetic energy, is unchanged by this decrease in a_3, but as the waves again attain their full amplitude a little later, the amplitude has to be greater in order to have the same total energy, now entirely potential in kind, since the virtual gravity $g + a_3$ has decreased. This also explains why it is possible to have instability when the frequency of the waves is only half that of the vessel. The wave frequency does vary with a_3, so that the explanation just given is only qualitative. However the effect of a small a_3 on wave frequency is small. Of course a variation of a_3 opposite to that just described would be stabilizing; but for a given variation of a_3 we have to consider waves with all time phases, and if waves with any particular time phase is unstable, the fluid is unstable. Even though the explanation given here is based on a very special stepwise schedule for a_3 and cannot replace the exact mathematical analysis, it does somehow satisfy our very natural wonder at the way a little periodic acceleration manages to destabilize the free surface.

11.2. Secular Instability of the Free Surface of a Liquid Layer Set in Periodic Motion by a Lower Boundary Oscillating in Its Own Plane

As another example of secular instability [Yih (1968b)], consider the flow of a viscous fluid of depth d with a free surface due to the horizontal oscillation of its lower boundary with the velocity $V \cos \omega_* t$. Using x, y, and τ as defined in (156), with d as the reference length, and writing

$$U(y,\tau) = \frac{\bar{u}(y,\tau)}{V} \tag{334}$$

in accordance with (156) and (160a), we have

$$\frac{\partial U}{\partial \tau} = \frac{1}{R} \frac{\partial^2 U}{\partial y^2}, \tag{335}$$

in which $R = Vd/\nu$. In (334), $\bar{u}$ is the dimensional velocity of the primary flow. The boundary conditions for U are

$$U(1,\tau) = \cos \omega_* t = \cos \omega \tau \quad \text{with } \omega = \frac{\omega_* d}{V},$$

and

$$\frac{\partial U}{\partial y} = 0 \quad \text{at } y = 0$$

if the origin of the coordinates is at the free surface and the coordinates are directed as shown in Fig. 14. The solution for U is

$$U = A[W + W^* - i \tanh \beta \tan \beta (W - W^*)], \tag{336}$$

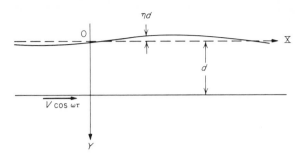

FIGURE 14. DEFINITION SKETCH FOR INSTABILITY OF THE FREE-SURFACE OF A LIQUID LAYER CAUSED TO FLOW BY AN OSCILLATING PLANE.

in which

$$W = \cosh\,[\beta(1+i)y]e^{i\omega t}, \qquad A = \frac{\cos\beta\cosh\beta}{2(\cos^2\beta + \sinh^2\beta)}, \quad \beta = \left(\frac{\omega R}{2}\right)^{1/2},$$

$$(337)$$

and the asterisk denotes the complex conjugate. The mean pressure is hydrostatic:

$$\bar{p} = g\rho y d.$$

Since U is now time-dependent, the stream function ψ and the perturbation pressure can no longer have their time dependence embodied exclusively in the exponential time factor. However, we can assume

$$\psi = \phi(y,\tau)e^{i\alpha x}, \qquad p' = f(y,\tau)e^{i\alpha x}. \tag{338}$$

The Orr-Sommerfeld equation now has the form

$$R\left[\left(\frac{\partial}{\partial\tau} + i\alpha U\right)(\phi'' - \alpha^2\phi) - i\alpha U''\phi\right] = \phi^{\text{iv}} - 2\alpha^2\phi'' + \alpha^4\phi \tag{339}$$

if primes still denote differentiation with respect to y. The nonslip conditions are

(i) $\phi(1,\tau) = 0,$

(ii) $\phi'(1,\tau) = 0.$

$$(340)$$

With

$$\eta = h(\tau)e^{i\alpha x}, \tag{341}$$

the equation corresponding to (295) is

$$\left[\frac{d}{d\tau} + i\alpha U(0,\tau)\right]h = -i\alpha\phi(0,\tau). \tag{342}$$

518   HYDRODYNAMIC STABILITY

The formulation of the free-surface conditions is similar to that presented in Sec. 9. These conditions are

(iii) The condition of zero shear stress:

$$U''(0,\tau)h + \phi''(0,\tau) + \alpha^2\phi(0,\tau) = 0. \tag{343}$$

(iv) The condition of zero normal stress:

$$i\alpha(F^{-2} + S\alpha^2)h + \frac{1}{R}(\phi''' - 3\alpha^2\phi') - \phi'_\tau - i\alpha U\phi' = 0, \tag{344}$$

in which
$$F^2 = \frac{V^2}{gd}, \qquad S = \frac{T}{\rho V^2 d},$$

F being the Froude number. The differential system governing stability consists of (339), (342), (340), (343), and (344).

All these equations have coefficients periodic in τ, and it can be shown that the solutions are of the Floquet form [Yih (1968b)]

$$\phi(y,\tau) = e^{\mu_1\tau}\chi(y,\tau),$$
$$h(\tau) = e^{\mu_1\tau}H(\tau), \tag{345}$$

in which μ_1 is a constant and χ and H are either periodic in τ with period τ_0 defined below or, in case a certain characteristic equation formable from the differential system has multiple roots, can involve polynomials as well as periodic functions in τ. In most cases χ and H are periodic in τ. In such cases (345) can perhaps be understood in the following way. If $\phi(y,\tau)$ and $h(\tau)$ are solutions of the system, so are $\phi(y, \tau + n\tau_0)$ and $h(\tau + n\tau_0)$ if n is an integer and

$$\tau_0 = \frac{2\pi}{\omega}$$

is the (dimensionless) period of the flow and hence of the coefficients of the differential system. At time τ, both $\phi(y,\tau)$ and $h(\tau)$, on the one hand, and $\phi(y,\tau + n\tau_0)$ and $h(\tau + n\tau_0)$, on the other, are solutions of the system. If we assume the system to have separable solutions or modes, then for each mode the latter, being time continuations of the former, must be, at time τ, proportional to them. If we take $n = 1$ for the moment and write

$$\phi(y, \tau + \tau_0) = e^{\mu_1\tau_0}\phi(y,\tau) \tag{346}$$

and
$$h(\tau + \tau_0) = e^{\mu_1\tau_0}h(\tau), \tag{347}$$

so that the constant of proportionality is $e^{\mu_1\tau_0}$, obviously by repeated application $\phi(y, \tau + n\tau_0)$ and $h(\tau + n\tau_0)$ are proportional to $\phi(y,\tau)$ and $h(\tau)$. Equations (346) and (347) state that

$$\phi(y, \tau + \tau_0)e^{-\mu_1(\tau+\tau_0)} = \phi(y,\tau)e^{-\mu_1\tau},$$
$$h(\tau + \tau_0)e^{-\mu_1(\tau+\tau_0)} = h(\tau)e^{-\mu_1\tau},$$

which means that

$$\phi(y,\tau)e^{-\mu_1\tau} \quad \text{and} \quad h(\tau)e^{-\mu_1\tau}$$

are periodic with period τ_0. Hence (345). We expect complications to arise if several modes coalesce; but this happens only if the characteristic equation mentioned above (and determined from the governing *partial* differential system) has multiple roots. If that happens, each of $\chi(y,\tau)$ and $H(\tau)$ has several forms, all but one of which involve polynomials of τ as well as periodic functions of τ. There is always one form of $\chi(y,\tau)$ and of $H(\tau)$ which is periodic in τ, and so we can treat $\chi(y,\tau)$ and $H(\tau)$ as periodic in τ and need only keep a watch over the possibility of degenerate cases (resulting from coalescence of modes), and then only for the neutral stability case, for if the real part of μ_1 is not zero the exponential growth or decay overshadows any polynomial growth.

We may reasonably expect to find long waves most unstable. The reason is as follows. If we let the plate accelerate with a constant acceleration a to the left and move the coordinate system with it, the latter will be a noninertial system in which there is a body force per unit mass equal to a and directed to the right. The problem then becomes the same as that for a liquid layer flowing down an inclined plane with the angle of inclination β determined from

$$\tan \beta = \frac{a}{g}.$$

The latter problem has already been treated in Sec. 9, where it was shown that long waves are most unstable. Although the actual acceleration of the lower boundary is not constant but is periodic, long waves are likely to remain least stable.

Thus we assume

$$\chi(\tau,y) = \phi_0(y,\tau) + \alpha\phi_1(y,\tau) + \alpha^2\phi_2(y,\tau) + \cdots,$$
$$H(\tau) = h_0(\tau) + \alpha h_1(\tau) + \alpha^2 h_2(\tau) + \cdots, \tag{348}$$
$$\mu_1 = \theta_0 + \alpha\theta_1 + \alpha^2\theta_2 + \cdots,$$

in which the ϕ's and h's are all periodic in τ and the θ's are constants. If (348) is substituted into (339) and (342) and terms of equal powers in α are collected, we obtain two series of equations in y and τ. The first of the series originating from (342) is

$$\frac{dh_0}{d\tau} + \theta_0 h_0 = 0.$$

Since h_0 must be periodic in τ, it follows immediately that

$$\theta_0 = 0,$$

and without loss of generality

$$h_0 = 1. \tag{349}$$

If θ_0 is imaginary, it must be multiples of $i\omega$, but then from (345) and (348) it can be seen that no new result is obtained. The other possibility is that $h_0 = 0$. But this would make $H(\tau) = O(\alpha)$, or the surface nearly flat for long waves. Judging from the results of a study [Yih (1963, pp. 331–334)] of the analogous flow of a liquid layer down an inclined plane, we can assume such waves to be much more stable than those for which $h_0 = 1$. Then for the first approximation the differential system is

$$R \frac{\partial}{\partial \tau} \phi_0'' = \phi_0^{\text{iv}}, \tag{350}$$

(ia) $\phi_0(1,\tau) = 0$,

(ii$a$) $\phi_0'(1,\tau) = 0$,

(iii$a$) $U''(0,\tau) + \phi_0''(0,\tau) = 0$,

(iv$a$) $\phi_0''' - R\phi_\tau' = 0$.

The solution of this system is

$$\phi_0 = -U(y,\tau) + B_0(\tau) + D(V + V^*) + iE(V - V^*), \tag{351}$$

in which

$$V(\tau) = \sinh \left[\beta(1 + i)y\right]e^{i\omega\tau}, \qquad B_0(\tau) = b_0 e^{i\omega\tau} + b_0^* e^{-i\omega\tau},$$

$$b_0 = \frac{A(1 - i\gamma)}{\cosh \beta(1 + i)}, \qquad \gamma = \tanh \beta \, \tan \beta,$$

$$D + iE = A(1 - i\gamma) \tanh \left[\beta(1 + i)\right],$$

D and E being real numbers. The τ dependence of ϕ_0 is dictated by (iii$a$).

For the next approximation, the second equation of the series corresponding to (342) is

$$\frac{dh_1}{d\tau} + \theta_1 = -iB_0(\tau).$$

Since B_0 contains only $\sin \omega t$ and $\cos \omega\tau$, and since h_1 is sinusoidally periodic with time, i.e., without a constant term,

$$\theta_1 = 0,$$

and
$$h_1 = -i \int B_0(\tau) \, d\tau = -\frac{1}{\omega} (b_0 e^{i\omega\tau} - b_0^* e^{-i\omega\tau}), \tag{352}$$

the constant of integration being chosen to be zero so that the term independent of τ in h is $h_0 \; (= 1)$ once and for all. This practice will not at all affect the stability criterion, and will be followed in calculating h_2. Equation (352) shows that even at the second approximation no instability is manifested. This is perhaps to be expected since the mean flow is periodic.

For ϕ_1, the governing system is, since $\theta_1 = 0$,

$$iR(U\phi_0'' - U''\phi_0) = \phi_1^{iv} - R\frac{\partial}{\partial\tau}\phi_1'',\tag{353}$$

(ib) $\phi_1(1,\tau) = 0$,

(ii$b$) $\phi_1'(0,\tau) = 0$,

(iii$b$) $U''(0,\tau)h_1 + \phi_1''(0,\tau) = 0$,

(iv$b$) $iF^{-2} + \dfrac{1}{R}\phi_1''' - \dfrac{\partial}{\partial\tau}\phi_1' - iU\phi_0' = 0$.

The τ dependence of ϕ_1 is dictated by this system. Instead of terms containing $\exp(i\omega\tau)$ and $\exp(-i\omega\tau)$, those containing $\exp(\pm i2\omega\tau)$ or no τ at all must be used. The particular solution of (353) is, with integration with respect to y understood,

$$\phi_{1p} = iR\left\{\frac{1}{2}\iint\left[\left(\phi_0'\int U\right) - U'\int(\phi_0 - B_0)\right] - J\right\},\tag{354}$$

in which J satisfies the equation

$$J^{iv} - R\frac{\partial}{\partial\tau}J'' = B_0U''.\tag{355}$$

The calculation of ϕ_{1p} involves some straightforward integrations. The result is

$$\phi_{1p} = iR(I_0 + I_1 + I_2 + I_0^* + I_1^* + I_2^*),\tag{356}$$

in which $I_0 = \dfrac{iA^2(1 + \gamma^2)}{4\omega R}\tanh[\beta(1 + i)](\sinh 2\beta y + i\sin 2\beta y)$,   (357)

$$I_1 = \frac{iA^2(1 + \gamma^2)}{\omega R\cosh[\beta(1 - i)]}\cosh[\beta(1 + i)y],\tag{358}$$

$$I_2 = -\frac{iA^2(1 - i\gamma)^2}{\omega R\cosh[\beta(1 + i)]}\cosh[\beta(1 + i)y]e^{i2\omega\tau}.\tag{359}$$

The complementary solution is

$$\begin{aligned}\phi_{1c} = A_1 + B_1y + C_1y^2 + D_1y^3 + E_1V_1\\+ F_1W_1 + E_1^*V_1^* + F_1^*W_1^* + G(\tau) + K(\tau)y,\end{aligned}\tag{360}$$

in which the eight coefficients are constants and

$$V_1 = \sinh[\beta(1 + i)y]e^{i2\omega\tau}, \qquad W_1 = \cosh[\beta(1 + i)y]e^{i2\omega\tau},$$

$$G(\tau) = g_1e^{i2\omega\tau} + g_1^*e^{-i2\omega\tau}, \qquad K(\tau) = k_1e^{i2\omega\tau} + k_1^*e^{-i2\omega\tau}.$$

Examination of (342) reveals that for prediction of stability for long waves, only the time-independent terms in

$$\phi_1 = \phi_{1p} + \phi_{1c}\tag{361}$$

and in $U(\theta,\tau)h_1$ need be calculated. For higher approximations at larger α the terms containing I_2, E_1, F_1, $G(\tau)$, and $K(\tau)$ should be included, but not at this stage. The time-independent part of ϕ_1 is

$$\Phi(y) = iR(I_0 + I_0^* + I_1 + I_1^*) + A_1 + B_1 y + C_1 y^2 + D_1 y^3, \quad (362)$$

which should satisfy the conditions

(ic) $\Phi(1) = 0$,

(iic) $\Phi'(1) = 0$,

(iiic) $\Phi''(0) + iA^2(1 + \gamma^2)R[\text{sech}\ \beta(1 + i) + \text{sech}\ \beta(1 - i)] = 0$,

(ivc) $iF^{-2} + \dfrac{1}{R}\Phi''' - iA^2\beta(1 + \gamma^2)[(1 + i)\tanh \beta(1 + i)$

$$+ (1 - i)\tanh \beta(1 - i)] = 0.$$

The solution is given by the equations

$$C_1 = 0,$$

$$D_1 = -\frac{i}{6}RF^{-2},$$

$$A_1 = 2D_1 - \frac{i2A^2(1 + \gamma^2)}{\omega}(R_1 - R_2), \qquad (363)$$

$$B_1 = -3D_1 - \frac{iRA^2(1 + \gamma^2)}{\beta}R_2,$$

in which R_1 and R_2 are respectively the real part of

$$\frac{i}{4}\tanh\ [\beta(1 + i)](\sinh 2\beta + i \sin 2\beta) + \frac{i}{\cosh \beta(1 - i)}\cosh \beta(1 + i),$$

$$\frac{i\beta}{2}\tanh\ [\beta(1 + i)](\cosh 2\beta + i \cos 2\beta) - \frac{2\beta}{(1 + i)\cosh \beta(1 - i)}\sinh \beta(1 + i).$$

The third equation of the series corresponding to (342) is

$$\frac{dh_2}{d\tau} = -\theta_2 - i[U(0,\tau)h_1 + \phi_1(0,\tau)]. \qquad (364)$$

The term in $U(0,\tau)h_1$ which is independent of τ is

$$-\frac{i2A^2(1 + \gamma^2)}{\omega}I_3 \qquad (365)$$

if I_3 denotes Im [sech $\beta(1 + i)$]. The part of $\phi_1(0,\tau)$ independent of τ is

$$\Phi(0) = A_1 + \frac{i2A^2(1 + \gamma^2)}{\omega}I_3. \qquad (366)$$

Thus the time-independent part on the right-hand side of (364) is simply

$-\theta_2 - iA_1$, and this must be zero since h_2 is periodic in τ. Hence

$$\theta_2 = -iA_1,$$

and the criterion sought is that the flow is unstable or stable according as $-iA_1$ is positive or negative, or according as

$$\frac{6A^2(1 + \gamma^2)}{\omega R} (R_2 - R_1) > \text{ or } < F^{-2}. \tag{367}$$

Table 2 shows the variation of L, the left-hand side of (367), with β. The maximum value of L is 0.279. When F^{-2} is greater than this value, there is stability. The data in Table 2 are shown graphically in Fig. 15. If L is positive, there can be instability even for small Reynolds numbers provided F is sufficiently large. It is interesting that for certain ranges of β the value of L is negative. In these ranges the motion stabilizes the free surface against the formation of long waves. For very small frequencies such that $\omega R = 2\beta^2 \ll 1$, the criterion derived from (367) is that the flow is unstable or stable according as

$$\tfrac{4}{5}A^2\omega^2R^2 > \text{ or } < F^{-2}, \tag{367a}$$

or, since $A = \tfrac{1}{2}$ for small β, according as

$$R'^2 > \text{ or } < 5F^{-2}, \tag{367b}$$

in which R' is the Reynolds number $\omega_* d^2/\nu$.

TABLE 2 L as function[†] of β

β	$L \times 10^n$	n	β	$L \times 10^n$	n
0.10	8.0042	5	3.20	-2.4914	3
0.20	1.2772	3	3.40	-1.7888	3
0.30	6.4093	3	3.60	-1.1007	3
0.40	1.9785	2	3.80	-5.7129	4
0.50	4.6005	2	4.00	-2.2517	4
0.60	8.7539	2	4.20	-3.1780	5
0.80	1.9975	1	4.40	5.5388	5
1.00	2.7859	1	4.60	7.9145	5
1.20	2.7075	1	4.80	7.1305	5
1.40	2.0792	1	5.00	5.2075	5
1.60	1.3790	1	5.20	3.2385	5
1.80	8.2447	2	5.40	1.6890	5
2.00	4.4619	2	5.60	6.6071	6
2.20	2.1067	2	5.80	8.0065	7
2.40	7.5952	3	6.00	-1.8288	6
2.60	7.1058	4	7.00	-5.2612	7
2.80	-2.1722	3	8.00	-7.6320	8
3.00	-2.8502	3	9.00	6.8866	10
			10.00	-1.6199	9

† From C.-S. Yih, *J. Fluid Mech.*, **31** (1968b), by permission of the Cambridge University Press.

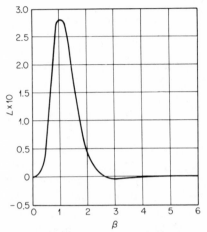

FIGURE 15.   THE VARIATION OF L WITH β.
WHEN $L > 1/F^2$, THERE IS INSTABILITY.
[*From C.-S. Yih, J. Fluid Mech.*, **31**
(1968b), *by permission of the Cambridge
University Press.*]

The result obtained is that

$$\mu_1 = -i\alpha^2 A_1 + O(\alpha^3).$$

For long waves, $\alpha \ll 1$, and the criterion (367) is valid.   As α increases
toward the order of 1, more terms are needed, but as long as $\alpha \ll 1$, the
stability or instability found is for any period of time, however long.
The problem of secular instability of the flow is solved explicitly for
long waves only.

The instability described above is a weak one, since μ_1 is of the
order of α^2.   Since a large F (or a small g) tends to make the flow unstable,
according to (367), it can be conjectured that the interface between two
layers of liquid differing only slightly in density (so that the gravity effect
is much reduced) may be much more unstable than the case studied here
if the liquids are set in motion by an oscillating plane.

PROBLEMS

1 Find the criterion for stability of a moisture-laden atmosphere, with
the possibility of condensation or evaporation taken into account.
This criterion will explain why clouds do not fall.   *Hint:* Ignore the
motion of any water particles relative to the surrounding gas, assume
an ideal-gas law for the dry air and one for water vapor, and consider

the isentropic change of state of a parcel of moisture-laden air, with the possibility of condensation or evaporation taken into account. The heat transfer between the liquid phase and the gaseous phase can be assumed reversible and uniform, but the heat transfer between the moist parcel of air as a whole and its surroundings is assumed zero. Latent heat must be taken into account. The criterion is obtained along the lines indicated in Sec. 2, though the situation is now more complicated.

2 A liquid of density ρ_1 and viscosity μ_1 occupies the space above $z = 0$, and a liquid of density $\rho_2 < \rho_1$ and viscosity μ_2 occupies the space below $z = 0$. Formulate the stability problem and outline one method for its solution.

3 A viscous fluid of thermal conductivity k and depth d has a uniform heat-source distribution of strength q (thermal units per unit volume). Show that the temperature distribution in the steady state is

$$\bar{T} = \frac{q}{2k} (x_3 d - x_3{}^2) + T_0,$$

the maximum value (over the depth d) of which is

$$\bar{T}_m = \frac{qd^2}{8k} + T_0.$$

Using the notation of Sec. 2.3, except that x_3 is now measured from the lower plate, for convenience, and assuming that unstable and neutrally stable disturbances are nonoscillatory, show that the relationship between the Rayleigh number

$$R = \frac{g\alpha d^5}{8\nu\kappa} \frac{q}{k}$$

and the wave number a at incipient instability is determined by the differential system

$$(D^2 - a^2)\theta = 4(1 - 2z)w,$$

$$(D^2 - a^2)^2 w = Ra^2\theta,$$

$$\theta = w = Dw = 0 \qquad \text{at } z = 0 \text{ and } z = 1.$$

Debler (1959) obtained, for $a = 4.0$, which was found to be the critical wave number for this problem, $R = 4{,}654$, $4{,}684$, and $4{,}669$ in the first, second, and third approximation, respectively. Check the figure $4{,}654$ in order to master Chandrasekhar's method.

4 Show that the stability of Couette flow in an annulus narrow compared with the radii is governed by

$$(D^2 - k^2 - \sigma R')(D^2 - k^2)u = 2k^2 R'\omega v,$$

$$(D^2 - k^2 - \sigma R')v = 2R' A u,$$

in which k is the wave number based on the gap width, that is,

$$k = \lambda\left(\frac{R_2}{R_1} - 1\right),$$

$$R' = \frac{\Omega_1(R_2 - R_1)^2}{\nu}, \qquad \omega = A + B\left[1 + \left(\frac{R_2}{R_1} - 1\right)\xi\right]^{-2},$$

$$\xi = \frac{r - R_1}{R_2 - R_1}, \qquad D = \frac{d}{d\xi},$$

and by the boundary conditions

$$u = \frac{du}{d\xi} = v = 0 \qquad \text{at } \xi = 0 \text{ and } \xi = 1.$$

5 Show that if Ω_1 and Ω_2 are nearly equal, the σ in the differential system in the preceding problem is real if the flow is unstable or neutrally stable.

6 For the problem treated in Sec. 4.1, show that if the motion is dependent on φ, so that Eq. (128) is replaced by

$$\phi = Ae^{\sigma t + in\varphi + imz}J_n(imr),$$

in which n is an integer and the real part is meant for ϕ, then (137) is replaced by

$$\sigma^2 = \frac{T}{\rho a^3}\frac{kI_n'(k)}{I_n(k)}(1 - n^2 - k^2),$$

in which

$$I_n(k) = e^{-in\pi/2}J_n(ik)$$

$$= \frac{k^n}{2^n n!}\left[1 + \frac{k^2}{2(2n + 2)} + \frac{k^4}{2(4)(2n + 2)(2n + 4)} + \cdots\right].$$

What conclusion can be drawn concerning the stability of unsymmetric disturbances?

7 If $m = 0$ for the preceding problem, the disturbance is two-dimensional. The degenerate form for ϕ is

$$\phi = Be^{\sigma t + in\varphi}r^n.$$

(This can be obtained by dividing the ϕ in the preceding problem by m^n, and then letting m approach zero. It can also be directly obtained by solving the two-dimensional Laplace equation in polar coordinates.) Show that

$$\sigma^2 = n(1 - n^2)\frac{T}{\rho a^3}.$$

Is there an obvious interpretation of why the fluid should be always stable?

8 A long viscous liquid hangs in air like a thread. Its undisturbed form is a circular cylinder if gravity effects are neglected. Formulate the differential system governing its stability and outline a method of solution. *Hint:* The curvature of the cylindrical surface is destabilizing.

9 A layer of inviscid and quiescent liquid of density ρ_2 occupies the space $-d \leq z \leq 0$. Below it is an infinite expanse of an inviscid liquid of density $\rho_3 > \rho_2$, also at rest. Above the finite layer is an infinite expanse of an inviscid liquid of density $\rho_1 < \rho_2$ moving horizontally with a velocity U. Study the stability of the system and compare the result with Helmholtz's result. Does the density difference $\rho_3 - \rho_2$ always stabilize the flow?

10 A liquid of viscosity μ_1, density ρ_1, and depth d_1 is in contact with a liquid below it, of viscosity μ_2, density $\rho_2 > \rho_1$, and depth d_2. The two-layer system is bounded above and below by rigid horizontal plates, the upper one of which moves with a velocity U_0 in its own plane relative to the lower one. Find the mean flow and formulate the problem of its stability. Suggest a method to solve this problem for long waves.

11 Show from (180) that

$$c_i \int_{y_1}^{y_2} (|\phi'|^2 + \alpha^2 |\phi|^2) \, dy = \frac{i}{2} \int_{y_1}^{y_2} U'(\phi\phi'^* - \phi^*\phi') \, dy$$

and that this is equivalent to

$$\frac{\partial}{\partial t} \int_{y_1}^{y_2} \int_0^{2\pi/\alpha} \frac{1}{2}(u'^2 + v'^2) \, dx \, dy = -\int_{y_1}^{y_2} \int_0^{2\pi/\alpha} U'u'v' \, dx \, dy.$$

Interpret this result.

12 Obtain from the Orr-Sommerfeld equation the equation of Synge:

$$-i\alpha c(I_1^2 + \alpha^2 I_0^2) = \alpha[Q + Q^* - (\alpha R)^{-1}(I_2^2 + 2\alpha^2 I_1^2 + \alpha^4 I_0^2)]$$

$$- i\alpha \int_{-1}^1 [U \, |\phi'|^2 + (\alpha^2 U + \frac{U''}{2}) \, |\phi|^2] \, dy,$$

in which

$$I_0^2 = \int_{-1}^1 |\phi|^2 \, dy, \qquad I_1^2 = \int_{-1}^1 |\phi'|^2 \, dy, \qquad I_2^2 = \int_{-1}^1 |\phi''|^2 \, dy,$$

$$Q = \frac{i}{2} \int_{-1}^1 U'\phi\phi'^* \, dy.$$

How can Synge's equation be used to establish bounds for c_i once

U is specified? *Hint:* The Schwartz inequality is

$$\left(\int fg \, dy \right)^2 < \int f^2 \, dy \int g^2 \, dy,$$

unless f is proportional to g [see Joseph (1968); his results have been uniformly improved by Yih (1969).]

13 A liquid with viscosity μ_1 and density ρ_1 is above, and in contact with, a liquid with viscosity μ_2 and ρ_2, both being in a porous medium of permeability k. If the liquids move vertically upward with a velocity V, show that waves at the interface of the form exp (inx), x being a horizontal Cartesian coordinate, grow or attenuate as exp (σt), in which t is the time and σ is determined by [Saffman and Taylor (1958)]

$$\frac{\sigma}{nk}(\mu_1 + \mu_2) = (\rho_1 - \rho_2)g + \frac{V}{k}(\mu_1 - \mu_2).$$

14 Show that the wave velocity c_r of neutral or unstable shear waves in plane Poiseuille flow or plane Couette-Poiseuille flow must be within the range of the velocity of the velocity of the flow, i.e., $U_{min} < c_r < U_{max}$. (Yih 1973.)

REFERENCES

BÉNARD, H., 1900: Tourbillons cellulaires dans une nappe liquide, pt. I, Description générale des phénomènes, pt. II, Procédés mécaniques et optiques d'examen; lois numériques des phénomènes, *Rev. Gen. Sci. Pur. Appl.*, **12**: 1261–1271 and 1309–1328.

BÉNARD, H., 1901: Les Tourbillons cellulaires dans une nappe liquide transportant de la chaleur par convection en régime permanent, *Ann. Chim. Phys.*, (7)**23**: 62–144.

BENJAMIN, T. B., 1957: Wave Formation in Laminar Flow down an Inclined Plane, *J. Fluid Mech.*, **2**: 554–574.

BENJAMIN, T. B., and F. URSELL, 1954: The Stability of the Plane Free Surface of a Liquid in Vertical Periodic Motion, *Proc. Roy. Soc. London Ser. A*, **225**: 505–515.

BENNEY, D. J., 1966: Long Waves in Liquid Films, *J. Math. Phys.*, **45**: 150–155.

CASE, K. M., 1960: Stability of Inviscid Plane Couette Flow, *Phys. Fluids*, **3**: 143–148.

CHANDRASEKHAR, S., 1954a: The Stability of Viscous Flow between Rotating Cylinders, *Mathematika*, **1**: 5–13.

CHANDRASEKHAR, S., 1954b: On Characteristic Value Problems in High Order Differential Equations Which Arise in Studies on Hydrodynamic and Hydromagnetic Stability, *Am. Math. Monthly*, **61**: 32–45.

CHANDRASEKHAR, S., 1955: The Character of the Equilibrium of an Incompressible Heavy Viscous Fluid of Variable Density, *Proc. Cambridge Phil. Soc.*, **51**: 162–178.

CHANDRASEKHAR, S., 1961: Adjoint Differential Systems in the Theory of Hydrodynamic Stability, *J. Math. Mech.*, **10**: 683–690.

CHANDRASEKHAR, S., and DONNA D. ELBERT, 1962: The Stability of Viscous Flow between Rotating Cylinders, pt. II, *Proc. Roy. Soc. London Ser. A*, **268**: 145–152.

CHRISTOPHERSON, D. G., 1940: Note on the Vibration of Membranes, *Quart. J. Math.*, **11**: 63–65.

DEBLER, W. R., 1959: The Onset of Natural Convection in a Fluid with Homogeneously Distributed Heat Sources, Ph.D. thesis, University of Michigan, Ann Arbor.

DEBLER, W. R., 1966: On the Analogy between Thermal and Rotational Hydrodynamic Stability, *J. Fluid Mech.*, **24**: 165–176.

DRAZIN, P. G., and L. N. HOWARD, 1966: Hydrodynamic Stability of Parallel Flow of Inviscid Fluid, *Advan. Appl. Mech.*, **9**: 1–89.

DUNN, D. W., and C. C. LIN, 1953: On the Role of Three-dimensional Disturbances in the Stability of Supersonic Boundary Layers, *J. Aeron. Sci.*, **19**: 491.

DUPREZ, F., 1851, 1854: Sur en cas particulier de l'équilibre des liquides, *Nouveaux Mém. Acad. Belg.*

ELIASSEN, A., E. HØILAND, and E. RIIS, 1953: Two-dimensional Perturbation of a Flow with Constant Shear of a Stratified Fluid, *Inst. Weather Climate Res.*, *Oslo. Publ.* 1.

FARADAY, M., 1831: On the Forms and States Assumed by Fluids in Contact with Vibrating Elastic Surfaces, *Phil. Trans.*, **1831**: 319–340. (This paper is an appendix to another paper with the title On a Peculiar Class of Acoustical Figures; and on Certain Forms Assumed by Group of Particles upon Vibrating Elastic Surfaces.)

FJØRTOFT, R., 1950: Application of Integral Theorems in Deriving Criteria of Stability for Laminar Flows and for the Baroclinic Circular Vortex, *Geofys. Publ.*, **17**: (5).

FRIEDRICHS, K. O., 1942: "Fluid Dynamics" (mimeographed lecture notes, Brown University, Providence), chap. 4, pp. 200–209.

GASTER, M., 1962: A Note on the Relation between Temporally-increasing and Spatially-increasing Disturbances in Hydrodynamic Stability, *J. Fluid Mech.*, **14**: 222–224.

GOLDSTEIN, S., 1931: On the Stability of Superposed Streams of Fluids of Different Densities, *Proc. Roy. Soc. London Ser. A*, **132**: 524–548.

GOREN, S. L., 1962: The Instability of an Annular Thread of Fluid, *J. Fluid Mech.*, **12**: 309–319.

GRAEF, M., 1961: Zur Instabilität ebener Riselfilme bei Berücksichtigung der Oberflächenspannung, *Max-Planck Institut Strömungsforschung*, *Bericht* 1.

GRAEF, M., 1966: Über die Eigenschaften zwei-und dreidimensionaler Störungen in Rieselfilmen an geneigten Wänden, *Mitt. Max-Planck Inst. Strömungsforschung Aeron. Versuchsanstalt*, 26.

HEISENBERG, W., 1924: Über Stabilität und Turbulenz von Flüssigkeitsströmen, *Ann. Phys. Leipzig*, (4)**74**: 577–627.

HELMHOLTZ, H. VON, 1868: Über discontinuirliche Flüssigkeitsbewegungen, *Monatsber. Akad. Wiss. Berlin*, **1868**: 215–228.

HØILAND, E., 1953: On Two-dimensional Perturbation of Linear Flow, *Geofys. Publ.*, **18**(9): 1–12.

HOWARD, L. N., 1961: Note on a Paper of John W. Miles, *J. Fluid Mech.*, **10**: 509–512.

JEFFREYS, H., 1928: Some Cases of Instability in Fluid Motion, *Proc. Roy. Soc. London Ser. A*, **118**: 195–208.

JOSEPH, D. D., 1968: Eigenvalue Bounds for the Orr-Sommerfeld Equation, *J. Fluid Mech.*, **33**: 617–621.

KAPITZA, P. L., 1948a: *Zh. Eksperim. i. Teor. Fiz.*, **18**: 3.

KAPITZA, P. L., 1948b: *Zh. Eksperim. i Teor. Fiz.*, **18**: 20.

KAPITZA, P. L., 1949: *Zh. Eksperim. i Teor. Fiz.*, **19**: 105.

KÁRMÁN, T. VON, 1934: Some Aspects of the Turbulence Problem, *Proc. 4th Intern. Congr. Appl. Mech., Cambridge*, pp. 54–91.

LIN, C. C., 1944: On the Stability of Two-dimensional Parallel Flows, *Proc. Natl. Acad. Sci. U.S.*, **30**: 316–323.

LIN, C. C., 1945: On the Stability of Two-dimensional Parallel Flows, pts. I–III, *Quart. Appl. Math.*, **3**: 117–142, 218–234, 277–301.

LIN, C. C., 1954a: Hydrodynamic Stability, *Proc. 5th Symp. Appl. Math. (AMS)*, pp. 1–18.

LIN, C. C., 1954b: Some Physical Aspects of the Stability of Parallel Flows, *Proc. Natl. Acad. Sci. U.S.*, **40**: 741–747.

LIN, C. C., 1955: "The Theory of Hydrodynamic Stability," Cambridge University Press, Cambridge.

LIN, C. C., and D. J. BENNEY, 1962: On the Instability of Shear Flows, *Proc. Symp. Appl. Math., Hydrodynamic Instability*, **13**: 1–24. (This paper contains an explanation of why observed unstable disturbances are often three-dimensional.)

MATTHIESSEN, L., 1868: Akustische Versuche, die kleinsten Transversalwellen der Flüssigkeiten betreffend, *Ann. Phys. Chem. Leipzig*, **134**: 107–117.

MATTHIESSEN, L., 1870: Über die Transversalschwingungen tönender tropfbarer und elastischer Flüssigkeiten, *Ann. Phys. Chem. Leipzig*, **141**: 375–393.

MAXWELL, J. C., 1890: "Scientific Papers," vol. II, Cambridge University Press, Cambridge.

MICHALKE, A., 1965: On Spatially Growing Disturbances in an Inviscid Shear Layer, *J. Fluid Mech.*, **23**: 521–544.

MILES, J. W., 1961: On the Stability of Heterogeneous Shear Flows, *J. Fluid Mech.*, **10**: 496–508.

ORR, W. MCF., 1907: The Stability or Instability of the Steady Motions of a Perfect Liquid and of a Viscous Liquid, pt. I, A Perfect Liquid, and pt. II, A Viscous Liquid, *Proc. Roy. Irish Acad.*, **27**: 9–68 and 69–138.

PEARSON, J. R. A., 1958: In Convection Cells Induced by Surface Tension, *J. Fluid Mech.*, **4**: 481–500.

PELLEW, A., and R. V. SOUTHWELL, 1940: On Maintained Convective Motion in a Fluid Heated from Below, *Proc. Roy. Soc. London Ser. A*, **176**: 312–343.

POTTER, M. C., 1966: Stability of Plane Couette-Poiseuille Flow, *J. Fluid Mech.*, **24**: 609–619.

RAYLEIGH, LORD, 1879: On the Capillary Phenomena of Jets, app. I, *Proc. Roy. Soc. London Ser. A*, **29**: 71.

RAYLEIGH, LORD, 1880: On the Stability or Instability of Certain Fluid Motions, "Scientific Papers," vol. 1, pp. 474–487, Cambridge University Press, Cambridge.

RAYLEIGH, LORD, 1883: On the Crispations of Fluid Resting upon a Vibrating Support, *Phil. Mag.*, (5)**16**: 50–58.

RAYLEIGH, LORD, 1892: On the Instability of a Cylinder of Viscous Liquid under Capillary Force, *Phil. Mag.*, **34**: 145.

RAYLEIGH, LORD, 1894: "Theory of Sound," 2d ed., Macmillan & Co., Ltd., London.

RAYLEIGH, LORD, 1916: On the Dynamics of Revolving Fluids, "Scientific Papers," vol. 6, pp. 447–453, Cambridge University Press, Cambridge.

ROBERTS, P. H., 1960: Characteristic Value Problems Posed by Differential Equations Arising in Hydrodynamics and Hydromagnetics, *J. Math. Anal. Appl.*, **1**: 195–214.

SAFFMAN, P. G., and G. I. TAYLOR, 1958: The Penetration of a Fluid into a Porous Medium or Hele-Shaw Cell Containing a More Viscous Fluid, *Proc. Roy. Soc. London Ser. A*, **245**: 312–329.

SCHLICHTING, H., 1933a: Zur Entstehung der Turbulenz bei der Platten-strömung, *Nachr. Ges. Wiss. Göttingen, Math. Phys. Kl.*, **1933**: 181–208.

SCHLICHTING, H., 1933b: Berechnung der Anfachung kleiner Störungen bei der Plattenströmung, *Z. Angew. Math. Mech.*, **13**: 171–174.

SCHLICHTING, H., 1935: Amplitudeverteilung und Energiebilanz der kleinen Störungen bei der Plattengrenzschicht, *Nachr. Ges. Wiss. Göttingen, Math. Phys. Kl.*, Fachgrappe I, **1**: 47–78.

SCHUBAUER, G. B., and H. H. SKRAMSTAD, 1947: Laminar Boundary-layer Oscillations and Transition on a Flat Plate, *J. Aeron. Sci.*, **14**: 69–78; see also *Natl. Advisory Comm. Aeron. Rept.* 909, 1948, originally issued as *Natl. Advisory Comm. Aeron. Advance Confidential Rept.*, 1943.

SEGEL, L. A., 1966: Nonlinear Hydrodynamic Stability Theory and Its Applications to Thermal Convection and Curved Flows, in Russell J. Donnelly (ed.), "Non-equilibrium Thermodynamics: Variational Techniques and Stability," The University of Chicago Press, Chicago. (This survey contains references to many papers on nonlinear stability theory of convection.)

SHEN, S. F., 1954: Calculated Amplified Oscillations in Plane Poiseuille and Blasius Flows, *J. Aeron. Sci.*, **21**: 62–64.

SOMMERFELD, A., 1908: Ein Beitrag zur hydrodynamischen Erklärung der turbulenten Flüssigkeitsbewegung, *Proc. 4th Intern. Congr. Math., Rome*, pp. 116–124.

SQUIRE, H. B., 1933: On the Stability for Three-dimensional Disturbances of Viscous Fluid Flow between Parallel Walls, *Proc. Roy. Soc., London Ser. A*, **142**: 621–628.

STERN, M. E., 1960: The "Salt-fountain" and Thermo-haline Convection, *Tellus*, **12**: 172–175.

STOMMEL, H., and H. G. FARMER, 1953: Control of Salinity in an Estuary by a Transition, *J. Marine Res.*, **12**: 13–20.

STOMMEL, H., A. B. ARONS, and D. BLAUCHARD, 1956: An Oceanographic Curiosity: The Perpetual Salt Fountain, *Deep-sea Res.* 3: 152–153.

SYNGE, J. L., 1938a: On the Stability of a Viscous Fluid Between Rotating Coaxial Cylinders, *Proc. Roy. Soc. London Ser. A*, **167**: 250–256.

SYNGE, J. L., 1938b: Hydrodynamic Stability, *Semi-centennial Publ. Am. Math. Soc.*, **2**: 227–269.

TAYLOR, G. I., 1915: Eddy Motion in the Atmosphere, *Phil. Trans. Roy. Soc. London Ser. A*, **215**: 1–26.

TAYLOR, G. I., 1923: Stability of a Viscous Liquid Contained between Two Rotating Cylinders, *Phil. Trans. Roy. Soc., London Ser. A*, **223**: 289-343.

TAYLOR, G. I., 1931: Effect of Variation in Density on the Stability of Superposed Streams of Fluid, *Proc. Roy. Soc. London Ser. A*, **132**: 499–523.

TAYLOR, G. I., 1950: The Instability of Liquid Surfaces when Accelerated in a Direction Perpendicular to Their Planes, pt. I, *Proc. Roy. Soc. London Ser. A*, **201**: 192–196.

TIETJENS, O., 1925: Beiträge zur Entstehung der Turbulenz, *Z. Angew. Math. Mech.* **5**: 200–217.

TOLLMIEN, W., 1929: Über die Entstehung der Turbulenz, *Nachr. Ges. Wiss. Göttingen, Math. Phys. Kl.*, **1929**: 21–44.

TOLLMIEN, W., 1935: Ein allgemeines Kriterium der Instabilität laminarer Geschwindigkeitsverteilungen, *Nachr. Ges. Wiss. Göttingen, Math. Phys. Kl.*, Fachgruppe I, **1**: 79–114.

TOMOTIKA, S., 1935: On the Instability of a Cylindrical Thread of a Viscous Liquid Surrounded by Another Viscous Liquid, *Proc. Roy. London Ser. Soc. A*, **150**: 322.

WATSON, G. N., "A Treatise on the Theory of Bessel Functions," 2d ed., Cambridge University Press, Cambridge.

WEBER, C., 1931: Zum Zerfall eines Flüssigkeitsstrahles, *Z. Angew. Math. Mech.*, **11**: 136.

YIH, C.-S., 1954: Stability of Parallel Laminar Flow with a Free Surface, *Proc. 2d U.S. Congr. Appl. Mech.*, pp. 623–628.

YIH, C.-S., 1955: Stability of Two-dimensional Parallel Flows for Three-dimensional Disturbances, *Quart. Appl. Math.*, **12**: 434–435.

YIH, C.-S., 1960: Instability of a Rotating Liquid Film with a Free Surface, *Proc. Roy. Soc. London Ser. A*, **258**: 63–86.

YIH, C.-S., 1961: Dual Role of Viscosity in the Instability of Revolving Fluids of Variable Density, *Phys. Fluids*, **4**: 806–811.

YIH, C.-S., 1963: Stability of Liquid Flow down an Inclined Plane, *Phys. Fluids*, **6**: 321–334.

YIH, C.-S., 1967a: Instability Due to Viscosity Stratification, *J. Fluid Mech.*, **27**: 337–352.

YIH, C.-S., 1967b: Instability of Laminar Flows Due to a Film of Adsorption, *J. Fluid Mech.*, **28**: 493–500.

YIH, C.-S., 1968a: Fluid Motion Induced by Surface Tension Variation, *Phys. Fluids*, **11**: 477–480.

YIH, C.-S., 1968b: Instability of Unsteady Flows or Configurations, pt. I, Instability of a Horizontal Liquid Layer on an Oscillating Plane, *J. Fluid Mech.*, **31**: 737–752.

YIH, C.-S., 1968c: Stability of a Horizontal Fluid Interface in a Periodic Vertical Electric Field, *Phys. Fluids*, **11**: 1447–1449.

YIH, C.-S., 1969, Note on Eigenvalue Bounds for the Orr-Sommerfeld Equation, *J. Fluid Mech.*, **38**: 273–278.

YIH, C.-S., 1972, Spectral Theory of Taylor Vortices, Part I. Structure of Unstable Modes, Archive for Rational Mech. and Analysis, *46*, pp. 218–240.

YIH, C.-S., 1972, Spectral Theory of Taylor Vortices, Part II. Proof of Nonoscillation, Archive for Rational Mech. and Analysis, *47*, pp. 288–300.

YIH, C.-S., 1973, Wave Velocity in Parallel Flows of a Viscous Fluid, *J. Fluid Mech.*, *58:* 703–708.

YUAN, C., 1966: Recalculation for the Stability of Free-surface Laminar Flow down an Inclined Plane, unpublished.

ADDITIONAL READING

CHANDRASEKHAR, S., 1961: "Hydrodynamic and Hydromagnetic Stability," Oxford University Press, London.

DAVEY, A., R. C. DI PRIMA, and J. T. STUART, 1968: On the Instability of Taylor Vortices, *J. Fluid Mech.*, **31**: 17–52. (Contains references to many modern research papers on the stability of Taylor vortices. From these and those given in Chandrasekhar's book a nearly complete list of references on Taylor vortices can be built by the star method.)

DI PRIMA, R. C., and J. T. STUART, 1964: Nonlinear Aspects of Instability in Flow between Rotating Cylinders, *Proc. 7th Intern. Congr. Appl. Mech.*, *Munich*. (Contains many references to papers on the nonlinear theory of the stability of Couette flow.)

GRAEBEL, W. P., 1966: On Determination of the Characteristic Equations for the Stability of Parallel Flows, *J. Fluid Mech.*, **24**: 497–508.

STUART, J. T., and J. WATSON, 1960: On the Nonlinear Mechanics of Wave Disturbances in Stable and Unstable Flows, *J. Fluid Mech.*, **9**: 353–370 [pt. 1, by Stuart] and 371–389 [pt. 2, by Watson]. (From this long paper references to works on nonlinear stability prior to 1960 can be found.)

CHAPTER TEN

TURBULENCE

1. INTRODUCTION

The study of hydrodynamic stability in the preceding chapter shows that if the Reynolds number or some other parameter governing stability becomes sufficiently large, the original laminar flow will become turbulent (very irregular) although, as in the case of Couette flow, the transition to turbulent flow may be accomplished through intermediate stages of more complicated laminar flows. Once the flow has become turbulent, its analysis becomes extremely difficult, although the equation of continuity still must be satisfied and the dynamics of the flow is still governed by the Navier-Stokes equations. One may consider the flow as established in the sense that at any location the mean quantities do not vary with time —and endeavor to determine the flow. With this approach, the difficulty is at least twofold. Often the simplifying assumptions concerning a flow under examination, e.g., the velocity being unidirectional in flow through pipes or the streamlines being circular in Couette flow, which obviate the specification of the conditions at infinity for these idealized flows, are no longer valid and corresponding assumptions cannot be specified in their stead. One may, of course, study the flow of an incompressible fluid entering a pipe through a smooth entrance, where the flow is laminar, and giving the flow an initial disturbance, study its subsequent development toward eventual turbulent flow if the Reynolds number is sufficiently high. In this way one can specify the upstream condition. Or one may consider the flow (which may be turbulent) caused by a body moving in an incompressible fluid bounded externally by a solid surface. The boundary conditions are then completely specified. But in both these cases, tracing the motion up to the point of transition to turbulence is no easy task, because of the difficulty presented by the nonlinear inertia terms in the Navier-Stokes equations. In short, complete specification of the boundary conditions is often impossible, and solution of the Navier-Stokes equations is always extremely difficult.

If one adopts the approach of specifying an initial condition together with the boundary conditions in a fluid of finite extent and studying the subsequent development of the flow, aside from the aforementioned difficulty of solution of the Navier-Stokes equations, one has to specify the highly irregular and complicated initial conditions of a turbulent flow encountered in nature or in the laboratory. This is neither possible nor indeed desirable. Thus analytical determination of turbulent flows in the same deductive way as analytical determination of laminar flows is not yet possible, and it is very doubtful that it will ever become so.

One may wonder why turbulent flows so defy complete analysis while the kinetic theory of gases is so successful in determining macroscopic properties of the gas whose molecules move about in as random a way as fluid particles in turbulent flow. The distinction lies in the crucial fact that gas molecules are very small in comparison with the empty space between them (or with their free path cubed) and that they can be treated as perfectly elastic balls in collision among themselves or against walls. No such simplification is possible for turbulent flows. In most cases (except very rarefied gases) the fluid can be treated as a continuum. The turbulence of the fluid can be represented by Fourier integrals, and thereby has a (continuous) spectrum describing its composition of *eddies* of various sizes. These eddies are very different from gas molecules. In the first place, they are not separated from each other. In fact, the small eddies are quite embedded in the larger ones. And they certainly cannot be treated as elastic balls whose collisions, against walls or among themselves, are governed by Newton's second law in its simplest form. When these eddies move about, if indeed they can be isolated in our imagination to permit such a statement, they affect the fluid around them. To state this more precisely, the fluctuating part of the velocity at any point is correlated statistically with that at a neighboring point. But the correlation falls off with distance, and becomes zero for two points far enough away. The distance at which this correlation falls to zero may be considered as an estimate of the size of the largest eddy, which contains eddies of smaller sizes. The finiteness of this eddy, the complexity of its structure, and the lack of separation of any fluid particle from its neighbors all conspire to render any application of the tools used in the kinetic theory of gases to the precise study of turbulent flows fruitless.

What, then, does one do if one wishes to study turbulent flows? From the point of view of an engineer who wishes to determine certain quantities of interest to him, e.g., the drag of a body moving with sufficient speed in a fluid to make its flow turbulent, the answer is straightforward. A dimensional analysis performed on the pertinent variables and subsequent systematic experiments according to the guide provided by this analysis will provide the answers he seeks. This in fact is what has been done wherever answers of practical importance are sought. But this does not help the person who wishes to *understand* turbulent flows. For him the task is more difficult and threatens to be perpetual. Whatever analysis he is able to apply is likely to rely on physical insight and on experiments either in making the appropriate hypothesis or at least for verification or disproof. Aside from the equation of continuity and the Navier-Stokes equations, his tools are statistical laws, demands of symmetry (such as exists in isotropic turbulence), the tensorial character of products of velocity components, and the Fourier integral, among others.

So great are the difficulties presented by turbulent flows and so large the literature that neither the scope nor the nature of this book permits an adequate treatment. The difficulty of selecting material for presentation becomes formidable, and we shall content ourselves with the following limited goals: (*a*) to see by one or two examples what turbulent flows are really like and (*b*) to see what analyses are possible, especially for isotropic turbulence. The treatment is far from exhaustive and far from sufficient, but it will at least suggest the nature of turbulent flows and indicate the spirit of their analysis. Within the limits of space and scope of this book such a treatment is perhaps all we can attempt.

2. TURBULENT FLOW IN PIPES

The most extensive measurement on turbulent flow which is steady and unidirectional *in the mean* was performed by Laufer (1953). Before presenting his results, we shall take this opportunity to demonstrate the problem of closure for turbulent flows. If incompressibility of the fluid is assumed, the equation of continuity is

$$\frac{\partial u_\alpha}{\partial x_\alpha} = 0, \tag{1}$$

and the Navier-Stokes equations are, in the notation of Chap. 2,

$$\rho\left(\frac{\partial u_i}{\partial t} + u_\alpha \frac{\partial u_i}{\partial x_\alpha}\right) = -\frac{\partial p}{\partial x_i} - \rho\frac{\partial \Omega}{\partial x_i} + \mu\,\nabla^2 u_i, \tag{2}$$

provided the viscosity μ is constant. Because of (1), (2) can be written as

$$\rho\frac{\partial u_i}{\partial t} + \frac{\partial(\rho u_i u_\alpha)}{\partial x_\alpha} = -\frac{\partial p}{\partial x_i} - \rho\frac{\partial \Omega}{\partial x_i} + \mu\,\nabla^2 u_i. \tag{3}$$

If a flow is steady in the mean, we can write

$$u_i = U_i + u_i', \tag{4}$$

in which U_i is the mean value of u_i taken over a long period of time. Thus, if a bar indicates that mean,

$$\overline{u_i'} = 0, \qquad \overline{U_j u_i'} = 0. \tag{5}$$

Substituting (4) into (3), taking the mean, and realizing that $\partial U_\alpha/\partial x_\alpha = 0$, we have

$$\rho U_\alpha \frac{\partial U_i}{\partial x_\alpha} + \frac{\partial \overline{\rho u_i' u_\alpha'}}{\partial x_\alpha} = -\frac{\partial \bar{p}}{\partial x_i} - \rho\frac{\partial \Omega}{\partial x_i} + \mu\,\nabla^2 U_i. \tag{6}$$

Comparing the terms $\partial\overline{\rho u_i' u_\alpha'}/\partial x_\alpha$ with the terms $\partial\tau_{\alpha i}/\partial x_\alpha$ in (2.4), we see that $-\overline{\rho u_i' u_\alpha'}$ is a component of a stress tensor. This is called the *Reynolds stress tensor*, and its components $-\overline{\rho u_i' u_\alpha'}$ are often called the *Reynolds stresses*. It can be seen that although (1) and (2) constitute a closed system of differential equations, i.e., a system with as many unknowns as equations, (6) and

$$\frac{\partial U_\alpha}{\partial x_\alpha} = 0 \tag{7}$$

do not constitute such a system, for written in the form of (6), the Reynolds stresses are unknowns in addition to U_i. Treating equation (3) in the mean therefore brings forth the problem of closure. We cannot attempt to solve (6). We can only use (6) to relate the Reynolds stresses to the mean quantities U_i. This is done by Laufer (1953) for pipe flows.

Laufer treated the turbulent flow of an incompressible fluid in a circular pipe, which is steady and unidirectional in the mean. Cylindrical coordinates are used, but Laufer calls them r, φ, x and denotes the corresponding velocity components by v', w', $U + u'$. The violation of convention is quite harmless, and we shall follow his usage in this section. The U is the mean velocity, and the letters with a prime indicate turbulent fluctuations. Using P to indicate the mean of the dynamic part of the pressure, Laufer obtained, upon taking the mean of the Navier-Stokes equations in cylindrical coordinates, the equations

$$\frac{1}{\rho}\frac{\partial P}{\partial x} = -\frac{1}{r}\frac{d}{dr}\,\overline{ru'v'} + \nu\left(\frac{d^2U}{dr^2} + \frac{1}{r}\frac{dU}{dr}\right),\tag{8}$$

$$\frac{1}{\rho}\frac{\partial P}{\partial r} = -\frac{1}{r}\frac{d}{dr}\,\overline{rv'^2} + \frac{\overline{w'^2}}{r}\,,\tag{9}$$

$$0 = -\frac{d}{dr}\,\overline{v'w'} - \frac{2\overline{v'w'}}{r}\,.\tag{10}$$

The left-hand side of (10) is zero because the flow is assumed to be axi-symmetric in the mean. Gravity effects are negligible and neglected.

Integrating (10), we have

$$\overline{v'w'} = \frac{C}{r^2}\,.$$

Since $\overline{v'w'} = 0$ at $r = a$ ($a =$ inner radius of the pipe), $C = 0$ and $\overline{v'w'} = 0$ for all values of r. Thus only (8) and (9) are left. Since all mean quantities are independent of x, (9) shows that $\partial P/\partial r$ is independent of x or $\partial P/\partial x$ is independent of r. We can then integrate (8) and (9) to produce

$$\frac{r^2}{2}\frac{1}{\rho}\frac{\partial P}{\partial x} = -r\left(\overline{u'v'} - \nu\frac{dU}{dr}\right) + A(x),\tag{11}$$

$$\frac{P}{\rho} = -\overline{v'^2} + \int_a^r \frac{\overline{w'^2} - \overline{v'^2}}{r}\,dr + B(x).\tag{12}$$

The functions $A(x)$ and $B(x)$ are to be determined by the boundary conditions (since $v = 0$ at $r = 0$ and the mean flow is axisymmetric)

$$\overline{u'v'} = 0 \quad \text{and} \quad \frac{dU}{dr} = 0 \quad \text{at } r = 0,\tag{13}$$

$$\overline{u'v'} = 0 = \overline{v'^2} \quad \text{and} \quad \nu\frac{dU}{dr} = -U_r^2 \quad \text{at } r = a,\tag{14}$$

because the velocity vanishes at the wall. The quantity U_r is simply defined by the second equation in (14), and is called the *shear velocity*. Thus (11) demands that $A(x) = 0$, and (from the condition at $r = a$)

$$\frac{1}{\rho}\frac{\partial P}{\partial x} = -\frac{2}{a}\,U_r^2,$$

or

$$\frac{P}{\rho} = -\frac{2}{a}\,U_r^2 x + C(r).\tag{15}$$

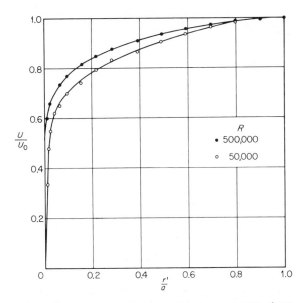

FIGURE 1. MEAN-VELOCITY DISTRIBUTION. IN FIGS. 1 TO 7, $r' = a - r$, a BEING RADIUS OF PIPE. [*From J. Laufer, Natl. Advisory Comm. Aeron. Tech. Note* 2954, 1953, *by permission of the National Aeronautics and Space Administrations.*]

Equations (12) and (15) give

$$B(x) = -\frac{2}{a} U_\tau^2 x$$

and also determine $C(r)$. Thus, finally, (11) and (12) become

$$\frac{r}{a} U_\tau^2 = \overline{u'v'} - \nu \frac{dU}{dr} \tag{16}$$

and

$$-\frac{P}{\rho} = \frac{2}{a} U_\tau^2 x + \overline{v'^2} + \int_a^r \frac{\overline{v'^2} - \overline{w'^2}}{r} \, dr. \tag{17}$$

Laufer's measurements for the mean velocity U are shown in Figs. 1 and 2. The dashed lines in Fig. 2 show agreement between the measured U and the U obtained from (11), with $A(x)$ and $\overline{u'v'}$ equal to zero (because the region under comparison is very near the wall), from measured $\partial P/\partial x$. Figures 3 to 5 show the distribution of u'', v'', and w'' defined by

$$(u'')^2 = \overline{u'^2}, \qquad (v'')^2 = \overline{v'^2}, \qquad (w'')^2 = \overline{w'^2}, \tag{18}$$

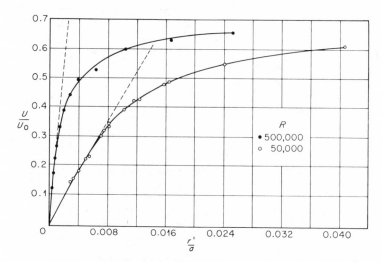

FIGURE 2. MEAN-VELOCITY DISTRIBUTION NEAR WALL. DASHED LINES ARE COMPUTED FROM PRESSURE-DROP MEASUREMENTS. (*From J. Laufer, Natl. Advisory Comm. Aeron. Tech. Note* 2954, 1953, *by permission of the National Aeronautics and Space Administration.*)

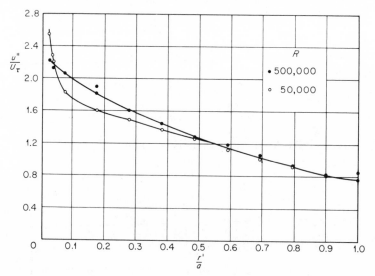

FIGURE 3. u'' DISTRIBUTION. (*From J. Laufer, Natl. Advisory Comm. Aeron. Tech. Note* 2954, 1953, *by permission of the National Aeronautics and Space Administration.*)

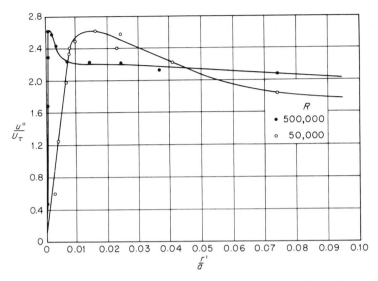

FIGURE 4. *u″* DISTRIBUTION NEAR THE WALL. (*From J. Laufer, Natl. Advisory Comm. Aeron. Tech. Note 2954, 1953, by permission of the National Aeronautics and Space Administration.*)

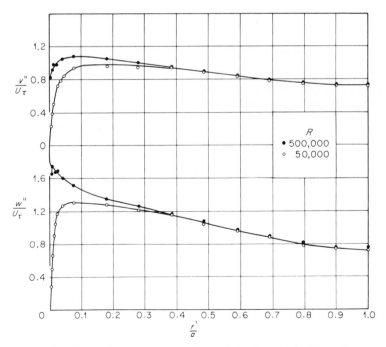

FIGURE 5. *v″* AND *w″* DISTRIBUTIONS. (*From J. Laufer, Natl. Advisory Comm. Aeron. Tech. Note 2954, 1953, by permission of the National Aeronautics and Space Administration.*)

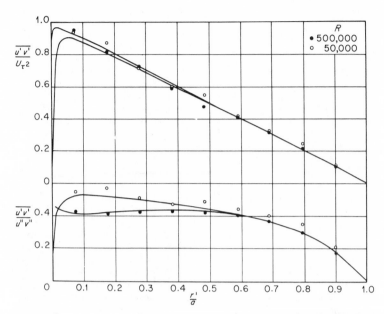

FIGURE 6.   REYNOLDS SHEARING STRESS AND DOUBLE-CORRELATION-COEFFICIENT
DISTRIBUTIONS.   CURVES CALCULATED FROM MEASURED dU/dr, u'', AND v''.
(*From J. Laufer, Natl. Advisory Comm. Aeron. Tech. Note 2954, 1953, by
permission of the National Aeronautics and Space Administration.*)

and Fig. 6 shows the distribution of $\overline{u'v'}$. In all cases the quantity in
question increases sharply from the wall to a maximum value near the wall
and then decreases toward a relative minimum at the center of the pipe.
This already indicates that most of the turbulence is produced near the
wall.   A further confirmation is obtained by considering the equation

$$\frac{r}{a} U_\tau^2 \frac{dU}{dr} = \overline{u'v'} \frac{dU}{dr} - \nu\left(\frac{dU}{dr}\right)^2, \tag{19}$$

obtained from (16).   Apart from a factor $2\pi r \rho \, dr$, this is the work done to
the fluid occupying the area $2\pi r \, dr$, per unit distance in the x direction, by
the pressure and shear forces on the annular space of thickness dr; for the
work W done to the fluid cylinder of radius r is, per unit distance along
the x axis,

$$W = -\frac{dP}{dx} \int_0^r 2\pi r U \, dr + 2\pi r \tau U,$$

in which τ is the total shear stress at the surface of the cylinder, and since,

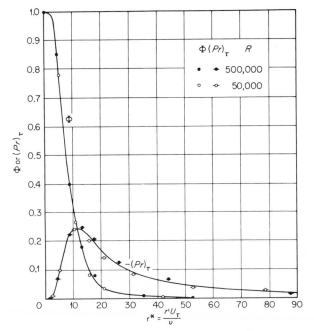

FIGURE 7. COMPARISON OF RATE OF TURBULENT-ENERGY PRO-
DUCTION WITH RATE OF DIRECT VISCOUS DISSIPATION NEAR
WALL. (*From J. Laufer, Natl. Advisory Comm. Aeron. Tech.
Note 2954, 1953, by permission of the National Aeronautical and
Space Administration.*)

according to (11),

$$r \frac{dP}{dx} = \frac{d}{dr} r\tau,$$

we have

$$dW = \frac{dW}{dr} dr = 2\pi r\tau \frac{dU}{dr} dr, \quad \text{with } \tau = \rho \left(\nu \frac{dU}{dr} - \overline{u'v'} \right). \quad (20)$$

The first term on the right-hand side of (19) represents the rate of turbulence production $(Pr)_t$, and the second term represents the rate of viscous
dissipation for the mean flow, called the *direct* viscous dissipation by
Laufer. Figure 7 shows that most of the direct viscous dissipation takes
place in a very narrow region (a = radius of the pipe)

$$r^* = \frac{(a - r)U_\tau}{\nu} < 15$$

and that at $r^* = 11.5$ the laminar shear stress is equal to the Reynolds
stress, so that turbulence production is equal to direct viscous dissipation.

The point $r^* = 11.5$ also happens to be the point of maximum turbulence production. It can be seen from Fig. 7 that most of the direct viscous dissipation and the turbulence production takes place near the wall. In a vague way this confirms our earlier realization that vorticity is created near the wall. In Fig. 7, $(Pr)_t$ refers to the first term on the right-hand side of (19), and Φ refers to the last term (without the minus sign) in (19). Laufer's measurements are all consistent with (16).

3. INTERMITTENCY OF TURBULENT BOUNDARY LAYERS

Corrsin (1943) discovered that when a laminar free stream passes along a flat plate and the boundary layer on the plate is turbulent, the level of turbulence does not decrease to zero smoothly but vanishes at some finite distance from the plate rather abruptly, leaving the flow beyond laminar. The surface at which the level of turbulence vanishes may be called the *interface*. Whatever vorticity has penetrated into the free stream by viscous diffusion seems to be very small, so that the fluid beyond the interface is not only in laminar flow but very nearly in irrotational flow. This interface, however, is by no means stationary. In fact it has random waves of lengths rather large compared with the scale of the turbulence or even with the thickness of the turbulent boundary layer. A hot-wire anemometer placed near the mean position of the interface shows turbulent fluctuations for a period of time, then shows little turbulence for a comparable period, and then shows turbulence again. The phenomenon (Figs. 8a and 8b) is called *intermittency*.

Intermittency is a very important phenomenon. In turbulent boundary layers the mean velocity in a region near the wall obeys the logarithmic law; i.e., aside from an additive constant the mean velocity in a turbulent region near the wall is proportional to the logarithm of the normal distance from the wall. Between this region and the wall there is a very thin region where the level of turbulence decreases to zero at the wall provided the wall is smooth. It is called the *laminar sublayer*. Measurement in this very thin layer is difficult, and it is certainly not well defined. From the wall region outward, the shear stress and the influence of the wall decrease, and the turbulence becomes increasingly influenced by the intermittency phenomenon occurring at the outer edge of the turbulent boundary layer. Recent researches show that at high Reynolds numbers and (surprisingly enough) especially at the convergent part of the flow, i.e., where the pressure decreases downstream, the intermittency penetrates very deeply into the boundary layer, sometimes right to the wall! The deep involvement of the interface with the boundary layer results in a "spotty" structure of the turbulence, rendering the use of spectrum analysis (Sec. 8) more difficult or even inadequate.

(a) r = 0 (on axis)

(b) r = 4.5 cm

(c) r = 7.5 cm

(d) r = 9.5 cm

(e) r = 14.5 cm

FIGURE 8a.   THE INTERMITTENCY SIGNAL IN A SHEAR FLOW OBTAINED BY CORRSIN (1943)
IN A SYMMETRIC JET.   [*Courtesy of Professor Stanley Corrsin.*]

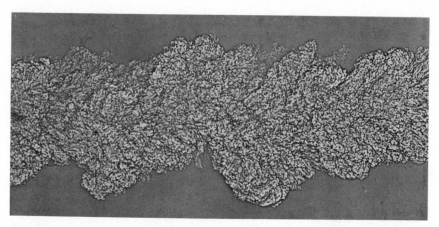

FIGURE 8*b*.   TURBULENT JET SURROUNDED BY POTENTIAL FLOW, THE MOVEMENT OF THE POSITION OF THE INTERFACE BEING THE CAUSE OF THE INTERMITTENCY PHENOMENON. [*Courtesy of Professor Stanley Corrsin.   The picture was supplied to him by the Aberdeen Proving Ground.*]

4. TAYLOR'S THEORY OF TURBULENT DIFFUSION

Turbulent diffusion was first investigated by Taylor (1921). *Diffusion is a process by which a fluid particle released at a certain point in the field of flow finds its way to other places.* Since the identification of the particle is then always kept in mind, it is not surprising that the phenomenon of turbulent diffusion is closely related to the correlation of the velocity of the particle at one instant with its velocity at another, or the Lagrangian correlation. This is brought out very clearly by Taylor's theory.

Suppose that we observe the values $p_1, p_2, \ldots, p_n$ of a quantity p at a large number of successive times $t_1, t_2, \ldots, t_n$ and suppose that the mean of the squares of $p_1, p_2, \ldots, p_n$, denoted by

$$\overline{p^2} = \frac{p_1{}^2 + p_2{}^2 + \cdots + p_n{}^2}{n},$$

as well as $\overline{(d^m p / dt^m)^2}$ for any n is constant. If further we observe the values $p_1 + \delta p_1, p_2 + \delta p_2, \ldots, p_n + \delta p_n$, at times $t_1 + \delta t, t_2 + \delta t, \ldots, t_n + \delta t$, where δt is a small interval of time, then to the first order we have

$$\overline{[p(t + \delta t)]^2}$$

$$= \frac{1}{n}\left[\left(p_1 + \frac{dp_1}{dt}\,\delta t\right)^2 + \left(p_2 + \frac{dp_2}{dt}\,\delta t\right)^2 + \cdots + \left(p_n + \frac{dp_n}{dt}\,\delta t\right)^2\right]$$

$$= \overline{p^2(t)} + \overline{2p\,\frac{dp}{dt}}\,\delta t.$$

It appears therefore that the time derivative of a mean of a quantity is the mean of the derivative of that quantity.

The constancy of p^2 then requires that

$$\overline{p \frac{dp}{dt}} = 0. \tag{21}$$

Differentiating (21), we obtain

$$\overline{p \frac{d^2p}{dt^2}} + \overline{\left(\frac{dp}{dt}\right)^2} = 0. \tag{22}$$

Hence the correlation of p and d^2p/dt^2 is, by definition and by virtue of (22),

$$R_1\left(p, \frac{d^2p}{dt^2}\right) = \frac{\overline{p(d^2p/dt^2)}}{\overline{p^2(d^2p/dt^2)^2}} = -\frac{\overline{(dp/dt)^2}}{\overline{p^2(d^2p/dt^2)^2}}, \tag{23}$$

which shows that the correlation between p and its second derivative with respect to time must be negative.

Similarly, since $\overline{(dp/dt)^2}$ is constant,

$$\overline{\frac{dp}{dt} \frac{d^2p}{dt^2}} = 0, \tag{24}$$

which yields, upon differentiation,

$$\overline{\frac{dp}{dt} \frac{d^3p}{dt^3}} + \overline{\left(\frac{d^2p}{dt^2}\right)^2} = 0. \tag{25}$$

Differentiation of (22) yields

$$\overline{p \frac{d^3p}{dt^3}} + 3\overline{\frac{dp}{dt} \frac{d^2p}{dt^2}} = 0.$$

Hence

$$\overline{p \frac{d^3p}{dt^3}} = 0, \tag{26}$$

which further yields, upon differentiation and use of (25),

$$\overline{p \frac{d^4p}{dt^4}} - \overline{\left(\frac{d^2p}{dt^2}\right)^2} = 0 \tag{27}$$

Proceeding in this way, it can be shown that

$$\overline{p \frac{d^{2n}p}{dt^{2n}}} = (-1)^n \overline{\left(\frac{d^np}{dt^n}\right)^2} \tag{28}$$

and

$$\overline{p \frac{d^{2n+1}p}{dt^{2n+1}}} = 0. \tag{29}$$

In analyzing an actual pt curve, it may be tedious to obtain the standard deviations of p and its derivatives. Taylor offers another method of defining the statistical properties of the curve which is equivalent to that given above but which is likely to be more manageable.

Suppose that one takes, as before, the values $p_1, p_2, \ldots, p_n$ at a large number of times $t_1, t_2, \ldots, t_n$. Let these values of p be correlated with the values $p_1', p_2', \ldots, p_n'$ at times $t_1 + \xi, t_2 + \xi, \ldots, t_n + \xi$, where ξ is a finite interval of time which may be positive or negative.

If $p(t)$ is the value of p at time t and $p(t + \xi)$ that at time $t + \xi$, then the Lagrangian correlation coefficient $R_1(\xi)$ is defined by

$$\overline{p(t)p(t + \xi)} = R_1(\xi)\left[\overline{p^2(t)}\overline{p^2(t + \xi)}\right]^{1/2}.$$

Since by hypothesis $\overline{p^2}$ is constant,

$$R_1(\xi) = \frac{\overline{p(t)p(t + \xi)}}{\overline{p^2}}. \tag{30}$$

Now $p(t + \xi)$ can be expanded in powers of ξ,

$$p(t + \xi) = p(t) + \xi \frac{dp}{dt} + \frac{\xi^2}{2!} \frac{d^2p}{dt^2} + \cdots. \tag{31}$$

Hence $\qquad \overline{p(t)p(t + \xi)} = \overline{p^2(t)} + \xi \overline{p\frac{dp}{dt}} + \frac{\xi^2}{2!} \overline{p\frac{d^2p}{dt^2}} + \cdots,$

and, by virtue of (28) and (29) and the definition of $R_1(\xi)$,

$$R_1(\xi) = 1 - \frac{\xi^2}{2!} \frac{\overline{(dp/dt)^2}}{\overline{p^2}} \cdots + (-1)^n \frac{\xi^{2n}}{2n!} \frac{\overline{(d^np/dt^n)^2}}{\overline{p^2}} + \cdots. \tag{32}$$

It will be seen that $R_1(\xi)$ is an even function of ξ, as might have been expected.

Now let the quantity p under consideration be the fluctuational velocity component v' in the y direction, which, in case $\bar{v} = 0$, is just the instantaneous velocity v. Then

$$\overline{v'(t)v'(\xi)} = R_1(\xi - t)\overline{[v'(t)]^2}.$$

If $\overline{v'^2}$ is constant, one has, since $R_1(\xi - t)$ is an even function $\xi - t$,

$$\int_0^t \overline{v'(t)v'(\xi)}\, d\xi = \overline{v'^2} \int_0^t R_1(\xi - t)\, d\xi = \overline{v'^2} \int_0^t R_1(t - \xi)\, d\xi$$

$$= \overline{v'^2} \int_t^0 - R_1(t - \xi)\, d(t - \xi) = \overline{v'^2} \int_0^t R_1(\xi)\, d\xi. \tag{33}$$

Also,

$$\int_0^t \overline{v'(t)v'(\xi)}\, d\xi = \overline{v'(t) \int_0^t v'(\xi)\, d\xi} = \overline{v'(t)Y} = \frac{1}{2}\frac{d}{dt}\overline{Y^2}, \tag{34}$$

so that

$$\overline{Y^2} = 2\overline{v'^2} \int_0^t \int_0^{t'} R_1(\xi) \, d\xi \, dt', \qquad (35)$$

where Y is the distance traversed by a particle in time t in the y direction, $\bar{v}$ being assumed to be zero.

When t is so small that $R_1(\xi)$ does not differ appreciably from 1 during the interval from 0 to t, (35) becomes

$$\overline{Y^2} = \overline{v'^2}t^2 \quad \text{or} \quad \sqrt{\overline{Y^2}} = v''t, \qquad (36)$$

where $(v'')^2 = \overline{v'^2}$. Equation (36) states that the standard deviation of a particle from its initial position is proportional to t when t is small.

The correlation coefficient $R_1(\xi)$ can be expected to fall to zero for large values of ξ. On the assumption that

$$\lim_{t \to \infty} \int_0^t R_1(\xi) \, d\xi = I \text{ (finite)},$$

a time T_1 can be defined such that for $t > T_1$, with unappreciable error,

$$\int_0^t R_1(\xi) \, d\xi = I.$$

Then, for $t > T_1$ after the beginning of the motion,

$$\frac{d}{dt} \overline{Y^2} = 2\overline{v'^2} I, \qquad (37)$$

according to (35), so that $\overline{Y^2}$ increases at a uniform rate. In the limit when $\overline{Y^2}$ is large,

$$\left(\overline{Y^2}\right)^{1/2} = v''(2IT)^{1/2}, \qquad (38)$$

so that the standard deviation of Y is proportional to the square root of T.

The trends of (36) and (38) agree with the observations of Richardson (1920), who performed some experiments on the diffusion of smoke emitted from a fixed point in a wind.

From (37)

$$\overline{Yv'} = \overline{v'^2} I.$$

Hence, utilizing (38), the correlation coefficient of Y and v' is, for very large T,

$$R_1(Y,v') = \frac{\overline{Yv'}}{\left(\overline{Y^2}\right)^{1/2}v''} = \left(\frac{I}{2T}\right)^{1/2}. \qquad (39)$$

If $\overline{Y^2}$ is measured, the equation

$$\frac{d^2}{dt^2} \overline{Y^2} = 2\overline{v'^2} R_1(\xi), \qquad (40a)$$

which is a consequence of (33) and (34), permits $R_1(\xi)$ to be computed.

5. TAYLOR'S THEORY OF ISOTROPIC TURBULENCE

Taylor's theory of turbulent diffusion shows the importance of $R_1(\xi)$. Further progress of the theory of turbulence was made by Taylor (1935, 1936) when he considered isotropic turbulence, or turbulence for which there are no privileged directions.

From (33) and (34),

$$\frac{1}{2}\frac{d}{dt}\,\overline{Y^2} = \overline{v'^2}\int_0^t R_1(\xi)\,d\xi, \tag{40b}$$

where bars again indicate mean values. We can define a length l_1 by

$$l_1 = v''\int_0^\infty R_1(\xi)\,d\xi, \tag{41}$$

which bears the same relationship to diffusion by turbulent motion that the mean free path does to molecular diffusion.

The length l_1 can be considered as the mean free path of particles in turbulent motion in the Lagrangian system.   It is also possible to define a length l_2 which will indicate the scale of turbulence in the Eulerian system.   If we imagine that the correlation R_2 between the values of u' (fluctuating velocity component in the x direction) at two points a distance y apart in the direction of y has been determined for various values of y and that R_2 falls to zero when $y \geq Y$, then l_2 can be defined as

$$l_2 = \int_0^\infty R_2(y)\,dy = \int_0^Y R_2(y)\,dy. \tag{42}$$

The length l_2 may be taken as a definition of the average size of the eddies.

The lengths l_1 and l_2 can be computed if $\int_0^t R_1(\xi)\,d\xi$ and $R_2(y)$ are measured, the former of which can be obtained by measuring $(d/dt)\,\overline{Y^2}$ and the latter by measuring $\overline{u'^2}$ and $\overline{u(0)'u'(y)}$, usually by hot-wire anemometers.

A method due to Prandtl is to pass the currents from the two hot wires (a distance y apart) through coils which cause deflections of a spot of light in two directions at right angles to each other.   If the two hot wires are identical and so close that the correlation is nearly 1.0, the spot of light moves over a very elongated elliptic area, the long axis of which is at $45°$ to the deflections caused by either of the wires in the absence of disturbances from the other.   By measuring the ratio of the principal axes of the elliptical blackened areas produced on a photographic plate by the moving spot of light during a prolonged exposure, it is possible to calculate R_2.   This method is specially suitable for measurements when the correlation is very high, that is, $1 - R_2(y)$ is small.

It may be mentioned that since at small t

$$\sqrt{\overline{Y^2}} = v''t, \tag{43}$$

and since $t = x/U$, where U is the mean velocity of flow,

$$\frac{\sqrt{\overline{Y^2}}}{x} = \frac{v''}{U} \tag{44}$$

at small x, which permits the measurement of the degree of turbulence by measuring the standard deviation of Y at distance x.

So far we have studied the diffusion phenomenon for constant v''. If the turbulence is decaying, v'' is not constant and the diffusion equation should be (34) instead of (40b) or, by an easy transformation,

$$\frac{1}{2} \frac{d}{dt} \overline{Y^2} = \overline{v'(t) \int_0^t v'(\xi)\, d\xi} = \overline{v'(t) \int_0^t v'(t - \xi)\, d\xi}, \tag{45}$$

where $Y = \int_0^t v'(\xi)\, d\xi = \int_0^t v'(t - \xi)\, d\xi$.

Writing $v''(t)$ for $\sqrt{\overline{[v'(t)]^2}}$ etc., we have

$$\frac{1}{2} \frac{d}{dt} \overline{Y^2} = v''(t) \int_0^t v''(t - \xi) R_1[v'(t), v'(t - \xi)]\, d\xi. \tag{46}$$

When v'' is not constant, it is not possible to proceed beyond (46), but experimental evidence seems to indicate that a satisfactory approximation can be obtained if we assume that turbulent fluctuations are proportional to the mean speed, so that if matter from a concentrated source is diffused over an area downstream from the source, an increase in the speed of the main flow leaves the relative distribution of matter in space unchanged though the absolute concentration is reduced. The necessary condition for this situation to exist is that $R_1[v'(t), v'(t - \xi)]$ be a function of ζ only, where

$$d\zeta = v''\, d\xi = \frac{v''}{U}\, dx \tag{47}$$

and $x = Ut$ is the distance downstream from the source. This can be seen by inspection of (46).

The equation which represents the lateral spread of matter or heat from a concentrated source is then

$$\frac{1}{2} \frac{U}{v''} \frac{d}{dx} \overline{Y^2} = \int_0^\zeta R_3(\zeta)\, d\zeta, \tag{48}$$

where

$$\zeta = \int_0^x \frac{v''}{U}\, dx$$

and $R_3(\zeta)$ is the correlation between the velocities of a particle at times t_1 and t_2, with ζ given by $\zeta = \int_{t_1}^{t_2} v''\,dt$. If $R_3(\zeta)$ falls to zero at a finite value of ζ, say $\zeta = \zeta_1$, and remains zero for all greater values of ζ, $\int_0^\zeta R_3(\zeta)\,d\zeta$ is finite. If l_ζ is written for $\int_0^{\zeta_1} R_3(\zeta)\,d\zeta$, then (48) becomes, for sufficiently large x,

$$\frac{1}{2}\frac{U}{v''}\frac{d}{dx}\,\overline{Y^2} = l_\zeta. \tag{49}$$

This has the same form as (40b) and (41), found for turbulence which is not decaying.

Equation (48) may be expressed in the form

$$\frac{1}{2}\frac{d}{d\zeta}\,\overline{Y^2} = \int_0^\zeta R_3(\zeta)\,d\zeta. \tag{50}$$

When ζ is small so that $R_3(\zeta)$ is approximately unity, (50) gives

$$\frac{1}{2}\frac{d}{d\zeta}\,\overline{Y^2} = \zeta, \tag{51}$$

from which, by integration,

$$\overline{Y^2} = \zeta^2 \qquad \text{or} \qquad \sqrt{\overline{Y^2}} = \zeta. \tag{52}$$

When the turbulence is constant, $\zeta = xv''/U$, so that (52) reduces to (44). If the turbulence is not constant, and if $\overline{Y^2}$ and v''/U are measured at a number of values of x, then both ζ and

$$\frac{1}{2}\frac{U}{v''}\frac{d}{dx}\,\overline{Y^2}$$

can be found. Thus $\int_0^\zeta R_3(\zeta)\,d\zeta$ can be plotted against ζ, and $R_3(\zeta)$ can be found graphically from the experimental curve so obtained.

We now proceed to study the dissipation of energy, the principal agents in which are the eddies of very small scale. The rate of dissipation of energy at any instant depends only on the viscosity and the instantaneous velocity in the following way [see (2.26)] if the mean velocity U is constant and is in the x direction:

$$\overline{\Phi} = \mu[2\overline{u_x'^2} + 2\overline{v_y'^2} + 2\overline{w_z'^2} + \overline{(v_x' + u_y')^2}$$
$$+ \overline{(w_y' + v_z')^2} + \overline{(u_z' + w_x')^2}]. \tag{53}$$

In this equation subscripts indicate partial differentiation, for brevity. If the representation of the essential statistical properties of the velocity field can be expressed by the R_2 curve and similar correlation curves, it should be possible to deduce from them the rate of dissipation of energy. This would in general involve a complicated analysis, but the problem can be much simplified if the field of turbulent flow is assumed to be isotropic, in which case

$$\overline{u_x'^2} = \overline{v_y'^2} = \overline{w_z'^2},$$

$$\overline{u_y'^2} = \overline{u_z'^2} = \overline{v_x'^2} = \overline{v_z'^2} = \overline{w_x'^2} = \overline{w_y'^2},$$

and

$$\overline{v_x'u_y'} = \overline{w_y'v_z'} = \overline{u_z'w_x'},$$

so that

$$\frac{\overline{\Phi}}{\mu} = 6(\overline{u_x'^2} + \overline{u_y'^2} + \overline{v_x'u_y'}). \tag{54}$$

Consider the most general possible expression for the mean value of any quadratic function of the nine first-order partial derivatives of the velocity components. The following table will show the number of terms in each of the 10 different groups, the total number of terms being $(9 \cdot 8)/2 + 9 = 45$:

Type	$\overline{u_x'^2}$	$\overline{u_x'u_y'}$	$\overline{u_y'^2}$	$\overline{u_y'u_z'}$	$\overline{u_z'v_x'}$	$\overline{u_x'v_y'}$	$\overline{u_x'v_z'}$	$\overline{u_y'v_x'}$	$\overline{u_y'v_z'}$	$\overline{u_z'v_z'}$
Symbol	a_1	a_2	a_3	a_4	a_5	a_6	a_7	a_8	a_9	a_{10}
No. of terms in group	3	6	6	3	6	3	6	3	6	3 = 45

From the equation of continuity and the condition of isotropy it can be obtained that

$$a_1 = -2a_6. \tag{55}$$

On rotating the reference axes by 45° in different ways, the condition of isotropy of turbulence furnishes

$$a_2 = a_4 = a_5 = a_7 = a_9 = a_{10} = 0, \tag{56}$$

$$a_1 - a_3 - a_6 - a_8 = 0. \tag{57}$$

One more equation is needed in order that all the nonvanishing a's can be expressed in terms of one of them, a_3, say. This can be obtained by noting that [Lamb (1932, p. 581)], by suitable combination and use of Green's theorem,

$$\int_V \frac{\overline{\Phi}}{\mu} \, dV = \int_V (\xi^2 + \eta^2 + \zeta^2) \, dV - \int_S \frac{\partial}{\partial n} q'^2 \, dS$$

$$+ 2 \int_S \begin{vmatrix} l' & m' & n' \\ u' & v' & w' \\ \xi & \eta & \zeta \end{vmatrix} dS, \tag{58}$$

where

$$\xi = w_y' - v_z', \quad \text{etc.},$$

$$q'^2 = u'^2 + v'^2 + w'^2,$$

n is measured in the direction of the normal to the surface S enclosing a control volume V and l', m', and n' are the direction cosines of this normal. If S is large compared with the scale of the turbulence, the surface integrals are small compared with the volume integrals and can be neglected. Hence, taking the mean values of all quantities in (58),

$$\frac{\overline{\Phi}}{\mu} = \overline{\xi^2} + \overline{\eta^2} + \overline{\zeta^2} = 6a_3 - 6a_8. \tag{59}$$

Equations (54) and (59) then give

$$6a_1 + 6a_3 + 6a_8 = 6a_3 - 6a_8,$$

or

$$a_1 + 2a_8 = 0. \tag{60}$$

Using (55), (57), and (60), we have

$$a_1 = a_3/2 = -2a_6 = -2a_8, \tag{61}$$

and

$$\frac{\overline{\Phi}}{\mu} = 7.5a_3 = 7.5 \overline{\left(\frac{\partial u'}{\partial y}\right)^2}. \tag{62}$$

From an equation analogous to (32), we have in this case

$$R_2 = 1 - \frac{1}{2!}\frac{y^2}{\overline{u'^2}}\overline{\left(\frac{\partial u'}{\partial y}\right)^2} + \frac{1}{4!}\frac{y^4}{\overline{u'^2}}\overline{\left(\frac{\partial^2 u'}{\partial y^2}\right)^2} + \cdots, \tag{63}$$

from which

$$\overline{\left(\frac{\partial u'}{\partial y}\right)^2} = 2\overline{u'^2}\lim_{y\to 0}\left(\frac{1 - R_2(y)}{y^2}\right). \tag{64}$$

Defining λ_1^2 as the radius of curvature of the R_2 curve at $y = 0$, we have

$$\frac{1}{\lambda_1^2} = 2\lim_{y\to 0}\left(\frac{1 - R_2(y)}{y^2}\right),$$

or, on putting $\lambda_1 = \lambda/\sqrt{2}$,

$$\frac{1}{\lambda^2} = \lim_{y\to 0}\left(\frac{1 - R_2(y)}{y^2}\right). \tag{65}$$

A physical interpretation of λ may be found by describing the parabola which touches the $R_2(y)$ curve of the origin. This parabola will cut the axis $R_2(y) = 0$ at the point $y = \lambda$. The λ may be regarded roughly as a measure of the diameters of the smallest eddies which are responsible for the dissipation of energy.

Combining (62), (64), and (65), we have

$$\overline{\Phi} = 15\mu\frac{\overline{u'^2}}{\lambda^2}. \tag{66}$$

If $\overline{\Phi}$, $\overline{u'^2}$, and λ^2 can be measured independently, the last equation offers a check on the theory. It may be noticed also that if the Reynold's stresses in geometrically similar fields of flow are proportional to $\overline{u'^2} = u''^2$, $\overline{\Phi}$ is proportional to u''^3, and λ to $(u'')^{-1/2}$. Since λ^{-2} is proportional to the curvature of the R_2 curve, the latter must be proportional to u''. In the limit of very high values of u'', the $R_2(y)$ curve may be expected to have a pointed top.

The question may be asked about the relation between λ and l, where l is some linear dimension defining the scale of the turbulence system. So far as changes in linear dimensions, velocity, and density are concerned,

$$\overline{\Phi} = \text{const} \cdot \left| \frac{\rho u''^3}{l} \right|. \tag{67}$$

Combination of (66) and (67) gives

$$\frac{\lambda^2}{l^2} = C \frac{\nu}{lu''} \, ,$$

so that if l is taken to be the mesh length M,

$$\frac{\lambda}{M} = A \sqrt{\frac{\nu}{Mu''}} \, , \tag{68}$$

where A is assumed by Taylor to be an absolute constant for all grids of a definite type, e.g., for all square-mesh grids or honeycombs.

One is now in a position to predict the way in which turbulence may be expected to decay when a definite scale has been given to it as the air stream passes through a regular grid or honeycomb.

The rate of loss of kinetic energy of the turbulence per unit volume is, in isotropic turbulence,

$$-\frac{\rho U}{2} \frac{d}{dx} (\overline{u'^2} + \overline{v'^2} + \overline{w'^2}) = -\frac{3\rho U}{2} \frac{d}{dx} (\overline{u'^2}).$$

This must be equal to the rate of dissipation $\overline{\Phi}$, so that

$$-\frac{3\rho U}{2} \frac{d}{dx} \overline{u'^2} = 15\mu \frac{\overline{u'^2}}{\lambda^2} \, ,$$

which becomes, by virtue of (68),

$$-\frac{U}{u''^3} \frac{d}{dx} u''^2 = \frac{10}{A^2 M} \, .$$

Integration gives

$$\frac{U}{u''} = \frac{5x}{A^2 M} + \text{const}, \tag{69}$$

which states that U/u'' increases linearly with x.

The space gradients of pressure fluctuations in isotropically turbulent flow will now be investigated. With ζ defined by (47), (32) gives, on substituting v' for the general quantity p, ζ for t, and $R_3(\zeta)$ for $R_1(\xi)$,

$$\overline{\left(\frac{Dv'}{D\zeta}\right)^2} = 2v''^2 \lim_{\zeta \to 0} \frac{1 - R_3(\zeta)}{\zeta^2}, \tag{70}$$

where $Dv'/D\zeta$ is the rate of change of v' with respect to ζ as the particle moves downstream with mean velocity U and v''^2 is written for $\overline{v'^2}$. If v'' is constant (nondecaying),

$$\frac{Dv'}{D\zeta} = \frac{Dv'}{D[(v''/U)x]} = \frac{1}{v''}\frac{Dv'}{Dt},$$

and

$$\overline{\left(\frac{Dv'}{Dt}\right)^2} = 2v''^4 \lim_{\zeta \to 0} \frac{1 - R_3(\zeta)}{\zeta^2} = 2\frac{v''^4}{\lambda_\zeta^2}. \tag{71}$$

Now Dv'/Dt is simply the acceleration of a particle in the direction y, since it is a derivative by pursuing the particle. The equation of motion in the y direction is, if viscous effects on p' are neglected,

$$-\frac{1}{\rho}\frac{\partial p'}{\partial y} = \frac{Dv'}{Dt}. \tag{72}$$

Thus the last two equations give

$$\overline{\left(\frac{\partial p'}{\partial y}\right)^2} = 2\rho^2\frac{v''^4}{\lambda_\zeta^2}. \tag{73}$$

It is natural to inquire about the relation of λ_ζ and λ. In the Eulerian system,

$$-\frac{1}{\rho}\frac{\partial p'}{\partial y} = \frac{\partial v'}{\partial t} + \frac{1}{2}\frac{\partial}{\partial y}(u'^2 + v'^2 + w'^2)$$

$$- w'\left(\frac{\partial w'}{\partial y} - \frac{\partial v'}{\partial z}\right) + u'\left(\frac{\partial v'}{\partial x} - \frac{\partial u'}{\partial y}\right).$$

Taylor made the assumption that $\dfrac{1}{\rho}\left[\overline{\left(\dfrac{\partial p'}{\partial y}\right)^2}\right]^{1/2}$ is of the same order of magnitude as

$$\frac{1}{2}\left[\overline{\left(\frac{\partial}{\partial y}q'^2\right)^2}\right]^{1/2},$$

which in isotropic turbulence is equal to

$$\frac{3}{2}\left[\overline{\left(\frac{\partial v'^2}{\partial y}\right)^2}\right]^{1/2}.$$

This in turn must be of the same order of magnitude as

$$\left[\overline{v'^2\left(\frac{\partial v'}{\partial y}\right)^2}\right]^{1/2}.$$

This led Taylor to assume that in isotropic turbulence

$$\left[\overline{\left(\frac{\partial p'}{\partial y}\right)^2}\right]^{1/2} = 3B\rho\left[\overline{v'^2\left(\frac{\partial v'}{\partial y}\right)^2}\right]^{1/2}, \tag{74}$$

where B is a constant which is expected to be of order of magnitude unity. From (73) and (74),

$$B^2 = \frac{2v''^4}{9\lambda_\zeta^2\overline{v'^2}(\overline{\partial v'/\partial y})^2}, \tag{75}$$

where $$\overline{\left(\frac{\partial v'}{\partial y}\right)^2} = a_1 = \frac{a_3}{2} = \frac{1}{2}\overline{\left(\frac{\partial u'}{\partial y}\right)^2} = \frac{v''^2}{\lambda^2} \tag{76}$$

by (61), (64), and (65). Thus

$$B^2 = \frac{2}{9}\frac{\lambda^2}{\lambda_\zeta^2}. \tag{77}$$

The assumption represented by (74) is therefore equivalent to the assumption that λ_ζ is a constant multiple of λ.

Experiments quoted in Taylor's paper (1935) showed that

$$l_1 = 0.1M, \qquad l_2 = 0.2M$$

and verified (66), (68), and (69), the value of A being found to be approximately 2. The set of experiments done by Schubauer (1935) gives $B = 0.94$, so that (77) is equivalent to $\lambda = 2\lambda_\zeta$ approximately. All these experimental results are subject to the restriction that lu''/ν, the Reynolds number of turbulence, is greater than some number which must be determined by experiment.

The essential features of diffusion in isotropic (nondecaying) turbulence, according to Taylor's theory, are as follows:

1 For time intervals which are small in comparison with the ratio of l_1 to u'', the diffusing quantity N spreads at a uniform rate proportional to u'', and the rate does not depend on l_1.

2 For time intervals which are large in comparison with the ratio of l_1 to u'', the diffusing quantity N spreads in accordance with the usual diffusion equation

$$\frac{DN}{Dt} = \frac{\partial}{\partial x}\left(K\frac{\partial N}{\partial x}\right) + \frac{\partial}{\partial y}\left(K\frac{\partial N}{\partial y}\right) + \frac{\partial}{\partial z}\left(K\frac{\partial N}{\partial z}\right)$$

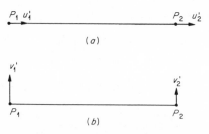

FIGURE 9. DEFINITION SKETCH.

with a constant coefficient of diffusion K equal to $l_1 u''$, l_1 being defined as $u'' \int_0^\infty R_1(\xi)\, d\xi$.

3 For intermediate intervals, the diffusion depends on $R_1(\xi)$.

If the turbulence is decaying, similar conclusions may be drawn, according to Taylor's theory, by replacing l_1 with l_ζ and by redefining K as $l_\zeta u''$, where u'' is now varying and l_ζ is defined as $\int_0^\infty R_3(\zeta)\, d\zeta$, ζ being defined as

$$\zeta = \int_0^x \frac{v''}{U}\, dx$$

and U being the mean velocity, which is in the x direction. [For later developments of the theory, see Taylor (1937, 1938a and b).]

6. VON KÁRMÁN'S THEORY

In isotropic turbulence the correlation tensor has spherical symmetry, and its components are functions only of the distance vector **r** between the two points and of the time t. Denote by u_1', v_1', w_1', and u_2', v_2', w_2' the components of the velocity fluctuations at the points P_1 and P_2 having coordinates $(x_1',0,0)$ and $(x_2',0,0)$, respectively. (*Caution:* The subscripts no longer distinguish coordinates: they now distinguish the locations.) Suppose that $\overline{u_1'^2}$, $\overline{v_1'^2}$, $\overline{w_1'^2}$, which by isotropy are equal, are independent of position and equal to $\overline{u'^2}$. Then

$$\overline{u_1'^2} = \overline{v_1'^2} = \overline{w_1'^2} = \overline{u_2'^2} = \overline{v_2'^2} = \overline{w_2'^2} = \overline{u'^2}.$$

The correlation coefficients $R_{22} = \overline{v_1' v_2'}/\overline{u'^2}$ and $R_{33} = \overline{w_1' w_2'}/\overline{u'^2}$ will be identical because of isotropy and will be a function $g(r,t)$ of r and t. The correlation coefficient $R_{11} = \overline{u_1' u_2'}/\overline{u'^2}$ will be a function $f(r,t)$ (see Fig. 9).

All the other correlation coefficients can be shown to be zero by rotations and reflections, remembering the isotropy of turbulence. Thus the correlation tensor is, for the particular points chosen,

$$
\begin{pmatrix}
f(r,t) & 0 & 0 \\
0 & g(r,t) & 0 \\
0 & 0 & g(r,t)
\end{pmatrix},
$$

which can be decomposed to

$$
[f(r,t) - g(r,t)]\begin{pmatrix}
1 & 0 & 0 \\
0 & 0 & 0 \\
0 & 0 & 0
\end{pmatrix} + g(r,t)\mathbf{I},
$$

where $\mathbf{I}$ is the tensor δ_{ij}. This is the starting point of von Kármán's theory (1937a and b). If a rotation is given to the coordinate axes, such that the points $(x_1',0,0)$ and $(x_2',0,0)$ assume their new coordinates (x_1,y_1,z_1) and (x_2,y_2,z_2), it can readliy be seen from the methods of tensor calculus (Appendix 2) that the correlation tensor will assume the form in the new coordinate system:

$$
R_{ij} = \frac{f(r,t) - g(r,t)}{r^2} X_i X_j + g(r,t)\delta_{ij}, \tag{78}
$$

where $X_1 = x_2 - x_1$, $X_2 = y_2 - y_1$, $X_3 = z_2 - z_1$, $r^2 = X_1^2 + X_2^2 + X_3^2$, and where the first subscript of R corresponds to the velocity component at (x_1,y_1,z_1), the second to that at (x_2,y_2,z_2), and subscripts 1, 2, 3 of the R's corresponding to the velocity components u', v', w'. The δ_{ij} is the Kronecker delta denoted by $\mathbf{I}$ above.

The equation of continuity is

$$
\frac{\partial u_2'}{\partial x_2} + \frac{\partial v_2'}{\partial y_2} + \frac{\partial w_2'}{\partial z_2} = 0.
$$

Multiplying this by $u_1'/\overline{u'^2}$, which is independent of x_2, y_2, and z_2, and averaging over time, we have

$$
\frac{\partial R_{11}}{\partial \overline{X}_1} + \frac{\partial R_{12}}{\partial \overline{X}_2} + \frac{\partial R_{13}}{\partial \overline{X}_3} = 0. \tag{79}
$$

Substitution of the R_{ij} given in (78) into (79) gives

$$
X_1\left[2(f - g) + r\left(\frac{\partial f}{\partial r}\right)\right] = 0.
$$

Since X_1 is arbitrary,

$$2f(r,t) - 2g(r,t) = -r\frac{\partial f(r,t)}{\partial r}.$$ (80)

Defining

$$L = \int_0^\infty R_y\, dx = \int_0^\infty g(r,t)\, dr,$$ (81)

$$L_x = \int_0^\infty R_x\, dx = \int_0^\infty f(r,t)\, dr,$$ (82)

where $R_y = g(r,t)$, $R_x = f(r,t)$, we have

$$L - L_x = \frac{1}{2}\int_0^\infty r\frac{\partial f}{\partial r}\, dr = -\frac{1}{2}\int_0^\infty f(r,t)\, dr = -\frac{L_x}{2},$$

or

$$2L = L_x.$$ (83)

Since f and g are even functions of r,

$$f = 1 + \frac{f_0'' r^2}{2} + \cdots,$$ (84)

$$g = 1 + \frac{g_0'' r^2}{2} + \cdots,$$ (85)

where the quantities f_0'', g_0'', etc., are functions of time only. From (80), $2f_0'' = g_0''$. Hence for small values of r,

$$R_{ij} = \left(1 + \frac{g_0''}{2}r^2\right)\delta_{ij} + \left(\frac{f_0'' - g_0''}{2}\right)X_iX_j$$

$$= (1 + f_0'' r^2)\delta_{ij} - \frac{f_0''}{2}X_iX_j.$$ (86)

The second derivatives are, for very small r and using X, Y, Z for X_1, X_2, and X_3,

$$\frac{\partial^2 R_{11}}{\partial X^2} = \frac{\partial^2 R_{22}}{\partial Y^2} = \frac{\partial^2 R_{33}}{\partial Z^2} = f_0'',$$ (87)

$$\frac{\partial^2 R_{11}}{\partial Y^2} = \frac{\partial^2 R_{11}}{\partial Z^2} = \text{similar terms by permutation} = 2f_0'',$$ (88)

$$\frac{\partial^2 R_{12}}{\partial X\, \partial Y} = \text{similar terms by permutation} = -\frac{f_0''}{2},$$ (89)

and all others, for example, $\partial^2 R_{12}/(\partial X\, \partial Z)$, are zero.

Von Kármán (1937a and b) points out that the correlation tensor is of the same form as the stress tensor for a continuous medium when there is spherical symmetry. In the analogy $f(r)$ corresponds to the principal radial stress at any point, $g(r)$ to the principal transverse stress, and the several R's to the stress components over planes normal to the coordinate axes. The relation between f and g given by the continuity equation corresponds to the condition for equilibrium of the stresses.

The correlations of the derivatives of the velocity components are found by noting that the velocity components at point 2 are independent of the coordinates of point 1, and vice versa. Thus, we have

$$\overline{v_2' \frac{\partial u_1'}{\partial x_1}} = \overline{\frac{\partial u_1' v_2'}{\partial x_1}} = \overline{u'^2} \frac{\partial R_{12}}{\partial x_1} = -\overline{u'^2} \frac{\partial R_{12}}{\partial X}$$

and

$$\overline{\frac{\partial u_1'}{\partial x_1} \frac{\partial v_2'}{\partial y_2}} = \overline{\frac{\partial}{\partial y_2} v_2' \frac{\partial u_1'}{\partial x_1}} = -\overline{u'^2} \frac{\partial^2 R_{12}}{\partial X \partial Y},$$

so that, upon letting points 1 and 2 coincide, we have

$$\overline{\frac{\partial u'}{\partial x} \frac{\partial v'}{\partial y}} = \overline{u_x' v_y'} = -\overline{u'^2} \left(\frac{\partial^2 R_{12}}{\partial X \partial Y} \right)_{X=Y=0} = -\frac{f_0''}{2} \overline{u'^2} = -\frac{f_0''}{2} u''^2 \qquad (90)$$

by virtue of (89). By similar reasoning, we obtain

$$\overline{\left(\frac{\partial u'}{\partial x} \right)^2} = \overline{\left(\frac{\partial v'}{\partial y} \right)^2} = \overline{\left(\frac{\partial w'}{\partial z} \right)^2} = -u''^2 f_0'', \qquad (91)$$

$$\overline{\left(\frac{\partial u'}{\partial y} \right)^2} = \text{similar terms by permutation} = -2u''^2 f_0'', \qquad (91a)$$

and

$$\overline{\frac{\partial v'}{\partial x} \frac{\partial u'}{\partial y}} = \overline{\frac{\partial w'}{\partial y} \frac{\partial v'}{\partial z}} = \overline{\frac{\partial u'}{\partial z} \frac{\partial w'}{\partial x}} = \frac{u''^2}{2} f_0''. \qquad (92)$$

Thus the relations between the quantities a_1, a_3, a_6, and a_8 in Taylor's theory have been found in a simpler and more elegant way.

Since g_0'' is the curvature of the g curve or the R_y curve, from (65) we see that

$$g_0'' = \frac{2}{\lambda^2}$$

and, since $2f_0'' = g_0''$, $\qquad\qquad f_0'' = \frac{1}{\lambda^2}, \qquad\qquad (93)$

so that all the mean values of the products of the derivatives can be expressed as multiples of u''^2/λ^2.

To investigate the propagation of the correlation with time, we recall that, there being no mean flow, the velocity fluctuations satisfy the Navier-Stokes equations of motion, the first of which is, for point 1,

$$\frac{\partial u_1'}{\partial t} + u_1' \frac{\partial u_1'}{\partial x_1} + v_1' \frac{\partial u_1'}{\partial y_1} + w_1' \frac{\partial u_1'}{\partial z'} = -\frac{1}{\rho} \frac{\partial p'}{\partial x_1} + \nu \nabla^2 u_1'. \qquad (94)$$

Multiplying by u'_2 and taking mean values, we have, on the assumption that the triple correlations and the term involving the pressure are zero,

$$\frac{\partial}{\partial t} R_{11} u''^2 = 2\nu u''^2 \nabla^2 R_{11}.$$

Identical equations for the other elements of the correlation tensor can be obtained, and all these equations including the one above can be replaced by the single equation

$$\frac{\partial}{\partial t} f u''^2 = 2\nu u''^2 \left(\frac{\partial^2 f}{\partial r^2} + \frac{4}{r} \frac{\partial f}{\partial r} \right). \tag{95}$$

Equations (84) and (95) entail

$$\frac{du''^2}{dt} = -10\nu u''^2 f''_0 = \frac{-10\nu u''^2}{\lambda^2}, \tag{96}$$

where d/dt has replaced $\partial/\partial t$ because u''^2 is a function of time only. Elimination of u''^2 from (95) and (96) gives

$$\frac{\partial f}{\partial t} = 2\nu \left(\frac{\partial^2 f}{\partial r^2} + \frac{4}{r} \frac{\partial f}{\partial r} + \frac{5}{\lambda^2} f \right). \tag{97}$$

This equation determines $f(r,t)$ for all subsequent times if f is given at $t = 0$ for all values of r.

If the initial shape of the correlation function $f(r,t)$ is arbitrary, its shape will in general change with time.   However, there are special cases in which the shape of the correlation function remains similar.   This will occur if $f(r,t)$ is a function of the dimensionless variable $\xi = r/\sqrt{\nu t}$ only. Then (97) is reduced to

$$\frac{d^2 f}{d\xi^2} + \left(\frac{4}{\xi} + \frac{\xi}{4} \right) \frac{df}{d\xi} + \frac{5\nu t}{\lambda^2} f = 0. \tag{98}$$

From this equation, on making $\xi \to 0$, we have, since f is an even function of ξ,

$$5 \left(\frac{d^2 f}{d\xi^2} \right)_{\xi=0} = -\frac{5\nu t}{\lambda^2}.$$

or

$$\frac{\nu t}{\lambda^2} = -\left(\frac{d^2 f}{d\xi^2} \right)_{\xi=0} = \alpha \text{ (say)}.$$

Thus

$$\lambda^2 = \frac{1}{\alpha} \nu t, \tag{99}$$

which means that λ^2 increases linearly with time if the shape of the correlation function remains similar.   The numerical factor α, which determines the rate of increase, is given by the initial shape of f.

We are now in a position to discuss the decay of turbulence, under the restriction that the shape of f remains similar. From (96) and (99),

$$\frac{du''^2}{dt} = -10\alpha\frac{u''^2}{t},$$

that is, $u''^2 = \frac{C}{t^{10\alpha}},$ $C = \text{const.}$ (100)

Now suppose that the fluid moves with uniform speed U in the direction of the x axis. If $t = t_0$ when $x = 0$ so that $t = t_0 + x/U$, (100) gives

$$u''^2 = u_0''^2/[1 + (x/Ut_0)]^{10\alpha}. \tag{101}$$

With λ_0 corresponding to t_0, (99) gives

$$\lambda^2 = \frac{\nu}{\alpha}\left(t_0 + \frac{x}{U}\right) = \lambda_0^2 + \frac{\nu}{\alpha}\frac{x}{U} = \lambda_0^2\left(1 + \frac{x}{Ut_0}\right). \tag{102}$$

Taylor's result contained in (69) can be written as

$$\frac{1}{u''} = \frac{1}{u_0''} + \text{const} \cdot \frac{x}{U}.$$

This is a special case of (101), with $\alpha = \frac{1}{5}$. Also (101) and (102) give

$$\frac{u''\lambda^2}{u_0''\lambda_0^2} = \left(1 + \frac{x}{Ut_0}\right)^{1-5\alpha}.$$

According to Taylor's assumption expressed in (68), $u''\lambda^2/\nu$ should be constant for the same grid. This is true again only if $\alpha = \frac{1}{5}$.

7. THE KÁRMÁN-HOWARTH THEORY

In von Kármán's theory just presented, triple correlations are neglected. Later, von Kármán (1938) and von Kármán and Howarth (1938) also considered the triple correlations

$$h(r,t) = \frac{\overline{v_1'^2 u_2'}}{u''^3}, \qquad h_1(r,t) = \frac{\overline{u_1'^2 u_2'}}{u''^3},$$

$$h_2(r,t) = \frac{\overline{u_1'v_1'u_2'}}{u''^3}, \tag{103}$$

where the two points 1 and 2 are aligned along the direction of u'. They then showed that the general triple correlation tensor $\mathbf{T}$ is a function of X, Y, Z, and t, that in isotropic turbulence $\mathbf{T}$ is expressible in terms of $h(r,t)$, $h_1(r,t)$, and $h_2(r,t)$, and that the development of these functions in powers of r begins with the r^3 term. The equation of continuity permits the expression of h_1 and h_2 in terms of h by the relations

$$h_1 = -2h, \tag{104}$$

$$h_2 = -h - \frac{r}{2}\frac{dh}{dr}. \tag{105}$$

Thus the tensor **T** can be expressed solely in terms of the scalar function $h(r,t)$. The development is similar to that leading to (80).

To investigate the propagation of the correlation with time, we again use (94) and the other two Navier-Stokes equations. Multiplying this equation by u_2' and introducing X, Y, and Z, and taking the mean, we obtain

$$\overline{u_2'\frac{\partial u_1'}{\partial t}} - \frac{\partial \overline{u_1'^2 u_2'}}{\partial X} - \frac{\partial \overline{u_1' v_1' u_2'}}{\partial Y} - \frac{\partial \overline{u_1' w_1' u_2'}}{\partial Z}$$

$$= -\frac{1}{\rho}\overline{\frac{\partial p'}{\partial x_1}u_2'} + \nu\left(\frac{\partial^2}{\partial X^2} + \frac{\partial^2}{\partial Y^2} + \frac{\partial^2}{\partial Z^2}\right)\overline{u_1' u_2'} \tag{106}$$

and, by a similar procedure,

$$\overline{u_1'\frac{\partial u_2'}{\partial t}} + \frac{\partial \overline{u_2'^2 u_1'}}{\partial X} + \frac{\partial \overline{u_2' v_2' u_1'}}{\partial Y} + \frac{\partial \overline{u_2' w_2' u_1'}}{\partial Z}$$

$$= -\frac{1}{\rho}\overline{\frac{\partial p'}{\partial x_2}u_1'} + \nu\left(\frac{\partial^2}{\partial X^2} + \frac{\partial^2}{\partial Y^2} + \frac{\partial^2}{\partial Z^2}\right)\overline{u_1' u_2'}. \tag{107}$$

Von Kármán and Howarth showed, by the use of symmetry arguments and the equation of continuity, that the pressure terms vanish. Adding the last two equations and introducing the correlation coefficient R_{11} (remembering that $\overline{u_1'^2 u_2'} = -\overline{u_2'^2 u_1'}$, etc.), we have

$$\frac{\partial}{\partial t}u''^2 R_{11} - 2\left(\frac{\partial}{\partial X}\overline{u_1'^2 u_2'} + \frac{\partial}{\partial Y}\overline{u_1' v_1' u_2'} + \frac{\partial}{\partial Z}\overline{u_1' w_1' u_2'}\right)$$

$$= 2\nu\left(\frac{\partial^2}{\partial X^2} + \frac{\partial^2}{\partial Y^2} + \frac{\partial^2}{\partial Z^2}\right)\overline{u_1' u_2'}. \tag{108}$$

This equation may be expressed in terms of the functions f, g, h_1, h_2, and h. Remembering the relation between h_1, h_2 and h, we obtain the relation between f and h

$$\frac{\partial f u''^2}{\partial t} + 2u''^3\left(\frac{\partial h}{\partial r} + \frac{4h}{r}\right) = 2\nu u''^2\left(\frac{\partial^2 f}{\partial r^2} + \frac{4}{r}\frac{\partial f}{\partial r}\right) \tag{109}$$

This equation expresses the change of f with t but cannot be solved without some knowledge of the function h, which illustrates that closure is the central problem in the theory of turbulence.

8. TAYLOR'S SPECTRUM ANALYSIS

The following words of Dryden (1943) serve as an excellent introduction to the spectral analysis of turbulence:

> *The description of turbulence in terms of intensity and scale resembles the description of the molecular motion of a gas by temperature and mean free path. A more detailed picture can be obtained by considering the distribution of energy among eddies of different sizes, or more conveniently the distribution of energy with frequency.*

As far as the present writer is aware, Taylor (1938a) was the first to develop a spectrum theory of turbulence. His analysis has been generalized to three-dimensional wave-number space by Batchelor (1949). For simplicity, Taylor's analysis for one-dimensional wave-number space will be presented.

We shall be concerned with the mean square value $\overline{u'^2}$. Expressing u' in the form of a Fourier integral,

$$u' = \frac{1}{\pi} \int_0^\infty \int_{-\infty}^\infty u' \cos \omega(t - t') \, dt' \, d\omega, \qquad (110)$$

where $\omega = 2\pi n$ is the angular velocity associated with n, and writing

$$I_1 = \frac{1}{2\pi} \int_{-T}^T u' \cos \omega t \, dt = \frac{1}{2\pi} \int_{-T}^T u' \cos 2n\pi t \, dt, \qquad (111)$$

$$I_2 = \frac{1}{2\pi} \int_{-T}^T u' \sin \omega t \, dt = \frac{1}{2\pi} \int_{-T}^T u' \sin 2n\pi t \, dt, \qquad (112)$$

we seek to express $\overline{u'^2}$ in terms of I_1 and I_2. This can be done in a simple manner by breaking $u'(t)$ into two parts, one being an even and the other an odd function of t. Thus,

$$u'(t) = E(t) + D(t),$$

where $\qquad E(t) = \dfrac{u'(t) + u'(-t)}{2} \qquad D(t) = \dfrac{u'(t) - u'(-t)}{2}. \qquad (113)$

The Fourier cosine transform for the even function $E(t)$ is

$$g(\omega) = \sqrt{\frac{2}{\pi}} \int_0^\infty E(t) \cos \omega t \, dt$$

$$= \sqrt{\frac{1}{2\pi}} \int_0^\infty [u'(t) + u'(-t)] \cos \omega t \, dt \qquad (114)$$

$$= \sqrt{\frac{1}{2\pi}} \int_{-\infty}^\infty u'(t) \cos \omega t \, dt = \sqrt{2\pi} \lim_{T \to \infty} I_1.$$

Similarly, the Fourier sine transform of the odd function $D(t)$ is

$$h(\omega) = \sqrt{2\pi} \lim_{T \to \infty} I_2. \tag{115}$$

Now, according to a theorem in Fourier integrals which corresponds to the Parseval's theorem in Fourier series,

$$\int_0^\infty E^2 \, dt = \lim_{T \to \infty} \int_0^T E^2 \, dt = \int_0^\infty g^2(\omega) \, d\omega = 2\pi \lim_{T \to \infty} \int_0^T I_1^2 \, d\omega, \tag{116}$$

Similarly, $$\int_0^\infty [D(t)]^2 \, dt = 2\pi \lim_{T \to \infty} \int_0^T I_2^2 \, d\omega. \tag{117}$$

But $$\lim_{T \to \infty} \frac{1}{T} \int_0^T (E^2 + D^2) \, dt = \lim_{T \to \infty} \frac{1}{T} \int_0^T \frac{[u'(t)]^2 + [u'(-t)]^2}{2} \, dt$$

$$= \lim_{T \to \infty} \frac{1}{2T} \int_{-T}^T [u'(t)]^2 \, dt = \overline{u''^2}. \tag{118}$$

So by virtue of the last three equations,

$$\overline{u''^2} = 2\pi \lim_{T \to \infty} \frac{1}{T} \int_0^T (I_1^2 + I_2^2) \, d\omega = 2\pi \int_0^\infty \lim_{T \to \infty} \frac{I_1^2 + I_2^2}{T} \, d\omega$$

$$= 4\pi^2 \int_0^\infty \lim_{T \to \infty} \frac{I_1^2 + I_2^2}{T} \, dn. \tag{119}$$

If we define $$F(n) = \frac{4\pi^2}{\overline{u''^2}} \lim_{T \to \infty} \frac{I_1^2 + I_2^2}{T}, \tag{120}$$

we can write (119) as $$\overline{u''^2} = \overline{u''^2} \int_0^\infty F(n) \, dn, \tag{121}$$

in which $F(n)$ appears as the probability density of the power spectrum, $F(n) \, dn$ is the contribution to $\overline{u''^2}$ from frequencies between n and $n + dn$, and

$$\int_0^\infty F(n) \, dn = 1. \tag{122}$$

Equation (120) gives the formula for obtaining $F(n)$ by measurement.

When the fluctuations are superposed on a stream of mean velocity U and are small compared with U, the changes in u' at a fixed point may be regarded as due to the passage of a fixed turbulent pattern over the point; i.e., it may be hypothesized that

$$u' = \phi(t) = \phi\left(\frac{x}{U}\right), \tag{123}$$

where x is measured upstream from the fixed point and $\phi(x/U)$ may be considered to be the instantaneous spatial distribution of u' at time $t = 0$.

The constancy of U implies that we are considering a turbulent flow far from solid boundaries. The correlation R_x between the fluctuations at time t and $t + x/U$ is defined by

$$R_x = \frac{\overline{\phi(t)\phi(t + x/U)}}{u''^2} = \lim_{T \to \infty} \frac{\int_{-T}^{T} \phi(t)\phi\left(t + \frac{x}{U}\right) dt}{2Tu''^2}. \tag{124}$$

A theorem in the theory of Fourier transform states that if $g(\omega)$ and $h(\omega)$ are the Fourier cosine transforms, respectively, of the even functions $E_1(t)$ and $E_2(t)$, then

$$\int_0^{\infty} g(\omega)h(\omega) \, d\omega = \int_0^{\infty} E_1(t)E_2(t) \, dt. \tag{125}$$

A similar theorem exists for the odd functions. By expressing $\phi(t)$ and $\phi(t + x/U)$ as sums of an even and an odd function, it can be proved by virtue of these theorems that (with $T = \infty$ in I_1 and I_2)

$$\int_{-\infty}^{\infty} \phi(t)\phi\left(t + \frac{x}{U}\right) dt = 8\pi^2 \int_0^{\infty} (I_1{}^2 + I_2{}^2) \cos\left(\frac{2\pi nx}{U}\right) dn, \tag{126}$$

from which we obtain, with (120) and (124),

$$R_x = \int_0^{\infty} F(n) \cos \frac{2\pi nx}{U} \, dn \tag{127}$$

and

$$F(n) = \frac{4}{U} \int_0^{\infty} R_x \cos \frac{2\pi nx}{U} \, dx. \tag{128}$$

A typical graph showing the variation of $F(n)$, obtained from (120) by measurement, with n is shown in Fig. 10. From (128) it can be seen that if R_x is independent of U, then $UF(n)$ must be a function of n/U. Figure 11 indicates that R_x is independent of U in the range of measurement. The value of $UF(n)$ calculated from (128) by measured values of R_x is also plotted in Fig. 10a, which shows the accuracy of the measurements. A portion of Fig. 10a is enlarged and shown in Fig. 10b. The measurements were made by L. F. G. Simmons.

The correlation coefficient R_x and the quantity $UF(n)/\sqrt{8\pi}$ are Fourier cosine transforms. If either is measured, the other can be computed. The correlation coefficient R_x is actually the function f in von Kármán's theory, so that the length λ is related to R_x by

$$\frac{1}{\lambda^2} = 2 \lim_{x \to 0} \frac{1 - R_x}{x^2}. \tag{129}$$

When n and x are small,

$$\cos \frac{2\pi nx}{U} = 1 - \frac{2\pi^2 n^2 x^2}{U^2}.$$

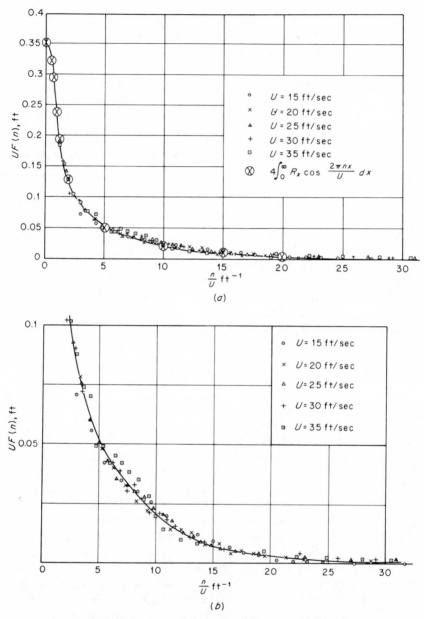

FIGURE 10.   (a) THE VARIATION OF $UF(n)$ WITH n/U, n BEING THE FREQUENCY, $F(n)$ THE
ENERGY DENSITY FUNCTION IN THE FREQUENCY SPECTRUM, AND U THE MEAN VELOCITY
OF FLOW.   (b) ENLARGEMENT OF A PORTION OF (a). [From G. I. Taylor, Proc. Roy. Soc.
London Ser. A, **164** (1938), by permission of the The Royal Society of London.]

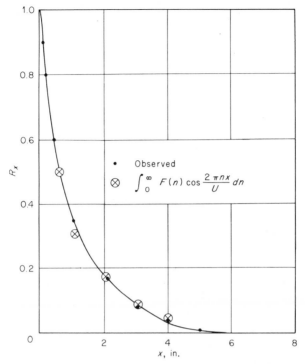

FIGURE 11. THE VARIATION OF R_x WITH x. [*From G. I. Taylor, Proc. Roy. Soc. London Ser. A,* **164** (1938), *by permission of The Royal Society of London.*]

Hence, using (122) and (127),

$$\frac{1}{\lambda^2} = \frac{4\pi^2}{U^2} \int_0^\infty n^2 F(n)\, dn. \tag{130}$$

Introducing the longitudinal scale

$$L_x = \int_0^\infty R_x\, dx$$

in (130) and (128), we obtain, respectively,

$$\frac{L_x^2}{\lambda^2} = 4\pi^2 \int_0^\infty \left(\frac{nL_x}{U}\right)^2 \frac{UF(n)}{L_x}\, d\frac{nL_x}{U} \tag{131}$$

and

$$\frac{UF(n)}{L_x} = 4 \int_0^\infty R_x \cos\left(\frac{2\pi nL_x}{U}\frac{x}{L_x}\right) d\frac{x}{L_x}, \tag{132}$$

both of which are expressed in dimensionless terms. If R_x is independent of U, (132) shows that $UF(n)/L_x$ is a function of nL_x/U alone. Typical curves of $UF(n)/L_x$ vs. nL_x/U obtained by the National Bureau of Standards and by the National Physical Laboratory show, however, that the relation between the two variables is not independent of U at large values of nL_x/U. Since from (131) the value of L_x/λ is determined chiefly by the value of $UF(n)/L_x$ at large values of nL_x/U, L_x/λ must not be independent of U, as is known to be true. Indeed it was from this established dependence of L_x/λ on U that the scatter of the $UF(n)/L_x$ vs. nL_x/U curves at large values of nL_x/U was attributed to the influence of U, for a given mesh length M and a given fluid.

When U is large, λ/L_x becomes small. Experimental measurements show that both R_x and R_y curves approach exponential curves, and (132) for the corresponding spectrum curve becomes[1]

$$\frac{UF(n)}{L_x} = \frac{4}{1 + 4\pi^2 n^2 L_x^2/U^2}.$$ (133)

This can be used to give a reference spectrum curve to compare with experimental data. As U decreases, λ increases, and the departures from (133) at large values of nL_x/U become greater. The change in the total energy due to the fluctuations associated with these departures in the spectrum at high frequencies is, however, very small.

Using (133), it is possible to compute the effect of varying the cutoff frequency of the measuring equipment on the measured value of the energy of the fluctuation. If the equipment passes high frequencies but cuts off sharply at a lower frequency n_0, the measured total energy is

$$\frac{\rho u''^2}{2} \int_{n_0}^{\infty} \frac{4(L_x/U)\,dn}{1 + 4\pi^2 n^2 L_x^2/U^2} = \left(1 - \frac{2}{\pi}\tan^{-1}\frac{2\pi n_0 L_x}{U}\right)\frac{1}{2}\rho u''^2.$$ (134)

Similarly, if the equipment passes low frequencies but cuts off sharply at a higher frequency n_h, the measured total energy is

$$\frac{1}{\pi}\tan^{-1}\frac{2\pi n_h L_x}{U}\,\rho u''^2.$$ (135)

In Taylor's discussion of discontinuous random motion (1921) the exponential function was found to be the limiting form of the correlation coefficient of small paths. Therefore, the fact that the correlation curves are of the exponential type at high values of U can be interpreted as meaning that at such high values of U, isotropic turbulence is a phenomenon of pure chance.

Figure 10 shows that the energy of turbulence is mainly carried by large eddies. When the turbulence decays, the energy-containing eddies must adjust to the decay. While the small eddies are mainly responsible for viscous dissipation because of the correspondingly larger rates of deformation for such eddies, the spectrum of the small eddies may actually

[1] The definition of L_x shows that if R_x is exponential, it must be of the form $\exp(-x/L_x)$. Then (132) gives (133).

be in equilibrium and serves, unchanging except in the last stages of decay, as a tunnel for the energy of the larger eddies to pass toward dissipation. This hypothesis, first made by Kolmogoroff (1941), seems to be very near the truth and has served as the starting point for many modern theories of turbulence.

9. DIMENSIONAL REASONING

Although the complete quantitative theory of turbulence is not yet available and there is some doubt of its eventual availability, some important qualitative results can be deduced by dimensional reasoning. We shall illustrate how this can be done by considering fully developed turbulence, or turbulence in flows at very large Reynolds numbers.

We recall that a length l_2 defined by (42) can be taken to be the definition of the average size of the eddies, the sizes of which vary.   For the purpose of the qualitative discussion to follow we shall define the sizes of eddies in the following simple way.   Consider the measurement of the velocity component v at a point as a function of time.   The result of this measurement for fully developed turbulence is a very irregular curve in the tv plane.   Although the simultaneous measurement of v along a straight line is impossible, it is reasonable to suppose that the variation of v with x (x being measured along that straight line) at any instant t is of the same irregular nature.   At any point P on this irregular xv curve we shall take an interval Δx defined by

$$x_P - \frac{\Delta x}{2} \leq x \leq x_P + \frac{\Delta x}{2}$$

and average v over Δx. We now take this average, which we call $v(x,\Delta x)$, and plot it against x for any chosen Δx.   The result will be a smoother curve than the original xv curve—a very smooth one indeed if Δx is large.   On this smoother curve we take the points at which the slope (or curvature, if preferable) is zero, take the average length between two consecutive points in this set of points, and define it as the size λ of the eddy, which is then a function of Δx.   For a large Δx, λ gives the order of magnitude of the sizes of the large eddies and for a small Δx, that of the sizes of the small eddies.   For flows at low Reynolds numbers, the xv curve may be fairly smooth to start with, so that λ does not decrease further as Δx reaches a lower limit, indicating that small eddies do not exist.   As the Reynolds number increases, smaller and smaller eddies appear.   The size of the largest eddies is of the order of the linear dimensions of the flow region, and we shall denote its magnitude by l.   The variation of v over l will be denoted by Δv.

To illustrate the dimensional reasoning mentioned before, consider ϵ, the mean rate of dissipation of energy per unit time *per unit mass* of fluid.   The energy dissipated derives from the largest eddies, as we have seen in the previous section, although eventually it is viscosity that is responsible for the dissipation.   Expressing ϵ in terms of l, Δv, and the density ρ, which are the only three possibly relevant quantities, we have

$$\epsilon \sim \frac{(\Delta v)^3}{l} \, . \tag{136}$$

As molecular viscosity owes its existence to molecular motion, turbulent fluctuations must also give rise to a viscosity which is generally called *eddy viscosity*, the idea of which originated with Boussinesq.   If we denote the kinematic eddy viscosity by ν_e, then a similar dimensional reasoning gives

$$\nu_e \sim l \, \Delta v. \tag{137}$$

We shall now consider the properties of turbulence regarding the small eddies, or small values of λ.   At places far enough away from solid boundaries it is reasonable to suppose that the turbulence *as regards small eddies* to be isotropic.   This is called *local isotropy* [Kolmogoroff (1941)].   The following important qualitative results are due to Kolmogoroff (1941) and Obukhoff (1941).

The size of λ below which viscous effects are important will be denoted by λ_0.   We shall consider eddy sizes in the interval $\lambda_0 \ll \lambda \ll l$. The variation of v over λ will be denoted by δv_λ, which must depend only on the kinematic quantities ϵ and λ.   A dimensional analysis then shows that

$$\delta v_\lambda \sim (\epsilon\lambda)^{1/3} \qquad \text{or} \qquad \delta v_\lambda \sim \Delta v \left(\frac{\lambda}{l}\right)^{1/3}. \tag{138}$$

The variation of v at any given point over the time interval τ, small compared with $T = l/u$, u being the mean velocity, can be obtained from (138) if the order of magnitude of λ therein is equated to $u\tau$.   Thus

$$\delta v_\tau \sim (\epsilon\tau u)^{1/3}. \tag{139}$$

We shall now give an estimate of λ_0.   Defining the Reynolds number for the eddy of size λ to be

$$R_\lambda = \frac{(\delta v_\lambda)\lambda}{\nu} \, , \tag{140}$$

we obtain from (138) that

$$R_\lambda \sim \frac{\Delta v \, \lambda^{1/3}}{\nu l^{1/3}} = R \left(\frac{\lambda}{l}\right)^{1/3}, \tag{141}$$

in which $R = (l \, \Delta v)/\nu$.   The order of magnitude of λ_0 is determined by the requirement that R_λ should be small, say of the order of unity.   Hence

$$\lambda_0 \sim lR^{-3/4}, \tag{142}$$

which could have been obtained by a dimensional analysis of the kinematic quantities λ_0, ϵ, and ν. By virtue of (142), (138) can be written as

$$\delta v_{\lambda_0} \sim \Delta v\, R^{-1/4}. \tag{143}$$

Finally, for $\lambda \ll \lambda_0$ the velocity distribution is smooth, and

$$\delta v_\lambda = \text{const} \cdot \lambda.$$

The order of magnitude of the constant must be $\delta v_{\lambda_0}/\lambda_0$, since $\delta v_\lambda \sim \delta v_{\lambda_0}$ when $\lambda \sim \lambda_0$. From (142) and (143) we have then

$$\delta v_\lambda \sim \frac{\Delta v\, R^{1/2}\lambda}{l}, \qquad \text{for } \lambda \ll \lambda_0. \tag{144}$$

PROBLEMS

1 Show that, in the notation of Sec. 9,

$$\epsilon \sim \nu_e \left(\frac{\Delta v}{l}\right)^2$$

and that

$$\Delta p \sim \rho(\Delta v)^2,$$

in which Δp is the pressure variation.

2 In the notation of Sec. 9, show that

$$\delta v_\tau \sim \Delta v \left(\frac{\tau}{T}\right)^{1/3}.$$

3 If $\delta v'_\tau$ is the variation of velocity v' of a fluid particle over the time interval τ as it moves about, show that

$$\delta v'_\tau \sim (\epsilon\tau)^{1/2}.$$

4 If at any point in a fully developed turbulent flow, steady in the mean, ω_0 is the order of magnitude of the frequencies of eddies in the range $\lambda \ll \lambda_0$, show that

$$\omega_0 \sim \frac{u}{\lambda_0} \qquad \text{and hence} \qquad \omega_0 \sim \frac{uR^{3/4}}{l},$$

in which u is the time-mean velocity at the point under question.

5 If the number of freedom n per unit volume of the fluid is defined to be the number of eddies of the size λ_0 per unit volume, show that the total number of freedom N of a turbulent flow varies as $R^{9/4}$, or

$$N \sim \left(\frac{R}{R_{\mathrm{cr}}}\right)^{9/4},$$

since N is of the order of unity when R is of the order of R_{cr}, which is the critical Reynolds number beyond which turbulence eventually occurs.

6 A point momentum source causes a turbulent jet, whose axis will be used as the polar axis. Assume that, as in the case of the laminar jet of Squire (Sec. 4.1 of Chap. 7),

$$u = \frac{F(\theta)}{R}, \qquad v = \frac{f(\theta)}{R},$$

in which spherical coordinates (R, θ, φ) are used and u and v are the velocity components corresponding to R and θ. Outside the jet region the flow can be assumed irrotational. Under these assumptions show that outside of the jet

$$F(\theta) = \mathrm{const} = -b \text{ (say)},$$

$$f(\theta) = -\frac{b(1 + \cos\theta)}{\sin\theta} = -b\cot\frac{\theta}{2}.$$

REFERENCES

BATCHELOR, G. K., 1949: The Role of Big Eddies in Homogeneous Turbulence, *Proc. Roy. Soc. London Ser. A*, **195**: 513.

CORRSIN, S., 1943: Investigation of Flow in an Axially Symmetric Heated Jet of Air, *Natl. Advisory Comm. Aeron. Advance Confidential Rept.* 3L23.

DRYDEN, H. L., 1943: A Review of the Statistical Theory of Turbulence, *Quart. Appl. Math.*, **1**: 7.

KÁRMÁN, T. VON, 1937a: On the Theory of Turbulence, *Proc. Natl. Acad. Sci. U.S.*, **23**: 98.

KÁRMÁN, T. VON, 1937b: The Fundamentals of the Statistical Theory of Turbulence, *J. Aeron. Sci.*, **4**: 131.

KÁRMÁN, T. VON, 1938: Some Remarks on the Statistical Theory of Turbulence, *Proc. 5th Intern. Congr. Appl. Mech., Cambridge, Mass.*, p. 347.

KÁRMÁN, T. VON, and L. HOWARTH, 1938: On the Statistical Theory of Isotropic Turbulence, *Proc. Roy. Soc. London Ser. A*, **164**: 192.

KOLMOGOROFF, A. N., 1941: On Degeneration of Isotropic Turbulence in an Incompressible Viscous Liquid, *Compt. Rend. Acad. Sci. U.R.S.S.*, **31**: 538.

LAMB, H., 1932: "Hydrodynamics," 6th ed., The Macmillan Company; reprinted by Dover Publications, Inc., New York, 1945 (with the same pagination).

LAUFER, J., 1953: The Structure of Turbulence in Fully Developed Pipe Flow, *Natl. Advisory Comm. Aeron. Tech. Note* 2954.

OBUKHOFF, A. M., 1941: On the Distribution of Energy in the Spectrum of Turbulent Flow, *Compt. Rend. Acad. Sci. U.R.S.S.*, **32**: 19, and *Izv. Acad. Nauk. S.S.S.R. Ser. Georg. Geofiz.*, **5**: 453.

RICHARDSON, L. F., 1920: Some Measurements of Atmospheric Turbulence, *Phil. Trans. Roy. Soc. London Ser. A*, **221**: 1.

SCHUBAUER, G. B., 1935: A Turbulence Indicator Utilizing the Diffusion of Heat, *Natl. Advisory Comm. Aeron. Rept.* 524.

TAYLOR, G. I., 1921: Diffusion by Continuous Movements, *Proc. London Math. Soc.*, **20A**: 196.

TAYLOR, G. I., 1935: Statistical Theory of Turbulence, pts. I–V, *Proc. Roy. Soc. London Ser. A*, **151**: 421.

TAYLOR, G. I., 1936: *idem.*, *Proc. Roy. Soc. London Ser. A*, **156**: 307.

TAYLOR, G. I., 1937: The Statistical Theory of Isotropic Turbulence, *J. Aeron. Sci.*, **4**: 311.

TAYLOR, G. I., 1938a: The Spectrum of Turbulence, *Proc. Roy. Soc. London Ser. A*, **164**: 476.

TAYLOR, G. I., 1938b: Some Recent Development in the Study of Turbulence, *Proc. 5th Int. Congr. Appl. Mech., Cambridge, Mass.*, p. 294.

ADDITIONAL READING

BATCHELOR, G. K., 1953: "The Theory of Homogeneous Turbulence," Cambridge University Press, London.

BATCHELOR, G. K., and A. A. TOWNSEND, 1948: Decay of Turbulence in the Final Period, *Proc. Roy. Soc. London Ser. A*, **194**: 527.

CORRSIN, S., 1953: The Turbulence Front at Free Stream Boundaries, *Proc. 3d Midwestern Conf. Fluid Mech., Univ. Minn.* (This and the following article deal with the intermittency phenomenon.)

CORRSIN, S., and A. L. KISTLER, 1955: Free-stream Boundaries of Turbulent Flows, *Natl. Advisory Comm. Aeron. Rept.* 1244.

HINZE, J. O., 1959: "Turbulence," McGraw-Hill Book Company, New York.

LIEPMANN, H. W., 1952: Aspects of the Turbulence Problem, pts. I and II, *Z. Angew. Math. Phys.*, **3**: 321–342 and 407–426.

TOWNSEND, A. A., 1956: "The Structure of Turbulent Shear Flow," Cambridge University Press, London.

APPENDIX ONE

BASIC
THERMODYNAMICS

1. THERMODYNAMIC SYSTEMS AND VARIABLES

A *thermodynamic system* is a quantity of matter separated from its environment by an enclosure, and the chief concern of thermodynamics is how, by means of thermodynamic systems, heat is transformed into mechanical energy or mechanical energy into heat.

The state of a thermodynamic system is described by *thermodynamic variables*, of which the most commonly used are its volume V, pressure p, and temperature t. No more than two of these variables are independent, and the equation specifying the dependence of a third variable on two independent variables is called the *equation of state*. In terms of V, p, and t, the equation of state, implying equilibrium, has the form

$$F(V,p,t) = 0. \tag{1}$$

The concept of volume needs no elaboration. The pressure can be defined to be the *force per unit area* that the system exerts on its enclosure. In the presence of an external field, such as the gravitational field of the earth, the pressure may vary from place to place in the system. But the nonuniformity of pressure can always be made as small as desired by dividing the system into sufficiently small portions. (For gases in the earth's gravitational field, the variation of pressure is only a small fraction of the mean pressure even if the height varies by as much as 500 ft.) As to the temperature, for the moment it can be considered as something two systems in thermal contact have in common after thermal equilibrium has been established, i.e., after the volume and pressure of either system no longer vary. It is often measured by the volume of a substance at constant pressure. The instrument which measures temperature is called a *thermometer*, of which the most common types are mercury, alcohol, or gas.

2. FIRST LAW OF THERMODYNAMICS

The first law of thermodynamics is the law of conservation of energy. If the energy added to a system in a form other than work is called *heat* and denoted by Q and the work done *by* the system is denoted by W, the first law states that

$$Q = \Delta U + W, \qquad (2)$$

in which ΔU is the increase of the internal energy U of the system. If work is done on the system rather than by it, W is negative and (2) states that the total energy input the system receives is equal to the increase of its internal energy.

Q is measured in calories or British thermal units (Btu). A calorie is the heat needed to raise the temperature of one gram of water at 14°C by one degree Celsius. A Btu is the heat needed to raise the temperature of one pound of water at 39°F by one degree Fahrenheit and is equal to 252.0 calories. In mechanical units Q is measured in joules (10^7 ergs) in the cgs system and in foot-pounds or watt-hours in the British system. The equivalence of the thermal and the mechanical units was established by Joule's paddle-wheel experiment[1] with water in an insulated container and is expressed by

$$1 \text{ calorie} = 4.19 \text{ joules.}$$

W is given by (1.77). If we neglect viscous effects and consider cases in which the normal stress, now equal to p, is uniform, then W can be evaluated in terms of p and V. If the surface of the substance constituting the thermodynamic system is denoted by A and the distance measured along the outward normal to it by n, and if the normal stress, equal to p since viscous effects are neglected, is uniform, then the work done by the substance as it expands or contracts from one configuration to

---

[1] In this experiment work was done on a paddle wheel revolving in a thermally insulated body of water, and the resulting increase in temperature was measured.

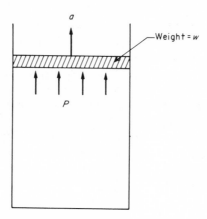

$PA - w = \frac{w}{g} a$

FIGURE 1. THE PRESSURE UNDER THE BLOCK IS NOT w/A IF THE BLOCK IS ACCELERATED BY THE EXPANSION OF GAS IN THE CYLINDER OF CROSS-SECTIONAL AREA A.

another nearby is

$$\int_A p \, dA \, dn = p \, dV,$$

in which dV is the difference between the initial and final values of the volume and is positive for an expansion and negative for a contraction. Thus

$$dW = p \, dV \qquad \text{and} \qquad W = \int_{V_2}^{V_2} p \, dV. \tag{3}$$

The differential form of (2) is therefore

$$dQ = dU + p \, dV \tag{2a}$$

under the assumption that viscous effects can be neglected and p is uniform.

The heat needed to raise the temperature of one unit mass of a substance one degree is called its *specific heat*. Since most substances expand on heating, (2a) shows that for most substances the specific heat at constant volume ($dV = 0$) is smaller than that at constant pressure ($dV > 0$), provided U is a function of temperature alone.

3. REVERSIBILITY

Although the first law, which is essentially a law of the energy budget, does not specify whether a process is reversible or not, the evaluation of the term W does depend on the nature of the process. For instance, if a gas (Fig. 1) contained in an insulated cylinder of cross-sectional area A

and covered by a frictionless sliding block of weight w is very slowly heated, the pressure in the gas is w/A, but if the heating is not infinitely slow, the block will be accelerated upward by the expanding gas and p will no longer be w/A. For a process to be infinitely slow, there must not be finite differences in pressure, temperature, etc., in the system. Such a process passes through a continuous sequence of quasi-equilibrium states, and is sometimes called *quasi-static*.

The infinite slowness of a process is a necessary condition for its reversibility; but it is not a sufficient condition. For instance, if a solid block is pushed slowly along the surface of another solid, the work done against friction (which remains finite no matter how slowly the block moves) is converted into heat, and the process is not reversible. Infinitely slow processes which are free from frictional or viscous dissipation are reversible. As subsequent developments will show, the concept of reversibility is a very important one in thermodynamics.

Since natural processes are not infinitely slow, the concept of reversibility would be quite useless if the requirement of "infinite slowness" were not further specified. When a material is subjected to a change of state, it takes some time, called the *relaxation time*, for that material to adjust to the new state. If the relaxation time is very small compared with the time during which a change of condition affecting its state (such as a change of pressure at its boundary) is imposed, the process is considered slow, and the process, in the absence of dissipative agents, will be reversible. Otherwise shock waves or finite waves with attendant viscous dissipation will be present, and the process will be irreversible.

4. IDEAL GASES

If a temperature θ is measured by the volume of an ideal gas at a particular pressure p_1, the equation

$$p_1 V = C_1(\theta + \theta_0)$$

holds exactly, because it is no more than the definition of θ. In this equation θ_0 is fixed arbitrarily, and the scale of θ is also chosen at will by choosing an arbitrary C_1. After the zero point and the scale of θ have been chosen (so that temperature is now measured by a standard gas thermometer), *all* gases at low densities are found to obey approximately an equation of the form (with different C's)

$$pV = C(\theta + \theta_0), \tag{4}$$

in which C varies only with the kind and amount of the gas used but not with the pressure p. As the density decreases indefinitely, the equations of state of all gases approach this form in the limit. All gases in this limiting state are called *ideal gases*. Their equations of state are therefore of the form (4).

For unit mass, (4) can be written as

$$\frac{p}{\rho} = R(\theta + \theta_0), \tag{4a}$$

in which R is constant for a particular gas and the temperature scale is a universal scale of a gas thermometer. If the freezing point of water under normal pressure is taken to be the zero point of θ,

$$\theta_0 = 273.16°C = 459.69°F$$

on the Celsius and Fahrenheit scale, respectively. This is not to say that the scale of θ is now identical with either the Celsius or Fahrenheit scale of a mercury thermometer, but the assignment of the value of θ_0 and of the zero point of θ does, for a certain gas, fix R and hence θ.

Although R is not a universal constant, the product mR, in which m is the molecular weight of the gas, is a universal constant for ideal gases. In Anglo-American units

$$mR = 49,700 \text{ ft}^2/(\text{F}°)(\text{sec}^2),$$

approximately. In the cgs system,

$$mR = 8.314 \times 10^7 \text{ ergs}/(\text{C}°)(\text{g}) = 8.314 \times 10^7 \text{ cm}^2/(\text{C}°)(\text{sec}^2).$$

If the absolute temperature (Kelvin scale) is defined to be

$$T = \theta + \theta_0,$$

(4) can be written as

$$\frac{p}{\rho} = RT \qquad \text{or} \qquad pv = RT, \tag{5}$$

in which $v = 1/\rho$ is the specific volume, or the volume of unit mass, and R is a constant derived from C in (4). The equation of state for a perfect gas of mass M is then

$$pV = MRT. \tag{5a}$$

Equation (5a) embodies:

BOYLE'S LAW   **For a definite amount of a gas, pV is constant at a constant temperature.**

CHARLES' LAW   **If temperature is measured by the volume of a gas of mass M at a constant pressure, then different gases give the same scale, provided M is proportional to the molecular weight m.**

Actually, if temperature is not measured by the volume but by the ratio of the volume to the volume at a reference temperature, different gases give the same temperature scale even if M is not proportional to m.

Equation (5a) can also be written in the form (see Sec. 9.2)

$$pV = NkT, \tag{5b}$$

in which k is a universal constant (the Boltzmann constant) and N the number of molecules in the gas. In this form the equation of state embodies:

AVOGADRO'S LAW N/V **has the same value for different gases at the same** p
   and T.

At low densities real gases behave almost like ideal gases and hence obey the three laws very closely.

4.1. Joule's Experiment

The internal energy U of a gas may depend on both T and V, but Joule showed with an experiment[1] that it is nearly independent of V. For that experiment Joule submerged in a calorimeter a container consisting of two chambers connected by a tube. Chamber A contained gas, but chamber B was empty. After the thermometer in the calorimeter indicated that thermal equilibrium had been reached, the stopcock in the connecting tube was opened and the thermometer was read again after equilibrium had been reestablished in the container. Joule found that the reading of the thermometer is nearly the same before and after expansion of the gas. Since the entire container consisting of the two chambers had not changed its volume, and since the readings of the thermometer indicated that there was no heat transfer to or from the container, the internal energy U of the gas was constant. Now since T was unchanged and V of the gas had increased, Joule's experiment showed that for a gas U is independent of V, or

$$\frac{\partial U}{\partial V} = 0, \tag{6}$$

and
$$U = U(T). \tag{7}$$

Equations (6) and (7) are nearly true for real gases at low densities. Moreover, they can be derived from (5) or (5a) for ideal gases. This will be done in Sec. 7.

In Joule's experiment, as the stopcock was opened, part of the gas rushed violently from chamber A to chamber B. The process was not quasi-static and hence by no means reversible. This experiment is thus an example of the application of the first law to irreversible processes.

4.2. Specific Heats

The heat needed to raise the temperature of one unit of mass of a substance by one degree is called its *specific heat*. If the volume of the substance is kept constant during the process, the specific heat is denoted by c_v and called the specific heat at constant volume. If the pressure is kept

---

[1] A similar and earlier experiment was done by Gay-Lussac.

constant during the process, it is called specific heat at constant pressure and denoted by c_p.

For the constant-volume process and unit mass of a substance,

$$dQ = c_v \, dT, \tag{8}$$

and, according to the first law,

$$dQ = dE, \tag{9}$$

in which E is U per unit mass.   Hence

$$dE = c_v \, dT. \tag{10}$$

For an ideal gas, c_v is a function of T alone because E is a function of T alone.   For real gases c_v may depend on both T and the volume, but in most cases of practical interest, it can be taken to be constant without committing too much error.

If a unit of mass is heated very slowly and expands at constant pressure, the first law states that

$$c_p \, dT = dQ = dE + p \, dv, \tag{11}$$

in which v is the specific volume, or volume of unit mass of a substance. For an ideal gas undergoing very slow expansion at constant pressure p, (5) gives

$$p \, dv = R \, dT, \tag{12}$$

and (11) becomes

$$c_p \, dT = dE + R \, dT. \tag{13}$$

Hence for an ideal gas c_p is a function of T alone, and, as a consequence of (10) and (13),

$$c_p = c_v + R. \tag{14}$$

For real gases (14) is nearly true at low densities, and in most cases of practical interest the dependence of c_p on T and v can be ignored without committing too much error.   In this book, c_p and c_v are treated as constants, for liquids as well as gases, unless otherwise stated.

The ratio of the specific heats c_p/c_v is usually denoted by γ.   In terms of γ, (14) can be written as

$$\frac{R}{c_v} = \gamma - 1. \tag{14a}$$

4.3.   Slow Adiabatic Transformations of an Ideal Gas

The relationships between p and ρ and between T and ρ of a gas as it undergoes "slow" adiabatic transformations are of great importance in the development of aerodynamics.   Consider an ideal gas of unit mass.

Its volume is then equal to the specific volume

$$v = \frac{1}{\rho}.$$

Since the transformations are adiabatic, the first law gives

$$c_v \, dT + p \, dv = 0.$$

By virtue of (5), this can be written as

$$\frac{dT}{T} + \frac{R}{c_v} \frac{dv}{v} = 0,$$

integration of which yields

$$\ln T + \frac{R}{c_v} \ln v = \text{const},$$

or, with R/c_v given by (14),

$$T\rho^{1-\gamma} = \text{const.} \tag{15}$$

This gives the relationship between T and ρ for slow adiabatic transformations. The constant is determined from the initial state. Slowness of the process has been specified because otherwise p cannot be evaluated from the equation of state (5).

The relationship between p and ρ now follows simply from (5) and (15) and is

$$\frac{p}{\rho^\gamma} = \text{const.} \tag{16}$$

5. THE CARNOT CYCLE

In preparation for the presentation of the second law of thermodynamics, the Carnot cycle will be discussed first. A Carnot cycle is a *reversible* cycle consisting of two isothermal processes at temperatures T_1 and T_2 and two adiabatic processes operating between T_1 and T_2, performed by a substance whose states can be represented on a Vp diagram. For instance, Fig. 2 describes the Carnot cycle performed by a gas. From A to B the gas expands slowly and isothermally at temperature T_2. From B to C the gas expands further, slowly and adiabatically. From C to D the gas is slowly compressed at constant temperature T_1 (less than T_2), and, finally, from D to A the slow compression is done adiabatically. The area of the figure $ABCD$ is the work done by the substance in one cycle. This is of course equal to the net heat input in the cycle, according to the first law. The net heat input is equal to the difference between the heat Q_2 absorbed from one reservoir at temperature T_2 during the expansion AB and the

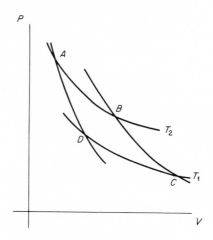

FIGURE 2. THE CARNOT CYCLE.

heat Q_1 transferred to another reservoir at temperature T_1 during the compression CD. The reason that a Carnot cycle should be reversible is that an irreversible process involves finite differences in pressure and temperature in the substance, whose states then cannot be meaningfully represented on a Vp diagram.

The efficiency of the Carnot cycle is defined to be the ratio of the work done to the heat absorbed at temperature T_2, that is,

$$\eta = \frac{Q_2 - Q_1}{Q_2} = 1 - \frac{Q_1}{Q_2}. \tag{17}$$

6. THE SECOND LAW OF THERMODYNAMICS

The second law is a law of nature supported by all available evidence. If the word *universe* is understood to mean a thermodynamic system and its environment, the second law of thermodynamics as stated by Thomson (Lord Kelvin) takes essentially the following form:

> **It is impossible to transform the heat absorbed from a reservoir at constant temperature into work without producing other changes in the universe.**

As stated by Clausius, the second law takes the form:

> **A transformation whose only final result is to transfer heat from a body at a lower temperature to another body at a higher temperature is impossible.**

Thomson's and Clausius' statements are equivalent, because the former follows from the latter and vice versa. If Thomson's statement were true and Clausius' false, the heat Q_1 transferred to the reservoir at

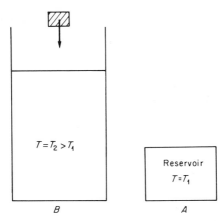

FIGURE 3. IF HEAT COULD BE ABSORBED FROM RESERVOIR A AND TURNED INTO WORK (REPRESENTED BY THE POTENTIAL ENERGY OF A MASS) WITHOUT PRODUCING ANY OTHER CHANGE IN THE UNIVERSE, HEAT COULD BE TRANSFERRED FROM RESERVOIR A TO RESERVOIR B AT A HIGHER TEMPERATURE BY LETTING THE MASS FALL INTO THE FLUID IN B AND PRODUCING HEAT BY VISCOUS DISSIPATION, WHICH THEN RESIDES IN B.

the lower temperature T_1 in the Carnot cycle could be transferred back to the reservoir at the higher temperature T_2. But then heat of the amount $Q_2 - Q_1$ would have been transformed into work without producing any other change in the universe, and Thomson's statement would be violated, leading to a contradiction. Conversely, if Clausius' statement were true and Thomson's false, the heat Q absorbed from a reservoir, after being turned into work, could be dissipated into heat (by friction or viscous dissipation, for instance) and transferred to a body at a higher temperature (see Fig. 3), violating Clausius' statement and leading to a contradiction.

7. THE ABSOLUTE TEMPERATURE SCALE

The temperature scales considered so far depend on the particular thermometers and the properties of the particular thermometric substances used. It is desirable to have an absolute temperature scale which is independent of the properties of the thermometric substance. To develop this scale, it is necessary to consider first the efficiency of heat engines.

Consider two heat engines which operate between the temperatures t_1 and t_2 ($> t_1$) and undergo adiabatic changes of state when not in thermal contact with the heat reservoirs. If the first engine operates in a reversible cycle, absorbs heat Q_2 at the temperature t_2, and delivers heat Q_1 to the reservoir at temperature t_1, then

$$\frac{Q_2}{Q_1} \geq \frac{Q_2'}{Q_1'}, \tag{18}$$

in which Q_2' is the heat input to the second engine at t_2 and Q_1' the heat delivered by it at t_1. To prove (18), it is convenient to consider a reversible engine identical to the first one *in every respect except size*, which is such that the heat delivered to the reservoir at temperature t_1 is Q_1'. Then simple proportionality demands that

$$\frac{Q_2}{Q_1} = \frac{Q_2''}{Q_1'}, \tag{19}$$

in which Q_2'' is the heat absorbed by the third engine at temperature t_2. Now if (18) were not true, one cycle of the second engine with one reverse cycle of the third engine would absorb the amount of heat $Q_2' - Q_2''$ from the reservoir at temperature t_2 and transform it into work, leaving everything else unchanged. This would violate the second law.

So far the second engine has not been specified to be reversible or irreversible. If it is also reversible, then the same argument which established (18), on interchanging the roles of the first and second engines, leads to

$$\frac{Q_2'}{Q_1'} \geq \frac{Q_2}{Q_1}, \tag{20}$$

and (18) and (20) give

$$\frac{Q_2}{Q_1} = \frac{Q_2'}{Q_1'}. \tag{21}$$

Since the efficiency is defined by (17), it follows that the efficiencies of all reversible engines are the same and that the efficiency of an irreversible engine can never exceed that of a reversible one.

Equation (21) states that the ratio Q_2/Q_1 is the same for all reversible engines and is independent of their special properties. It must then depend only on the temperatures t_1 and t_2, that is,

$$\frac{Q_2}{Q_1} = f(t_1, t_2),$$

in which the function $f(t_1, t_2)$ is a universal function of the two variables t_1 and t_2. Now consider a heat engine that operates at the temperatures t_2 and t_0 ($< t_1 < t_2$). The ratio of heat input to heat delivery is

$$\frac{Q_2}{Q_0} = f(t_0, t_2). \tag{22}$$

The corresponding ratio for an engine operating at the temperatures t_1 and t_0 is

$$\frac{Q_1}{Q_0} = f(t_0, t_1), \qquad (23)$$

in which Q_0 is assumed to be the same as for the first engine, for convenience of discussion. Division of (22) by (23) yields

$$\frac{Q_2}{Q_1} = \frac{f(t_0, t_2)}{f(t_0, t_1)}. \qquad (24)$$

If now t_0 is taken to be a fixed reference temperature, (24) can be written as

$$\frac{Q_2}{Q_1} = \frac{f(t_2)}{f(t_1)} = \frac{\theta(t_2)}{\theta(t_1)}, \qquad (25)$$

in which

$$\theta(t) = Kf(t). \qquad (26)$$

Equation (25) permits the use of θ as the temperature instead of the empirical temperature t indicated by a particular thermometer. Of course, the constant K in (26) can be arbitrarily chosen, so that the magnitude of the unit of θ can be chosen at will. The usual choice is to make the difference in θ at the boiling and freezing temperatures of water at atmospheric pressure equal to 100 degrees on one scale, or 180 degrees on another. But it should be remembered that the resulting absolute temperature scale does not necessarily agree with the empirical scale of an arbitrary thermometer, except at the end points, even if that thermometer has the same number of divisions between the two end points. However, it will now be shown that the temperature scale of an ideal-gas thermometer agrees exactly with the absolute temperature scale if the units are chosen to be the same. Thus real-gas thermometers at low densities give a good approximation to the absolute temperature scale.

Consider the Carnot cycle described in Fig. 2. As the gas expands isothermally from A to B, the internal energy U, being a function of T alone, does not change. Hence according to the first law and (5a)

$$Q_2 = W = \int_{V_A}^{V_B} p \, dV = \int_{V_A}^{V_B} MRT_2 \frac{dV}{V} = MRT_2 \ln \frac{V_B}{V_A}. \qquad (27)$$

Similarly, the heat delivery at t_1 is

$$Q_1 = MRT_1 \ln \frac{V_C}{V_D}. \qquad (28)$$

But since BC and DA are adiabatics and (15) can be written as

$$TV^{\gamma-1} = \text{const},$$

the volumes are further related by the equations

$$T_2 V_B^{\gamma-1} = T_1 V_C^{\gamma-1} \quad \text{and} \quad T_1 V_D^{\gamma-1} = T_2 V_A^{\gamma-1}.$$

Hence

$$\frac{V_B}{V_A} = \frac{V_C}{V_D}, \qquad (29)$$

and it follows from (27) and (28) that

$$\frac{Q_2}{Q_1} = \frac{T_2}{T_1}, \tag{30}$$

which, together with (25), proves the equivalence of the absolute temperature scale and the ideal-gas temperature scale of Kelvin, provided the units are taken to be the same.

8. ENTROPY

Consider a thermodynamic system operating in a cycle which involves adiabatic processes and isothermal ones at the temperatures $T_1, T_2, \ldots, T_n$, absorbing heat $Q_1, Q_2, Q_3, \ldots, Q_n$ when in contact with heat reservoirs at these temperatures. It will be shown that

$$\sum_{i=1}^{n} \frac{Q_i}{T_i} \leq 0, \tag{31}$$

in which Q_i is considered to be negative if the system delivers heat to the reservoir at temperature T_i instead of absorbing heat from it. The equality sign in (31) is valid if the cycle is reversible.

Imagine n heat engines operating in Carnot cycles between the temperatures T_1 and T_0, T_2 and T_0, $\ldots$, T_n and T_0, in which T_0 is a new temperature level. The heat *delivered* to the n heat reservoirs will be assumed to be precisely $Q_1, Q_2, \ldots, Q_n$, respectively, and the heat *absorbed* by the engines from the reservoir at temperature T_0 will be denoted by $Q_{01}, Q_{02}, \ldots, Q_{0n}$, respectively. Then the original thermodynamic system and these n engines, after each has performed a cycle, will have *absorbed* the amount of heat

$$Q_0 = \sum_{i}^{n} Q_{0i} \tag{32}$$

from the reservoir at temperature T_0 and transformed it into work. Now the Carnot cycles being reversible, the following equations hold:

$$\frac{Q_{0i}}{T_0} = \frac{Q_i}{T_i}, \qquad i = 1, 2, \ldots, n. \tag{33}$$

Hence
$$Q_0 = T_0 \sum_{i=1}^{n} \frac{Q_i}{T_i}.$$

T_0 is positive (because the volume and pressure of an ideal gas are positive, and Q_0 has to be negative, for otherwise an amount of heat would have been taken from the reservoir at constant temperature T_0 and transformed into work, leaving everything else unchanged, and this would violate the second law. Hence (31) holds.

If the cycle of the original thermodynamic system is reversible, for

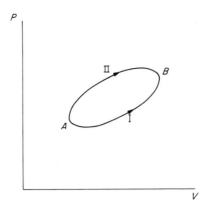

FIGURE 4. SKETCH SHOWING THAT dQ/T
IS AN EXACT DIFFERENTIAL.

the reversed cycle Q_i becomes $-Q_i$ and the equation corresponding to (31) is

$$-\sum_{i=1}^{n} \frac{Q_i}{T_i} \leq 0, \tag{34}$$

which, together with (31), demands that

$$\sum_{i=1}^{n} \frac{Q_i}{T_i} = 0 \tag{35}$$

for a reversible process.

So far it has been assumed that the number of temperature levels at which the system operates is finite (and equal to n). This restriction can evidently be removed since the number of temperature levels can be indefinitely increased. In the limit, the variation of T can be continuous, and (35) becomes

$$\oint \frac{dQ}{T} = 0, \tag{36}$$

provided that the cycle is *reversible* and the line integral is along a closed path.

Equation (36) is of tremendous importance in thermodynamical considerations. It states that (Fig. 4)

$$\int_{I} \frac{dQ}{T} - \int_{II} \frac{dQ}{T} = 0,$$

or

$$\int_{I} \frac{dQ}{T} = \int_{II} \frac{dQ}{T}.$$

This means that the integral

$$\int_{A}^{B} \frac{dQ}{T} \tag{37}$$

is independent of path and is a function only of the limits of integration. If a standard state is indicated by the subscript 0,

$$S(A) = \int_0^A \frac{dQ}{T}, \qquad (38)$$

and, since the integral (37) is independent of path,

$$\left(\int_A^B \frac{dQ}{T} \right)_R = \int_0^B \frac{dQ}{T} - \int_0^A \frac{dQ}{T} = S(B) - S(A), \qquad (39)$$

in which the subscript R emphasizes that the process is reversible. The quantity dQ/T is therefore an exact differential, which, in consistency with (39), will be denoted by dS. The quantity S is called the *entropy*. The standard state at which the entropy is zero has been assigned *arbitrarily* in (38). Hence the entropy is defined only up to an additive constant. As long as only the differences in entropy are of interest, it does not matter what this constant is, but in the kinetic theory of gases it is sometimes important to know what this constant is or to know what state can be taken as the standard state of zero entropy. The third law of thermodynamics, or Nernst's theorem, can be stated in the following form:

The entropy of every system at absolute temperature zero can always be taken to be zero.

In essence this theorem means that *all possible states* of a thermodynamic system at absolute zero have the same entropy, which can therefore be taken to be any constant whatever, and in particular zero.

It must be remembered that (39) is valid only for a reversible process. For a process which may not be reversible, (36) must be replaced by

$$\oint \frac{dQ}{T} \leq 0, \qquad (40)$$

which is the limiting form of (31). Suppose now that a system reaches a state B from a state A by *any process* whatever and that it then returns to state A by a reversible one. Equation (40) can be written as

$$\int_A \frac{dQ}{T} + \left(\int_B^A \frac{dQ}{T} \right)_R \leq 0, \qquad (41)$$

in which the subscript R indicates reversible. By virtue of (39), (41) can be written as

$$\int_A^B \frac{dQ}{T} \leq S(B) - S(A). \qquad (42)$$

The equality sign holds if the process pertaining to the integral on the left-hand side is reversible, as (39) shows. Although it does not follow from the development of (42) that the inequality sign must hold for irreversible processes, in fact it does.

Equation (42) shows that the entropy can increase even if there is no heat input to a system. For instance, in Joule's gas-expansion experiment no heat was absorbed by the gas, but after it rushed violently through to chamber B and equilibrium was finally reestablished, the entropy had been increased. The increase in entropy caused by irreversible processes is evaluated by replacing it by a reversible one having the same end states and by using (39), in which the left-hand side is understood to be an integral for a reversible process.

8.1. Entropy of an Ideal Gas

Since the evaluation of entropy of an ideal gas is useful in aerodynamics, it will be given here. Consider an ideal gas of unit mass, for which the equation of state is (5). According to the first law,

$$dQ = c_v \, dT + p \, dv, \tag{43}$$

in which dQ is the heat input for a reversible process. If p is eliminated by the use of (5) and (43) is divided by T,

$$dS = \frac{dQ}{T} = \frac{c_v}{T} \, dT + \frac{R}{v} \, dv,$$

and one integration gives

$$S = c_v \ln T + R \ln v + \text{const} = c_v \ln T - R \ln \rho + \text{const}.$$

If this equation is divided by c_v, (14a) used, and the exponential of the resulting equation taken, one has, finally,

$$\frac{T}{\rho^{\gamma-1}} = \text{const } e^{S/c_v}, \tag{44}$$

or, by virtue of (5),

$$\frac{p}{\rho^\gamma} = \text{const } e^{S/c_v}. \tag{45}$$

A process of transformation which preserves the entropy is called *isentropic*. A system with uniform entropy everywhere is called *homentropic*. Thus isentropy is historical in nature, whereas homentropy is geographic. A body of gas originally homentropic or nonhomentropic remains homentropic or nonhomentropic, respectively, after all portions of it have undergone isentropic changes of state. An originally homentropic gas may become nonhomentropic after its parts have gone through nonisentropic changes of state.

9. RUDIMENTS OF THE KINETIC THEORY OF GASES

So far the thermodynamic principles have been presented from the macroscopic point of view. Some results in the kinetic theory of gases will now be presented, in order to:

1 Have some detailed information about a gas considered not merely as a thermodynamic body but a collection of molecules.

2 Gain more insight into the meaning of temperature, even if some of the pertinent results are merely stated and not derived.

3 Derive the equation of state of an ideal gas, instead of taking it as an empirical result.

4 Derive the value for the ratio γ of the specific heats.

5 Acquaint the reader with the spirit of the kinetic theory of gases.

In particular, the derivation of the equation of state and the value of γ for ideal gases will illustrate how the kinetic theory of gases can produce results not analytically obtainable by the methods of macroscopic thermodynamics.

9.1. Velocity Distribution in Ideal Gases at Equilibrium

One important question concerning ideal gases in equilibrium is the distribution of velocities among the molecules. This question was answered by Maxwell. Let the velocity components in the directions of increasing Cartesian coordinates x, y, and z be denoted by u, v, and w. Maxwell made the important and fruitful assumption that the fraction of molecules with u in the range $(u, u + du)$ is $F(u^2)\,du$, in which $F(u^2)$ is called the *probability-density function* for the distribution of u. Similarly, he assumed that the fraction of molecules with v in the range $(v, v + dv)$ is $F(v^2)\,dv$ and that the fraction of molecules with w in the range $(w, w + dw)$ is $F(w^2)\,dw$. It is implied in this assumption that what happens in one direction is independent of what happens in the other directions and that only the magnitudes of u, v, and w are important, their signs being immaterial. Although the truth of Maxwell's assumption cannot be proved a priori, it leads to correct conclusions and is now generally accepted.

The fraction of molecules with velocities in the range

$$(u, u + du; v, v + dv; w, w + dw)$$

is then $\qquad\qquad F(u^2)F(v^2)F(w^2)\,du\,dv\,dw.$

Note that the three F's signify the same function, because of the *isotropy of space*, or the lack of privileged direction or directions. Now $du\,dv\,dw$ is a volume element in the velocity space (Fig. 5). The fraction of molecules with velocities represented in any other volume element $d\kappa$ containing the point $P(u,v,w)$ in the velocity space is

$$F(u^2)F(v^2)F(w^2)\,d\kappa. \tag{46}$$

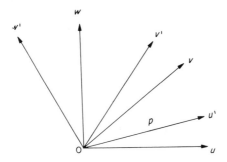

FIGURE 5. THE VELOCITY SPACE.

If the new set of Cartesian coordinates x', y', and z' is introduced, with x' measured along OP, and if u', v', and w' are the velocity components in the directions of increasing x', y', and z', the fraction of molecules with velocities represented in $d\kappa$ is

$$F(0)F(0)F(q^2)\,d\kappa, \tag{47}$$

in which $q^2 = u'^2 = u^2 + v^2 + w^2.$

Comparison of (46) and (47) reveals that $F(u^2)F(v^2)F(w^2)$ must be a function of q^2 only, and so must its logarithm; that is,

$$\ln F(u^2) + \ln F(v^2) + \ln F(w^2) = \text{function of } (u^2 + v^2 + w^2).$$

But this can be true only if

$$\ln F(u^2) = \alpha - \beta u^2, \qquad \text{etc.,}$$

α and β being constant. Thus the fraction of molecules with u in the range $(u, u + du)$ is proportional to $e^{-\beta u^2}\,du$ and is equal to[1]

$$\frac{e^{-\beta u^2}\,du}{\displaystyle\int_{-\infty}^{\infty} e^{-\beta u^2}\,du} = \left(\frac{\beta}{\pi}\right)^{1/2} e^{-\beta u^2}\,du. \tag{48}$$

---

[1] Recall that with r and θ denoting polar coordinates introduced in the xy plane,

$$r^2 = x^2 + y^2,$$

and the area element is $r\,dr\,d\theta$. Then

$$\left(\int_{-\infty}^{\infty} e^{-\beta x^2}\,dx\right)^2 = \int_{-\infty}^{\infty}\int_{-\infty}^{\infty} e^{-\beta x^2 - \beta y^2}\,dx\,dy = \int_{0}^{\infty}\int_{0}^{2\pi} e^{-\beta r^2}\,r\,dr\,d\theta$$

$$= 2\pi \int_{0}^{\infty} e^{-\beta r^2}\,r\,dr = -\frac{\pi}{\beta} e^{-\beta r^2}\Big|_{0}^{\infty} = \frac{\pi}{\beta}.$$

Hence

$$\int_{-\infty}^{\infty} e^{-\beta x^2}\,dx = \left(\frac{\pi}{\beta}\right)^{1/2}.$$

Thus β must be positive. The average kinetic energy of a molecule for motion in the x direction is

$$\frac{\int_{-\infty}^{\infty} \frac{1}{2} mu^2 e^{-\beta u^2} \, du}{(\pi/\beta)^{\frac{1}{2}}} = \frac{m}{4\beta}, \tag{49}$$

and the average kinetic energies for motion in the other directions have exactly the same value. The average kinetic energy of a molecule is therefore $3m/4\beta$ if only translational energy is counted.

The probability density functions

$$\left(\frac{\beta}{\pi}\right)^{\frac{1}{2}} e^{-\beta u^2}, \qquad \text{etc.}$$

describe a velocity distribution called the *Maxwellian distribution*. They have the same form as the probability-density function describing the well-known Gaussian distribution of errors.

In books[1] on the kinetic theory of gases it is shown that in a gas mixture β for each constituent gas is proportional to its molecular weight. For instance, if there are two gases A and B in the mixture,

$$\beta_A = \frac{m_A}{2\theta}, \qquad \beta_B = \frac{m_B}{2\theta}, \tag{50}$$

in which θ is common to both gas constituents. From (49) and (50) it follows that the kinetic energy for motion of each gas in any direction is $\theta/2$. Since θ is what the two gases at equilibrium in a mixture have in common no matter what their molecular weights and their masses are, it can be identified as a temperature. Thus the temperature θ is two-thirds of the translational kinetic energy. If rotational kinetic energies are also present, as for gases with more than one atom in each molecule, θ is still a measure of the total kinetic energy.

9.2. Equation of State of an Ideal Gas

Consider an ideal gas whose molecules have no interactions with one another. To simplify the discussion, it will be assumed to be contained in a rectangular box (Fig. 6), although the shape of the container is really immaterial in the following discussion. Since interaction among molecules is neglected, molecules with velocity component u situated at a distance less than ut from the wall normal to the x axis and of area A will hit that wall in time t. The number of such molecules that will hit the wall is then proportional to the volume Aut. If there are N molecules in

---

[1] See, for instance, E. A. Guggenheim, "Elements of the Kinetic Theory of Gases," Pergamon Press, New York, 1960. The brief sketch given here follows the development given in that book.

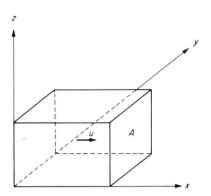

FIGURE 6. SKETCH FOR THE CALCULA-
TION OF PRESSURE ON A.

the box, and if the volume of the box is denoted by V, the number of molecules in the volume Aut with the x component of the velocity in the range $(u, u + du)$ is, by proportion and according to (48),

$$N \frac{Aut}{V} \left(\frac{\beta}{\pi}\right)^{1/2} e^{-\beta u^2} du, \tag{51}$$

in which Aut may be supposed less than V, although this restriction can be removed.   Now these molecules, after hitting the wall under consideration, will bounce back with a velocity component $-u$.   Thus the change of momentum of each of these molecules is $-2mu$, and the impulse received by the wall is

$$\frac{N}{V} Aut(2mu)f(u)\, du = \frac{N}{V} At2mf(u)u^2\, du,$$

in which $$f(u) = \left(\frac{\beta}{\pi}\right)^{1/2} e^{-\beta u^2}.$$

Integrating over all *positive* values of u, one has

$$\frac{N}{V} At2m\int_0^\infty f(u)u^2\, du = \frac{N}{V} Atm\int_{-\infty}^\infty f(u)u^2\, du \tag{52}$$

as the total impulse received by the wall in time t.   The pressure on the wall is this impulse divided by At, and is

$$\frac{N}{V} \frac{m}{2\beta} = \frac{N}{V} \theta, \tag{53}$$

according to (50), in which the subscripts can be dropped for this discussion.   Thus [note that this θ is not the θ in (4)]

$$p = \frac{N}{V} \theta, \tag{54}$$

in which, as stated before, θ can be identified with the temperature. Equation (54) is the equation of state of an ideal gas.

The temperature θ is related to T by

$$\theta = kT,$$

in which k is the Boltzman constant and has the value 1.3803×10^{-16} erg/degree. Equation (54) then is identical with (5b).

9.3. Evaluation of γ for Ideal Gases

For monatomic gases, the translational kinetic energy is the total kinetic energy and is

$$\tfrac{3}{2}N\theta = U, \tag{55}$$

in which the kinetic energy has been identified with the internal energy U. Since

$$pV = N\theta = MRT,$$

(55) can be written as

$$U = \tfrac{3}{2}MRT.$$

For unit mass,

$$E = \tfrac{3}{2}RT,$$

and

$$c_v = \left(\frac{\partial E}{\partial T}\right)_V = \frac{3}{2}R$$

Equation (14a) then gives

$$\gamma = \tfrac{5}{3}$$

for monatomic gases.

For diatomic gases, there are five degrees of freedom, three translational and two rotational. The kinetic energy corresponding to rotation about the axis joining the atoms can be neglected. The principle of equipartition of energy in all degrees of freedom holds, so that $N\theta/2$ is now only one-fifth of the internal energy. Hence

$$U = \tfrac{5}{2}N\theta,$$

and a similar development shows that

$$c_v = \tfrac{5}{2}R \quad \text{and} \quad \gamma = \tfrac{7}{5}.$$

For gases with more than two atoms in a molecule, the total number of degrees of freedom is five if the atoms are collinear (as in the case of CO_2) and six otherwise. If the number of freedom is six,

$$c_v = 3R \quad \text{and} \quad \gamma = \tfrac{4}{3}.$$

The number of degrees of freedom increases for diatomic and multiatomic gases by 2 or more as the temperature becomes high enough for vibrational energy to be important. Thus at high temperatures γ for such gases decreases below $\tfrac{7}{5}$ or $\tfrac{4}{3}$.

PROBLEM

1 An insulated cylinder contains an ideal gas of volume V_0, with a frictionless piston of weight w_0 on top. The cross-sectional area of the cylinder is A. Find the entropy change and final volume of the gas in each of the following two cases:

a Iron filings of total weight w_1 are added gradually, compressing the gas to a new volume V.

b A weight w_1 is released suddenly but from rest just above the piston, compressing the gas to a new volume V'.

If S_0 is the initial entropy, S the final entropy for case **a**, and S' the final entropy for case **b**, is S' greater than S_0? Is S greater than S_0? Is V' greater than V?

ADDITIONAL READING

GUGGENHEIM, E. A., 1960: "Elements of the Kinetic Theory of Gases," Pergamon Press, New York.

JEANS, J., 1946: "An Introduction to the Kinetic Theory of Gases," Cambridge University Press, London.

TRUESDELL, C., 1966: Thermodynamics of Deformation, in S. Eskinazi (ed.), "Modern Developments in the Mechanics of Continua," Academic Press, Inc., New York.

ZEMANSKY, M. W., 1957: "Heat and Thermodynamics," 5th ed., McGraw-Hill Book Company, New York.

APPENDIX TWO

CURVILINEAR COORDINATES

1. TENSORS

If $(x^1, x^2, \ldots, x^N)$ is any set of coordinates (not necessarily Cartesian) in an N-dimensional space and $(y^1, y^2, \ldots, y^N)$ another, and if

$$a^i_j = \frac{\partial y^i}{\partial x^j} \quad \text{and} \quad b^i_j = \frac{\partial x^i}{\partial y^j}, \tag{1}$$

a general mixed tensor of order $m + n$ is a set of N^{m+n} quantities $t^{i_1 \cdots i_m}_{j_1 \cdots j_n}$ obeying the laws of transformation

$$\bar{t}^{i_1 \cdots i_m}_{j_1 \cdots j_n} = a^{i_1}_{\alpha_1} \cdots a^{i_m}_{\alpha_m} b^{\beta_1}_{j_1} \cdots b^{\beta_n}_{j_n} t^{\alpha_1 \cdots \alpha_m}_{\beta_1 \cdots \beta_n}, \tag{2a}$$

$$t^{i_1 \cdots i_m}_{j_1 \cdots j_n} = b^{i_1}_{\alpha_1} \cdots b^{i_m}_{\alpha_m} a^{\beta_1}_{j_1} \cdots a^{\beta_n}_{j_n} \bar{t}^{\alpha_1 \cdots \alpha_m}_{\beta_1 \cdots \beta_n}, \tag{2b}$$

in which the quantities with a bar denote tensor components in the y system, and in which summation over the α's and β's on the right-hand sides is understood. This tensor is contravariant to the order m and covariant to the order n. The laws (2a) and (2b) are not independent, for the one can be derived from the other.

A tensor of order 1 is composed either of the N quantities t^i or of the N quantities t_i, and is a vector. In the former case it is called a *contravariant vector* and in the latter a *covariant vector*. A tensor of order zero is a scalar, which does not change with a change of coordinates. In the following sections the space considered will be a Euclidean space of three dimensions ($N = 3$), that is, a three-dimensional space in which Cartesian coordinates can be introduced.

With differential calculus it can readily be established that

$$a_\alpha^i b_j^\alpha = \delta_j^i \quad \text{and} \quad a_i^\alpha b_\alpha^j = \delta_i^j, \tag{3}$$

in which

$$\delta_j^i = \delta_i^j = \begin{cases} 0 & \text{if } i \neq j, \\ 1 & \text{if } i = j. \end{cases}$$

From (3) it can be deduced that if a mixed tensor is summed over any number of its contravariant indices which are numerically and respectively identical with an equal number of its covariant indices, a new tensor results whose order is lower than that of the original tensor by twice the number of indices over which the summation is performed, with the order of contravariance and covariance reduced equally.

For instance,

$$\bar{t}_\alpha^{\alpha i} = a_\beta^\alpha a_\gamma^i b_\alpha^\delta t_\delta^{\beta\gamma} = \delta_\beta^\delta a_\gamma^i t_\beta^{\beta\gamma} = a_\gamma^i t_\beta^{\beta\gamma}. \tag{4}$$

Thus $t_\beta^{\beta\gamma}$ is a tensor of order 1, and in fact a contravariant vector. The process of summation just described is called *contraction*. Tensors formed by contraction are called *contracted tensors*.

For Cartesian coordinates,

$$a_j^i = b_i^j, \tag{5}$$

so that in calculations the distinction between contravariance and covariance disappears for a Cartesian tensor. Therefore contraction of a Cartesian tensor can be performed over any number of pairs of indices. However, the distinction between contravariance and covariance still is a useful concept. In fact, only by dint of this conceptual distinction can the transformation of tensors and differential equations be most systematically carried out.

2. PSEUDO-TENSORS

The determinant

$$\Delta = |a_j^i| = \begin{vmatrix} a_1^1 & a_2^1 & a_3^1 \\ a_1^2 & a_2^2 & a_2^3 \\ a_1^3 & a_2^3 & a_3^3 \end{vmatrix} \tag{6}$$

is very useful. There exist sets of quantities which transform like tensors except for the presence of a power in Δ in the formula of transformation. Such quantities form a *pseudo-tensor*. A pseudo-tensor is sometimes composed of some of the elements (or of an element) of a tensor of a higher order.

An example of pseudo-tensor of order zero is furnished by the contravariant tensor of the third order σ^{ijk} with the following property

$$\sigma^{ijk} = [i,j,k]\sigma, \tag{7}$$

in which

$$[i,j,k] = \begin{cases} 1 & ijk \text{ of same cyclic order as } 1, 2, 3, \\ -1 & ijk \text{ of same cyclic order as } 3, 2, 1, \\ 0 & i, j, \text{ and } k \text{ not all different.} \end{cases}$$

With
$$\bar{\sigma}^{ijk} = a_\alpha^i a_\beta^j a_\gamma^k \sigma^{\alpha\beta\gamma},$$
it follows that

$$\bar{\sigma}^{ijk} = [\alpha,\beta,\gamma]a_\alpha^i a_\beta^j a_\gamma^k \sigma = [i,j,k]\Delta \cdot \sigma, \tag{8}$$

since
$$[\alpha,\beta,\gamma]a_\alpha^i a_\beta^j a_\gamma^k = [i,j,k]\Delta \tag{9}$$

by definition of the determinant Δ. Thus $\bar{\sigma}^{ijk}$ has the same property as σ^{ijk}, that is, $\bar{\sigma}^{ijk} = [i,j,k]\bar{\sigma}$, where

$$\bar{\sigma} = \Delta \cdot \sigma. \tag{10}$$

The quantity σ obeying the transformation law (10) is a *pseudo-scalar*, or a pseudo-tensor of order zero, because after a transformation of coordinates it is multiplied only by a power of Δ. Since this power is Δ itself, σ is called a *capacity scalar*. The meaning of the word capacity here can be explained by the following example. The determinant representing the volume of the parallelepiped formed by the N infinitesimal vectors $d^{(1)}x^i$, $d^{(2)}x^i$, $d^{(N)}x^i$ is

$$d\sigma = |d^{(i)}x^j| = \begin{vmatrix} d^{(1)}x^1 & d^{(1)}x^2 & \cdots & d^{(1)}x^N \\ \hline d^{(N)}x^1 & d^{(N)}x^2 & \cdots & d^{(N)}x^N \end{vmatrix} \tag{11}$$

and transforms according to the law

$$d\bar{\sigma} = \frac{\partial(y^1 \cdots y^N)}{\partial(x^1 \cdots x^N)} d\sigma = \Delta \, d\sigma. \tag{12}$$

Since $d\sigma$ is a volume element, and since volume is roughly synonymous with capacity, any quantity which transforms like $d\sigma$ (and is therefore a pseudo-scalar) is called a capacity scalar.

Similarly, a covariant tensor of order three $\bar{P}_{ijk}$ with the property

$$P_{ijk} = [i,j,k]P \tag{13}$$

has the component
$$P_{123} = P, \tag{14}$$

which obeys the transformation law

$$\bar{P} = \frac{1}{\Delta}P. \tag{15}$$

P is therefore a pseudo-scalar, and is called a *density scalar*. Since the power of Δ in (15) is Δ^{-1}, the product of a capacity scalar and a density scalar is a scalar invariant in the same way that the product of a volume and a density is equal to a scalar invariant (mass). This fact accounts for the name density scalar for p.

An example of a covariant pseudo-vector is one whose components are three distinct components of an antisymmetric contravariant tensor of the second order t^{ij} in three-dimensional space. This tensor has the property

$$t^{ij} = -t^{ji},$$

and hence can be represented by

$$
\begin{vmatrix} t^{11} & t^{12} & t^{13} \\ t^{21} & t^{22} & t^{23} \\ t^{31} & t^{32} & t^{33} \end{vmatrix}
=
\begin{vmatrix} 0 & \xi_3 & -\xi_2 \\ -\xi_3 & 0 & \xi_1 \\ \xi_2 & -\xi_1 & 0 \end{vmatrix}.
\tag{16}
$$

Because of antisymmetry

$$\bar{t}^{ij} = a_\alpha^i a_\beta^j t^{\alpha\beta} = (a_2^i a_3^j - a_3^i a_2^j)t^{23} + (a_3^i a_1^j - a_1^i a_3^j)t^{31} + (a_1^i a_2^j - a_2^i a_1^j)t^{12},
\tag{17}$$

from which it is clear that $\bar{t}^{ij}$ is also antisymmetric and can be represented by an equation similar to (16). But since according to (3),

$$
\begin{bmatrix} a_1^1 & a_2^1 & a_3^1 \\ a_1^2 & a_2^2 & a_3^2 \\ a_1^3 & a_2^3 & a_3^3 \end{bmatrix}
\begin{bmatrix} b_1^1 & b_2^1 & b_3^1 \\ b_1^2 & b_2^2 & b_3^2 \\ b_1^3 & b_2^3 & b_3^3 \end{bmatrix}
=
\begin{bmatrix} 1 & 0 & 0 \\ 0 & 1 & 0 \\ 0 & 0 & 1 \end{bmatrix},
$$

the matrix $[b_\beta^\alpha]$ is the inverse of $[a_\beta^\alpha]$, so that

$$b_\beta^\alpha = \frac{A(\beta,\alpha)}{\Delta},
\tag{18}$$

in which $A(\beta,\alpha)$ is the cofactor of a_α^β. Thus, with (16), (17) can be written as

$$\bar{t}^{ij} = \Delta(b_k^1\xi_1 + b_k^2\xi_2 + b_k^3\xi_3) = \Delta \cdot b_k^\alpha\xi_\alpha,$$

or

$$\bar{\xi}_k = \Delta \cdot b_k^\alpha\xi_\alpha,
\tag{19}$$

if (i,j,k) is cyclic with $(1,2,3)$. Equation (19) shows that the three distinct elements of the antisymmetric tensor of second order t^{ij} form a pseudo-vector—a covariant capacity vector. Similarly, those of a covariant tensor t_{ij} form a contravariant density vector. In Cartesian coordinates, Δ is equal to 1, and the ξ's in (19) form a Cartesian vector, as was shown in connection with the vorticity tensor (Chap. 1).

One important theorem concerning pseudo-vectors is that the divergence of a contravariant density vector is a density scalar, i.e., if

$$\bar{A}^i = \frac{1}{\Delta} a_\alpha^i A^\alpha, \tag{20}$$

then

$$\frac{\partial A^i}{\partial x^i} = P \tag{21}$$

is a density scalar.   A straightforward calculation shows that

$$\bar{P} = \frac{\partial \bar{A}^i}{\partial y^i} = \frac{1}{\Delta} a_\alpha^i \frac{\partial A^\alpha}{\partial x^\beta} b_i^\beta + A^\alpha b_i^\beta \left(\Delta^{-1} \frac{\partial a_\alpha^i}{\partial x^\beta} - \Delta^{-2} a_\alpha^i \frac{\partial \Delta}{\partial x^\beta} \right).$$

But

$$\frac{1}{\Delta} a_\alpha^i b_i^\beta \frac{\partial A^\alpha}{\partial x^\beta} = \frac{1}{\Delta} \delta_\alpha^\beta \frac{\partial A^\alpha}{\partial x^\beta} = \frac{1}{\Delta} \frac{\partial A^\alpha}{\partial x^\alpha} = \frac{1}{\Delta} P,$$

and, by the rules of differentiation of a determinant,

$$\frac{\partial \Delta}{\partial x^\beta} = \frac{\partial a_\alpha^i}{\partial x^\beta} A(i,\alpha) = \frac{\partial a_\alpha^i}{\partial x^\beta} \Delta b_i^\alpha,$$

so that

$$\Delta^{-1} \frac{\partial a_\alpha^i}{\partial x^\beta} - \Delta^{-2} a_\alpha^i \frac{\partial \Delta}{\partial x^\beta} = \Delta^{-1} \frac{\partial a_\alpha^i}{\partial x^\beta} - \Delta^{-1} \frac{\partial a_\alpha^i}{\partial x^\beta} = 0.$$

Thus (24) becomes

$$\bar{P} = \frac{1}{\Delta} P,$$

and the theorem is proved.

The following three facts are important for later developments: the product of a capacity scalar and a density scalar is a scalar, that of a capacity scalar and a density tensor is a tensor, and that of a density scalar and a capacity tensor is a tensor.   The proofs of these statements are self-evident.   The general law of transformation of pseudo-tensors is illustrated in (10), (15), (19), and (20), and need not be written out in full.

3.   THE FUNDAMENTAL METRIC TENSOR

If x^i are the coordinates of an N-dimensional space, the square of the shortest distance between two points infinitesimally separated in that space is calculated by

$$(ds)^2 = g_{ij}\, dx^i\, dx^j, \tag{22}$$

in which dx^i are the differences in coordinates between the two points and g_{ij} is symmetric ($g_{ij} = g_{ji}$).   The quantities g_{ij} specify the geometry of the space considered and constitute the metric tensor of that space. Since ds^2 is a pure scalar (or scalar invariant) and dx^i form a contravariant vector, g_{ij} form a covariant tensor of order 2.   This tensor is also called the *fundamental tensor*.   A Euclidean space is one for which the tensor g_{ij}

is equal to (or can be reduced to) the Kronecker tensor δ_{ij}, which is equal to 1 or zero according as i is or is not equal to j. For convenience, the determinant formed by g_{ij} is denoted by g. In the y coordinates the fundamental tensor is denoted by $\bar{g}_{ij}$, with a determinant $\bar{g}$. From

$$\bar{g}_{ij} = b_i^\alpha b_j^\beta g_{\alpha\beta}$$

it follows that

$$\bar{g} = |\bar{g}_{ij}| = |b_i^\alpha| \, |g_{\alpha\beta}| \, |b_j^\beta| = \frac{1}{\Delta^2} g,$$

or

$$\bar{g}^{1/2} = \frac{1}{\Delta} g^{1/2}. \tag{23}$$

Thus $\sqrt{g}$ is a density scalar.

If the quantities g^{ij} are defined by either of the relationships

$$g_{i\alpha} g^{l\alpha} = \delta_i^l \qquad \text{or} \qquad g_{\alpha i} g^{\alpha l} = \delta_i^l, \tag{24}$$

in which δ_i^l is the Kronecker delta, and if G_{ij} is the cofactor of g_{ij} in g, then

$$g^{ij} = g^{ji} = \frac{G_{ij}}{g}. \tag{25}$$

Since $g^{ij} \, dx_i \, dx_j$ is a scalar but is equal to $(ds)^2$ in Cartesian coordinates,

$$(ds)^2 = g^{ij} \, dx_i \, dx_j \tag{26}$$

in any coordinates. From (24) it can be deduced that the determinant of the contravariant fundamental tensor g^{ij} is $1/g$. Since $g^{1/2}$ is a density scalar, the quantity

$$|g^{ij}|^{1/2} = g^{-1/2}$$

is a capacity scalar because its product with $g^{1/2}$ is a pure scalar. The pseudo scalars $g^{1/2}$ and $g^{-1/2}$ are extremely important, for multiplication by them permits the formation of density or capacity vectors from vectors and vice versa.

Just as the square of a distance can be calculated with (22), the square of the magnitude of a vector u^i or u_i is expressed by the scalar invariant

$$|\mathbf{u}|^2 = g_{ij} u^i u^j, \tag{27a}$$

or

$$|\mathbf{u}|^2 = g^{ij} u_i u_j. \tag{27b}$$

These formulas are very important for correlating the physical components of vectors with their mathematical components.

4. COVARIANT DIFFERENTIATION OF VECTORS AND TENSORS

A space is called *Euclidean* if Cartesian coordinates can be introduced therein. Thus a plane is Euclidean, whereas the two-dimensional space of a spherical surface is non-Euclidean. The surface of a sheet of paper is still a two-dimensional Euclidean space even after the paper has been bent in a three-dimensional space without stretching or tearing, for the Cartesian coordinates introduced on it when it was a plane still exist and still govern the geometry in the two-dimensional space.

In the differentiation of a vector it is necessary to find the difference between a vector at a point P and a vector at a neighboring point P'. To find this difference one must transport the former vector to P' in such a fashion (*parallel transport*) that in its new position it represents the vector at P. The difference between the vector originally at P' and the vector transported from P to P' can then be taken at one point, namely, P'. In Euclidean spaces, the parallel transport of a vector presents no problem, since the notion of Euclidean parallelism is well known. In particular, if Cartesian coordinates are used, a vector parallel to a given one is simply one with identical components. (The same is not true in other coordinates.) In non-Euclidean space the law of parallelism (or rather of parallel transport of a vector) had to be invented such that it reduces to the Euclidean law of parallelism in a Euclidean space. This law is:

A vector is parallel-transported from one point to the other if it makes the same angle with the line of shortest distance (a geodesic) connecting the two points throughout the transport.

In non-Euclidean spaces, a vector, after being parallel-transported around a closed circuit composed of geodesics, may return to its initial position a different vector. This is due to the curvature of the space in which this transport is effected. For instance, on the surface of a sphere a vector situated on the Equator at longitude 0° and pointing east can be parallel-transported along the equator to a new location at longitude 90°E, at which it still points east. Then it can be parallel-transported to the North Pole along that longitude, pointing east all the way. After it reaches the North Pole, it can be parallel-transported along the longitude 0° pointing *north* all the way until it reaches its original position on the Equator. But now, at its original position, it is a different vector!

It is a much simpler matter to find the difference between two vectors in a Euclidean space, whether Cartesian coordinates are used or not. Since in such a space Cartesian coordinates can be introduced, and since the law of parallelism is particularly simple in those coordinates, the vector difference desired can be obtained first in Cartesian coordinates, then in other coordinates by a transformation. Fortunately, in the study of the transformation of vectors, tensors, and equations appearing in hydrodynamics, only the change of coordinates is involved—the space under consideration is always Euclidean.

With y^m ($m = 1, 2, 3$) denoting Cartesian coordinates, if the vector $\bar{u}^l$ situated at a point P with coordinates y^m is parallel-transported to another point P' with coordinates $y^m + dy^m$, its components simply retain their magnitudes:

$$\bar{u}(y^m + dy^m) = \bar{u}(y^m). \tag{28}$$

In another set of coordinates, for example, x^k ($k = 1, 2, 3$), the components u^i of the same vector at P are no longer identical with those of the same vector at P' (after parallel transport). In fact, the vector at P has the components

$$u^i(x^k) = b_l^i(x^k)\bar{u}^l, \tag{29}$$

and the same vector at P' has the components (the subscript pt means parallel transport)

$$u^i(x^k + dx^k)_{pt} = b_l^i(x^k + dx^k)\bar{u}^l = u^i(x^k) + \frac{\partial b_l^i}{\partial x^k} dx^k \bar{u}^l$$

$$= u^i(x^k) + \frac{\partial b_l^i}{\partial y^m} a_k^m a_j^l u^j dx^k = u^i(x^k) - \Gamma_{jk}^i u^j dx^k, \tag{30}$$

in which x^k are the x coordinates of P and $x^k + dx^k$ those of P' and $\bar{u}^l$ is evaluated either at P or at P'. By virtue of (28), it makes no difference where it is evaluated. The quantity

$$\Gamma_{jk}^i = - \frac{\partial b_l^i}{\partial y^m} a_k^m a_j^l = - \frac{\partial^2 x^i}{\partial y^l\, \partial y^m} a_k^m a_j^l \tag{31}$$

is clearly symmetric in j and k, that is,

$$\Gamma_{jk}^i = \Gamma_{kj}^i. \tag{32}$$

According to (30), after the parallel transport of a vector from P to P', the components of the same vector in the x coordinates are changed. The absolute differential of a field vector u^i is equal to the difference between u^i evaluated at P' and $u^i(x^k + dx^k)_{pt}$ calculated with (30) and representing at P' the vector $u^i(x^k)$ at P, to which it is parallel:

$$Du^i = u^i(x^k + dx^k) - u^i(x^k) + \Gamma_{jk}^i u^j dx^k. \tag{33}$$

Thus the covariant derivative of u^i with respect to x^k is

$$\frac{Du^i}{Dx^k} = \frac{\partial u^i}{\partial x^k} + \Gamma_{jk}^i u^j. \tag{34}$$

Since Du^i are just the transforms of the ordinary differentials $d\bar{u}^l$ in Cartesian coordinates, they are the components of a differential vector and transform as such. Thus, the covariant derivative of a vector is a tensor, as can also be shown by a direct demonstration.

To find the covariant derivative of a covariant vector, it can be assumed that

$$\frac{Dv_i}{Dx^k} = \frac{\partial v_i}{\partial x^k} + B_{ik}^h v_h, \tag{35}$$

in which the B's are to be determined. Since $u^i v_i$ is a scalar, its absolute gradient is exactly equal to its ordinary gradient and no correction terms are needed. Thus

$$\frac{\partial}{\partial x^k}(u^i v_i) = \frac{D(u^i v_i)}{\partial x^k} = \frac{\partial(u^i v_i)}{\partial x^k} + v_i \Gamma_{jk}^i u^j + u^i B_{ik}^h v_h,$$

and, with i replaced by j and h by i in the last sum,

$$(\Gamma^i_{jk} + B^i_{jk})v_i u^j = 0.$$

Since v_i and u^j are arbitrary, it follows that

$$B^i_{jk} = -\Gamma^i_{jk}. \tag{36}$$

Thus the formula for the covariant derivative of a covariant vector is

$$\frac{Dv_i}{Dx^k} = \frac{\partial v_i}{\partial x^k} - \Gamma^h_{ik}v_h. \tag{37}$$

The covariant derivative of a mixed tensor t^i_j can be found by considering the scalar formed by contraction:

$$S = t^i_j u^j v_i.$$

The covariant derivative of S is just its ordinary derivative, so that

$$\frac{DS}{Dx^k} = \frac{Dt^i_j}{Dx^k} u^j v_i + t^i_j \frac{D}{Dx^k} u^j v_i$$

$$= \frac{\partial S}{\partial x^k} = \frac{\partial t^i_j}{\partial x^k} u^j v_i + t^i_j \frac{\partial}{\partial x^k} u^j v_i, \tag{38}$$

From this it follows that

$$u^j v_i \left(\frac{Dt^i_j}{Dx^k} - \frac{\partial t^i_j}{\partial x^k} \right) + t^i_j \left[\frac{D(u^j v_i)}{Dx^k} - \frac{\partial(u^j v_i)}{\partial x^k} \right] = 0,$$

or $\qquad u^j v_i \left(\frac{Dt^i_j}{Dx^k} - \frac{\partial t^i_j}{\partial x^k} \right) + t^i_j (\Gamma^j_{hk} u^h v_i - \Gamma^h_{ik} u^j v_h) = 0, \tag{39}$

since, from (34) and (37),

$$\frac{D(u^j v_i)}{Dx^k} = u^j \frac{Dv_i}{Dx^k} + v_i \frac{Du^j}{Dx^k} = \frac{\partial(u^j v_i)}{\partial x^k} + \Gamma^j_{hk} u^h v_i - \Gamma^h_{ik} u^j v_h.$$

The repeating indices in (39) can be changed so that the equation becomes

$$u^j v_i \left(\frac{Dt^i_j}{Dx^k} - \frac{\partial t^i_j}{\partial x^k} + \Gamma^h_{jk} t^i_h - \Gamma^i_{hk} t^h_j \right) = 0.$$

Since u^j and v_i are arbitrary, it follows that

$$\frac{Dt^i_j}{Dx^k} = \frac{\partial t^i_j}{\partial x^k} + \Gamma^i_{hk} t^h_j - \Gamma^h_{jk} t^i_h. \tag{40}$$

By similar procedures, it can be demonstrated that

$$\frac{Dt^{ij}}{Dx^k} = \frac{\partial t^{ij}}{\partial x^k} + \Gamma^j_{kh} t^{ih} + \Gamma^i_{kh} t^{kj} \tag{41}$$

and $\qquad \dfrac{Dt_{ij}}{Dx^k} = \dfrac{\partial t_{ij}}{\partial x^k} - \Gamma^h_{kj} t_{ih} - \Gamma^h_{ki} t_{hj}. \tag{42}$

That Dt^i_j / Dx^k is a tensor follows from the equation

$$\frac{DS}{Dx^k} = \frac{Dt^i_j}{Dx^k} u^j v_i + t^i_j \frac{Du^j}{Dx^k} v_i + t^i_j u^j \frac{Dv_i}{Dx^k}$$

and from the fact that

$$\frac{DS}{Dx^k}, \qquad t^i_j \frac{Du^j}{Dx^k} v_i, \qquad \text{and} \qquad t^i_j u^j \frac{Dv_i}{Dx^k}$$

all are covariant vectors. Similarly the quantities on the left-hand sides of (41) and (42) are tensors. In fact, covariant derivatives of tensors are automatically tensors, with the order of covariance increased by 1.

5. CHRISTOFFEL SYMBOLS

Although the symbol Γ^i_{jk} has been defined by (31), a simple expression for it in terms of the components of the fundamental tensor g_{ij} is desirable. Since

$$|\mathbf{u}|^2 = g_{ij} u^i u^j$$

is an invariant and is therefore unchanged by a parallel transport of u^i,

$$0 = \frac{D|\mathbf{u}|^2}{Dx^k} = u^i u^j \frac{Dg_{ij}}{Dx^k},$$

since

$$\frac{Du^i}{Dx^k} = \frac{Du^j}{Dx^k} = 0$$

for a parallel transport. Thus

$$\frac{Dg_{ij}}{Dx^k} = \frac{\partial g_{ij}}{\partial x^k} - \Gamma^h_{jk} g_{ih} - \Gamma^h_{ki} g_{jh} = 0. \tag{43}$$

If

$$\Gamma_{i,jk} = \Gamma^h_{jk} g_{ih}, \tag{44}$$

then

$$\Gamma^h_{jk} = \Gamma^l_{jk} g_{il} g^{ih} = g^{ih} \Gamma_{i,jk}, \tag{45}$$

and

$$\frac{\partial g_{ij}}{\partial x^k} = \Gamma_{i,jk} + \Gamma_{j,ki}. \tag{46}$$

By symmetry,

$$\frac{\partial g_{jk}}{\partial x^i} = \Gamma_{j,ki} + \Gamma_{k,ij}, \tag{47}$$

and

$$\frac{\partial g_{ki}}{\partial x^j} = \Gamma_{k,ij} + \Gamma_{i,jk}. \tag{48}$$

Thus

$$\Gamma_{i,jk} = \frac{1}{2} \left(\frac{\partial g_{ij}}{\partial x^k} + \frac{\partial g_{ki}}{\partial x^j} - \frac{\partial g_{jk}}{\partial x^i} \right), \tag{49}$$

and

$$\Gamma^h_{jk} = \frac{1}{2} g^{ih} \left(\frac{\partial g_{ij}}{\partial x^k} + \frac{\partial g_{ki}}{\partial x^j} - \frac{\partial g_{jk}}{\partial x^i} \right). \tag{50}$$

The symbols $\Gamma_{i,jk}$ and Γ^h_{jk} are called the *Christoffel symbols* of the first and second kind, respectively.

6. PHYSICAL COMPONENTS OF VECTORS AND TENSORS

The mathematical components of a vector or a tensor do not necessarily have the dimensions of the physical quantities (such as velocity, acceleration, rate of deformation, and stress) which they stand for, since the quantities a_j^i or b_j^i are not necessarily dimensionless. Therefore, if these components have the desired physical dimensions in Cartesian coordinates, they do not necessarily retain them in other coordinates. Thus, it is necessary to find the expressions for the physical components of vectors and tensors, with the desired physical dimensions.

If the coordinates are orthogonal, the square of an infinitesimal displacement is

$$(ds)^2 = g_{11}(dx^1)^2 + g_{22}(dx^2)^2 + g_{33}(dx^3)^2,$$

from which it may be deduced, according to the theorem of Pythagoras, that the physical components of the displacement vector can be defined by

$$dx(i) = \sqrt{g_{ii}}\, dx^i, \qquad i \text{ unsummed.} \tag{51}$$

This definition is adequate even if the coordinates are not orthogonal, for (22) can be rewritten in terms of the physical components just defined without any difficulty.   In fact, $(ds)^2$ can be expressed as a quadratic form of the physical components $dx(i)$, with unity as the coefficient of the square terms, for $(ds)^2$ must be equal to the square of $dx(1)$, for instance, if $dx(2)$ and $dx(3)$ are zero. If the components g_{ii} (unsummed) are determined from geometrical considerations, and if Eqs. (51) are substituted into this quadratic form, (22) is obtained in the process.

Since velocity is the time rate of change of the displacement and acceleration is the time rate of change of the velocity, and since time is a scalar quantity in Newtonian mechanics, the physical components of velocity and of acceleration are defined by

$$u(i) = \sqrt{g_{ii}}\, u^i, \qquad i \text{ unsummed,} \tag{52}$$

$$a(i) = \sqrt{g_{ii}}\, a^i, \qquad i \text{ unsummed.} \tag{53}$$

So far only contravariant vectors have been considered.   Since

$$dx^i = g^{ij}\, dx_j,$$

(51) becomes

$$dx(i) = \sqrt{g_{ii}}\, g^{ij}\, dx_j, \qquad i \text{ unsummed,}$$

and similar equations can be immediately written for the velocity and acceleration components.   If the coordinates are orthogonal,

$$g^{11} = \frac{g_{22}g_{33}}{g} = \frac{1}{g_{11}},$$

and in general $\quad g^{ii} = \dfrac{1}{g_{ii}}, \qquad i$ unsummed,

so that

$$dx(i) = \frac{dx_i}{\sqrt{g_{ii}}}, \qquad u(i) = \frac{u_i}{\sqrt{g_{ii}}}, \qquad \text{and } a(i) = \frac{a_i}{\sqrt{g_{ii}}}, \qquad i \text{ unsummed.}$$

$$(54)$$

Physical components of a tensor can be defined to be its mathematical components in the Cartesian coordinates y^i. Indeed, from (51) to (54) it can be deduced that the same statement applies to vectors, as it should. If u^j and v_i are two arbitrary vectors and t^i_j is a mixed tensor, and if quantities in Cartesian coordinates are denoted by a bar,

$$t^j_j v_i u^j = \bar{t}^i_j \bar{v}_i \bar{u}^j = t(i,j)v(i)u(j),\tag{55}$$

in which $t(i,j)$ denotes the physical components of t^i_j. But from (52) and (54), (55) can be written as

$$t(i,j)v(i)u(j) = \sqrt{\frac{g_{ii}}{g_{jj}}}\, t^i_j v(i)u(j).$$

Since u^j and v_i are arbitrary, it follows that

$$t(i,j) = \sqrt{\frac{g_{ii}}{g_{jj}}}\, t^i_j, \qquad \text{unsummed.}\tag{56}$$

Similarly,

$$t(i,j) = \sqrt{g_{ii}g_{jj}}\, t^{ij}, \qquad \text{unsummed,}\tag{57}$$

$$t(i,j) = \frac{1}{\sqrt{g_{ii}g_{jj}}}\, t_{ij}, \qquad \text{unsummed.}\tag{58}$$

The physical components of tensors of higher order can be similarly defined. The law of forming them from the mathematical components is now clear. The equations relating the physical components of a pseudo-tensor to its mathematical components are similar to (56) to (58), with the only difference that a power of the determinant Δ now occurs on the right-hand sides.

7. GRADIENT, DIVERGENCE, AND LAPLACIAN

Since the ordinary derivatives of a scalar function ϕ are exactly its absolute derivatives, the quantities $\partial\phi/\partial x^i$ form a covariant vector. The physical components of this vector can be derived from (54) to be

$$\frac{1}{\sqrt{g_{ii}}}\frac{\partial\phi}{\partial x^i}, \qquad i \text{ unsummed,}\tag{59}$$

which have the expected form, since $\sqrt{g_{ii}}\, dx^i$ are the physical displacements $dx(i)$,

Since the divergence of a contravariant density vector is a density scalar, as proved by (21), and since $\sqrt{g}$ is a density scalar,

$$\frac{1}{\sqrt{g}} \frac{\partial \sqrt{g}\, u^i}{\partial x^i}$$

is a pure scalar. Thus, in general coordinates, the divergence of a vector is calculated by the following formula in terms of its physical components:

$$\operatorname{div} \mathbf{u} = \frac{1}{\sqrt{g}} \frac{\partial \sqrt{g/g_{ii}}\, u(i)}{\partial x^i}. \tag{60}$$

The contracted tensor

$$g^{i\alpha} \frac{\partial \phi}{\partial x^\alpha}$$

is a contravariant vector. Therefore

$$\nabla^2 \phi = \frac{1}{\sqrt{g}} \frac{\partial}{\partial x^i}\left(\sqrt{g}\, g^{i\alpha} \frac{\partial \phi}{\partial x^\alpha}\right) \tag{61}$$

is a pure scalar and is the Laplacian of ϕ in general coordinates, because in Cartesian coordinates it reduces to the sum of the three second derivatives of ϕ with respect to the three coordinates.

For orthogonal coordinates, g_{ij} is zero if i is not equal to j, and

$$g_{11} = h_1^2, \qquad g_{22} = h_2^2, \qquad \text{and} \qquad g_{33} = h_3^2.$$

The physical components of the gradient are then

$$\frac{1}{h_1} \frac{\partial \phi}{\partial \alpha}, \qquad \frac{1}{h_2} \frac{\partial \phi}{\partial \beta}, \qquad \text{and} \qquad \frac{1}{h_3} \frac{\partial \phi}{\partial \gamma} \tag{62}$$

if x^1, x^2, and x^3 are replaced by α, β, and γ. Since now

$$g = (h_1 h_2 h_3)^2 \qquad \text{and} \qquad g^{ii} = \frac{1}{g_{ii}}, \qquad i \text{ unsummed,}$$

(60) and (61) become, respectively,

$$\operatorname{div} \mathbf{u} = \frac{1}{h_1 h_2 h_3}\left[\frac{\partial(h_2 h_3 u)}{\partial \alpha} + \frac{\partial(h_3 h_1 v)}{\partial \beta} + \frac{\partial(h_1 h_2 w)}{\partial \gamma}\right] \tag{63}$$

and

$$\nabla^2 = \frac{1}{h_1 h_2 h_3}\left[\frac{\partial}{\partial \alpha}\left(\frac{h_2 h_3}{h_1} \frac{\partial}{\partial \alpha}\right) + \frac{\partial}{\partial \beta}\left(\frac{h_3 h_1}{h_2} \frac{\partial}{\partial \beta}\right) + \frac{\partial}{\partial \gamma}\left(\frac{h_1 h_2}{h_3} \frac{\partial}{\partial \gamma}\right)\right], \tag{64}$$

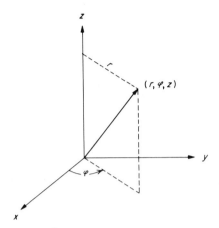

FIGURE 1. CYLINDRICAL COORDINATES.

in which ϕ has been omitted, and

$$u,v,w = u(1), u(2), u(3), \tag{65}$$

respectively.

In cylindrical coordinates (r,φ,z) (Fig. 1),

$$h_1 = 1, \qquad h_2 = r, \qquad \text{and} \qquad h_3 = 1,$$

and the physical gradient of a scalar function ϕ has the components

$$\frac{\partial \phi}{\partial r}, \qquad \frac{1}{r}\frac{\partial \phi}{\partial \varphi}, \qquad \frac{\partial \phi}{\partial z}. \tag{66}$$

The divergence of a (physical) vector is

$$\text{div } \mathbf{u} = \frac{1}{r}\frac{\partial}{\partial r}(ru) + \frac{1}{r}\frac{\partial v}{\partial \varphi} + \frac{\partial w}{\partial z}, \tag{67}$$

and the Laplacian operator is

$$\nabla^2 \equiv \frac{\partial^2}{\partial r^2} + \frac{1}{r}\frac{\partial}{\partial r} + \frac{1}{r^2}\frac{\partial^2}{\partial \varphi^2} + \frac{\partial^2}{\partial z^2}. \tag{68}$$

In spherical coordinates (R,θ,φ) (Fig. 2),

$$h_1 = 1, \qquad h_2 = R, \qquad h_3 = R \sin \theta,$$

so that the gradient of ϕ has the components

$$\frac{\partial \phi}{\partial R}, \qquad \frac{1}{R}\frac{\partial \phi}{\partial \theta}, \qquad \frac{1}{R \sin \theta}\frac{\partial \phi}{\partial \varphi}. \tag{69}$$

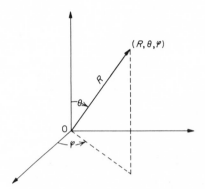

FIGURE 2.   SPHERICAL COORDINATES.

The divergence of a vector is

$$\text{div } \mathbf{u} = \frac{1}{R^2} \frac{\partial}{\partial R} (R^2 u) + \frac{1}{R \sin \theta} \frac{\partial}{\partial \theta} (v \sin \theta) + \frac{1}{R \sin \theta} \frac{\partial w}{\partial \varphi}, \quad (70)$$

and the Laplacian operator is

$$\nabla^2 = \frac{1}{R^2} \frac{\partial}{\partial R} \left(R^2 \frac{\partial}{\partial R} \right) + \frac{1}{R^2 \sin \theta} \frac{\partial}{\partial \theta} \left(\sin \theta \frac{\partial}{\partial \theta} \right) + \frac{1}{R^2 \sin^2 \theta} \frac{\partial^2}{\partial \varphi^2}. \quad (71)$$

8.   THE VORTICITY AND RATE-OF-DEFORMATION TENSORS

From (37) and (32) it follows that

$$\frac{\partial u_i}{\partial x^k} - \frac{\partial u_k}{\partial x^i} = \frac{Du_i}{Dx^k} - \frac{Du_k}{Dx^i}$$

if u is written instead of v. Thus the vorticity components formed by partial differentiation form an antisymmetric covariant tensor of order 2, the physical components of which are defined by (58). In general orthogonal coordinates, these components, calculated from (54) and (58), are

$$\xi = \frac{1}{h_2 h_3} \left[\frac{\partial (h_3 w)}{\partial \beta} - \frac{\partial (h_2 v)}{\partial \gamma} \right], \qquad \eta = \frac{1}{h_3 h_1} \left[\frac{\partial (h_1 u)}{\partial \gamma} - \frac{\partial (h_3 w)}{\partial \alpha} \right],$$

$$\zeta = \frac{1}{h_1 h_2} \left[\frac{\partial (h_2 v)}{\partial \alpha} - \frac{\partial (h_1 u)}{\partial \beta} \right], \quad (72)$$

which are the three physical components of the curl of the velocity in orthogonal coordinates. In cylindrical coordinates,

$$\xi = \frac{1}{r}\left[\frac{\partial w}{\partial \varphi} - \frac{\partial(rv)}{\partial z}\right], \qquad \eta = \frac{\partial u}{\partial z} - \frac{\partial w}{\partial r}, \qquad \zeta = \frac{1}{r}\left[\frac{\partial(rv)}{\partial r} - \frac{\partial u}{\partial \varphi}\right],$$

(73)

and in spherical coordinates,

$$\xi = \frac{1}{R^2 \sin\theta}\left[\frac{\partial(R\sin\theta\, w)}{\partial\theta} - \frac{\partial(Rv)}{\partial\varphi}\right], \qquad \eta = \frac{1}{R\sin\theta}\left[\frac{\partial u}{\partial\varphi} - \frac{\partial(R\sin\theta\, w)}{\partial R}\right],$$

$$\zeta = \frac{1}{R}\left[\frac{\partial(Rv)}{\partial R} - \frac{\partial u}{\partial\theta}\right].$$

(74)

The rate-of-deformation tensor, in covariant form, has the components

$$\frac{Du_i}{Dx^k} + \frac{Du_k}{Dx^i}.$$

The physical components are calculated from (37), (54), and (58). After a lengthy but straightforward calculation it can be shown that, for orthogonal coordinates, the physical components of the rate-of-deformation tensor are

$$\frac{1}{2}e_{\alpha\alpha} = \frac{1}{h_1}\frac{\partial u}{\partial\alpha} + \frac{v}{h_1 h_2}\frac{\partial h_1}{\partial\beta} + \frac{w}{h_3 h_1}\frac{\partial h_1}{\partial\gamma},$$

$$\frac{1}{2}e_{\beta\beta} = \frac{1}{h_2}\frac{\partial v}{\partial\beta} + \frac{w}{h_2 h_3}\frac{\partial h_2}{\partial\gamma} + \frac{u}{h_1 h_2}\frac{\partial h_2}{\partial\alpha},$$

$$\frac{1}{2}e_{\gamma\gamma} = \frac{1}{h_3}\frac{\partial w}{\partial\gamma} + \frac{u}{h_3 h_1}\frac{\partial h_3}{\partial\alpha} + \frac{v}{h_2 h_3}\frac{\partial h_3}{\partial\beta},$$

$$e_{\beta\gamma} = \frac{h_3}{h_2}\frac{\partial}{\partial\beta}\frac{w}{h_3} + \frac{h_2}{h_3}\frac{\partial}{\partial\gamma}\frac{v}{h_2},$$

$$e_{\gamma\alpha} = \frac{h_1}{h_3}\frac{\partial}{\partial\gamma}\frac{u}{h_1} + \frac{h_3}{h_1}\frac{\partial}{\partial\alpha}\frac{w}{h_3},$$

$$e_{\alpha\beta} = \frac{h_2}{h_1}\frac{\partial}{\partial\alpha}\frac{v}{h_2} + \frac{h_1}{h_2}\frac{\partial}{\partial\beta}\frac{u}{h_1},$$

(75)

in which we have used $e_{\alpha\beta}$ instead of $e(\alpha,\beta)$, etc., for brevity. The physical stress components for an incompressible fluid are simply

$$\tau_{ij} = p\delta_{ij} + \mu e_{ij},$$

(76)

in which δ_{ij} is the ordinary Kronecker delta, really just the physical components of the Kronecker tensor.

In cylindrical coordinates,

$$\frac{1}{2}e_{rr} = \frac{\partial u}{\partial r}, \qquad \frac{1}{2}e_{\varphi\varphi} = \frac{1}{r}\frac{\partial v}{\partial \varphi} + \frac{u}{r}, \qquad \frac{1}{2}e_{zz} = \frac{\partial w}{\partial z},$$

$$e_{\varphi z} = \frac{1}{r}\frac{\partial w}{\partial \varphi} + \frac{\partial v}{\partial z}, \qquad e_{zr} = \frac{\partial u}{\partial z} + \frac{\partial w}{\partial r}, \qquad e_{r\varphi} = r\frac{\partial}{\partial r}\frac{v}{r} + \frac{1}{r}\frac{\partial u}{\partial \varphi},$$

(77)

and in spherical coordinates,

$$\frac{1}{2}e_{RR} = \frac{\partial u}{\partial R}, \qquad \frac{1}{2}e_{\theta\theta} = \frac{1}{R}\frac{\partial v}{\partial \theta} + \frac{u}{R},$$

$$\frac{1}{2}e_{\varphi\varphi} = \frac{1}{R\sin\theta}\frac{\partial w}{\partial \varphi} + \frac{u}{R} + \frac{v\cot\theta}{R},$$

(78)

$$e_{\theta\varphi} = \frac{\sin\theta}{R}\frac{\partial}{\partial\theta}\frac{w}{\sin\theta} + \frac{1}{R\sin\theta}\frac{\partial v}{\partial\varphi}, \qquad e_{\varphi R} = \frac{1}{R\sin\theta}\frac{\partial u}{\partial\varphi} + R\frac{\partial}{\partial R}\frac{w}{R},$$

$$e_{R\theta} = R\frac{\partial}{\partial R}\frac{v}{R} + \frac{1}{R}\frac{\partial u}{\partial\theta}.$$

9. THE EQUATION OF CONTINUITY AND THE NAVIER-STOKES EQUATIONS

The equation of continuity is simply

$$\frac{\partial \rho}{\partial t} + \operatorname{div}\rho\mathbf{u} = 0, \tag{79}$$

in which the divergence of a vector is expressed by (63) in general coordinates and by (67) and (70) in particular coordinates.

The Navier-Stokes equations can be written as

$$\frac{\partial u^i}{\partial t} + u^\alpha \frac{Du^i}{Dx^\alpha} = X^i - \frac{1}{\rho}g^{i\alpha}\frac{D}{Dx^\alpha}(p - \lambda\theta) + \frac{g^{\beta\alpha}}{\rho}\frac{D}{Dx^\beta}\left[\mu g^{\gamma i}\left(\frac{Du_\alpha}{Dx^\gamma} + \frac{Du_\gamma}{Dx^\alpha}\right)\right],$$

(80)

which for Cartesian coordinates reduces to the ones given in Chap. 2 and yet is tensorially homogeneous, since

$$\theta = \operatorname{div}\mathbf{u} = \frac{1}{\sqrt{g}}\frac{\partial\sqrt{g}\,u^i}{\partial x^i} \tag{81}$$

is a pure scalar. If u^i, u_α, and u_γ are expressed in terms of the physical components and the rule of covariant differentiation is followed (a scalar covariant differentiation is simply a partial differentiation), the Navier-Stokes equations can be formed explicitly in general coordinates and in any particular set of coordinates. The calculation, though laborious, is straightforward, but the results are voluminous for general coordinates—at least a whole page is needed for writing the equations. Since Eqs. (80), though not immediately explicit, are already the Navier-Stokes equations in general coordinates, the lengthy equations in terms of h_i will not be presented. Instead, only the Navier-Stokes equations for a homogeneous fluid with constant viscosity μ and for cylindrical and spherical coordinates will be given. These are, for cylindrical coordinates,

$$\frac{\partial u}{\partial t} + u\frac{\partial u}{\partial r} + \frac{v}{r}\frac{\partial u}{\partial \varphi} + w\frac{\partial u}{\partial z} - \frac{v^2}{r} = -\frac{1}{\rho}\frac{\partial p}{\partial r} + X_r + \nu\left(\nabla^2 u - \frac{u}{r^2} - \frac{2}{r^2}\frac{\partial v}{\partial \varphi}\right)$$

$$\frac{\partial v}{\partial t} + u\frac{\partial v}{\partial r} + \frac{v}{r}\frac{\partial v}{\partial \varphi} + w\frac{\partial v}{\partial z} + \frac{uv}{r}$$

$$= -\frac{1}{\rho r}\frac{\partial p}{\partial \varphi} + X_\varphi + \nu\left(\nabla^2 v + \frac{2}{r^2}\frac{\partial u}{\partial \varphi} - \frac{v}{r^2}\right), \quad (82)$$

$$\frac{\partial w}{\partial t} + u\frac{\partial w}{\partial r} + \frac{v}{r}\frac{\partial w}{\partial \varphi} + w\frac{\partial w}{\partial z} = -\frac{1}{\rho}\frac{\partial p}{\partial z} + X_z + \nu\,\nabla^2 w,$$

in which the X's are the body-force components and the Laplacian operator is defined by (68). In spherical coordinates, the corresponding equations are

$$\frac{\partial u}{\partial t} + u\frac{\partial u}{\partial R} + \frac{v}{R}\frac{\partial u}{\partial \theta} + \frac{w}{R\sin\theta}\frac{\partial u}{\partial \varphi} - \frac{v^2 + w^2}{R}$$

$$= -\frac{1}{\rho}\frac{\partial p}{\partial R} + X_R + \nu\left(\nabla^2 u - \frac{2u}{R^2} - \frac{2}{R^2}\frac{\partial v}{\partial \theta} - \frac{2v\cot\theta}{R^2} - \frac{2}{R^2\sin\theta}\frac{\partial w}{\partial \varphi}\right),$$

$$\frac{\partial v}{\partial t} + u\frac{\partial v}{\partial R} + \frac{v}{R}\frac{\partial v}{\partial \theta} + \frac{w}{R\sin\theta}\frac{\partial v}{\partial \varphi} + \frac{uv}{R} - \frac{w^2\cot\theta}{R} \qquad (83)$$

$$= -\frac{1}{\rho}\frac{1}{R}\frac{\partial p}{\partial \theta} + X_\theta + \nu\left(\nabla^2 v + \frac{2}{R^2}\frac{\partial u}{\partial \theta} - \frac{v}{R^2\sin^2\theta} - \frac{2\cot\theta}{R^2\sin\theta}\frac{\partial w}{\partial \varphi}\right),$$

$$\frac{\partial w}{\partial t} + u\frac{\partial w}{\partial R} + \frac{v}{R}\frac{\partial w}{\partial \theta} + \frac{w}{R\sin\theta}\frac{\partial w}{\partial \varphi} + \frac{wu}{R} + \frac{vw\cot\theta}{R}$$

$$= -\frac{1}{\rho}\frac{1}{R\sin\theta}\frac{\partial p}{\partial \varphi} + X_\varphi + \nu\left(\nabla^2 w - \frac{w}{R^2\sin^2\theta}\right.$$

$$\left. + \frac{2}{R^2\sin\theta}\frac{\partial u}{\partial \varphi} + \frac{2\cot\theta}{R^2\sin\theta}\frac{\partial v}{\partial \varphi}\right).$$

Note that in (79) and (80) ρ, p, λ, and μ are treated as scalars.

PROBLEMS

1. Show that

$$\frac{Dg^{id}}{Dx^k} = 0.$$

Then does it matter whether the $g^{\gamma i}$ in (80) is inside the bracket or outside of it?

2. Verify (72) to (74).

3. Verify (75), (77), and (78).

4. Verify (82) by the tensorial method.

5. Verify (83).

6. Derive the Navier-Stokes equations in toroidal coordinates.

ADDITIONAL READING

BRILLOUIN, L., 1938: "Les tenseurs en mécanique et en elasticité," Masson et Cie, Paris; reprinted by Dover Publications, Inc., New York, 1946.

INDEX

This book was set in Times Roman by The Universities Press, printed on permanent paper and bound by The Maple Press Company. The designer was Barbara Ellwood; the drawings were done by J. & R. Technical Services, Inc. The editors were B. J. Clark and J. W. Maisel. Morton I. Rosenberg supervised the production.